1st International Conference on Science, Mathematics, Environment and Education (ICoSMEE 2017)

Journal of Physics: Conference Series Volume 1022

Solo Baru, Indonesia
16 September 2017

Editors:

Nurma Yunita Indriyanti
Murni Ramli
Puguh Karyanto
Getut Pramesti

ISBN: 978-1-5108-6322-4
ISSN: 1742-6588

Printed from e-media with permission by:

Curran Associates, Inc.
57 Morehouse Lane
Red Hook, NY 12571

Some format issues inherent in the e-media version may also appear in this print version.

Printed by Curran Associates, Inc. (2018)

For permission requests, please contact the Institute of Physics
at the address below.

Institute of Physics
Dirac House, Temple Back
Bristol BS1 6BE UK

Phone: 44 1 17 929 7481
Fax: 44 1 17 920 0979

techtracking@iop.org

Additional copies of this publication are available from:

Curran Associates, Inc.
57 Morehouse Lane
Red Hook, NY 12571 USA
Phone: 845-758-0400
Fax: 845-758-2633
Email: curran@proceedings.com
Web: www.proceedings.com

Table of contents

Volume 1022

The 1st International Conference on Science, Mathematics, Environment and Education

16 September 2017, Solo Baru, Indonesia

Accepted papers received: 1 May 2018
Published online: 15 May 2018

Science education

Physics, instrumentation, and energy

Physics education

Chemistry and material science

Chemistry education

Biology

IOP Conf. Series: Journal of Physics: Conf. Series **1022** (2018) 011001 doi:10.1088/1742-6596/1022/1/011001

Preface

Global framework of sustainable development proposed by UN has led to several in-depth scientific discussions focused on studies and reviews of the strategies and progress made so far. The main objective of the performed discussions is to examine the progress and seeking appropriate and more operational strategies addressing the challenge of better global sustainability achievements. In the discussion, university has an important position in providing solution from both, in result-chain area and in causal-chain area.

It cannot be denied that natural, environmental and education science play as a major backbone underpinning effort in seeking the solution of achieving the goal of sustainable development. Scholarly research, reviews and reports from these majors are of valuable scientific databases in achieving the goal of sustainable development. The spirit of having contribution in the aforementioned UN goal has led the UNS and the committee to organize this conference. This conference involving Science, Mathematics, Environment and Education in order to address the challenge of fulfilling holistic approaches towards the improvement of the quality of human life without ignoring the sustainability of the environmental system.

This first International Conference on Science, Mathematics, Environment and Education (ICoSMEE) aims at bringing together researchers, educators, scientists, and scholar students in the area of Science, Mathematics, Environment and Education to exchange and share their experiences, ideas, and findings and to discuss practical challenges encountered and the solutions to develop humanity and the quality of human life in a sustainable manner. The conference was held in Surakarta, Indonesia on September 16th, 2017. Hundred of mathematics and science education experts and practitioners joined the event. There were six keynote speakers who came from Indonesia, United Kingdom, Australia, and United State of America. Moreover, more than 200 articles were presented in the conference and 59 articles are selected to be publishing in the present conference proceeding.

The Editors,

Dr.paed Nurma Yunita Indriyanti
Murni Ramli, Ed.D
Puguh Karyanto, Ph.D
Getut Pramesti, M.Si.

IOP Conf. Series: Journal of Physics: Conf. Series **1022** (2018) 011001 doi:10.1088/1742-6596/1022/1/011001

List of Committee

International Advisory Commitee
Prof. Dr. Hans-Dieter Barke
Prof Bob Bucat
Prof. Lilia Halim
Prof. Dr. Karl-Heinz Van Pee
Benny Tjahjono, Ph.D
Prof. Tugrul U Daim, Ph.D.
Dr. Dani Harmanto BEng (Hons) ,MSc, PhD ,FHEA PgCert C.Eng MIED AMIMechE
Bagus Putra Muljadi
Heri Setya Kusuma

Organizing committee

Chair
Puguh Karyanto, Ph.D
Dr.rer.nat Sri Mulyani

Conference Secretariat & Website
Dr. Maria Ulfa
Daru Wahyuningsih, M.Pd.
Arum Wulandari, M.Pd.
Delisma Wisnu Adi, M.Pd.

Conference Treasury
Nurmiyati, M.Si.
Budi Utami, M.Pd

Paper and Proceeding
Dewanto Harjunowibowo, M.Sc
Maria Ciptaning Sabdo Kawedhar, S.Pd
Shinta Maharani, S.Pd

Conference Reception
Dr. Muzzazinah
Dr. Sri Yamtinah

Publications, Transportation and Documentation
Doni Wahyu Prabowo, S.Pd
Suwarno, M.Pd.

Conference Banquet
Bakti Mulyani, M.Si.
Dr. Sri Widoretno

Peer review statement

All papers published in this volume of *Journal of Physics: Conference Series* have been peer reviewed through processes administered by the proceedings Editors. Reviews were conducted by expert referees to the professional and scientific standards expected of a proceedings journal published by IOP Publishing.

IOP Conf. Series: Journal of Physics: Conf. Series **1022** (2018) 012001 doi:10.1088/1742-6596/1022/1/012001

Erlang circular model motivated by inverse stereographic projection

G Pramesti[1]

[1]Department of Mathematic Education, Universitas Sebelas Maret, Surakarta, Indonesia

Corresponding email: getutpramesti@staff.uns.ac.id

Abstract. The Erlang distribution is a special case of the Gamma distribution with the shape parameter is an integer. This paper proposed a new circular model used inverse stereographic projection. The inverse stereographic projection which is a mapping that projects a random variable from a real line onto a circle can be used in circular statistics to construct a distribution on the circle from real domain. From the circular model, then can be derived the characteristics of the Erlang circular model such as the mean resultant length, mean direction, circular variance and trigonometric moments of the distribution.

Key words: Inverse stereographic projection, Erlang, Circular

1. Introduction

Erlang which is a member of Exponential family is a special distribution of Gamma distribution. Erlang distribution is a Gamma distribution with parameters which are positive integers. Erlang distribution is used nowadays in the field of stochastic processes and biomathematics. In addition, the implementation of the Erlang distribution can be used in the field of insurance. For example, if the amount of insurance claim X have distribution of Erlang. In this illustration, Erlang works on real domain. Directional statistics is sub statistics field where measurement in direction ways. This paper will be proposed a new model circular distribution using inverse stereographic projection in circular measurement. As we know, at least there are four methods of obtaining circular distribution [7]. Using stereographic projection, we will apply Erlang distribution with real line domain then identify such real random variable on the circumference of the circle. There are many paper talk about circular stereographic distribution. Phani, Girija and Dattatreya [10] proposed circular model induced by inverse stereographic projection on extreme value distribution. Abe, Shimizu and Pewsey [1] also introduced a symmetric unimodal models such as t distribution motivated by inverse stereographic projection. Getut [4] also derived exponential circular distribution, Chi Square circular and Chi Square semi circular [5] and the latest paper is Beta circular [6].

2. Materials and Methods

2.1 Circular and Semicircular Data

Mostly, measurement in any field is regarded as real number. Actually, in many diverse field, any observation can be measured as a direction. For example, circular data. Wind direction or direction of migrating birds as a circular data can be measured the compass and the clock [8]. Other example, such as

IOP Conf. Series: Journal of Physics: Conf. Series **1022** (2018) 012001 doi:10.1088/1742-6596/1022/1/012001

component lifetime, orientations of certain organism, direction of a pollutant also can be regarded as directional observation. A set of such observation on direction is referred to as directional data [5]. Direction can be represented as points on the circumference of a unit circle centre or a unit vectors connecting the origin to these points [6]. Thus, observation on two dimensional directions are called as circular data and in three dimensions are called as spherical data. Circular data which can be represented by an angle $\theta \in [-\pi, \pi)$ or $\theta \in [0, 2\pi)$. The angle is periodic, i.e. $\theta = \theta + 2\pi m, m \in Z$. Beside circular data, a random variable which having values on a semicircle and have the domain in $\left(-\frac{\pi}{2}, \frac{\pi}{2}\right)$ or $[0, \pi)$ can be called as axial or semi-circular data [2].

2.2 Circular Models

Circular models is the model of circular random variable. The circular models can be modelled by its distribution function. Mardia and Jupp [8] describe some families of circular models such as Poisson, Uniform, Von Mises, Cardioid, Wrapped Normal, Wrapped Poisson and Wrapped Cauchy. The distribution function of circular is defined as the function on the whole real line and the distribution can be regarded as that of a random angle θ. As other word we can say that a circular distribution can be represent as a probability distribution. The distribution of this circular variable also can be described by its characteristic function which is defined only at integer values, the next term it's called as p^{th} trigonometric moment of θ and can be write as (1) [7].

$$\varphi_\theta(p) = \int_0^{2\pi} e^{ip\theta} dF(\theta), \quad p = 0, \pm 1, \pm 2, \cdots \qquad (1)$$

Circular models can be generated from any known probability distribution, [7] describe a few methods such as: a) By wrapping a linear distribution around the unit circle. Rao, Girija and Devaraaj also derived the characteristics of wrapped gamma [3]. Shogo Kato and M.C Jones make an extended family of circular distribution related to wrapped Cauchy distributions via Brownian motion [12], b) By characterizing properties, c) offset distribution, and d) stereographic projection

A continuous random variable, X on the real line is said to have the Erlang distribution [11] with scale parameter $\lambda > 0$ and shape parameter r, with notation $X \sim ERL(\alpha, r)$ and the probability density function of random variable, $f(x)$ defined as (2)

$$f(x, \lambda, r) = \frac{1}{\lambda^r (r-1)!} x^{r-1} e^{-\frac{x}{\lambda}}, \ x > 0 \qquad (2)$$

The random variable X that equals the interval length until r counts occur in a Poisson has an Erlang random variable for $r = 1, 2, \ldots$

2.3 Measure of Location and Dispersion

The distribution of the circular data can be modelled as the probability density function (pdf) and the cumulative density function (cdf). If $f(\theta)$ denote the pdf of the random variable θ and $F(\theta)$ denote the cdf of the random variable θ then for finding measure of location and dispersion of the distribution, we can work with the characteristic function [2]. As on the real line, directional data also have any model that be described via it's distribution. We can call it then as a circular distribution. Thus, circular distribution is a probability distribution whose total probability is concentrated on the circumference of a unit circle [7]. The trigonometric moment write by (1) which can be wrote as (3).

$$\varphi_p = \alpha_p + i\beta_p \qquad (3)$$

Where

$\alpha_p = E[\cos p\theta]$, $|\alpha_p| \le 1$

And

$\beta_p = E[\sin p\theta]$, $|\beta_p| \le 1$

The characteristic function of a random angle θ is the doubly infinite sequence of complex numbers [7]. For $p \ge 0$, (3) can be write as (4)

$$\varphi_p = \rho_p + e^{i\mu_p} \tag{4}$$

For the value of $p = 1$. The $\rho_1 = \rho$ is called the mean resultant length and the $\mu_1 = \mu$ is called the mean direction [8]. Mean resultant length and the mean direction can be obtained from (5).

$$\rho_p = \sqrt{\alpha_p^2 + \beta_p^2} \tag{5}$$

And

$$\mu_p = \arctan\left(\frac{\beta_p}{\alpha_p}\right) \tag{6}$$

The dispersion measure of circular variable random can be called as circular variance, υ. The circular variance of a random angle θ is defined as [7]:

$$v = 1 - \rho \tag{7}$$

2.4 Stereographic Projection

Stereographic projection or Mobius (bilinear) transformation was proposed by Minh and Farnum [9]. Stereographic projection is defined as (8).

$$\theta = 2\tan^{-1}\left(\frac{x-u}{v}\right) \tag{8}$$

Stereographic projection is used to induce distribution on the circle. if x is random variable chosen on real line and $x \in (-\infty, \infty)$ then the inverse stereographic projection defined by a one to one mapping given by (9)

$$x = u + v\tan\left(\frac{\theta}{2}\right) \tag{9}$$

Abe *et al* [1] defined θ as random variable angle on the circle of radius v, $u \in Z$ and $v > 0$, $v = \sqrt{m}$, $m = 2n + 1, n = 0,1,2,\ldots$

Radhika *et al* [11] proposed a model for periodic data whose density is given by (10).

$$g(\theta) = vl\left[\frac{1+\tan^2\left(\frac{l\theta}{2}\right)}{2}\right] f\left(u + v\tan\frac{\theta}{2}\right) \tag{10}$$

For $v > 0, l \in N$

3. Results

3.1 The Stereographic Erlang Circular Distribution

If $g(\theta)$ denote the probability density function (pdf) of the random angle θ on the unit circle. Let $F(x), f(x)$ denote the cumulative density function and (pdf) of the random variable x on the real line. The probability density function (pdf) of the random variable Erlang circular can be induced by an inverse stereographic projection follows with Radhika *et al* [11] methodology. Later we can called the model as stereographic Erlang circular.

Let the pdf of Erlang random variable X on real line, $f(x)$. Then applying inverse stereographic projection defined one to one mapping (8), (9) and (10) we can find the pdf of stereographic Erlang circular random variable, $g(\theta)$.

$$g(\theta) = v\left[\frac{1+\tan^2\left(\frac{\theta}{2}\right)}{2}\right] f\left(u + v\tan\left(\frac{\theta}{2}\right)\right)$$

$$g(\theta) = \frac{v}{2}\sec^2\left(\frac{\theta}{2}\right) f\left(u + v\tan\left(\frac{\theta}{2}\right)\right)$$

First, we looking for the value of $f\left(u + v\tan\left(\frac{\theta}{2}\right)\right)$

i.e $f(x) = \frac{1}{\lambda^r(r-1)!}x^{r-1}e^{-\frac{x}{\lambda}}$

$$f\left(u + v\tan\left(\frac{\theta}{2}\right)\right) = \frac{1}{\lambda^r(r-1)!}\left(u + v\tan\left(\frac{\theta}{2}\right)\right)^{r-1} e^{-\frac{\left(u+v\tan\left(\frac{\theta}{2}\right)\right)}{\lambda}}, \lambda > 0, r = 1,2,\ldots$$

Let $u = 0$ we can find,

$$f\left(v\tan\left(\frac{\theta}{2}\right)\right) = \frac{1}{\lambda^r(r-1)!}\left(v\tan\left(\frac{\theta}{2}\right)\right)^{r-1} e^{-\frac{\left(v\tan\left(\frac{\theta}{2}\right)\right)}{\lambda}}, \lambda > 0, r = 1,2,\ldots$$

So,

$$g(\theta) = \frac{v}{2}\sec^2\left(\frac{\theta}{2}\right)\frac{1}{\lambda^r(r-1)!}\left(v\tan\left(\frac{\theta}{2}\right)\right)^{r-1} e^{-\frac{\left(v\tan\left(\frac{\theta}{2}\right)\right)}{\lambda}}, \lambda > 0, r = 1,2,\ldots$$

Let $k = \frac{v}{\lambda}$

Because of $v > 0$ and $\lambda > 0$ then $k > 0$ and the pdf of random variable as follow (11)

$$g(\theta) = \frac{k^r}{2}\sec^2\left(\frac{\theta}{2}\right)\tan^{r-1}\left(\frac{\theta}{2}\right)e^{-k\tan\left(\frac{\theta}{2}\right)} \quad (11)$$

Where

$k > 0, r = 1,2,\ldots, \theta \in [-\pi,\pi)$ or $\theta \in [0,2\pi)$

If r =1, k =1/4, ½ and ¾ then the probability density function from stereographic Erlang circular can be graphed as figure 1.

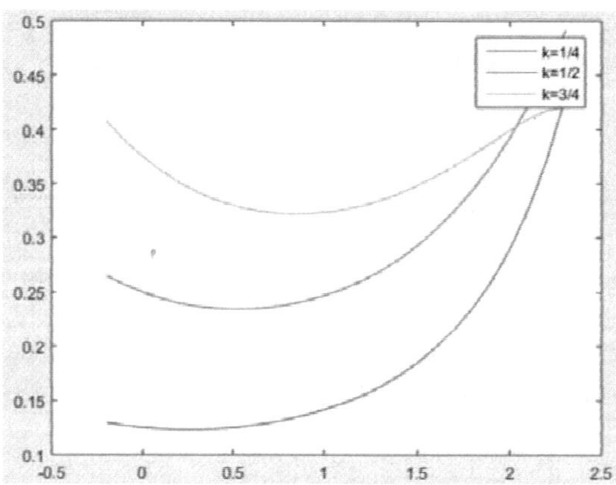

Figure 1. r =1, k=1/4, ½, ¾ for $\boldsymbol{\theta} \in \left[-\frac{3}{4}\pi, \frac{3}{4}\pi\right)$

Figure 1 is pdf of Erlang circular with different scale parameters, k. From figure 1, we can that the probability density function of Erlang circular will be going up as well as it's scale.

If k =1/2, r =1,2,3 then the probability density function can be graphed as figure 2.

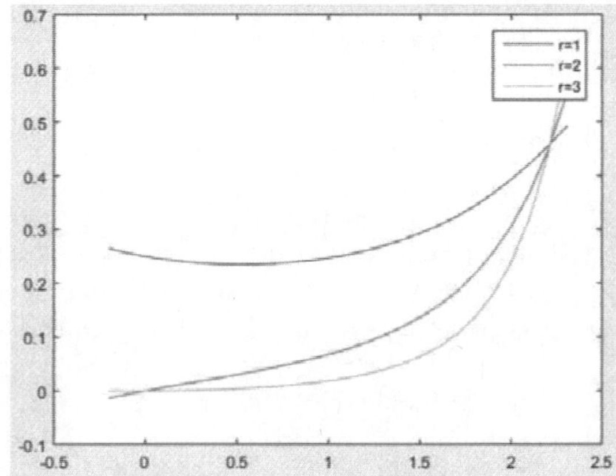

Figure 2. k=1/2, r =1, 2 and 3

Figure 2 is the pdf of Erlang circular with different shape parameters, r. Opposite with the figure 1, the probability density function will be down as the shape goes up. From figure 1 and figure 2, the shape of probability density function of stereographic Erlang circular looks opposite with the Exponential circular stereographic [4].

The descriptive measure of any random variable can be measured from the center and dispersion of measurements. And in circular random variable, both measurement can showed by each mean resultant length, mean direction and measure of dispersion. For k=0.1, 0.3, 0.5, 0.75, 1 and r=1, we can find the mean resultant length ρ , mean direction, μ and measure of dispersion υ as Table 1.

Table 1. Descriptive Measure of Stereographic Erlang Circular

k	α_1	β_1	ρ	μ	υ
0.1	-580.442	-22.8789	580.8927	0.039396	-579.893
0.3	-1.97E-08	-6.5415E-06	6.54E-06	1.567792	0.999993
0.5	-6.6475E-13	-2.1593E-12	2.26E-12	1.272149	1
0.75	-5.4213E-20	-1.741E-19	1.82E-19	1.268923	1
1	-4.4213E-27	-1.412E-26	1.48E-26	1.267344	1

From the Table 1 above, we can see that the mean resultant length of the circular random variable will be going small whenever k is large. Thus, the mean direction looks fluctuate when the measure of dispersion is large as the value of k.

Thus, for k =0.1, 0.3, 0.5 and 1, we can graph the doubly infinite sequence of complex numbers (α_p, β_p) as Figure 3.

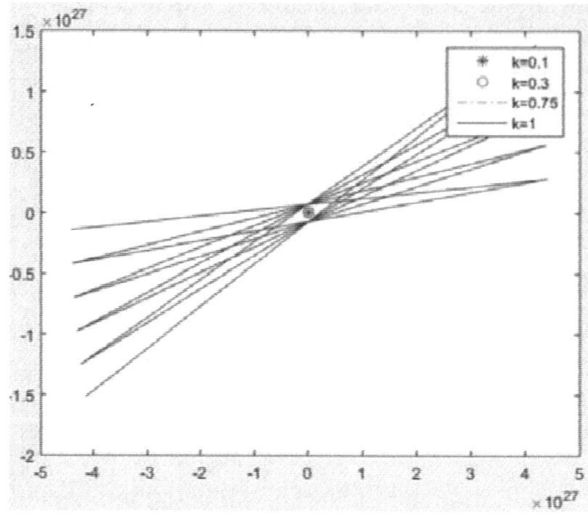

Figure 3. Doubly Infinite Sequence for k =0.1, 0.3, 0.5, 1

In same domain $\boldsymbol{\theta} \in \left[-\frac{3}{4}\pi \frac{3}{4}\pi\right)$, we also can graph the trigonometric moment with same value of k =0.1, 0.3, 0.5, 1 as figure 4.

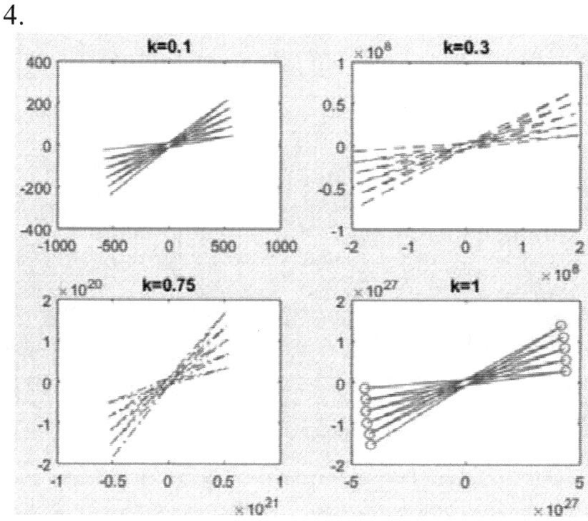

Figure 4. Trigonometric Moment for k =0.1, 0.3, 0.5, 1

From figure 3 and 4 we can see that the trigonometric moment can graphed as doubly infinity sequence and the trigonometric moment will be large as the value of k .

4. Conclusion

From the discussion above, it can be concluded that using stereographic projection, we can derive a new circular model, viz stereographic Erlang circular. From the descriptive measure of the model, we can conclude that the mean resultant length will be small whenever $\boldsymbol{k}$ is large. Thus, the mean direction looks fluctuate when the measure of dispersion will be large as the value of $\boldsymbol{k}$. Thus, the trigonometric moments will be large as the value of the scale parameter of the stereographic Erlang circular.

5. References

[1] Abe T, Shimizu K and Pewsey A 2010 Symmetric unimodal models for directional data motivated by inverse stereographic projection *Journal of the Japan Statistical Society* **40**(1) 45–61

[2] Byoung J A and Hyoung M K A 2008 New family of semicircular models: The semicircular Laplace Distribution *Communications of the Korean Statistical Society* **15**(5) 775-81

[3] Dattatreya Rao, S V S Girija and V J Devaraaj 2013 On Characteristics of Wrapped Gamma Distribution IRACST- Engineering Science and Technology ; *An International Journal (ESTIJ)*, ISSN : 2250-349 **3**(2) April
[4] Getut Pramesti and Yanming Jin Exponential circular distribution motivated by inverse stereographic projection 2016 *International Journal of Applied Mathematics and Statistics* ISSN:0973-7545 **54**(2) 114-22
[5] Getut Pramesti The Stereographic Semicircular Chi Square Models 2015 *Far East Journal of Theoretical Statistics* **51**(3) 49-58
[6] Getut Pramesti The Characteristics of Stereographic Beta Circular Model 2017 *Far East Journal of Probability and Statistic* **53** (2), 77-88
[7] Jammalamadaka SR and SenGupta A 2001 *Topics in Circular Statistics* (New York: World Scientific)
[8] Mardia K V and Jupp PE 2000 *Directional Statistics* (England: John Wiley & Sons)
[9] Minh D L P and Farnum N R 2000 *Commun. Stat.-Theory Meth.* **32** 1-9
[10] Phani Y, Girija S V S and Dattatreya R Circular Model Induced by Inverse Stereographic On Extreme – Value Distribution 2012 *Engineering Science and Technology An International Journal* **2**(5) 881-88
[11] A J V Radhika, Y Phani, S V S Girija and A V Dattatreya Rao On Stereographic Semicircular Gamma 2013 *Proceedings of the National Conference on Recent Trends in Mathematical Computing*- NCRTMC. ISBN 978-93-82338-68-0
[12] Shogo Kato and M C Jones 2013 An Extended family of circular distributions related to wrapped Cauchy distribution via Brownian motion *Bernoulli* **19**(1) 154-171 doi : 10.3150/11-BEJ397

IOP Conf. Series: Journal of Physics: Conf. Series **1022** (2018) 012002 doi:10.1088/1742-6596/1022/1/012002

Parameter estimation of multivariate multiple regression model using bayesian with non-informative Jeffreys' prior distribution

D R S Saputro[1], F Amalia[2], P Widyaningsih[3] and R C Affan[4]

[1,2,3]Department of Mathematics, Universitas Sebelas Maret

Corresponding email: dewiretnoss@staff.uns.ac.id, firdaamalia021@gmail.com, purnami_w@staff.uns.ac.id

Abstract. Bayesian method is a method that can be used to estimate the parameters of multivariate multiple regression model. Bayesian method has two distributions, there are prior and posterior distributions. Posterior distribution is influenced by the selection of prior distribution. Jeffreys' prior distribution is a kind of Non-informative prior distribution. This prior is used when the information about parameter not available. Non-informative Jeffreys' prior distribution is combined with the sample information resulting the posterior distribution. Posterior distribution is used to estimate the parameter. The purposes of this research is to estimate the parameters of multivariate regression model using Bayesian method with Non-informative Jeffreys' prior distribution. Based on the results and discussion, parameter estimation of β and Σ which were obtained from expected value of random variable of marginal posterior distribution function. The marginal posterior distributions for β and Σ are multivariate normal and inverse Wishart. However, in calculation of the expected value involving integral of a function which difficult to determine the value. Therefore, approach is needed by generating of random samples according to the posterior distribution characteristics of each parameter using Markov chain Monte Carlo (MCMC) Gibbs sampling algorithm.

1. Introduction

Multivariate multiple regression model is used to modeling the relationship between m responses and a single set of predictor variables (Johnson and Wichern [8]). Multivariate multiple regression model has unkown parameters. Parameters are the characteristic of population. The parameters value is obtained from parameter estimation. According to Bolstad [4], there are two statistical approaches that generally are used to estimate the parameters. The first is the classical approach which based on all information from the random sample. The second approach is Bayesian method which combine the information from the random sample and the information of previous research to estimate the parameters. In Bayesian method the information from the random sample is likelihood function and the information of previous research we get prior distribution. According to Harris et al. [6], basic of Bayesian method is to multiplying between likelihood function and prior distribution which proportional with posterior distribution. Posterior distribution is used to estimate the parameters and it is influenced by the selection of prior distribution.

Generally, selecting prior is based on the parameter information available or not. If the parameter information is available, we use informative prior. Informative prior has significant effect on posterior distribution and more subjective (Gelman et al. [5]). If the parameter information is not available, we use Non-informative prior. Non-informative prior are more objective than most classical analyzes (Yang and Berger [15]). Jeffreys' prior is a kind of Non-informative prior which commonly used in multivariate multiple Bayesian regression model. This prior is good to overcome invariant parameters.

Basically, Bayesian method states that future events can be predicted with terms of previous events that have occurred (Saputro et al. [11]). This research discuss about parameters estimation of multivariate multiple regression model using Bayesian method with Non-informative Jeffreys' prior.

2. The Previous Research

Here are some previous research about this material: Sun and Berger [13] research about the kind of Non-informative prior that could be used for normal multivariate models. Robert and Rousseau [10] research about Non-informative Jeffreys' prior in 2010. Iswari et al. [7] apply Bayesian method on simple linear regression model with Non-informative prior in 2014, while Prasdika et al. [9] in 2016 also examined apply Bayesian methods on simple regression model with informative prior. In the same year, Sinay and Hsu [12] apply the Bayesian method on multivariate regression model with uniform prior. While research about the application of Non-informative Jeffreys' prior on regression model is done by Amalia et al. [1].

3. Results and Discussion

3.1 Multivariate Multiple Regression Model

According to Johnson and Wichern [8], each response variable $\mathbf{Y_1, Y_2, \dots, Y_m}$ is assumed to follow its own regression model, so that

$$\mathbf{Y_1 = \beta_{01} + \beta_{11}X_1 + \beta_{21}X_2 + \cdots + \beta_{r1}X_r + \varepsilon_1}$$
$$\mathbf{Y_2 = \beta_{02} + \beta_{12}X_1 + \beta_{22}X_2 + \cdots + \beta_{r2}X_r + \varepsilon_2}$$
$$\vdots$$
$$\mathbf{Y_m = \beta_{0m} + \beta_{1m}X_1 + \beta_{2m}X_2 + \cdots + \beta_{rm}X_r + \varepsilon_m}.$$

the error term $\boldsymbol{\varepsilon}^T = [\boldsymbol{\varepsilon_1}, \boldsymbol{\varepsilon_2}, \dots, \boldsymbol{\varepsilon_m}]$ has $E[\boldsymbol{\varepsilon}] = 0$, and $Var[\boldsymbol{\varepsilon}] = \Sigma$. Σ is matrix variance ordo m × m. Let $\boldsymbol{X_j} = [X_{j0}, X_{j1}, \dots, X_{jr}]$ denote the value of predictor variables for the jth trial, let $\boldsymbol{Y_j^T} = [Y_{j1}, Y_{j2}, \dots, Y_{jm}]$ be the responses, and $\boldsymbol{\beta_j} = [\beta_{0j}, \beta_{1j}, \dots, \beta_{jr}]$ is regression parameter. $\boldsymbol{Y_j}$ has multivariate normal distribution $\boldsymbol{Y_j} \sim N_m(\boldsymbol{X_j\beta}, \boldsymbol{\Sigma})$. The density function for normal multivariate is

$$f(\boldsymbol{Y_j}|\boldsymbol{\beta}, \boldsymbol{\Sigma}) = (2\pi)^{-\frac{m}{2}} |\Sigma|^{-\frac{1}{2}} exp\left(-\frac{1}{2}(\boldsymbol{Y_j} - \boldsymbol{X_j\beta})^T \Sigma^{-1}(\boldsymbol{Y_j} - \boldsymbol{X_j\beta})\right).$$

3.2 Bayesian Method

Basic of Bayesian method is to determine the likelihood function and prior distribusion, after that determine the posterior distribution from multiplying the likelihood function and prior distribusion. Parameter estimation is the expected value from posterior marginal distribution. Prior distribution function denote by $f(\theta)$ with θ is parameter. Likelihood function is joint probability function from Y as the random sampel, if θ is known and denote by $f(Y|\theta)$, and the posterior distribution function denote by $f(\theta|Y)$ as follows

$$f(\theta|Y) = \frac{f(\theta, Y)}{f(Y)} = \frac{f(\theta)f(Y|\theta)}{f(Y)} \propto f(\theta)f(Y|\theta).$$

3.3 Likelihood Function

The joint density function of m random variable $\mathrm{Y_{j1}, Y_{j2}, \dots, Y_{jm}}$ evaluated at $\mathrm{y_{j1}, y_{j2}, \dots, y_{jm}}$, with j=1,2,...,n say f(Y|θ), is referred to as a likelihood function, then

$$\mathrm{f(Y|\theta)} = \prod_{j=1}^{n} \mathrm{f(Y_j|\theta) = f(Y_1|\theta)f(Y_2|\theta). \dots . f(Y_n|\theta)}.$$

The likelihood function with random variable has multivariate normal distribution $\mathbf{Y_j} \sim \mathrm{N_m}(\mathbf{X_j}\beta, \boldsymbol{\Sigma})$, is

$$f(\boldsymbol{Y}\,|\boldsymbol{\beta},\boldsymbol{\Sigma}) = \prod_{j=1}^{n}(2\pi)^{-\frac{m}{2}}\,|\Sigma|^{-\frac{1}{2}}exp\left(-\frac{1}{2}(\boldsymbol{Y}_j - \boldsymbol{X}_j\boldsymbol{\beta})^T\,\Sigma^{-1}(\boldsymbol{Y}_j - \boldsymbol{X}_j\boldsymbol{\beta})\right)$$

$$= (2\pi)^{-\frac{nm}{2}}\,|\Sigma|^{-\frac{n}{2}}exp\left(tr\left(\sum_{j=1}^{n}-\frac{1}{2}(\boldsymbol{Y}_j - \boldsymbol{X}_j\boldsymbol{\beta})^T\,\Sigma^{-1}(\boldsymbol{Y}_j - \boldsymbol{X}_j\boldsymbol{\beta})\right)\right)$$

$$= (2\pi)^{-\frac{nm}{2}}\,|\Sigma|^{-\frac{n}{2}}exp\left(tr\left(-\frac{n}{2}(\overline{\boldsymbol{Y}} - \boldsymbol{X}_j\boldsymbol{\beta})^T\,\Sigma^{-1}(\overline{\boldsymbol{Y}} - \boldsymbol{X}_j\boldsymbol{\beta}) - \frac{1}{2}\Sigma^{-1}\boldsymbol{S}\right)\right)$$

with $\overline{\boldsymbol{Y}} = \frac{1}{n}\sum_{j=1}^{n}\boldsymbol{Y}_j$ and $\boldsymbol{S} = \sum_{j=1}^{n}(\boldsymbol{Y}_j - \overline{\boldsymbol{Y}})^T\,(\boldsymbol{Y}_j - \overline{\boldsymbol{Y}})$.

3.4 Non-informative Jeffreys' prior

According to Harris et al. [6] prior distribution is previous distribution information. Usually, knowledge of the prior will be available and included in the analysis as informative prior. Knowledge of the prior may be obtained from the opinions of experts or because it reuses the posterior distribution from the previous research, but if the knowledge of the prior is uncertain, lost, or ignored, non-informative prior is used.

Jeffreys' prior is kind of Non-informative prior. Jeffreys' prior is square root of Fisher information written by $f(\boldsymbol{\beta}) = [\boldsymbol{I}(\boldsymbol{\beta})]^{\frac{1}{2}}$ with $\boldsymbol{I}(\boldsymbol{\beta})$ is Fisher's information. According to Bain and Engelhardt [3], Fisher's information is written as

$$\boldsymbol{I}(\boldsymbol{\beta}) = -E_\beta\left(\frac{\partial^2\,log\,f(\boldsymbol{Y}|\boldsymbol{\beta})}{\partial\boldsymbol{\beta}^2}\right)$$

If $\boldsymbol{\theta} = (\theta_1, \theta_2, \ldots, \theta_p)^T$ is a vector, it is used $f(\boldsymbol{\theta}) = [\det\mathbf{I}(\boldsymbol{\theta})]^{\frac{1}{2}}$ with $\mathbf{I}(\boldsymbol{\theta})$ is Fisher's information matrix ordo (pxp). The Fisher Information is

$$\mathbf{I}_{ij}(\boldsymbol{\theta}) = -E_\theta\left(\frac{\partial^2\log f(\mathbf{Y}|\boldsymbol{\theta})}{\partial\theta_i\,\partial\theta_j}\right)$$

with $i = 1,2,\ldots,p$ and $j = 1,2,\ldots,p$.

Non-informative prior of multivariate multiple regression is $f(\boldsymbol{\theta})$ with $\boldsymbol{\theta} = (\boldsymbol{\beta}, \Sigma)$, assumsed $\boldsymbol{\beta}$ and Σ is independent so $f(\boldsymbol{\theta}) = f(\boldsymbol{\beta})f(\Sigma)$. Furthermore, a Non-informative prior distribution $f(\boldsymbol{\theta})$ is determined from the following steps.

Non-informative Jeffreys' prior for $\boldsymbol{\beta}$ is

$$\mathbf{I}(\boldsymbol{\beta}) = -E\left(\frac{\partial^2}{\partial\boldsymbol{\beta}^2}\log f(\mathbf{Y}|\boldsymbol{\theta})\right) = -E(-\Sigma^{-1}) = \Sigma^{-1}$$

$$f(\boldsymbol{\beta}) = \sqrt{\mathbf{I}(\boldsymbol{\beta})} = \sqrt{\Sigma^{-1}} = \Sigma^{-\frac{1}{2}}.$$

Multivariate multiple regression model with *m* dimensi has Non-informative Jeffreys' prior for $\boldsymbol{\beta}$

$$f(\boldsymbol{\beta}) = f(\beta_1)\,\times\,f(\beta_2)\,\times \ldots \times f(\beta_m) = \Sigma^{-\frac{1}{2}} \times \Sigma^{-\frac{1}{2}} \times \ldots \times \Sigma^{-\frac{1}{2}} = \Sigma^{-\frac{m}{2}}$$

Non-informative Jeffreys' prior for Σ is

$$\mathbf{I}(\Sigma) = -E\left(\frac{\partial^2}{\partial\Sigma^2}\log f(\mathbf{Y}|\boldsymbol{\theta})\right) = -E(-(\mathbf{Y}_j - \mathbf{X}_j\boldsymbol{\beta})^T\,\Sigma^{-3}(\mathbf{Y}_j - \mathbf{X}_j\boldsymbol{\beta})) = \Sigma^{-2}$$

$$f(\Sigma) = \sqrt{\mathbf{I}(\Sigma)} = \sqrt{\Sigma^{-2}} = \Sigma^{-1}.$$

Non-informative Jeffreys' prior for $\boldsymbol{\theta} = (\boldsymbol{\beta}, \sum)$ is

$$f(\boldsymbol{\theta}) = f(\beta_1, \beta_2, \dots, \beta_m) f(\boldsymbol{\Sigma}) = \boldsymbol{\Sigma}^{-\frac{m}{2}} \boldsymbol{\Sigma}^{-1} = \boldsymbol{\Sigma}^{-\frac{m+2}{2}}$$

Thus Non-informative Jeffreys' prior for multivariate multiple regression models is $f(\boldsymbol{\beta}, \sum) = \sum^{-\frac{m+2}{2}}$.

3.5 Posterior Distribution

After obtaining likelihood function and prior distribution, we determine the posterior distribution. The posterior distribution function for multivariate multiple regression function is given below

$$\begin{aligned} f(\boldsymbol{\beta}, \boldsymbol{\Sigma}|\mathbf{Y}) &\propto |\boldsymbol{\Sigma}|^{-\frac{m+2}{2}} (2\pi)^{-\frac{nm}{2}} |\boldsymbol{\Sigma}|^{-\frac{n}{2}} \exp\left(\operatorname{tr}\left(-\frac{n}{2}(\overline{\mathbf{Y}} - \mathbf{X}_j\boldsymbol{\beta})^T \boldsymbol{\Sigma}^{-1}(\overline{\mathbf{Y}} - \mathbf{X}_j\boldsymbol{\beta}) - \frac{1}{2}\boldsymbol{\Sigma}^{-1}\mathbf{S}\right)\right) \\ &\propto |\boldsymbol{\Sigma}|^{-\frac{1}{2}} |\acute{0}|^{-\frac{m+n+1}{2}} \exp\left(\operatorname{tr}\left(-\frac{n}{2}(\overline{\mathbf{Y}} - \mathbf{X}_j\boldsymbol{\beta})^T \boldsymbol{\Sigma}^{-1}(\overline{\mathbf{Y}} - \mathbf{X}_j\boldsymbol{\beta})\right)\right) \exp\left(\operatorname{tr}\left(-\frac{1}{2}\boldsymbol{\Sigma}^{-1}\mathbf{S}\right)\right) \\ &\propto \left(|\boldsymbol{\Sigma}|^{-\frac{1}{2}} \exp\left(\operatorname{tr}\left(-\frac{n}{2}(\overline{\mathbf{Y}} - \mathbf{X}_j\boldsymbol{\beta})^T \Sigma^{-1}(\overline{\mathbf{Y}} - \mathbf{X}_j\boldsymbol{\beta})\right)\right)\right) \left(|\boldsymbol{\Sigma}|^{-\frac{m+n+1}{2}} \exp\left(\operatorname{tr}\left(-\frac{1}{2}\Sigma^{-1}\mathbf{S}\right)\right)\right) \\ &\propto \left(\left|\frac{\boldsymbol{\Sigma}}{n}\right|^{-\frac{1}{2}} \exp\left(\operatorname{tr}\left(-\frac{1}{2}(\overline{\mathbf{Y}} - \mathbf{X}_j\boldsymbol{\beta})^T \left(\frac{\Sigma}{n}\right)^{-1} (\overline{\mathbf{Y}} - \mathbf{X}_j\boldsymbol{\beta})\right)\right)\right) \\ &\quad \left(|\boldsymbol{\Sigma}|^{-\frac{m+n+1}{2}} \exp\left(\operatorname{tr}\left(-\frac{1}{2}\mathbf{S}\,\boldsymbol{\Sigma}^{-1}\right)\right)\right) \end{aligned}$$

The posterior marginal distribution for $\boldsymbol{\beta}$ is $N_m\left(\overline{\boldsymbol{Y}}, \frac{\Sigma}{n}\right)$ and posterior marginal distribution for $\boldsymbol{\Sigma}$ is $W^{-1}(\boldsymbol{S}^{-1}, n)$.

3.6 Parameter Estimation

The parameter estimation is determined from the expected value or the average value of marginal posterior distribution for the mean $\boldsymbol{\beta}$ and variance $\boldsymbol{\Sigma}$. Given $\boldsymbol{\beta}$ is the matrix of regression parameter of size (r+1)×m so the parameter estimate for $\boldsymbol{\beta}$ is

$$\begin{aligned} \widehat{\boldsymbol{\beta}} &= E[\boldsymbol{\beta}] = E[\beta_{01}, \beta_{11}, \dots, \beta_{(r+1)\times m}] \\ &= \int_{-\infty}^{\infty} \int_{-\infty}^{\infty} \dots \int_{-\infty}^{\infty} \boldsymbol{\beta} \left|\frac{\Sigma}{n}\right|^{-\frac{1}{2}} exp\left(tr\left(-\frac{1}{2}(\overline{\boldsymbol{Y}} - \boldsymbol{X}_j\boldsymbol{\beta})^T \left(\frac{\Sigma}{n}\right)^{-1} (\overline{\boldsymbol{Y}} - \boldsymbol{X}_j\boldsymbol{\beta})\right)\right) d\beta_{(r+1)\times m} \dots d\beta_{11} d\beta_{01} \end{aligned}$$

and the parameter estimate for $\sum$ is

$$\begin{aligned} \widehat{\boldsymbol{\Sigma}} &= E[\boldsymbol{\Sigma}] = E[\boldsymbol{\Sigma}_{11}, \boldsymbol{\Sigma}_{12}, \dots, \boldsymbol{\Sigma}_{m\times m}] \\ &= \int_{-\infty}^{\infty} \int_{-\infty}^{\infty} \dots \int_{-\infty}^{\infty} \boldsymbol{\Sigma} |\boldsymbol{\Sigma}|^{-\frac{m+n+1}{2}} exp\left(tr\left(-\frac{1}{2}\boldsymbol{S}\,\boldsymbol{\Sigma}^{-1}\right)\right) d\boldsymbol{\Sigma}_{m\times m} \dots d\boldsymbol{\Sigma}_{12} d\boldsymbol{\Sigma}_{11}. \end{aligned}$$

When we estimate $\boldsymbol{\beta}$ and $\boldsymbol{\Sigma}$ using integral dimensions with complex functions that are difficult to determine the solution. It is necessary to approach the complex functions by generating random samples using MCMC Gibbs sampling algorithm. According to Walsh [14] MCMC is used to simulate sampling from posterior distributions. MCMC algorithms has two kind of algorithms, there are Metropolis-Hastings and Gibbs sampling. The Gibbs sampling algorithm is a special occurrence of Metropolis-Hastings, since the parameter is estimated to be more than one (Asriadi [2]). In the Bayesian multivariate regression model there are two parameters that are estimated so that the algorithm used is Gibbs sampling algorithm.

The next step is to generate a random sample of posterior marginal distribution using MCMC Gibbs sampling algorithm. The parameter estimate is obtained after calculating the mean value of the random sample of the generated result.

The Gibbs sampling algorithm for multivariate multiple regression model is

(1) Construck a multivariate multiple regression model to obtain $\boldsymbol{\beta}$ and $\boldsymbol{\Sigma}$.

(2) Initialize value $\boldsymbol{\beta}^{(0)}$ and $\boldsymbol{\Sigma}^{(0)}$ based on the information from first step.

(3) Obtain $\mathbf{Y}_j^{(1)} \sim N_m(\mathbf{X}_j\boldsymbol{\beta}^{(0)}, \boldsymbol{\Sigma}^{(0)})$.

(4) Generate $\boldsymbol{\beta}^{(1)}|\mathbf{Y}_j^{(1)} \sim N_m\left(\overline{Y}, \frac{\boldsymbol{\Sigma}}{n}|\boldsymbol{\beta}^{(0)}\right)$ with $\overline{Y} = \frac{1}{n}\sum_{j=1}^{n} \mathbf{Y}_j$.

(5) Generate $\boldsymbol{\Sigma}^{(1)}|\mathbf{Y}_j^{(1)} \sim \mathrm{IW}(\mathbf{S}^{-1}, \mathrm{n}|\boldsymbol{\Sigma}^{(0)})$.

(6) Repeat step 3,4, and 5 as much as M repetition until we get $\boldsymbol{\beta}^{(M)}|\mathbf{Y}_j^{(M)} \sim \mathrm{N_m}\left(\bar{\mathrm{Y}}, \frac{\Sigma}{\mathrm{n}}|\boldsymbol{\beta}^{(M-1)}\right)$ and $\boldsymbol{\Sigma}^{(M)}|\mathbf{Y}_j^{(M)} \sim \mathrm{IW}(\mathbf{S}^{-1}, \mathrm{n}|\boldsymbol{\Sigma}^{(M-1)})$.

The Gibbs sampling algorithm generates a sample sequence $(\boldsymbol{\beta}^{(1)}, \boldsymbol{\beta}^{(2)}, .., \boldsymbol{\beta}^{(M)})$ dan $(\boldsymbol{\Sigma}^{(1)}, \boldsymbol{\Sigma}^{(2)}, \ldots, \boldsymbol{\Sigma}^{(M)})$. The parameter estimation for $\boldsymbol{\beta}$ is $\widehat{\boldsymbol{\beta}} = \frac{1}{M}\sum_{i=1}^{M}\boldsymbol{\beta}^{(i)}$ and the parameter estimation for $\boldsymbol{\Sigma}$ is $\widehat{\boldsymbol{\Sigma}} = \frac{1}{M}\sum_{i=1}^{M}\boldsymbol{\Sigma}^{(i)}$.

3.7 Application

Bayesian method with Non-informative Jeffreys' prior is used to estimate multivariat multiple regression model from the data. The data is generated from multivariate normal distribution with number of data n = 88, has two dimensions (m = 2), the mean $\bar{y}^T = (19715,\ 6937)$, and the variance $\Sigma = \begin{pmatrix} 2495993 & 11264928 \\ 11264928 & 5329880 \end{pmatrix}$. Data generated using R software. Here are the steps in estimating the parameters.

From the mean and variance we can determine the generate data of posterior distribution function using MCMC Gibbs sampling, so we get $\boldsymbol{\beta}^{(i)}|\boldsymbol{Y}_j^{(i)} \sim N_m\left(\bar{Y}, \frac{\Sigma}{n}|\boldsymbol{\beta}^{(i-1)}\right)$ and $\boldsymbol{\Sigma}^{(i)}|\boldsymbol{Y}_j^{(i)} \sim IW(\boldsymbol{S}^{-1}, n|\boldsymbol{\Sigma}^{(i-1)})$, with i=1,2,3,...,M. The parameters estimation in determine from average of the generate data of posterior distribution function. Parameters estimation is shown by Tabel 1.

Table 1. Parameter estimation

Parameter	Mean	percentil 0.25	Percentil 0.975	Explanation
$\widehat{\beta_{01}}$	0.0056794	0.0046285	0.0068708	Significant
$\widehat{\beta_{11}}$	0.0055101	0.0050309	0.0059481	Significant
$\widehat{\beta_{21}}$	79.69864	29.7474	129.1843	Significant
$\widehat{\beta_{02}}$	-0.000897	-0.0008824	-0.001864	Significant
$\widehat{\beta_{12}}$	0.0031068	0.0028224	0.003394	Significant
$\widehat{\beta_{22}}$	82.15982	40.84126	124.8749	Significant
$\widehat{\Sigma_{11}}$	807413.4	598997.2	1089146	Significant
$\widehat{\Sigma_{21}}$	211616.5	100467.5	337936.4	Significant
$\widehat{\Sigma_{22}}$	314092.5	233559.6	423550.0	Significant

The parameter estimation of Bayesian multivariate multiple regression model given below.

$$\hat{\beta} = \begin{pmatrix} \beta_{01} & \beta_{02} \\ \beta_{11} & \beta_{12} \\ \beta_{21} & \beta_{22} \end{pmatrix} = \begin{pmatrix} 0.0056794 & -0.000897 \\ 0.0055101 & 0.0031068 \\ 79.69864 & 82.15982 \end{pmatrix}$$

$$\hat{\Sigma} = \begin{pmatrix} \Sigma_{11} & \Sigma_{21} \\ \Sigma_{21} & \Sigma_{22} \end{pmatrix} = \begin{pmatrix} 807413.4 & 211616.5 \\ 211616.5 & 314092.5 \end{pmatrix}$$

IOP Conf. Series: Journal of Physics: Conf. Series **1022** (2018) 012002 doi:10.1088/1742-6596/1022/1/012002

4. Conclusions

Based on results and discussion, the following conclusions is obtained. Parameter estimation of multivariate multiple regression model using Bayesian method was obtained by determining the expected value of the posterior marginal distribution f (β |Y) and f (Σ |Y). However, in calculation of parameter estimation, involving integral function which is difficult to determine the solution, so we use sampling with Markov chain Monte Carlo (MCMC) Gibbs sampling algorithm as approach in estimating its parameters. The result of MCMC Gibbs sampling algorithm is obtained by the matrix random sample $\left(\beta^{(1)}, \beta^{(2)}, \ldots, \beta^{(M)}\right)$ and $\left(\Sigma^{(1)}, \Sigma^{(2)}, \ldots, \Sigma^{(M)}\right)$. The parameter estimation for β is $\widehat{\beta} = \frac{1}{M}\sum_{i=1}^{M} \beta^{(i)}$ and the parameter estimation for Σ is $\widehat{\Sigma} = \frac{1}{M}\sum_{i=1}^{M} \Sigma^{(i)}$.

5. References

[1] Amalia F, D R S Saputro and T J Parmaningsih 2016 *Estimasi Parameter Model Regresi Linier Sederhana Bayes dengan Distribusi Prior Noninformatif* Seminar Nasional 2016 Matematika dan Pendidikan Matematika

[2] Asriadi A 2007 Simulasi Stokastik menggunakan Algoritma Metropolis Hastings *JMAP (Jurnal Matematika dan Pembelajaran) Jurusan Matematika FMIPA UNJ* **6**(2) 45-52

[3] Bain L J and M Engelhardt 1992 *Introduction to probability and mathematical statistics* Second ed. (Buxbury Press, Inc., California)

[4] Bolstad W M 2007 *Introduction to Bayesian statistics* Second ed. (New Jersey: Wiley)

[5] Gelman A, J B Carlin, H S Stern, D B Dunson, A Vehtari and D B Rubin, *Bayesian Data Analysis* Second ed. (New York: Chapman and Hall)

[6] Harris P, M Cox, C Matthews, I Smith and L Wrigh 2015 *a guide to Bayesian inference for regression problems* European Metrologu Research Programme (EMRP)

[7] Iswari A A I A C, I W Sumarjaya and I G A M Srinadi 2014 Analisis Regresi Bayes Linear Sederhana dengan Prior Noninformatif *E-Jurnal Matematika* **3**(2) 38-44

[8] Johnson R and D W Wichern 2007 *Applied multivariate statistical analysis* Sixth ed. (Pearson Education, Inc., America)

[9] Prasdika D A, D R S Saputro and T J Parmaningsih 2016 *Estimasi parameter model regresi linier sederhana bayes dengan distribusi prior informative* Seminar Nasional 2016 Matematika dan Pendidikan Matematika

[10] Robert C P and J Rousseau and Harold Jeffreyss 2009 Theory of Probability Revisited *Statistical Science* **24**(2) 141-72

[11] Saputro D R S, P Widyaningsih, F Handayani, and N A Kurdhi 2016 Prior and Posterior Dirichlet Distribution on Bayesian Networks (BNs) *AIP Publishing* **1827**(36)

[12] Sinay M S and J S J Hsu 2014 Bayesian Inferensi of a multivariate regression model (Hindawi Publishing Corporation)

[13] Sun D and J O Berger 2006 *Objective priors for the multivariate normal model* (Benidorm)

[14] Walsh B 2004 Markov Chain Monte Carlo and Gibbs sampling *Lecture Notes for EEB* 581

[15] Yang R and J O Berger 1998 A *catalog of non-informative priors* Technical Report, Dake University, ISDS 97-42

IOP Conf. Series: Journal of Physics: Conf. Series **1022** (2018) 012003 doi:10.1088/1742-6596/1022/1/012003

On the Locating Edge Domination Number of Comb Product of Graphs

Dafik[1,2], I. H. Agustin[1,3], Moh. Hasan[1,3], R. Adawiyah[1,2], R. Alfarisi[1,4] and D.A.R. Wardani[1]

[1]CGANT University of Jember, Jember, Indonesia
[2]Mathematics Edu. Depart. University of Jember, Jember, Indonesia
[3]Mathematics Depart, University of Jember, Jember, Indonesia
[4]Elementary Sch. Teacher Edu. Depart., University of Jember, Indonesia

E-mail: d.dafik@unej.ac.id, ikahestiagustin@gmail.com, robiatul@unej.ac.id, alfarisi.fkip@unej.ac.id

Abstract. Let $G = (V, E)$ be a graph. For $e_1 \in E$, $N(e_1)$ denote the neighborhoods of e_1 in G. A set $D \subseteq E$ is a locating edge dominating set if every two distinct edges $e_1, e_2 \in E(G) \setminus D$ satisfy that $\emptyset \neq N(e_1) \cap D \neq N(e_2) \cap D \neq \emptyset$. The locating edge domination number $\gamma'_L(G)$ is the minimum cardinality of locating edge dominating set. The comb product between G and H, denoted by $G \rhd H$, is a graph obtained by one copy of G and $|V(G)|$ copies of H, and grafting the i vertex of G to the u_i in i-th copy of H. In this paper, we will analyze the locating edge dominating number of comb product of graphs and also find its best lower bound.

Keywords: Locating edge dominating set, locating edge domination number, comb product of graphs.

1. Introduction

Let G be a nontrivial, finite, simple, undirected and connected graphs, with vertex set $V(G)$, edge set $E(G)$ and with no isolated vertex. The graph G is a pair of the set $V(G)$ and $E(G)$, for more detail definition of graph see [6, 2, 10, 11].

An edge in graph G dominates it self and its adjacent edges. A subset S' of $E(G)$ is called an edge dominating set of G if every edge not in S' is adjacent to some edge in S'. The edge dominating number of G, denoted by $\gamma'(G)$, of G is the minimum cardinality taken over all edge dominating sets of G. The concept of edge dominating was introduced by Hedetniemi and Mitchel [7]. Arumagam and Velammal [1] studied the edge domination number of graphs. Furthermore, a new development in edge dominating set is locating edge dominating set. The locating edge dominating set is a new concept in edge dominating set study. A set $D \subseteq E$ is a locating edge dominating set if every two edges $e_1, e_2 \in E(G) \setminus D$ satisfy that $\emptyset \neq N(e_1) \cap D \neq N(e_2) \cap D \neq \emptyset$. The locating edge domination number $\gamma'_L(G)$ is the minimum cardinality of locating edge dominating set. The concept of a locating dominating set was firstly introduced and studied by Slater [8, 9, 3, 4, 5, 12].

In this paper, we mainly discuss about the locating edge domination number of comb product of graph. The comb product between G and H, denoted by $G \rhd H$, is a graph obtained by one copy of G and $\mid V(G) \mid$ copies of H, and grafting the i vertex of G to the u_i in i-th copy of H.

Some special graphs that we use in this paper are cycle graph, path graph, and star graph. Cycle is a single vertex with a self-loop or simple graph C with $|V_c| = |E_c|$ that is drawn such that all of its vertices and edges lie on a single circle. An $n - vertex$ of cycle graph is denoted by C_n [3]. A path is a simple graph with $|V_p| = |E_p| + 1$ that is drawn such that all of its vertices and edge lie on a single straight line. A path graph with m vertices and $m - 1$ edges is denoted by P_m [3].These three graphs are operated by using comb product operation. Then, we analyze their locating edge domination set as well as we determine their locating edge domination number. We also analyze the lower bound of locating edge domination number on comb product of graph G and H $(G \rhd H)$ [13].

2. Main results

From now, we will show the properties of the locating edge dominating number of comb product of graphs. Firstly, we discuss about the locating edge dominating number of star and cycle. The following theorems are useful for determining the locating edge domination number of comb product of the next theorem. The results are as follows.

Theorem 2.1 *Let S_n be a star graph with order $n \geq 2$, then the locating edge domination number of star graph is $\gamma'_L(S_n) = n - 1$*

Proof. The star graph S_n is a connected graph with vertex set $V(S_n) = \{x, x_i; 1 \leq i \leq n\}$ and edge set $E(S_n) = \{xx_i; 1 \leq i \leq n\}$. The cardinality of vertex set and edge set, respectively, are $|V(S_n)| = n + 1$ and $|E(S_n)| = n$. In order to prove $\gamma'_L(S_n) = n - 1$ with $n \geq 2$, we should show $\gamma'_L(S_n) \leq n - 1$ and $\gamma'_L(S_n) \geq n - 1$. First, we need to prove that the lower bound of the locating edge domination number of star graph is $\gamma'_L(S_n) \geq n-1$. Suppose $\gamma'_L(S_n) < n-1$. The locating edge dominating set in S_n is $D = \{xx_i; 1 \leq i \leq n - 2\}$. Thus, the cardinality of the edge dominating set is $|D| = n - 2$. The edge set in S_n without the element of edge dominator is $E - D = \{xx_i : n - 1 \leq i \leq n\}$. The intersection between the edge neighborhood $E - D$ with D are as follows:

$$N(xx_i : i = n - 1) \cap D = \{xx_i; 1 \leq i \leq n - 2\}$$
$$N(xx_i : i = n) \cap D = \{xx_i; 1 \leq i \leq n - 2\}$$

It is clearly to see that the intersection between the edge neighborhood and the dominator $N(xx_{n-1}) \cap D = \{xx_i; 1 \leq i \leq n - 2\}$ and $N(xx_n) \cap D = \{xx_i; 1 \leq i \leq n - 2\}$ is the same intersection. It contradicts the property of local edge dominating set. Therefore, the lower bound of locating edge domination number of S_n is $\gamma'_L(S_n) \geq n - 1$.

Furthermore, we will show that the upper bound of locating edge domination number of $\gamma'_L(S_n) \leq n - 1$. Let $D = \{xx_i; 1 \leq i \leq n - 1\}$ be the locating edge dominating set in S_n. The non-dominator set in S_n $E(G) - D = \{xx_i : i = n - 1\}$. Thus, we get the intersection between the neighborhood $N(E)$ and $E \in E(G) - D$ and the dominator set D is as follow:

$$N(xx_i : i = n - 1) \cap D = \{xx_i; 1 \leq i \leq n - 2\}$$

Based on the results above, it can be seen that the intersection between the neighborhood $N(E)$ with $E \in E(G) - D$ and the dominator D are different and it is not empty set. Thus, it can be concluded that $\gamma'_L(S_n) \leq n-1$. Since the lower bound and upper bound of locating edge domination number respectively, are $\gamma'_L(S_n) \geq n - 1$ and $\gamma'_L(S_n) \leq n - 1$. It can be concluded that the locating edge domination number of S_n is $\gamma'_L(S_n) = n - 1$. □

Theorem 2.2 *Let C_n be a cycle graph with order $n \geq 3$, then the locating edge domination number of cycle is $\gamma'_L(C_n) = \lceil \frac{2n}{5} \rceil$*

Proof. The cycle graph C_n is a connected graph with vertex set $V(C_n) = \{x_i : 1 \leq i \leq n\}$ and edge set $E(C_n) = \{x_i x_i + 1 : 1 \leq i \leq n-1\} \cup \{x_i x_1 : i = n\}$. The cardinality of vertices set and edges set, respectively, are $|V(C_n)| = n$ $|E(C_n)| = n$. We will describe the proof into five cases.
Case 1: For $n \equiv 0(\text{mod}5)$, we will show that the lower bound of the locating edge domination number of C_n is $\gamma'_L(C_n) \geq \frac{2n}{5}$. Suppose $\gamma'_L(C_n) < \frac{2n}{5}$. The locating edge dominating set of C_n is $D = \{x_i x_{i+1}; 1 \leq i \leq n-1, i \equiv 1(\text{mod}5)\} \cup \{x_i x_{i+1}; 1 \leq i \leq n-2, i \equiv 4(\text{mod}5)\}$. Thus, the cardinality of the edge dominating set is $|D| = \frac{2n}{5} - 1$. The edge set in C_n without the element of edge dominator is $E - D$. The intersection between the edge neighborhood $E - D$ with D are as follow:

$$N(x_i x_{i+1} : i \equiv 2(\text{mod}5)) \cap D = \{x_{i-1} x_i : i \equiv 2(\text{mod}5), 1 \leq i \leq n-2\}$$
$$N(x_i x_{i+1} : i \equiv 3(\text{mod}5)) \cap D = \{x_{i+1} x_{i+2} : i \equiv 3(\text{mod}5), 1 \leq i \leq n-3\}$$
$$N(x_i x_{i+1} : i \equiv 0(\text{mod}5)) \cap D = \{x_{i+1} x_{i+2}, x_{i-1} x_i : i \equiv 0(\text{mod}5), 1 \leq i \leq n-2\}$$
$$N(x_1 x_n) \cap D = \{x_1 x_2\}$$
$$N(x_n x_{n-1}) \cap D = \emptyset$$
$$N(x_{n-2} x_{n-1}) \cap D = \emptyset$$

It is clearly to see that the intersection between the edge neighborhood and the dominator $N(x_n x_{n-1}) \cap D = \emptyset$ and $N(x_{n-2} x_{n-1}) \cap D = \emptyset$ is the same intersection and an empty set. It contradicts the property of locating edge dominating set. Therefore, the lower bound of locating edge domination number of C_n is $\gamma'_L(C_n) \geq \frac{2n}{5}$.

Furthermore, we will show that the upper bound of locating edge domination number of C_n is $\gamma'_L(C_n) \leq \frac{2n}{5}$. Let a locating edge dominating set in C_n be $D = \{x_i x_{i+1}; 1 \leq i \leq n-1, i \equiv 1(\text{mod}5)\} \cup \{x_i x_{i+1}; 1 \leq i \leq n-1, i \equiv 4(\text{mod}5)\}$. The non-dominator set in C_n $E(G) - D$. Thus, We get intersection between the neighborhood $N(E)$ with $E \in E(G) - D$ and dominator set D is as follow:

$$N(x_i x_{i+1} : i \equiv 2(\text{mod}5)) \cap D = \{x_{i-1} x_i : i \equiv 2(\text{mod}5), 1 \leq i \leq n-2\}$$
$$N(x_i x_{i+1} : i \equiv 3(\text{mod}5)) \cap D = \{x_{i+1} x_{i+2} : i \equiv 3(\text{mod}5), 1 \leq i \leq n-2\}$$
$$N(x_i x_{i+1} : i \equiv 0(\text{mod}5)) \cap D = \{x_{i+1} x_{i+2}, x_{i-1} x_i : i \equiv 0(\text{mod}5), 1 \leq i \leq n-2\}$$
$$N(x_1 x_n) \cap D = \{x_1 x_2, x_n x_{n-1}\}$$

Based on the results above, it can be seen that the intersection between the neighborhood $N(E)$ with $E \in E(G) - D$ and dominator D is unique and it is not empty set. Thus, it can be concluded that $\gamma'_L(C_n) \leq n - 1$. Since, the lower bound and upper bound of locating edge domination number respectively, are $\gamma'_L(C_n) \geq \frac{2n}{5}$ and $\gamma'_L(C_n) \leq \frac{2n}{5}$. It can be concluded that the locating edge domination number of C_n is $\gamma'_L(C_n) = \frac{2n}{5}$.
Case 2: For $n \equiv 4(\text{mod}5)$, we will show that the lower bound of the locating edge domination number of C_n is $\gamma'_L(C_n) \geq \frac{2(n+1)}{5}$. Suppose $\gamma'_L(C_n) < \frac{2(n+1)}{5}$. The locating edge dominating set in C_n is $D = \{x_i x_{i+1}; 1 \leq i \leq n-1, i \equiv 1(\text{mod}5)\} \cup \{x_i x_{i+1}; 1 \leq i \leq n-2, i \equiv 4(\text{mod}5)\}$. Thus, the cardinality of the edge dominating set is $|D| = \frac{2(n+1)}{5} - 1$. The edge set in C_n without the element of edge dominator is $E - D$. The intersection between the edge neighborhood $E - D$ with D are as follow:

$$N(x_i x_{i+1} : i \equiv 2(\text{mod}5)) \cap D = \{x_{i-1} x_i : i \equiv 2(\text{mod}5), 1 \leq i \leq n-2\}$$
$$N(x_i x_{i+1} : i \equiv 3(\text{mod}5)) \cap D = \{x_{i+1} x_{i+2} : i \equiv 3(\text{mod}5), 1 \leq i \leq n-3\}$$
$$N(x_{n-1} x_n) \cap D = \emptyset$$
$$N(x_i x_{i+1} : i \equiv 0(\text{mod}5)) \cap D = \{x_{i+1} x_{i+2}, x_{i-1} x_i : i \equiv 0(\text{mod}5), 1 \leq i \leq n-2\}$$
$$N(x_n x_1) \cap D = \{x_1 x_2\}$$

IOP Conf. Series: Journal of Physics: Conf. Series **1022** (2018) 012003 doi:10.1088/1742-6596/1022/1/012003

It is clearly to see that the intersection between the edge neighborhood and the dominator $N(x_nx_{n-1}) \cap D = \emptyset$ is an empty set. It contradicts to property of locating edge dominating set. Therefore, the lower bound of locating edge domination number of C_n is $\gamma'_L(C_n) \geq \frac{2(n+1)}{5}$.

Furthermore, we will show that the upper bound of locating edge domination number of C_n is $\gamma'_L(C_n) \leq \frac{2(n+1)}{5}$. Let a locating edge dominating set of C_n be $D = \{x_ix_{i+1}; 1 \leq i \leq n-1, i \equiv 1(\text{mod}5)\} \cup \{x_1x_n, x_ix_{i+1}; 1 \leq i \leq n-2, i \equiv 4(\text{mod}5)\}$. The non-dominator set in C_n $E(G) - D$. Thus, we get intersection between the neighborhood $N(E)$ with $E \in E(G) - D$ and dominator set D is as follow:

$$N(x_ix_{i+1} : i \equiv 2(\text{mod}5)) \cap D = \{x_{i-1}x_i : i \equiv 2(\text{mod}5), 1 \leq i \leq n-2\}$$
$$N(x_ix_{i+1} : i \equiv 3(\text{mod}5)) \cap D = \{x_{i+1}x_{i+2} : i \equiv 3(\text{mod}5), 1 \leq i \leq n-3\}$$
$$N(x_{n-1}x_n) \cap D = \{x_1x_n\}$$
$$N(x_ix_{i+1} : i \equiv 0(\text{mod}5)) \cap D = \{x_{i+1}x_{i+2}, x_{i-1}x_i : i \equiv 0(\text{mod}5), 1 \leq i \leq n-2\}$$

Based on the results above, it can be seen that the intersection between the neighborhood $N(E)$ with $E \in E(G) - D$ and dominator D is unique and it is not empty set. Thus, it can be concluded that $\gamma'_L(C_n) \leq n-1$. Since, the lower bound and upper bound of locating edge domination number respectively, are $\gamma'_L(C_n) \geq \frac{2(n+1)}{5}$ and $\gamma'_L(C_n) \leq \frac{2(n+1)}{5}$. It can be concluded that the locating edge domination number of C_n is $\gamma'_L(C_n) = \frac{2(n+1)}{5}$.

Case 3: For $n \equiv 3(\text{mod}5)$, we will show that the lower bound of locating edge domination number of C_n is $\gamma'_L(C_n) \geq \frac{2(n+2)}{5}$. Suppose $\gamma'_L(C_n) < \frac{2(n+2)}{5}$. The locating edge dominating set in C_n is $D = \{x_ix_{i+1}; 1 \leq i \leq n-1, i \equiv 1(\text{mod}5)\} \cup \{x_ix_{i+1}; 1 \leq i \leq n-2, i \equiv 4(\text{mod}5)\}$. Thus, the cardinality of the edge dominating set is $|D| = \frac{2(n+2)}{5} - 1$. The edge set in C_n without the element of edge dominator is $E - D$. The results of intersection between the edge neighborhood $E - D$ with D are as follow:

$$N(x_ix_{i+1} : i \equiv 2(\text{mod}5)) \cap D = \{x_{i-1}x_i : i \equiv 2(\text{mod}5), 1 \leq i \leq n-2\}$$
$$N(x_{n-1}x_n) \cap D = \{x_{n-2}x_{n-1}\}$$
$$N(x_ix_{i+1} : i \equiv 3(\text{mod}5)) \cap D = \{x_{i+1}x_{i+2} : i \equiv 3(\text{mod}5), 1 \leq i \leq n-2\}$$
$$N(x_ix_{i+1} : i \equiv 0(\text{mod}5)) \cap D = \{x_{i+1}x_{i+2}, x_{i-1}x_i : i \equiv 0(\text{mod}5), 1 \leq i \leq n-2\}$$
$$N(x_nx_1) \cap D = \{x_2x_1\}$$

It is clearly to see that the intersection between the edge neighborhood and the dominator $N(x_nx_1) \cap D = \emptyset$ and $N(x_2x_3) \cap D = \emptyset$ is the same intersection. It contradicts the property of locating edge dominating set. Therefore, the lower bound of locating edge domination number of C_n is $\gamma'_L(C_n) \geq \frac{2(n+2)}{5}$.

Furthermore, we will show that the upper bound of locating edge domination number of C_n is $\gamma'_L(C_n) \leq \frac{2(n+2)}{5}$. Let a locating edge dominating set of C_n be $D = \{x_ix_{i+1}; 1 \leq i \leq n-1, i \equiv 1(\text{mod}5)\} \cup \{x_1x_n, x_ix_{i+1}; 1 \leq i \leq n-1, i \equiv 4(\text{mod}5)\}$. The non-dominator set in C_n $E(G) - D$. Thus, we get intersection between the neighborhood $N(E)$ with $E \in E(G) - D$ and dominator set D is as follow:

$$N(x_ix_{i+1} : i \equiv 2(\text{mod}5)) \cap D = \{x_{i-1}x_i : i \equiv 2(\text{mod}5), 1 \leq i \leq n-2\}$$
$$N(x_{n-1}x_n) \cap D = \{x_1x_2, x_{n-2}x_{n-1}\}$$
$$N(x_ix_{i+1} : i \equiv 3(\text{mod}5)) \cap D = \{x_{i+1}x_{i+2} : i \equiv 3(\text{mod}5), 1 \leq i \leq n-2\}$$
$$N(x_ix_{i+1} : i \equiv 0(\text{mod}5)) \cap D = \{x_{i+1}x_{i+2}, x_{i-1}x_i : i \equiv 0(\text{mod}5), 1 \leq i \leq n-2\}$$

IOP Conf. Series: Journal of Physics: Conf. Series **1022** (2018) 012003 doi:10.1088/1742-6596/1022/1/012003

Based on the results above, it can be seen that the intersection between the neighborhood $N(E)$ with $E \in E(G) - D$ and dominator D is unique and it is not empty set. Thus, it can be concluded that $\gamma'_L(C_n) \leq n-1$. Since, the lower bound and upper bound of locating edge domination number respectively, are $\gamma'_L(C_n) \geq \frac{2(n+2)}{5}$ and $\gamma'_L(C_n) \leq \frac{2(n+2)}{5}$. It can be concluded that the locating edge domination number of C_n is $\gamma'_L(C_n) = \frac{2(n+2)}{5}$.

Case 4: For $n \equiv 2(\text{mod}5)$, we will show that the lower bound of locating edge domination number of C_n is $\gamma'_L(C_n) \geq \frac{2n+1}{5}$. Suppose $\gamma'_L(C_n) < \frac{2n+1}{5}$. The locating edge dominating set in C_n is $D = \{x_i x_{i+1}; 1 \leq i \leq n-1, i \equiv 1(\text{mod}5)\} \cup \{x_i x_{i+1}; 1 \leq i \leq n-2, i \equiv 4(\text{mod}5)\}$. Thus, the cardinality of the edge dominating set is $|D| = \frac{2n+1}{5} - 1$. The edge set in C_n without the element of edge dominator is $E - D$. The results of intersection between the edge neighborhood $E - D$ with D are as follow:

$$N(x_i x_{i+1} : i \equiv 2(\text{mod}5)) \cap D = \{x_{i-1}x_i : i \equiv 2(\text{mod}5), 1 \leq i \leq n-2\}$$
$$N(x_{n-1}x_n) \cap D = \emptyset$$
$$N(x_1 x_n) \cap D = \{x_1 x_2\}$$
$$N(x_i x_{i+1} : i \equiv 3(\text{mod}5)) \cap D = \{x_{i+1}x_{i+2} : i \equiv 3(\text{mod}5), 1 \leq i \leq n-2\}$$
$$N(x_i x_{i+1} : i \equiv 0(\text{mod}5)) \cap D = \{x_{i+1}x_{i+2}, x_{i-1}x_i : i \equiv 0(\text{mod}5), 1 \leq i \leq n-3\}$$
$$N(x_{n-1}x_{n-2}) \cap D = \{x_{n-3}x_{n-2}\}$$

It is clearly to see that the intersection between the edge neighborhood and the dominator $N(x_n x_{n-1}) \cap D = \emptyset$ and $N(x_2 x_3) \cap D = N(x_n x_1) \cap D = \{x_1 x_2\}$ and $N(x_{n-2}x_{n-1}) \cap D = N(x_{n-4}x_{n-3}) \cap D = \{x_{n-2}x_{n-3}\}$ is the same intersection. It contradicts property of locating edge dominating set. Therefore, the lower bound of locating edge domination number of C_n is $\gamma'_L(C_n) \geq \frac{2n+1}{5}$.

Furthermore, we will show that the upper bound of locating edge domination number of C_n is $\gamma'_L(C_n) \leq \frac{2n+1}{5}$. Let a locating edge dominating set in C_n be $D = \{x_i x_{i+1}; 1 \leq i \leq n-1, i \equiv 1(\text{mod}5)\} \cup \{x_i x_{i+1}; 1 \leq i \leq n-1, i \equiv 4(\text{mod}5)\}$. The non-dominator set in C_n $E(G) - D$. Thus, we get intersection between the neighborhood $N(E)$ with $E \in E(G) - D$ and dominator set D is as follow:

$$N(x_i x_{i+1} : i \equiv 2(\text{mod}5)) \cap D = \{x_{i-1}x_i : i \equiv 2(\text{mod}5), 1 \leq i \leq n-2\}$$
$$N(x_1 x_n) \cap D = \{x_1 x_2, x_n x_{n-1}\}$$
$$N(x_i x_{i+1} : i \equiv 3(\text{mod}5)) \cap D = \{x_{i+1}x_{i+2} : i \equiv 3(\text{mod}5), 1 \leq i \leq n-2\}$$
$$N(x_i x_{i+1} : i \equiv 0(\text{mod}5)) \cap D = \{x_{i+1}x_{i+2}, x_{i-1}x_i : i \equiv 0(\text{mod}5), 1 \leq i \leq n-2\}$$

Based on the results above, it can be seen that the intersection between the neighborhood $N(E)$ with $E \in E(G) - D$ and dominator D is unique and it is not empty set. Thus, it can be concluded that $\gamma'_L(C_n) \leq n-1$. Since, the lower bound and upper bound of locating independent domination number respectively, are $\gamma'_L(C_n) \geq \frac{2n+1}{5}$ and $\gamma'_L(C_n) \leq \frac{2n+1}{5}$. It can be concluded that the locating edge domination number of C_n is $\gamma'_L(C_n) = \frac{2n+1}{5}$.

Case 5: For $n \equiv 1(\text{mod}5)$, we will show that the lower bound of locating edge domination number of C_n is $\gamma'_L(C_n) \geq \frac{2n+3}{5}$. Suppose $\gamma'_L(C_n) < \frac{2n+3}{5}$. Thus, the locating edge dominating set in C_n is $D = \{x_i x_{i+1}; 1 \leq i \leq n-1, i \equiv 1(\text{mod}5)\} \cup \{x_i x_{i+1}; 1 \leq i \leq n-1, i \equiv 4(\text{mod}5)\}$. Thus, the cardinality of the edge dominating set is $|D| = \frac{2n+1}{5} - 1$. The edge set in C_n without the element of edge dominator is $E - D$. The results of intersection between the edge neighborhood $E - D$ with D are as follow:

$$N(x_i x_{i+1} : i \equiv 2(\text{mod}5)) \cap D = \{x_{i-1}x_i : i \equiv 2(\text{mod}5), 1 \leq i \leq n-2\}$$
$$N(x_i x_{i+1} : i \equiv 3(\text{mod}5)) \cap D = \{x_{i+1}x_{i+2} : i \equiv 3(\text{mod}5), 1 \leq i \leq n-2\}$$

IOP Conf. Series: Journal of Physics: Conf. Series **1022** (2018) 012003 doi:10.1088/1742-6596/1022/1/012003

$$N(x_ix_{i+1} : i \equiv 0(\text{mod}5)) \cap D = \{x_{i+1}x_{i+2}, x_{i-1}x_i : i \equiv 0(\text{mod}5), 1 \leq i \leq n-2\}$$
$$N(x_{n-1}x_n) \cap D = \{x_{n-2}x_{n-1}\}$$
$$N(x_1x_n) \cap D = \{x_2x_1\}$$

It is clearly to see that the intersection between the edge neighborhood and the dominator $N(x_2x_3) \cap D = N(x_nx_1) \cap D = \{x_1x_2\}$ and $N(x_nx_{n-1}) \cap D = N(x_{n-2}x_{n-3}) \cap D = \{x_{n-2}x_{n-1}\}$ is the same intersection. It contradicts property of locating edge dominating set. Therefore, the lower bound of locating independent domination number of C_n is $\gamma'_L(C_n) \geq \frac{2n+3}{5}$.

Furthermore, we will show that the upper bound of locating edge domination number of C_n is $\gamma'_L(C_n) \leq \frac{2n+3}{5}$. Let a locating edge dominating set in C_n $D = \{x_ix_{i+1}; 1 \leq i \leq n-1, i \equiv 1(\text{mod}5)\} \cup \{x_1x_n, x_ix_{i+1}; 1 \leq i \leq n-1, i \equiv 4(\text{mod}5)\}$. The non-dominator set in C_n $E(G) - D$. Thus, we get intersection between the neighborhood $N(E)$ with $E \in E(G) - D$ and dominator set D is as follow:

$$N(x_ix_{i+1} : i \equiv 2(\text{mod}5)) \cap D = \{x_{i-1}x_i : i \equiv 2(\text{mod}5), 1 \leq i \leq n-2\}$$
$$N(x_ix_{i+1} : i \equiv 3(\text{mod}5)) \cap D = \{x_{i+1}x_{i+2} : i \equiv 3(\text{mod}5), 1 \leq i \leq n-2\}$$
$$N(x_ix_{i+1} : i \equiv 0(\text{mod}5)) \cap D = \{x_{i+1}x_{i+2}, x_{i-1}x_i : i \equiv 0(\text{mod}5), 1 \leq i \leq n-2\}$$
$$N(x_{n-1}x_n) \cap D = \{x_1x_n, x_{n-2}x_{n-1}\}$$

Based on the results above, it can be seen that the intersection between the neighborhood $N(E)$ with $E \in E(G) - D$ and dominator D is unique and it is not empty set. Thus, it can be concluded that $\gamma'_L(C_n) \leq n-1$. Since, the lower bound and upper bound of locating independent domination number respectively, are $\gamma'_L(C_n) \geq \frac{2n+3}{5}$ and $\gamma'_L(C_n) \leq \frac{2n+3}{5}$. It can be concluded that the locating edge domination number of C_n is $\gamma'_L(C_n) = \frac{2n+3}{5}$.

Based on five cases in above that we obtain locating edge domination number of C_n is $\gamma'_L(C_n) = \frac{2n}{5}$ for $n \equiv 0(\text{mod}5)$, $\gamma'_L(C_n) = \frac{2n+2}{5}$ for $n \equiv 4(\text{mod}5)$, $\gamma'_L(C_n) = \frac{2n+4}{5}$ for $n \equiv 3(\text{mod}5)$, $\gamma'_L(C_n) = \frac{2n+1}{5}$ for $n \equiv 2(\text{mod}5)$ and $\gamma'_L(C_n) = \frac{2n+3}{5}$ for $n \equiv 5(\text{mod}5)$. Thus, it can be combined as $\gamma'_L(C_n) = \lfloor \frac{2n}{5} \rfloor$ for $n \geq 3$. □

Before we continue to discuss about locating edge domination number on comb product of graph, let we analyze the lower bond of locating edge domination number of $\gamma'_L(G \rhd H)$. The result is as follow:

Lemma 1 *Let G, H be any two connected graphs. Let P_G respectively be the number of vertices in G and $\gamma'_L(H)$ respectively be the locating edge domination number of graph H. The locating edge domination number of comb product between graph G and graph H is $\gamma'_L(G \rhd H) \geq P_G \times \gamma'_L(H)$*

Proof. The comb product between G and H is a graph obtained by one copy of G and $\mid V(G) \mid$ copies of H, and grafting the i vertex of G to the u_i in i-th copy of H. Let one of en edge in H that adjacent with edge in G as the result of comb product operation (suppose E_i), be an element of locating edge dominating set. E_i can dominate all edge in G. So, we only need to consider the locating edge domination number of graph H. Because graph G consist of P_G vertices, we need $P_G \times \gamma'_L(H)$. By choosing $P_G \times \gamma'_L(H)$, all edge are dominated by the edge dominating set. The neighborhood of $E(G \rhd H) - D$ are also different. Because the neighborhood of edge which is not edge locating dominating set in graph H precisely different based on the definition of $\gamma'_L(H)$. The intersection between edge in G with D will also be different. It can be concluded that the locating edge domination number of comb product between graph G and graph H is $\gamma'_L(G \rhd H) \geq P_G \times \gamma'_L(H)$. □

The following theorems show the properties of the locating edge domination number on comb product of path and cycle, cycle and star, star and cycle. Path order n, star order $n+1$, and

IOP Conf. Series: Journal of Physics: Conf. Series **1022** (2018) 012003 doi:10.1088/1742-6596/1022/1/012003

cycle graph order n, respectively, are denoted by P_n, S_n, and C_n. Here are some results of locating edge dominating number of comb product graphs.

Theorem 2.3 *Let $P_n \rhd S_m$ be a comb product of path order n and star order $m+1$. For $n \geq 2$ and $m \geq 2$, then the locating edge domination number of $P_n \rhd S_m$ is $\gamma'_L(P_n \rhd S_m) = nm - n$*

Proof. Let graph $P_n \rhd S_m$ be a comb product of path order n and star order $m+1$. The vertex set $V(P_n \rhd S_m) = \{c_i; 1 \leq i \leq n\} \cup \{x_i^j : 1 \leq i \leq n 1 \leq j \leq m\}$ and edge set $E(P_n \rhd S_m) = \{c_i x_i^j : 1 \leq i \leq n 1 \leq j \leq m\} \cup \{x_i^1 x_{i+1}^1 : 1 \leq i \leq n-1\}$. The cardinality of vertices set and edge set, respectively, are $|V(P_n \rhd S_m| = n(m+1)$ and $|E(P_n \rhd S_m)| = nm+n-1$. In order to prove γ'_L $(P_n \rhd S_m) = nm - n$ with $n \geq 2$ and $m \geq 2$, we should show $\gamma'_L(P_n \rhd S_m) \leq nm - n$ and $\gamma'_L(P_n \rhd S_m) \geq nm - n$.

First, we need to prove that the lower bound of the locating edge domination number of $P_n \rhd S_m$ is $\gamma'_L(P_n \rhd S_m) \geq nm - n$, we first analyze the locating edge domination number of star graph order m. Based on Theorem 2.1, we know that $\gamma'_L(S_n) = n-1$. Based on Lemma 1, we have $\gamma'_L(G \rhd H) \geq P_G \times \gamma'_L(H)$. It can be seen that the lower bond of the locating domination number of $P_n \rhd S_m$ is $\gamma'_L(G \rhd H) \geq n \times \gamma'_L(S_m)$. We can conclude that $\gamma'_L(G \rhd H) \geq n \times (m-1) = nm - n$. However, it is attain the sharpest lower bound. Furthermore, we will show that the upper bound of locating edge domination number of $P_n \rhd S_m$ is $\gamma'_L(P_n \rhd S_m) \leq nm - n$. Let a locating edge dominating set in $P_n \rhd S_m$ be $D = \{c_i x_i^j : 1 \leq i \leq n-1 1 \leq j \leq m\}$. The non-dominator set in $P_n \rhd S_m$ $E(G) - D = \{c_i x_i^j : i = n 1 \leq j \leq m\} \cup \{x_i^1 x_{i+1}^1 : 1 \leq i \leq n-1\}$. Thus, we get intersection between the neighborhood $N(E)$ with $E \in E(G) - D$ and dominator set D are as follow:

$$N(c_i x_i^j : i = n 1 \leq j \leq m) \cap D = \{c_i x_i^j, c_i x_{i+1}^j, c_i x_{i+2}^j, ..., c_i x_{n-1}^j : 1 \leq j \leq m\}$$
$$N(x_i^1 x_{i+1}^1) \cap D = \{c_i x_i^j, c_i x_i^{j+1} : i = 1 : 1 \leq j \leq m-1\}$$

Based on the results above, it can be seen that the intersection between the neighborhood $N(E)$ with $E \in E(G) - D$ and dominator D is unique and not empty set. Thus, it can be concluded that $\gamma'_L(P_n \rhd S_m) \leq nm - n$. Since, the lower bound and upper bound of locating edge domination number respectively, are $\gamma'_L(P_n \rhd S_m) \geq nm - n$ and $\gamma'_L(P_n \rhd S_m) \leq nm - n$. It can be concluded that the locating edge domination number of $P_n \rhd S_m$ is $\gamma'_L(P_n \rhd S_m) = nm - n$.
□

Theorem 2.4 *Let $C_n \rhd S_m$ be a comb product of cycle order n and star order $m+1$. For $n \geq 3$ and $m \geq 2$, the locating edge domination number of cycle and star $C_n \rhd S_m$ is $\gamma'_L(C_n \rhd S_m) = nm - n$*

Proof. Let graph $C_n \rhd S_m$ be a comb product of cycle order n and star order $m+1$. The vertex set $V(C_n \rhd S_m) = \{c_i\} \cup \{x_i^j : 1 \leq i \leq n 1 \leq j \leq m\}$ and edge set $E(C_n \rhd S_m) = \{c_i x_i^j : 1 \leq i \leq n 1 \leq j \leq m\} \cup \{x_i^1 x_{i+1}^1 : 1 \leq i \leq n-1\} \cup \{x_1^1 x_n^1\}$. The cardinality of vertices set and $|V(C_n \rhd S_m)| = n(m+1)$ and the cardinality of edges $|E(C_n \rhd S_m)| = nm + n$. In order to prove γ'_L $(C_n \rhd S_m) = nm - n$ with $n \geq 3$ and $m \geq 2$, we should show $\gamma'_L(C_n \rhd S_m) \leq nm - n$ and $\gamma'_L(C_n \rhd S_m) \geq nm - n$.

First, we need to prove that the locating edge domination number of $C_n \rhd S_m$ is $\gamma'_L(C_n \rhd S_m) \geq nm - n$, we first analyze the locating edge domination number of star graph order m. Based on Theorem 2.1, we know that $\gamma'_L(S_n) = n-1$. Based on Lemma 1, we have $\gamma'_L(G \rhd H) \geq P_G \times \gamma'_L(H)$. It can be seen that the lower bound of $(C_n \rhd S_m)$ is $\gamma'_L(G \rhd H) \geq n \times \gamma'_L(S_m)$. We can conclude that $\gamma'_L(G \rhd H) \geq n \times (m-1) = nm - n$. However, it is attain the sharpest lower bound. Furthermore, we will show that the upper bound of locating edge domination number of $C_n \rhd S_m$ is $\gamma'_L(C_n \rhd S_m) \leq nm - n$. Let a locating edge dominating set of

$C_n \rhd S_m$ be $D = \{c_i x_i^j : 1 \leq i \leq n 1 \leq j \leq m-1\}$. The non-dominator set in $C_n \rhd S_m$ is $E(G) - D = \{c_i x_i^m : 1 \leq i \leq n\} \cup \{x_1^1 x_n^1, x_i^1 x_{i+1}^1 : 1 \leq i \leq n-1\}$. Thus, we get intersection between the neighborhood $N(E)$ with $E \in E(G) - D$ and dominator set D are as follow:

$$N(c_i x_i^m : 1 \leq i \leq n) \cap D = \{c_i x_i^j ; 1 \leq i \leq n, 1 \leq j \leq m-1\}$$
$$N(x_i^1 x_{i+1}^1) \cap D = \{c_i x_i^1, c_{i+1} x_{i+1}^1 ; 1 \leq i \leq n-1\}$$
$$N(x_1^1 x_n^1) \cap D = \{c_1 x_1^1, c_n x_n^1\}$$

Based on the results above, it can be seen that the intersection between the neighborhood $N(E)$ with $E \in E(G) - D$ and dominator D is unique and not empty set. Thus, it can be concluded that $\gamma'_L(C_n \rhd S_m) \leq nm - n$. Since, the lower bound and upper bound of locating domination number, respectively, are $\gamma'_L(C_n \rhd S_m) \geq nm - n$ and $\gamma'_L(C_n \rhd S_m) \leq nm - n$. It can be concluded that the locating edge domination number of $C_n \rhd S_m$ is $\gamma'_L(C_n \rhd S_m) = nm - n$.
□

3. Concluding Remarks

In this paper, we have determined the exact values of locating edge domination number on comb product of graph, namely $(P_n \rhd S_m)$ and $(C_n \rhd S_m)$. We also analyze the lower bound of locating edge domination number. This research is a new topic in edge dominating set study. Thus, it still has many open problems.

Open Problem 1 *Let G and H be any connected graph, determine a sharp upper bound of* $\gamma'_L(G \rhd H)$ *in term of the number of edge in graph.*

Open Problem 2 *Let G and H be any connected graph, determine the locating edge domination number of the other operations in graph.*

4. Acknowledgments

We gratefully acknowledge the support from CGANT - University of Jember of year 2017.

5. References

[1] Arumugam S and Velammal S 1998 Edge domination in graphs *Taiwanese journal of Mathematics* 173-179.
[2] Chartrand G and Lesniak L 2000 *Graphs and digraphs 3rd* ed (London: Chapman and Hall)
[3] Colbourn C J, Slater P J, Stewart L K 1987 Locating-dominating sets in series-parallel networks *Congr. Numer.* **56** 135-162.
[4] Finbow A, Hartnell B L 1988 On locating dominating sets and well-covered graphs *Congr. Numer.* **65** 191-200.
[5] Foucaud F, Henning M A, Löwenstein C and Sasse T 2016 Locating dominating sets in twin-free graphs *Discrete Applied Mathematics* **200** 52-58.
[6] Gross J L, Yellen J and Zhang P 2014 *Handbook of graph Theory* Second Edition CRC Press Taylor and Francis Group
[7] Hedetniemi S T and Mitchell S 1977 Edge domination in trees *In Proc. 8th SE Conf. Combin., Graph Theory and Computing, Congr. Numer* **19** 489-509.
[8] Slater P J 1987 Dominating and location in acyclic graphs, *Networks* **17** 55-64.
[9] Slater P J 1988 Dominating and reference sets in graphs, *J. Math. Phys. Sci.* **22** 445-455
[10] Dafik, Purnapraja A K and Hidayat R 2015 Cycle-Super Antimagicness of Connected and Disconnected Tensor Product of Graphs *Procedia Computer Science* **74** pp 9399
[11] Dafik, Slamin, Tanna D, Semanicova-Fenovckova A and Baca M 2017 Constructions of H-antimagic graphs using smaller edge-antimagic graphs *Ars Combin.* **133** pp 233-245
[12] Gembong A W, Slamin, Dafik, and Agustin I H 2017 Bound of Distance Domination Number of Graph and Edge Comb Product Graph *J. Phys.: Conf. Ser.* **855** 012014
[13] Alfarisi R, Darmaji, Dafik 2017 On the star partition dimension of comb product of cycle and complete graph *Journal of Physics: Conference Series* **855** No. 1 p. 012005

Marked point process for modelling seismic activity (case study in Sumatra and Java)

Hasih Pratiwi[1], **Lia Sulistya Rini**[2] **and I Wayan Mangku**[3]

[1]Statistics Department, Universitas Sebelas Maret, Surakarta, Indonesia
[2]Mathematics Department,Universitas Sebelas Maret, Surakarta, Indonesia
[3]Mathematics Department, Institut Pertanian Bogor, Bogor, Indonesia

Corresponding email: [1]hpratiwi@mipa.uns.ac.id, [2]liasulistyarini@gmail.com, [3]wayan.mangku@gmail.com

Abstract. Earthquake is a natural phenomenon that is random, irregular in space and time. Until now the forecast of earthquake occurrence at a location is still difficult to be estimated so that the development of earthquake forecast methodology is still carried out both from seismology aspect and stochastic aspect. To explain the random nature phenomena, both in space and time, a point process approach can be used. There are two types of point processes: temporal point process and spatial point process. The temporal point process relates to events observed over time as a sequence of time, whereas the spatial point process describes the location of objects in two or three dimensional spaces. The points on the point process can be labelled with additional information called marks. A marked point process can be considered as a pair (x, m) where x is the point of location and m is the mark attached to the point of that location. This study aims to model marked point process indexed by time on earthquake data in Sumatra Island and Java Island. This model can be used to analyse seismic activity through its intensity function by considering the history process up to time before t. Based on data obtained from U.S. Geological Survey from 1973 to 2017 with magnitude threshold 5, we obtained maximum likelihood estimate for parameters of the intensity function. The estimation of model parameters shows that the seismic activity in Sumatra Island is greater than Java Island.

1. Introduction

Indonesia is in a position that is vulnerable to the threat of natural disasters. This is due to the geographical position of Indonesia which is at the meeting of three large plates of earth namely the Eurasian plate, the Indo-Australian plate and the Pasific plate. The encounter and movement of three large earth's plates creates an active volcano ranks and earthquake potentials in most of the Indonesian archipelago.

Sumatra Island and Java Island are two areas of Indonesia that are often affected earthquake. Among the various regions in Sumatra, the areas with high earthquake potential are Aceh and West Sumatra. Based on data from United State Geological Survey (USGS) in 1960 to June 2017 in Aceh Province area, there are at least 2400 earthquakes with magnitude size above 4 on Richter scale. As for the area of West Sumatra at the same time interval, there are at least 900 earthquakes with magnitude measurements above 4 on the Richter scale. Since 1990 until June 2017, there have been recorded 473 earthquakes that occurred in Java area with magnitude grater than or equal to 4 on Richter scale. Examples of major earthquakes that occur on the Java Island include the Yogyakarta earthquake which occurred on May 2006 with magnitude 5.9 on the Richter scale. A month later, an earthquake accompanied by a tsunami occurred in Pangandaran, Ciamis Regency, West Java. Then in September 2009, a major earthquake also occurred in Tasikmalaya Regency, West Java.

Point process is a major subject in statistical seismology. Point process is a part of stochastic process that can be used to describe events in a particular pattern. In this process, earthquakes are viewed as a random collection of points in a space, where each point represents the time or location of an event [2]. The occurrences of earthquakes in certain locations can be viewed as points, whereas the size associated with the occurrence of an earthquake is the magnitude or depth [10]. In this process, the occurrence of the event is memory less and mutually independent from occurrences of other events.

In seismological research, some models that can be used to describe the aftershock pattern are Omori's law and Omori-Utsu's law. These mathematical models are not always accurate in modeling earthquakes for long periods of time. The relation between mainshock and aftershocks can be illustrated by other mathematical models which are the development of Omori-Utsu's Law [5,8,9]. Further research related to stochastic modeling to analyze seismic activity in a region with point process approach has also been done by some experts [7,12]. In this article we discuss the marked point process which is indexed by time. This process is presented by an intensity function which considering the occurrence of mainshock and its aftershocks. Then we will estimate this function through maximum likelihood method with considering the history process.

2. Marked Point Process

The point process is defined as a random collection of points located on a particular region. The points of this process can be expressed as events, time of events, location of events, or time and location of events [6]. The construction of a point process can be done by several approaches, one of them is through a counting process. The state space and the parameter space of this counting process are non negative integer sets and subsets of R^d with $d \geq 1$. A counting process $X(t)$ is defined as the number of events occurring during time t. The counting process $X(t)$ satisfies the following properties:

1) $X(t) \geq 0$

2) $X(t)$ is integer

3) If $s<t$, then $X(s) \leq X(t)$

4) For $s<t$, then $X(t) - X(s)$ denotes the number of events occurring in the interval $(s, t]$.

The counting process has independent increments property. It means that the number of occurrences between time s and t, $X(t) - X(s)$, independent of the number of events occurred up to time s. The counting process has stationary increments property if the distribution of the number of occurrences at an interval depends only on the length of the interval. There are two types of point processes, i.e. the temporal point process and spatial point process. The temporal point process relates to events observed at a sequence of time, while the spatial point process relates to the location of the object in a two or three-dimensional space [11].

IOP Conf. Series: Journal of Physics: Conf. Series **1022** (2018) 012004 doi:10.1088/1742-6596/1022/1/012004

The points of the point process are often labeled with additional information called marks. A marked point process can be considered as a pair *(X, m)* where *x* is the point of location and *m* is the mark or quantity attached to the point.

Definition 1. A marked point process in a space *S* with the mark in a space *M* is a point process *Y* at $S \times M$ such that $N_Y(K \times M) < \infty$, for all compact sets $K \subset S$.

This definition describes that the projected process (from unmarked points) is finite local [2].

Theorem 2. Let *Y* be the marked point process on *S* with marks in *M*. Let *X* be projected at *S* (from unmarked points), then

1) *X* is a Poisson process in *S* with intensity μ, and given *X*, the mark attached to point in *X* independently and identically distributed with joint distribution *Q* on *M*;

2) *Y* is a Poisson process in *S x M* with intensity measure $\mu \otimes Q$.

3. Conditional Intensity Function

An important concept of this temporal marked point process is the conditional intensity function on the history of a time-dependent process. This concept describes the probability of an event in the future which depends on previously information up to the present time. In seismology this function can describe seismic activity in a region for a certain period of time. The ground intensity function is the rate of events depend on time, and not only influenced by the present time but also influenced by the previous events, or history process [3]. This function describes the instant Poisson rate as a function that depends only on time. Let $N_\delta(t)$ be the number of occurrences at the time interval $[t, t + \delta]$, the ground intensity function can be written as

$$\lambda(t,\theta_1,\ldots,\theta_m \mid H_t) = \lim_{\delta\to 0}\frac{1}{\delta}P(N_\delta(t) > 0 \mid H_t)$$

where $N_\delta(t)$ is the number of occurrences at the time interval $[t, t + \delta)$.

Furthermore we discuss the intensity function on the marked point process indexed by time with the magnitude as its marked. According to Kagan [4], the rate of aftershocks is defined as

$$N(t) = \frac{C_1}{t} \tag{1}$$

where C_1 is parameter and *t* is time measured from the beginning of a mainshock. Nowadays a more complex equation is used to estimate the rate of aftershocks:

$$N(t) = \frac{C_1}{(C_2 + t)^p} \tag{2}$$

where C_1, C_2, and *p* are parameters. Equation (2) is a modification from Omori's Law (1) and called modified Omori's Law or Omori-Utsu's Law.

In the epidemic model, the number of individuals living at time *t* is controlled by the rate of immigration, birth rate and death rate. In the case of earthquake, immigration refers to the occurrence of independent events, whereas birth is associated with the sequence of events triggered by the previous event. The process of birth and death depends on age for each individual *x* living at time *t*.

IOP Conf. Series: Journal of Physics: Conf. Series **1022** (2018) 012004 doi:10.1088/1742-6596/1022/1/012004

For the next interval $(t, t + dt)$, there are probability of one birth, $g(x)dt$, and probability of one death, $h(x)dt$. Birth and death are independent of each other. If the birth process considers immigration at rate μ per unit time and there is no death, $h(x) = 0$ then the process has the following conditional intensity:

$$\lambda(t) = \mu + \Sigma_{ti<t} g(t-t_i) \quad (3)$$

where t_i is the ith occurrence time and $N(t)$ is the number of occurrences, $\{t_i\}$, in $(0, t)$.

Equation (3) can be extended to the multivariate point process, $\{t_i^m\}$, and the conditional intensity function for the discrete magnitude value of j and m is

$$\lambda_j(t) = \mu_j + \Sigma_m \Sigma_{ti<t} g^j_m (t-t_i^m) \quad (4)$$

If we assume that $g_{jm}(t) = c(m)g_j(t)$ and the point process $N(t) = \Sigma_m N_m(t)$ then the conditional intensity is given by $\lambda(t) = \Sigma_j \lambda_j(t)$ and

$$\lambda(t) = \mu + \Sigma_{ti<t} c(m_i)\, g(t-t_i) \quad (5)$$

where t_i is the occurrence time of $N(t)$, m_i is the magnitude corresponding to t_i, and $g(t) = \Sigma_j g_j(t)$.

Based on the Omori-Utsu law (2) and the epidemic type model (5), we require the following assumptions:

1) The ground seismic rate in a region is constant μ, which means that the ground seismic occurs according to the stationary Poisson process with constant rate μ.

2) All earthquake events, including aftershocks, may cause subsequent aftershocks, and any subsequent aftershocks generate the second subsequent independently. The number of aftershocks of a major earthquake with magnitude M follows Poisson distribution with average $c(M)$:

$$c(M) = \kappa e^{\alpha(M-M_0)} \quad (6)$$

where M_0 is the magnitude threshold while κ and α are parameters.

3) Aftershocks of a mainshock with magnitude M occurring at certain time intervals appropriate with Poisson process with rate $c(M)g(t-t_0)$ with $g(t)$ is a normalized Omori function at a certain time so that $\int g(t)dt = 1$ and is expressed as

$$g(t) = \frac{p-1}{c}(1+\frac{t}{c})^{-p} \quad (7)$$

where c and p are parameters, while t_0 is the occurrence time of the mainshock.

4) Magnitude distribution is independent with rate of occurrence. By using the explicit form of the Gutenberg-Richter relationship, the probability density function of the magnitude is

$$f(M) = \beta e^{\beta(M-M0)}$$

where β corresponds with b, $\beta = 2.30b$.

By substituting $c(m_i)$ with equation (6) and $g(t - t_i)$ with equation (7) in equation (5), the conditional intensity function of marked point process is

IOP Conf. Series: Journal of Physics: Conf. Series **1022** (2018) 012004 doi:10.1088/1742-6596/1022/1/012004

$$\lambda(t) = \mu + \kappa \sum_{t_i<t} e^{\alpha(M_i - M_0)} (1 + \frac{t - t_i}{c})^{-p} \tag{8}$$

4. Seismicity Modeling for Sumatra Island and Java Island

In this section we estimate the conditional intensity function for earthquake data in Sumatra and Java. The earthquake data is a secondary data sourced from the United States Geological Survey (USGS). The earthquake data contains times of ith earthquake occurrence, t_i, and magnitude of ith earthquake occurrence, m_i. We use earthquake data occurred from January 1973 to April 2017 with a magnitude of ≥ 5 and a depth of ≤ 70 km.

4.1 Sumatra Earthquake Data

In the Sumatra Island for the period time January 1973 to April 2017 with magnitude of ≥ 5 and depth of ≤ 70 km there are 2,114 earthquakes. Based on the maximum likelihood estimate method, the conditional intensity function (8) for Sumatra earthquake data can be written as

$$\lambda(t) = 0.0291 + 0.224 \; \Sigma_{t_i<t} \; e^{2.4417(M_i - M_0)} (1 + \frac{t - t_i}{0.0199})^{0.993}$$

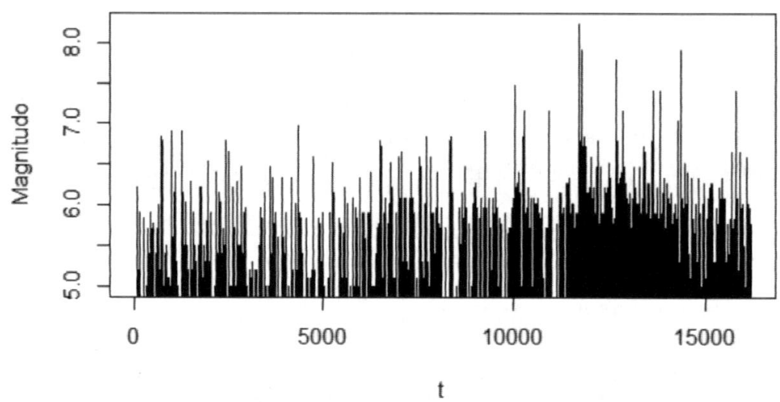

Figure 1. Plot of Magnitude and Time of Sumatra Earthquake Data

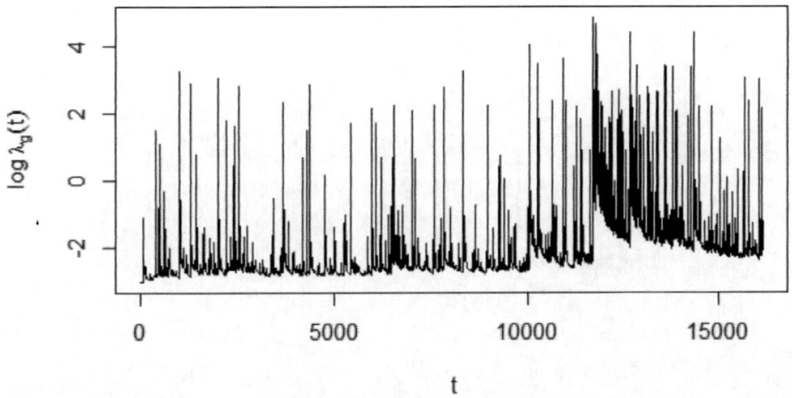

Figure 2. Plot of Logarithmic of Conditional Intensity Function for Sumatra Earthquake Data

Plot of magnitude and time of Sumatra earthquake data is presented in Figure 1 while the plot of logarithmic of conditional intensity function is presented in Figure 2. Based on Figure 1 it can be seen that on March 11, 2012, September 12, 2007, March 28, 2005, March 11, 2012 and December 26, 2004 there was an earthquake with large magnitude occurring in off the west coast of northern Sumatra, southern Sumatra, northern Sumatra, and off the west coast of northern Sumatra. Based on Figure 2 it appears that these areas also have high intensity functions.

4.2 Java Earthquake Data

In Java Island for the time period from January 1973 to April 2017 with magnitude ≥ 5 and depth ≤ 70 km there are 488 earthquakes. Based on maximum likelihood estimate method, the intensity function of marked point process for Java earthquake data can be written as:

$$\lambda(t) = 0.0152 + 10.8592\ \Sigma_{t_i<t}\ e^{0.0875(M_i-M_0)}(1+\frac{t-t_i}{0.0043})^{-1.0484}$$

Plot of magnitude and time of Java earthquake data is presented in Figure 3 whereas plot of the logarithmic of conditional intensity function is presented in Figure 4. Based on Figure 3 it can be seen that on September 7, 1974, April 3, 2011, October 25, 2000, September 2, 2009 and July 17, 2006, there was an earthquake with large magnitude occurring in Southern Java and the Sunda Strait. Based on Figure 4, it is seen that in those areas the conditional intensity function for Java earthquake data is also high.

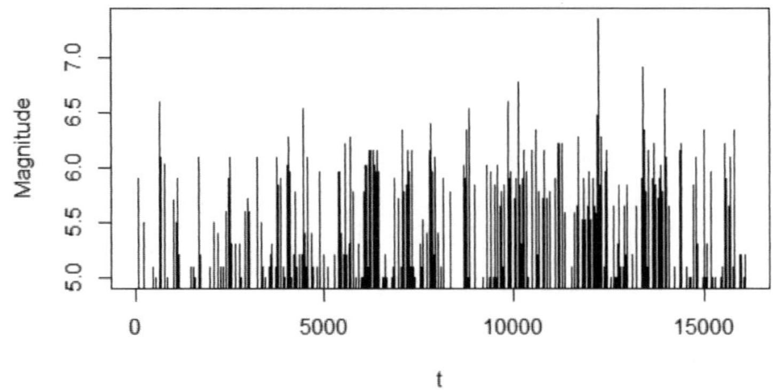

Figure 3. Plot of Magnitude and Time for Java Earthquake Data

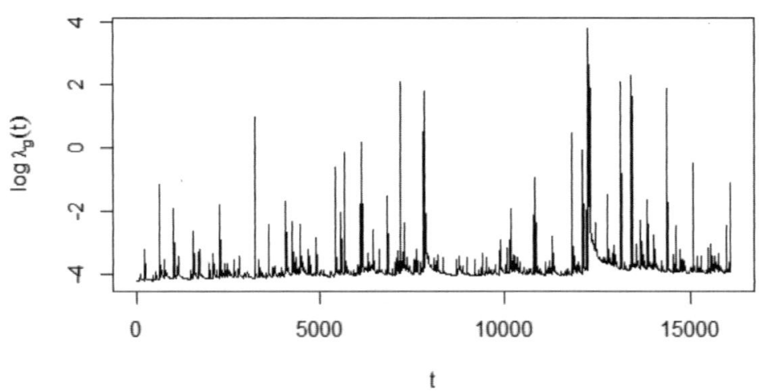

Figure 4. The Plot of Logarithmic of Conditional Intensity Function for Java Earthquake Data

4.3 Seismic Activity

In this section seismic activity on Sumatra and Java islands is compared by considering all parameters of their intensity function. Table 1 shows that parameter value μ of Sumatra earthquake data is greater than Java which means that the base seismic rate of Sumatra Island greater than Java Island.

The parameter value κ of Java earthquake data is greater than Sumatra which means that the productivity of aftershocks in Java is higher than Sumatra, while the parameter value α of Sumatra earthquake data is greater than Java which means that the efficiency of earthquake with certain magnitude in producing aftershocks in Sumatra is higher than Java. Furthermore, the parameter value c of Sumatra earthquake data is greater than Java which means that in time scale, the decay rate of aftershocks in Sumatra is higher than Java, and parameter p of Java earthquake data is greater than Sumatra which means that the decay rate of aftershocks in Java is higher than Sumatra.

Table 1. Parameter Estimate of Conditional Intensity Function for Earthquake Data in Sumatra and Java

Parameter	Sumatra	Java
μ	0.0291	0.0152
κ	0.2240	10.8592
α	2.4417	0.0875
c	0.0199	0.0043
p	0.9930	1.0484

5. Conclusion

Intensity function on a marked point process can modeling seismic activity in a region. By using maximum likelihood estimate method we can apply this model to determine earthquake characteristic in Sumatra and Java islands. Base seismic rate in Sumatra is greater than the one in Java. Although decay rate of aftershocks in time in Sumatra is higher than the one in Java, but impact of earthquake occurrences in Sumatra is greater than the one in Java. This process have not yet considered the spatial component for its intensity function, so the improvement of intensity function (5) can be done to model seismic activity in a region by considering magnitude, time occurrence, and location.

6. References

[1] Console R, M Murru and G Falcone 2010 Retrospective forecasting of m≥4.0 earthquakes in new zealand, *Pure and Applied Geophysics* 167

[2] Daley D J and D Vere-Jones, 2007 A*n introduction to the theory of point process, volume II: general theory and structure*, (Springer: New York)

[3] Harte D 2010 Pt process: An R package for modelling marked point processes indeed by time, *J. of Stat. Soft.* **35**(8)

[4] Kagan Y Y 2014 *Errthquakes: models, statistics, testable forecasts* (John Wiley and Sons, Ltd)

[5] Kumazawa T and Y Ogata 2014 Nonstationary ETAS models for nonstandard earthquakes *The Annals of Applied Statistics* **8**(3)

[6] Ogata Y 1988 Statistical models for earthquake occurrences and residual analysis for point processes *J. of the Amer. Stat. Assoc.* **83** 9–27

[7] Ogata Y 1999 Seismicity analysis through point process modelling: A review *Pure and Applied Geophysics* **155** 471–507

[8] Ogata Y and H Tsuruoka 2016 Statistical monitoring of aftershock sequences: A sase study of the 2015 Mw 7.8 Gorkha, Nepal *Earthquake, Earth, Planets and Space*

[9] Omi T, Y Ogata, Y Hirata and K Aihara 2014 Estimating the ETAS model from an early aftershock sequence *Geophys.Res. Lett.* **50** 850–57

[10] Pratiwi H, I Slamet, D R S Saputro and Respatiwulan 2017 Self-exciting point process in modeling earthquake occurrences *Journal of Physics: Conference Series* **855** 1-7

[11] Schoenberg F P 2000 *Introduction to Point Processes* (UCLA: New York)

[12] Zhuang J 2000 Statistical modelling of seismicity patterns before and after the 1990 Oct 5 Cape Palliser Earthquake, New Zealand *New Zealand Journal of Geology & Geophysics* **43** 447–60

Acknowledgments
The authors would like to thank Ministry of Research, Technology, and Higher Education of the Republic of Indonesia and Universitas Sebelas Maret for supporting this research through Grant of Pasca Doctor 2017.

Zero point and zero suffix methods with robust ranking for solving fully fuzzy transportation problems

P T B Ngastiti[1], Bayu Surarso[2] and Sutimin[2]

[1]Magister Program of Mathematics FSM Diponegoro University
[2]Department of Magister Mathematics FSM Diponegoro University
Jl. Prof. H. Soedarto, S.H. Tembalang Semarang

[*]Corresponding email: tetralian1010@gmail.com, bayusurarso@yahoo.com, sutimin@undip.ac.id

Abstract. Transportation issue of the distribution problem such as the commodity or goods from the supply tothe demmand is to minimize the transportation costs. Fuzzy transportation problem is an issue in which the transport costs, supply and demand are in the form of fuzzy quantities. Inthe case study at CV. Bintang Anugerah Elektrik, a company engages in the manufacture of gensets that has more than one distributors. We use the methods of zero point and zero suffix to investigate the transportation minimum cost. In implementing both methods, we use robust ranking techniques for the defuzzification process. The studyresult show that the iteration of zero suffix method is less than that of zero point method.

Keywords: Zero suffix method, Zero point method, Fully fuzzy transportation problems

1. Introduction

The current business competition does notonly involve between companies but also the networks of the supply chain. Supply chain is a network of companies that works together to create and deliver a product into theend users or consumers[1]. A company that maintaines the continuity of production, need to runthe production process more efficient and effective. Inventories can be in the form of raw materials that are stored for processing,processed components, processed materials in manufacturing processes and finished items that are stored for sale. Inventories play an important role for the company to run well [2].

The last part of supply chain management in a company. It is related to thetransportation problems. The transportation problemisthe studyon distribution of a commodity or items from a supply quantity to a demand quantity to minimize the transportation cost. The purpose of the fuzzy transportation problem is to determine the delivery schedule by minimizing the total of fuzzy cost in order to keep the inventory quantity and fuzzy demand.

CV. Bintang Anugrah Elektrikis a company that engages in the manufacture of the gensets, it has more than one distributors. In the delivery activity at CV. Bintang Anugrah Elektrik, the company sends the genset to various cities, therefore CV. Bintang Anugrah Elektrik requires a policy on shipping costs to minimize the cost so that it must be a proper method to solve transportation problem. Transportation problems can besolved by usingzero point method and zero suffix method to get the minimum cost.

IOP Conf. Series: Journal of Physics: Conf. Series **1022** (2018) 012005 doi:10.1088/1742-6596/1022/1/012005

Ismail Mohideen S. and Senthil Kumar P. used the zero point method of multiplication operation and concluded that it is better than both Vogel's Approximation method and the modified distribution method [3]. Sharma Gaurav *et.al* .explained that the zero point method as the symmetric procedure for transportation problems, is easy to be applied for any type of transportation problem with an objective function (maximum or minimum). This methodprovidethe decisions for an optimal solution of the logistic problems in the transportation problem [4]. In solving transportation problem, Pandian P. and Natarajan G usedzero point method with trapezoidal fuzzy number to find the optimal value of the objective function. This method is the systematic procedure that can be used to make the decisions, when we solve the various types of logistics problems involving fuzzy parameters [5]. Concerning with the fuzzy transportation problem, Samuel E. also studied the method of the increasing of zero point, and concluded that this method was simpler and more efficient compared to VAM, SVAM, GVAM, RVAM, BVAM. He stated that this method was easy to be understood that providedthe optimal solution [6].

In the currently, a number of researcher investigate regarding to the transportation solution using zero point method or zero suffix method. Fegade *et.al.* applied the zero suffix method to solve this problem. They inferred that the optimal solution was converted by fuzzy problem to crisp using robust ranking method.Based on fuzzy problem, the total fuzzy cost reaches optimal more effective [7]. Chandrasekaran, S.*et. al.* studied the fuzzy transportation problem using heptagon fuzzy number with the zero suffix method and they found the optimal solution in solving fuzzy problem [8]. Nirmala. G and Anju R used the zero suffix method to solve an optimal solution and they obtained the optimal solution with fewer iterations [9]. Annie Christi M.S and Kumari Shoba. K.applied the robust ranking with the zero suffix method and they got the optimal solution of fuzzy problems accurately and effectively [10].The aims of the research is to investigate the number of iterations, allocations, optimized solutions of fully fuzzy transportation problems by using Zero Suffix methods and Zero Point methods with robust ranking technique.

2. Preliminaries

In this section, some basic definitions of trapezoidal fuzzy numbers are presented.

2.1 Definition [11]**:**

Let X denote a universal set. A fuzzy subset $\tilde{A}$ of X is defined as ordered set that its members are as follows:

$$\mu_{\tilde{A}} : X \rightarrow [0,1]$$

which assign to each element $x \in X$ to a real number $\mu_{\tilde{A}}(x)$ in the interval [0,1].

2.2 Definition [5]**:**

Afuzzy number $\tilde{A}$ is a trapezoidal fuzzy number that is expressed by (a_1,a_2,a_3,a_4) where a_1,a_2,a_3 and a_4 arereal number and the membership function $\mu_{\tilde{A}}(x)$ is given by,

$$\mu_{\tilde{A}}(x;a_1,a_2,a_3,a_4) = \begin{cases} 0 & \text{for } x \le a_1 \\ \dfrac{(x-a_1)}{(a_2-a_1)} & \text{for } a_1 \le x \le a_2 \\ 1 & \text{for } a_2 \le x \le a_3 \\ \dfrac{(a_4-x)}{(a_4-a_3)} & \text{for } a_3 \le x \le a_4 \\ 0 & \text{for } x \ge a_4 \end{cases}$$

2.3 Definition [12]**:**

Fully fuzzy transport problem (FFTP) is a linear programming problem with the specific structure. If in a transportation issue, all parameters and variables are fuzzy, then it is includedas full fuzzy transport issues. Thefull fuzzy transport issue (FFTP) is formulated as follows:

$$\text{Min}\tilde{z} \approx \sum_{i=1}^{m}\sum_{j=1}^{n}\tilde{c}_{ij}\tilde{x}_{ij},$$

subject to

$\sum_{j=1}^{n}\tilde{x}_{ij} = \tilde{a}_i \text{for} i = 1,2,\dots,m$

$\sum_{i=1}^{m}\tilde{x}_{ij} = \tilde{b}_j \text{for} j = 1,2,\dots,n$

$$\sum_{j=1}^{n}\tilde{x}_{ij} = \tilde{a}_i \approx \sum_{i=1}^{m}\tilde{x}_{ij} = \tilde{b}_j \text{for } i = 1,2,\dots,m \text{ and } j = 1,2,\dots,n$$

$\tilde{x}_{ij} \succcurlyeq 0$for alli and j

3. Robust Ranking Techniques in Trapezoidal Fuzzy

If $\widetilde{A}$ is the trapezoidal fuzzy number, then the robust ranking can be defined as follows [7]:

$$R(\tilde{A}) = \int_0^1 (0,5)(\alpha_\alpha^L, \alpha_\alpha^U)d\alpha$$

where$\tilde{A} = a_1, a_2, a_3, a_4$and

$$a_a^L, a_a^U = (a_2 - a_1)\alpha + a_1, a_4 - (a_4 - a_3)\alpha$$

4. Zero Point Method and Zero Suffix Method

4.1 The Steps for Solving the Fully Fuzzy Transportation Problem

The steps to solve the fully fuzzy transportation problem byusing the zero suffix method and the zero point method with the robust ranking on the trapezoidal fuzzy number can be presented as in the following figure 1:

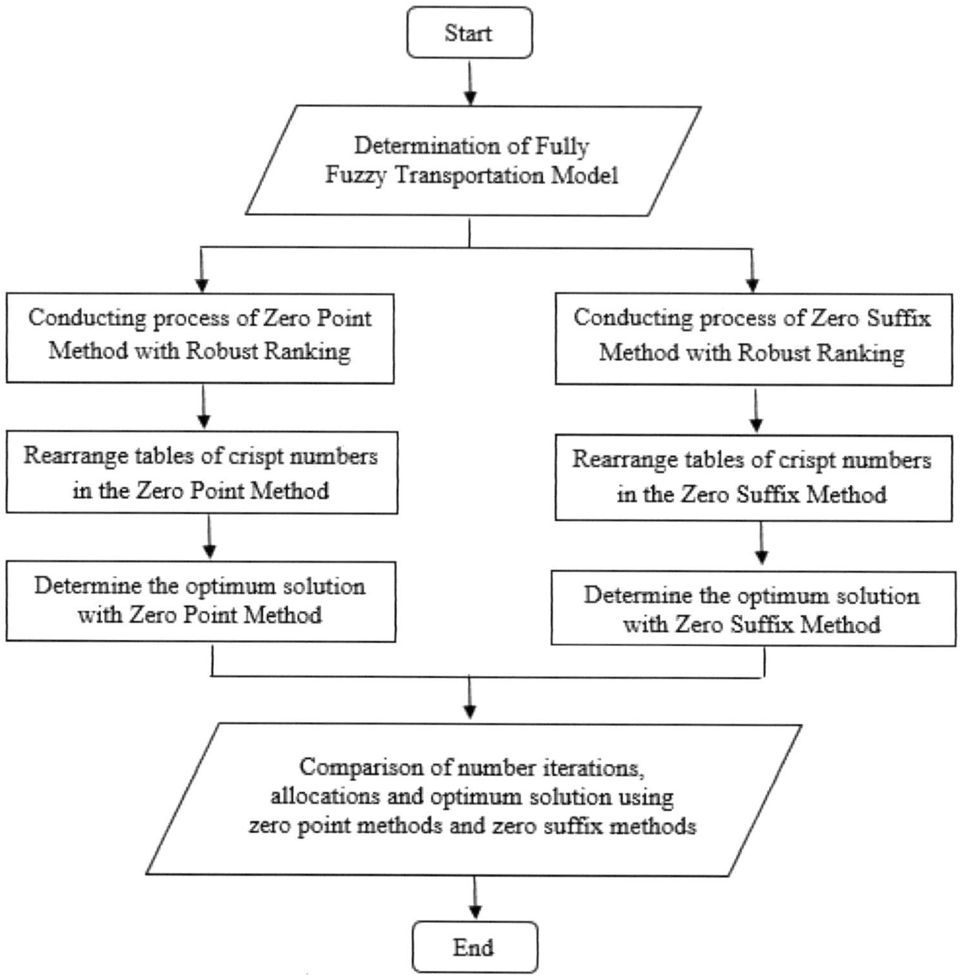

Figure 1. The Fully Fuzzy Transportation Problem for Zero Suffix Method And Zero Point Method

4.2 Zero Point Method

Steps of Zero Point Method used to find the optimal solution of fuzzy balanced transportation problem [5].

Step 1 : Construct the fuzzy transportation table of the fuzzy transportation problem, then convert it into the balancedfuzzy transportation table.

Step 2 : Subtract each row entries of the fuzzy transportation table from the row that hasthe minimum value.

Step 3 : Subtract each column entries of the result of fuzzy transportation table after using the Step 2 from the column with minimum value.

Step 4 : Check if each the fuzzy demand column (b_j) is less than to the sum of the fuzzy supplies(a_i)the cost reduction result in the column was zero. Also, check if each the fuzzy supply row (a_i) was less than to sum of the fuzzy demands column (b_j) in which the raw value iszero ($c_{ij} = 0$). If it isfulfilled, go to step 7, if it is not,go to step 5.

Step 5 : Closethe minimum values with the horizontal line and vertical line that have zero values from the reduction result of the fuzzy tranportation tableso that some rows or columns are not covered, when they do not fufillthe condition in step 4.

Step 6 : Construct the revision of the new fuzzy transportation table as follows:

IOP Conf. Series: Journal of Physics: Conf. Series **1022** (2018) 012005 doi:10.1088/1742-6596/1022/1/012005

(i) Find the smallest entry of the reducedw_{ij} of fuzzy cost matrix which was not covered by any lines.
(ii) Reduce the smallest reduced costsw_{ij} to all the uncovered costsc_{ij} and add the costs c_{ij} that is covered by the intersection of two lines, then return to step 4.

Step 7 : Select a reduced cell in the fuzzy transportation table that has the largest of the reduced cost, called (α, β) . If there are more one value, choose one of this value maximum cost. Say (α, β). If there were more than one, then select one.

Step 8 : Select a cell in the -α row or and $-\beta$ column of the reduced fuzzy transportation table whose the reduced cost zero ($c_{ij} = 0$) and then, select the maximum allocation to the cell. If such cell does not occured for the maximum value, find the next maximum so that such a cell occurs. If such cell does not occur for any value, we select any cell in the reduced fuzzy transportation table in which was reduced cost zero.

Step 9 : Reform the reduced fuzzy transportation table after deleting on the supply row and demand column.

Step 10 : Repeat step 7 to step 9 until all the fuzzy supply points are fully used and all the fuzzy demand points are fully received.

Step 11 : The allocation generates the fuzzy solutions on fuzzy transport issues.

4.3 Zero Suffix Method

The zero suffix method is a method for finding the optimal solution of the transportation problems[7]. The steps to solve the transportation problemsby using this method can be described as follows.

Step 1 : Construct the transportation table.

Step 2 : Subtract each row enteries of the transportation table from the minimum row . Then,subtract each column enteries of the transportation table on a certain minimum coloumn.

Step 3 : In the matrix cost,there is an entry with zero valuein each row and column, then find the suffix value denoted by S.

$$S = \frac{\text{Add the cost of nearest adjacent sides of zero}}{\text{Additional Cost}}$$

Step 4 : Choose the maximum of S if it has oneof themaximum value. Ifit has two or more of the same value, then choose just one, the cost becomes the allocation of goods by regarding to the demand and supply.

Step 5 : After step 4, select minimum of $\{a_i, b_j\}$, then allocate it into transportation table. The resulting table should has at least one cost of 0 in each row and column, otherwise repeat step 2.

Step 6 : Repeat step 3 to step5 until the optimal solution is obtained. The optimal cost is obtained if the column or row is saturated (suffix values = 0)

4.4 Numerical Case

In this section give numerical solution of take a transportation problem for CV. Bintang Anugerah Elektrikusing zero point method and zero suffix method. We use robust ranking techniquesfor the defuzzification process. Next, we give comparison of results of the implication of Zero Point Method and Zero Suffix Method.

IOP Conf. Series: Journal of Physics: Conf. Series **1022** (2018) 012005 doi:10.1088/1742-6596/1022/1/012005

Table 1. Fully Fuzzy Transportation Table

From \ To		Destination: Demak	Destination: Purwodadi	Destination: Kendal	Supply
Source	West Semarang	(35000, 50000, 60000, 75000)	(45000, 60000, 70000, 90000)	(30000, 45000, 55000, 70000)	(2,3,5,7)
	Temanggung	(55000, 65000, 80000, 100000)	(55000, 70000, 85000, 115000)	(45000, 60000, 75000, 95000)	(0,1,3,6)
	East Semarang	(40000, 45000, 55000, 80000)	(45000, 55000, 65000, 85000)	(40000, 50000, 55000, 70000)	(1,3,4,5)
	Demand	(1,2,4,6)	(2,4,5,8)	(1,3,5,7)	

Table 1 is a full fuzzy transportationtablewith form of trapezoidal numbers which transportation costs, supply, and demand are quantities fuzzy. Now apply robust ranking technique for the above fuzzy transportation problem

$$R(\tilde{A}) = \int_0^1 (0{,}5)(\alpha_\alpha^L, \alpha_\alpha^U) d\alpha$$

Where $a_\alpha^L, a_\alpha^U = \{(a_2 - a_1)\alpha + a_1, a_4 - (a_4 - a_3)\alpha\}$

R $(\tilde{c}_{11})$ =R$(35000, 50000, 60000, 75000)$

$$= \int_0^1 (0{,}5)(15000\alpha + 35000 + 75000 - 15000\alpha) da$$

$$= \int_0^1 (0{,}5)(110000) da$$

$$= 55000$$

Similarly

$R(\tilde{c}_{12})$=66250, $R(\tilde{c}_{13})$=50000,$R(\tilde{c}_{21})$ =75000, $R(\tilde{c}_{22})$=81250, $R(\tilde{c}_{23})$=68750, $R(\tilde{c}_{31})$=55000, $R(\tilde{c}_{32})$=62500, $R(\tilde{c}_{33})$=53750

Rank of all supply

$R(\tilde{a}_{11})$=4,25 $R(\tilde{a}_{12})$=2,5,$R(\tilde{a}_{13})$ =3,25

Rank of all demand

$R(\tilde{b}_{11})$=3,25 $R(\tilde{b}_{12})$=4,75,$R(\tilde{b}_{13})$ =4

Now, from the robust ranking technique, we have the following crisp value table.

IOP Conf. Series: Journal of Physics: Conf. Series **1022** (2018) 012005 doi:10.1088/1742-6596/1022/1/012005

Table 2. Crisp Transportation Table

From \ To		Destination			Supply
		Demak	Purwodadi	Kendal	
Source	West Semarang	55000	66250	50000	4,25
	Temanggung	75000	81250	68750	2,5
	East Semarang	55000	62500	53750	3,25
Demand		3,25	4,75	4	

From table 2, we get the total supply and total demand as follows:

$$\sum_{j=1}^{n} \tilde{a}_i = 4{,}25 + 2{,}5 + 3{,}25 = 10$$

$$\sum_{i=1}^{m} \tilde{b}_j = 3{,}25 + 4{,}75 + 4 = 12$$

It was found that$\sum_{j=1}^{n} \tilde{a}_i < \sum_{i=1}^{m} \tilde{b}_j$So the transportation problem was not balanced. Next, to be balanced,$\sum_{j=1}^{n} \tilde{a}_i = \sum_{i=1}^{m} \tilde{b}_j$,Then we have the following Table 3.

Table 3. Balanced Full Fuzzy Transportation Table

From \ To		Destination			Supply
		Demak	Purwodadi	Kendal	
Source	West Semarang	55000	66250	50000	4,25
	Temanggung	75000	81250	68750	2,5
	East Semarang	55000	62500	53750	3,25
	Dummy	0	0	0	2
Demand		3,25	4,75	4	

Table 3 represents a transportation table has been balanced by adding a dummy row.

4.4.1 Implementation of Zero Point Method

In this section we apply zero point method to get optimum solution of cost transportation problem.

Table 4. Optimal Solution with Zero Point

From \ To		Destination			Supply
		Demak	Purwodadi	Kendal	
Source	West Semarang	0,25	3750	4	-
	Temanggung	1250	2,5	0	-
	East Semarang	3	0,25	3750	-
	Dummy	7500	2	12500	-
Demand		-	-	-	

Table 4 is a transportation table with an optimal value, obtained using the zero point method. By solving zero point method we get the following allocation as follows:
$x_{11} = 0{,}25, x_{13} = 4, x_{22} = 2{,}5, x_{31} = 3, x_{32} = 0{,}25, x_{42} = 2$

We get the minimum transportation costas follows:

$$Z = \sum_{i=1}^{m} \sum_{j=1}^{n} c_{ij} x_{ij}$$

$Z = 55000(0{,}25) + 50000(4) + 81250(2{,}5) + 55000(3) + 62500(0{,}25) + 0(2)$
$Z = 597500$

4.4.2 Implementation of Zero suffix method
In this section we apply zero suffix method to get optimum solution of cost transportation problem.

Table 5. Optimal Solution with Zero Suffix

From \ To		Destination			Supply
		Demak	Purwodadi	Kendal	
Source	West Semarang	0,25 55000	66250	4 50000	-
	Temanggung	75000	2,5 81250	68750	-
	East Semarang	3 55000	0,25 62500	53750	-
	Dummy	0	2 0	0	-
Demand		-	-	-	

Table 5 is a transportation table with an optimal value using the zero suffix method. By solving zero suffix method we get the following allocation as follows:
$x_{11} = 0{,}25, x_{13} = 4, x_{22} = 2{,}5, x_{31} = 3, x_{32} = 0{,}25 , x_{42} = 2$

We get the minimum transportation costas follows:
$Z = 55000(0{,}25) + 50000(4) + 81250(2{,}5) + 55000(3) + 62500(0{,}25) + 0(2)$
$Z = 597.500$

The results of the implementation of zero point and zero suffix methods are shown in Table 6 as follows.

IOP Conf. Series: Journal of Physics: Conf. Series **1022** (2018) 012005 doi:10.1088/1742-6596/1022/1/012005

Table 6. Comparison of the Result of Zero Point Method and Zero Suffix Method

No	Name of method	Number of iteration	Allocations	Optimal Solution
1	Zero Point Method	5	$x_{11} = 0{,}25$ $x_{13} = 4$ $x_{22} = 2{,}5$ $x_{31} = 3$ $x_{32} = 0{,}25$ $x_{42} = 2$	597500
2	Zero Suffix Method	4	$x_{11} = 0{,}25$ $x_{13} = 4$ $x_{22} = 2{,}5$ $x_{31} = 3$ $x_{32} = 0{,}25$ $x_{42} = 2$	597500

Table 6 is a comparison table of the zero point method and the zero suffix method by comparing iteration, allocation and optimal solution

5. Conclusions

The comparison result shows that in Zero Point Method, the total iteration to reach the optimum solution after 5 times with allocation 6 items with the optimal solution is 597.500. While in Zero Suffix Method, the number of iterations to reach the optimum solution are needed 4 times with allocation 6 items, and the optimal solution is 597.500, it is the same value if compared to the first method. It can be concluded that the Zero Suffix Method usesless iterationsthan Zero Point Method, whereas allocation and minimum optimum solution are the same.

6. References

[1] Kristiawan, Evelyn and Tarigan Zeplin Jiwa Husada 2014 *Penerapan inbound logistik pada PT. Mekar Armada Jaya di Magelang dengan pendekatan konsep supply chain management.* **2** 1-12

[2] Merdiani, Riska Wulan, Intan Berlianty and Puryani 2015 *Pengembangan model persediaan single vendormulti buyer dengan kebijakan pengiriman* **4** 19-26

[3] Pandian P dan Natarajan 2010 *A new algorithm for finding a fuzzy optimalsolution for fuzzy transportation problems* **4** 79 – 90

[4] Ismail Mohideen S and Senthil Kumar P 2010*A Comparative study on transportation problem in fuzzy environment* **2** 151-8

[5] Sharma Gaurav, Abbas S H and Gupta V K 2012 *Optimum solution of transportation problem with the help of zero point method* **1** 1-6

[6] Samuel A Edward 2012 *Improved Zero Point Method (IZPM) for the transportation problems* **6** 5421-26

[7] Fegade M R, Jadhav V A and Muley A A 2012 *Solving fuzzy transportation problem using zero suffix and robust ranking methodology* **2** 36-9

[8] Chandrasekaran S, Kokila G and Saju Junu 2013 *Ranking of heptagon number using zero suffix method* **4** 2256-57

[9] Nirmala G and Anju R 2014 *An application of fuzzy quantifier in fuzzy transportation problem* **3** 175-7

[10] Annie Christi M S and Kumari Shoba K 2015 *Two stage fuzzy transportation problem using symmetric trapezoidal fuzzy number* **4** 07-10

[11] Sakawa Masatoshi 1993 *Fuzzy sets and interactive multiobjective optimization* (New York: Plenum Press)

[12] A Khoshnava and M R Mozaffari 2015 *Fully fuzzy transportation problem* **1** 42-54

IOP Conf. Series: Journal of Physics: Conf. Series **1022** (2018) 012006 doi:10.1088/1742-6596/1022/1/012006

Several comparison result of two types of equilibrium (Pareto Schemes and Stackelberg Scheme) of game theory approach in probabilistic vendor – buyer supply chain system with imperfect quality

R Setiawan

Mathematics Education, Faculty of Teacher Training and Education Sciences
Sebelas Maret University, Ir. Sutami No. 36 A, Kentingan, Jebres, Surakarta

Corresponding email: rubono.matematika@staff.uns.ac.id

Abstract. In this paper, Economic Order Quantity (EOQ) of the vendor-buyer supply-chain model under a probabilistic condition with imperfect quality items has been analysed. The analysis is delivered using two concepts in game theory approach, which is Stackelberg equilibrium and Pareto Optimal, under non-cooperative and cooperative games, respectively. Another result is getting acomparison of theoptimal result between integrated scheme and game theory approach based on analytical and numerical result using appropriate simulation data.

1. Introduction

Supply chain analysis is one of the interesting research topics in operation research in which industrial engineering problem as a primary reference. The objective of supply chain analysis is to find the mathematical model, often called by inventory model, that represents the relationship between a firm in supply chain system and to find the optimal condition for all of the decision variables. Based on previous references until the present, supply chain management analysis is lead to multiplayer in such system. Firstly, two-level supply chain, known as a vendor –buyer system. In two level supply-chain system, refer to the literature, is known about the integrated scheme which is used by many authors in that field. In the integrated scheme, there is only one possible strategy between each party in the system that is cooperative policy, cooperative in price coordination and other policy like a discount. The aim of the integrated scheme is to get long-term mutual benefits. Goyal [11] introduced an integrated scheme for getting EOQ of inventory problem in his paper. He considered the joint optimization problem of a single vendor and a single buyer, in which he assumed that the vendor's rate is infinite. Then, after the Goyal's work, many authors are also interested in integrated analysis. For example, Banerjee [4] assumed a finite rate of production and developed a joint economic-lot-size (JELS) model for a product with a lot-for-lot shipment policy. He proved that through JELS model, that combine both jointly ordering policy and appropriate price adjustment. Furthermore, there are several results in integrated vendor – buyer system like Goyaland Szendrovits [12], Goyal [15], Goyal and Gupta [13], Goyal [14], Hill (1997), Hill [18], Hoque and Goyal [19], Setiawan and Triyanto [28]. In general, the integrated scheme only applicable in a system which each firm agree to join their economic move.

In order to get the more realistic model, the development of inventory model always concerns about the crucial topics that represent the actual problem in an industrial process. One of those concerning topic is defective items in a lot of production. Some researcher argues that product quality is not always perfect in quality, but is directly affected by the reliability of the production process used. Therefore, the process may deteriorate and produce defectives or poor-quality item. There are many author that explain the effect of imperfect quality items in various assumption of inventory model in their result, for example Salameh and Jaber [27], Huang [21] in his result that extended the integrated vendor – buyer inventory model by accounting for imperfect quality items and considered the situation where the delivery quantity sent to the buyer was identical for each shipment, Goyal et al. [16], Huang [21] developed an inventory model to determine an optimally integrated vendor –buyer inventory policy for unreliable production process in a JIT (Just In Time) manufacturing environment. Another result in supply chain management which includes imperfect quality is proposed by several authors like Lo et al. [24], Wee et al [30], Yoet al. [32], Chang and Ho [6], Sana [26], Hsu and Hsu [20]. Another concerning in current analysis of inventory model is uncertainty condition. One condition that under consideration with uncertainty condition is lead time demand length. In the face of reality, the information about that time is quite limited and most often provided in first two moments of demand only. Lead time length in supply chain management maybe is one of the crucial problems that inflict many other problems in supply chain process, cost, and shortage condition, if they not handled properly. According to the Lin [22], lead time can be reduced by additional crushing cost, customers service level improved, inventory in safety stocks reduced, and the competitive edge in business increased, in other words, it is controllable. Lin [22] study an integrated vendor-buyer inventory policy for a continuous review model with a random number of defective items and screening process gradually. Shortages are allowed and partially backlogged on the buyer's side and that the lead time demand distribution is unknown, except its first two moments.

In recent of industrial competition, only thinking about the integrated scheme is no longer suitable for any case of supply chain management. Each member in supply chain system may play a different strategy rather than integrated scheme. That chosen one depends on the economic purposes of each member. They can choose the cooperative or non-cooperative type of strategy to each other. According to the current literature, games theory is the best mathematical tool for getting an analysis of this condition and also finding the optimal result. There are two types of games related to the strategy analysis in supply chain management that cooperative (centralized approach) and non-cooperative (decentralized approach) games. In cooperative games condition, there are no dominant players, the players choose their strategies simultaneously, for instances, an integrated scheme in the inventory system. While, in a non-cooperative game, this is not exactly true. There are some circumstances that each member in inventory system can dominate each other. It can be classified as leader and follower. There are two types of non-cooperative games that are dynamics non-cooperative games and static non-cooperative games. According to the literature by Alaei et al [1], Stackelberg game is a non-cooperative game. Each firm makes an optimal decision given the behavior of the other firm, and therefore, none has an incentive to deviate unilaterally from the equilibrium. On the other hand, for the cooperative game, according to literature by Peters [25], the results of non-cooperative approaches are not Pareto efficient. In order to achieve this solution, which is Pareto-efficient, a cooperative game can be applied by Abad and Jaggi [2]. In this game, the weighted sum of the player's objectives which agree on the result of the Nash equilibrium is optimized. There are many kinds of literature about the optimal analysis of supply chain model using game theory approach. But, according to current literature, all of them are still limited to the deterministic model and most of them did not consider the model with the probabilistic condition and imperfect process condition such as lead time demands, defective items, inspection error rate, shortage and partial back-ordering. According to the history of game theory literature, the first complete analysis and basic theory regarding the application of game theory in thesupply management system were analyzed by Cachon and Netessine [7]. They focus on basic result about various equilibrium solution properties of cooperative and non-cooperative games in supply chain model. Another literature about the application of game theory in order to strategic behavior modeling in the industrial and economic field had been written by Geçkil and Anderson [10]. Esmaelli et al [9] have analyzed two level seller and buyer model, they use a basic deterministic model that proposed by Abad [2]. They use Stackelberg strategy solution concept to analyze the noncooperative game and Pareto optimal concept to analyze the cooperative game. But, this paper still meets with the assumption that shortage is not permitted and no back ordering or partial back ordering process allowed. Elyasi et al [8] study modified EOQ for deterministic two – echelon (a manufacturer and a supplier) model with imperfect quality items. They use just in time (JIT) concept (centralized model) in inventory problem to formulate total cost of inventory. In their paper, the optimal solution also analyzed by game theory approach: non-cooperative static game using and cooperative game that first introduced by Abad and Jaggi [2]. A cooperative game is applied in order to get Pareto – efficient solution which refers to literature by Peters [25]. In their result, they proved that the result of the cooperative model is the nearest one to the result of the centralized model. Wu et al. [31] consider a supply chain with one supplier and two competing retailers and develop six different scenarios of power imbalance that leads to different sequences of moving among the members of the supply chain. They implement both Nash and Stackelberg equations and compare the gained results.

2. Material and Method

2.1. General Assumptions

There are single-vendor and single-buyer. A single type of product is considered. Both the vendor and the buyer play the non-cooperative game under the monopolistic condition, which addresses to the vendor's move. Because of that condition, then the Buyer becomes the leader and both the buyer as a follower in that system. According to the current theory in applied game theory in strategic behavior, we will use Stackelberg Equilibrium approach to analyzing that situation. It's assumed that defective items only occur in the production process. Follow to the random number of defective items in retailer arriving order lot. In the buyer's side, the defective items exist in an arriving lot of size Q with defective percentage γ and probability function $f(\gamma)$. Vendor's production rate for non-defective items is greater than buyer's demand rate. Shortage conditions are allowed in the relationship between the vendor and the buyer. Shortages are partially back ordered with afraction of the demand $\beta, \beta \in [0,1]$ during the stock out period that will be backordered. We ignore the inspecting time in buyer's side, so defective items can be inspected immediately by 100 % screening process of the lot. The Buyer will return all defective items to the manufacturer upon receipt of the next lot. The reorder point r = expected demand during lead time + safety stock (SS) and SS = kx (standard deviation of lead time), so $r = DL + k\sigma\sqrt{L}$, where k is asafety factor.

2.1.1. Notations

In this paper, we use some following parameter with some specific notation

Q: The size of the shipments from the vendor to the buyer, a decision variable

N: The number of lots in which the product is delivered from the vendor to the buyer.
a positive integer, a decision variable.

D: Expected demand per unit time on the Buyer for deteriorating items

P: The production rate at the vendor

F: The freight (transportation) cost per shipment.

S_b: Buyer's ordering cost per order.

S_v: Vendor's set up cost per production.

h_{b1}: Buyer's holding cost for non-defective item per unit per unit time.

h_{b2}: Buyer's holding cost for defective item per unit per unit time.

h_v: Vendor's holding cost per unit time.

π: Buyer's shortage cost per unit short.

γ: The probability that an item produced is defective.

π_0: Buyer's marginal profit (cost of demand lost) per unit.

IOP Conf. Series: Journal of Physics: Conf. Series **1022** (2018) 012006 doi:10.1088/1742-6596/1022/1/012006

β :The fraction of the demand during the stock-out period.
c_{wv}: The vendor's unit warranty cost per defective items.
c_{ib} : Screening cost at the buyer.
R : Reorder point for deteriorating items.
X: The lead time demand which has a p.d.f. f(x) with finite demand DL and standard deviation $\sigma\sqrt{L}$. DL is anexpectation of demand when thelead time occurs. Then, σ is standardof deviation of demand per unit time.
x: Buyer's unit screening cost.

3. Result and Discussions

3.1. *Buyer Total Cost*

The buyer inventory cost per cycle per unit time is consists of components due to placing an order, transportation cost, screening cost, holding cost, expected shortage cost and lead time crushing cost. The mathematical formula of buyer's total cost is given by the following equation:

$$TC_b(Q,k,n,L) = S_b + nF + c_{ib}Q + [\pi + \pi_0(1-\beta)]E[(X-r)^*] + C(L)$$
$$+ \frac{h_{b1}Q(1-\gamma)(1-\theta)\left[\frac{Q\gamma\theta}{2x(1-\gamma)(1-\theta)} + \frac{Q(1-\gamma)(1-\theta)}{2} + k\sigma\sqrt{L} + (1-\beta)E[(X-r)^*]\right]}{D}$$
$$+ h_{b2}\left[\frac{Q^2(1-\gamma)(1-\theta)}{D} - \frac{Q^2\gamma\theta}{2x}\right] \qquad (3.1)$$

Using renewal-reward theorem, the expected average total cost per unit time for the buyer follows the following formula

$$ETC_b(Q,k,n,L)$$
$$= \frac{D\big(S_b + nF + c_{ib}Q + [\pi + \pi_0(1-\beta)]E[(X-r)^*] + C(L)\big)}{Q(1-\gamma)(1-\theta)}$$
$$+ h_{b1}\left[\frac{Q\gamma\theta}{2x(1-\gamma)(1-\theta)} + \frac{Q(1-\gamma)(1-\theta)}{2} + k\sigma\sqrt{L} + (1-\beta)E[(X-r)^*]\right]$$
$$+ h_{b2}\left[\frac{Q^2(1-\gamma)(1-\theta)}{D} - \frac{Q^2\gamma\theta}{2x}\right] \qquad (3.2)$$

or equivalent with

$$ETC_b(Q,k,n,L)$$
$$= \frac{D\big(S_b + nF + c_{ib}Q + [\pi + \pi_0(1-\beta)]E[(X-r)^*] + C(L)\big)}{Q(1-\gamma)(1-\theta)}$$
$$+ h_{b1}\left[\frac{Q(1-\gamma)(1-\theta)}{2} + k\sigma\sqrt{L} + (1-\beta)E[(X-r)^*]\right] + h_{b2}Q\gamma\theta$$
$$+ \frac{(h_{b1}D - h_{b2})QD\gamma\theta}{2x(1-\gamma)(1-\theta)} \qquad (3.3)$$

where $\pi + \pi_0(1-\beta) = \bar{\pi}$. Based on the equation (3.2), we get the worst distribution of $ETC_b(Q,k,L)$

$$ETC_b(Q,k,n,L) = \frac{D\left(S_b + nF + c_{ib}Q + \frac{\overline{\pi}\sigma\sqrt{L}}{2}\left(\sqrt{1+k^2}-k\right) + C(L)\right)}{Q(1-\gamma)(1-\theta)}$$
$$+ h_{b1}\left[\frac{Q(1-\gamma)(1-\theta)}{2} + k\sigma\sqrt{L} + (1-\beta)\frac{\sigma\sqrt{L}}{2}\left(\sqrt{1+k^2}-k\right)\right] + h_{b2}Q\gamma\theta$$
$$+ \frac{(h_{b1}D - h_{b2})QD\gamma\theta}{2x(1-\gamma)(1-\theta)} \quad (3.4)$$

3.2. *Vendor Total Cost*

The vendor's inventory production per unit time can be obtained by subtracting the accumulated buyer's inventory level from the accumulated vendor's inventory level as follow

$$\left[nQ\left(\frac{Q}{P} + (n-1)\right) - \frac{nQ\left(\frac{nQ}{P}\right)}{2}\right] - T(Q + 2Q + \cdots + (n-1)Q)$$
$$= \frac{nQ^2}{P} - \frac{n^2Q^2}{2P} + \frac{n(n-1)Q^2(1-\gamma)(1-\theta)}{2D} \quad (3.5)$$

Thereby, holding cost for the vendor is formulated by

$$h_v\left(\frac{nQ^2}{P} - \frac{n^2Q^2}{2P} + \frac{n(n-1)Q^2(1-\gamma)(1-\theta)}{2D}\right) \quad (3.6)$$

$$ETC_v(Q,n) = \frac{D}{nQ(1-\gamma)(1-\theta)}\left[S_v + c_{vw}Q\gamma\theta\right.$$
$$\left.+ h_v\left(\frac{nQ^2}{P} - \frac{n^2Q^2}{2P} + \frac{n(n-1)Q^2(1-\gamma)(1-\theta)}{2D}\right)\right] \quad (3.7)$$

3.3. *EOQ by Integrated Scheme*

In this paper, we will use JIT (Just in Time) system concept to analyse the optimum policy in theintegrated inventory system. In practically, JIT system is suitable for the integrated inventory system. This system focuses primarily on purchasing and manufacturing required for immediate consumption. It requires a spirit of cooperation and same long-term strategic partnership between the buyer and the vendor. They can coordinate their product, inventory strategies and share information with each other to determine the best policy and also profit sharing for both parties. Following the work by Banerjee [4] and Lin [23] about concept of joint optimization for the buyer and the vendor, we get the joint total expected average cost per cycle per unit time as follows

$$JETC(Q,k,n,L) = \frac{D\left(S_b + nF + c_{ib}Q + \frac{\bar{\pi}\sigma\sqrt{L}}{2}\left(\sqrt{1+k^2}-k\right) + C(L)\right)}{Q(1-\gamma)(1-\theta)} + h_{b1}\left[\frac{Q(1-\gamma)(1-\theta)}{2} + k\sigma\sqrt{L} + (1-\beta)\frac{\sigma\sqrt{L}}{2}\left(\sqrt{1+k^2-k}\right)\right] + h_{b2}Q\gamma\theta + \frac{(h_{b1}D - h_{b2})QD\gamma\theta}{2x(1-\gamma)(1-\theta)} + \frac{D}{nQ(1-\gamma)(1-\theta)}\left[S_v + c_{vw}Q\gamma\theta + h_v\left(\frac{nQ^2}{P} - \frac{n^2Q^2}{2P} + \frac{n(n-1)Q^2(1-\gamma)(1-\theta)}{2D}\right)\right] \quad (3.8)$$

It is easy to prove the JETC is a convex function for Q and k. Thereby, for fixed n and $L \in [L_i, L_{i-1}]$, the minimum value of $JETC(Q,k,n,L)$ occurs at the critical point (Q,k). Using Karush-Kuhn-Tucker (KKT) conditions for optimal condition, e.g. first derivative criteria, we get EOQ and optimal value of the others as follow:

$$Q = \sqrt{\frac{h_{b1}\frac{(1-\gamma)(1-\theta)}{2} + h_{b2}\gamma\theta + \frac{(h_{b1}D-h_{b2})D\gamma\theta}{2x(1-\gamma)(1-\theta)} + \frac{Dh_v}{(1-\gamma)(1-\theta)P}\left(\frac{1}{P} - \frac{n}{2P} + \frac{(n-1)(1-\gamma)(1-\theta)}{2D}\right)}{\left[\frac{D\left(S_b+nF+\frac{\bar{\pi}\sigma\sqrt{L}}{2}(\sqrt{1+k^2}-k)+C(L)\right)}{(1-\gamma)(1-\theta)} + \frac{DS_v}{n(1-\gamma)(1-\theta)}\right]}}$$

$$k^* = 1 + \frac{1}{{A_1}^2 + 2A_1}, \text{ where } A_1 = \frac{2h_{b1}}{\frac{D\bar{\pi}}{Q(1-\gamma)(1-\theta)} + h_{b1}(1-\beta)}$$

3.4. EOQ by Stackelberg Scheme

The second analysis of our research is aboutthenon-cooperative result of our inventory model.We consider the situation that the buyer and the vendor prefer to play non-cooperative strategies. Mathematically, those conditions lead to Stackelberg scheme in games theory approach. The vendor as leader chooses his move first and then buyer as follower reacts by observing respective decision. The buyer gets his best own move consistent with available information. Finally, the leader determines his optimal move as Stackelberg's optimal move consider the optimal reaction from the follower. According to the Basar and Olsder in Esmaelli, et al. [9], the objective of the leader in inventory system is to design his move in such ways as to maximize his revenue after considering all rational moves that the follower can devise. By applying Karush-Kuhn-Tucker (KKT) conditions for optimum condition, we get expected average total cost of buyer as follower for fixed value of lead time $L \in [L_{i-1,}L_i]$.

$$Q_{fl} = \sqrt{\frac{D\left(S_b + nF + c_{ib}Q + \frac{1}{2}\bar{\pi}\sqrt{L}\left(\sqrt{1+k^2}-k\right) + C(L)\right)}{\frac{1}{2}h_{b1}(1-\gamma)^2(1-\theta)^2 + \frac{1}{2x}(h_{b1}D - h_{b2})D\gamma\theta + h_{b2}\gamma\theta(1-\gamma)(1-\theta)}} \quad (3.9)$$

On the other hand, if we apply KKT condition to the vendor's expected average total cost, we get the vendor optimal move:

$$Q_{ld} = \sqrt{\frac{2DS_v}{n_{ld}(n_{ld}-1)(1-\gamma)(1-\theta)}} \quad (3.10)$$

$$n_{ld} = \sqrt{\frac{2(DS_v + c_{vw}Q\gamma\theta)}{h_vQ^2(1-\gamma)(1-\theta)}} \quad (3.11)$$

Thereby, optimal decision of the leader, that called by Stackelberg optimal solution, is determined based on all reasonable reaction of the buyer as follower $Q^S = Q^*_{ld},\ n^S = n^*_{ld},\ k^S = \frac{1-A_2{}^2}{2A_2}$ where,

$$A_2 = \frac{\frac{4DS_vB_1}{n_{ld}(n_{ld}-1)(1-\gamma)(1-\theta)} - 2D\big(S_b + nF + c_{ib}Q + C(L)\big)}{D\bar{\pi}\sqrt{L}}$$

$$B_1 = \frac{1}{2}h_{b1}(1-\gamma)^2(1-\theta)^2 + \frac{1}{2x}(h_{b1}D - h_{b2})D\gamma\theta + h_{b2}\gamma\theta(1-\gamma)(1-\theta) \qquad (3.12)$$

Equation (3.12) is proposed in implicit form. Using this result is often difficult and not applicable in practice. According to the almost researcher in supply chain management, numerical simulation for the set of equation (3.12) using suitable procedure algorithm and appropriate data may be the best way to describe the approximation value of the optimal solution in a numerical sense and tell us about the condition in the inventory system. Thereby, solution procedure algorithm and also numerical simulation will be provided in next chapter.

3.5. EOQ by Pareto Optimal Solution

According to the Abad and Jaggi [2] and Elyasi et al. [8], the optimal result of the non-cooperative approaches is not Pareto – efficient. To get the Pareto-efficient solution, a cooperative game is applied. We use the cooperative game introduced by Abad and Jaggi [2]and also used by Elyasi et.al. [8]. The weighted sum of the players' objectives is optimized with assumption that the players (vendor and buyer) agree on the result of the Nash equilibrium

$$Z = \lambda_p(ETC_b) + (1-\lambda_p)ETC_v,\quad 0 < \lambda < 1 \qquad (3.13)$$

so

$$Z = \lambda_p\left[\frac{D\left(S_b + nF + c_{ib}Q + \frac{\bar{\pi}\sigma\sqrt{L}}{2}\left(\sqrt{1+k^2}-k\right) + C(L)\right)}{Q(1-\gamma)(1-\theta)} + h_{b_1}\sigma\sqrt{L}\left(k + \frac{(1-\beta)\left(\sqrt{1+k^2}-k\right)}{2}\right) + h_{b_2}Q\gamma\theta + \frac{(h_{b_1}D - h_{b_2})QD\gamma}{2x(1-\gamma)} + \frac{h_{b_1}Q(1-\gamma)(1-\theta)}{2}\right] + (1-\lambda_p)\left[\frac{D}{nQ(1-\gamma)(1-\theta)}\left[S_v + c_{vw}Q\gamma\theta + h_v\left[\frac{n(n-1)Q^2(1-\gamma)(1-\theta)}{2D}\right]\right]\right] \qquad (3.14)$$

To get optimal solution of decision variables, we take the first partial derivative of Z respect to n and then set this equal to zero to get optimal value of parameter λ_p as follow

$$n^* = \sqrt{\frac{(1-\lambda_p{}^*)D(S_v + c_{vw}Q^*\gamma\theta)}{Q^*(1-\gamma)(1-\theta)\left(\frac{\lambda_p{}^*F}{Q^*(1-\gamma)(1-\theta)} + \frac{(1-\lambda_p{}^*)h_v Q^{*2}(1-\gamma)(1-\theta)}{2D}\right)}} \tag{3.15}$$

$$\lambda_p{}^* = \frac{1}{\left(1+\frac{F}{D}\right)} \tag{3.16}$$

For fixed value of $L_i \in [L_{i-1}, L_i], i = 1,2,3,..,N,$ optimal shipment lot size Q and it's safety factor, which yield the minimum of Z are solutions to the set of following equations $\frac{\partial Z}{\partial Q} = 0,\ \frac{\partial Z}{\partial k} = 0$ corresponding to the optimal value of $\lambda_p{}^*$. Thereby, we have

$$Q^* = \sqrt{\frac{\frac{\lambda_p{}^*D}{(1-\gamma)}\left(S_b + n^*F + \frac{\bar{\pi}\sigma\sqrt{L}}{2}\left(\sqrt{1+k^{*2}} - k^*\right) + C(L)\right) - \frac{(1-\lambda_p{}^*)DS_v}{n^*(1-\gamma)(1-\theta)}}{\lambda_p{}^*\left(h_{b_2}\gamma + \frac{(h_{b_1}D - h_{b_2})D\gamma}{2x(1-\gamma)} + \frac{h_{b_1}(1-\gamma)}{2}\right) + \frac{(1-\lambda_p{}^*)h_v(n-1)}{2}}} \tag{3.17}$$

$$k^* = \sqrt{\frac{(B_2-1)^2}{(1-(B_2-1)^2)}},\quad B_2 = \frac{2h_{b_1}}{\bar{\pi}\left(h_{b_1}(1-\beta) + \frac{D}{2Q(1-\gamma)}\right)} \tag{3.18}$$

As we know from (3.17) and (3.18), anexplicit form of *k*, *Q*, and *n* are not possible, so we use a numerical example to get an optimal approximation of decision variables.

3.6. Numerical Simulation and Comparison Result

We use that proposed by Lin [22] to find the approximation of the optimal value decision variables and also the total cost of inventory system by Pareto optimal concept, Stackelberg's scheme and integrated system. For numerical example, we consider an integrated inventory with single vendor, single buyer and single product with following example data: D = 1200 unit/year, σ = 8 unit /week, S_b =\$ 400/order, S_v = \$ 3000/setup, P = 3000 units/year, h_v : \$ 14/unit/ year, h_{b1} = \$ 15/unit /year, h_{b2} = \$ 11/unit/year, F = \$ 20/shipment, c_{wv}: \$ 5/unit, c_{ib} = \$0.5/unit, x= 175200 unit/year, π_0= \$ 50/unit. The lead time has three components. We follow the lead time data that proposed in reference Lin [21]. We propose the numerical result of simulation data with vary in parameter β and γ value. Then we investigate the effects of parameters β and γ to the to the inventory total cost and decision variables. The optimum value result of decision variable and total cost function value as objective function by Pareto optimal scheme and also integrated scheme as comparison result. According to the numerical example result, we get some valuable information for our analytical result. Total cost of inventory system based on Pareto optimal scheme is more economically profitable to each party rather than integrated scheme. Although, they are included in the same category of strategy that is cooperative strategy, according to the numerical result about the effect of defective item rate (γ), the larger value of defective item rate, then it will be result in smaller value of optimum order quantity *Q*and it's total cost. But, there is an interesting result about the value of defective item rate, we can say that an "bifurcation value" in the value of about 0.05. If the defective item rate is greater than those critical point, then the total value will increase than before. From this result, it's recommended that tolerable value of defective item rate of this supply chain model is less than 0.05. In the other-hand, for fixed value of defective item, then the changes value of β also resulted in changes in the optimal value of total inventory cost. Based on numerical example, overall, in average, the greater value fraction of the demand in the stock out period β then it will result to the smaller value of total cost of inventory system.

4. Conclusion and Remarks

In this research, we propose the mathematical model of probabilistic inventory under probabilistic condition using three different schemes of the optimum solution. First, one is integrated scheme, thesecond one is anon-cooperative game using Stackelberg optimum and then the third ones are cooperative schemes using Pareto Optimal scheme. Because of the complexity of the model, the result all of the optimal solution from the all of the scheme is proposed in implicit form. We investigate the optimal order quantity under two condition of game theory. While the vendor and the buyer agree to play cooperative strategy, then Pareto optimal scheme is used to get the optimal decision of each party.

Based on the analytical result, it's difficult to get an explicit form of each decision variable. To get more information about that variable we can use numerical simulation with some appropriate data. Regarding the numerical example result, we get the result that total cost of inventory system based on Pareto Optimal scheme is more economically profitable to each partyrather than integrated scheme. On the other hand, if both of vendor and buyer does not agree with cooperative schemes, or rather prefer to determine their own optimal solution, then Stackelberg games is used to analyze this situation. Based on the numerical result for optimal result, the Stackelberg Scheme is also more economically profitable than integrated scheme.

5. References

[1] Alaei S, Hajji A, Alaei R and Behvaresh M 2015*Acta Polytech. Hung.* **12** 221–39
[2] Abad P L and Jaggi C K 2003 *Int. J. Prod. Econ.* **83** 115-22
[3] Abad P L 1994 *Eur. J. Oper. Res.* **78** 334-54
[4] Banerjee A 1986 *Decis. Sci.* **3** 292-311
[5] Bylka S 2003 *Int. J. Prod. Econ.* **82** 533-44
[6] Chang H C and Ho C H 2010 Omega **38** 233-37
[7] Cachon G and Netessine S 2004 *Handbook of Qualitative Supply Chain Analysis* vol 4, ed S D Wu and Z J Shen (US: Springer) pp 200-33
[8] Elyasi M, Khoshalhan F, and Khanmirzaee M 2014 *Int. J. Eng. Comp.* **5** 211-22
[9] Esmaeili M, Aryanezhad M B and Zeephongsekul P 2009 *Eur. J. Oper. Res.* **195**
[10] Geçkil I K and Anderson P L 2010 *Applied Game Theory and Strategic Behaviour* (Boca Raton: Taylor and Francis)
[11] Goyal S K 1977 *Int. J. Prod. Res.* **15** 107-11
[12] Goyal S K and Szendrovits A Z 1986 *Eng. Cost. Prod. Econ.* **10** 203-10
[13] Goyal S K and Gupta Y P 1989 *Eur. J. Oper. Res.* **41** 261-69
[14] Goyal S K 1995 *Eur. J. Oper. Res.* **82** 209-10
[15] Goyal S K 1988 *Decision Sciences* **19** 236-41
[16] Goyal S K, Huang C K and Chen K C 2003 *Prod. Plan. Control* **14** 596-602
[17] Hill R M 1997 *Eur. J. Oper. Res.* **9** 493-99
[18] Hill R M 1999 *Int. J. Prod. Res.* **37** 2463-75
[19] Hoque M A and Goyal S K 2000 *Int. J. Prod. Econ.* **65** 305-15
[20] Hsu J T and Hsu L F 2012 *Int. J. Eng. Comp.* **3** 703-20
[21] Huang C K 2002 *Prod. Plan. Control* **13** 355-61
[22] Lin H J 2012 *Yugosl. J. Oper. Res.* **23** 87-109
[23] Lin T Y 2009 *J. Oper. Res. Soc. Jpn.* **52** 307-20
[24] Lo S T, Wee H M and Huang W C 2007 *Int. J. Prod. Econ.* **106** 248-60
[25] Peters H. 2008 *Game Theory:A Multi-Levelled Approach* (Berlin: Springer)
[26] Sana S S 2011 *Decis. Support. Syst.* **50** 539-47
[27] Salameh M K and Jaber M Y 2000 *Int. J. Prod. Econ.* **64** 59-64
[28] Setiawan R and Triyanto 2016 *Far East J. Math. Sci.* **99** 109-32
[29] Setiawan R 2016 *Far East J. Math. Sci.* **100** 1695-704
[30] Wee H M, Yu J and Chen M C 2007 *Omega* **35** 7-11
[31] Wu C H, Chen C W and Hsieh C C 2012 *Int. J. Prod. Econ.* **135** 339-50

[32] Yo S H, Kim D and Park M S 2009 Int. J. Prod. Econ. **121** 255-65

Acknowledgement
The author would say to thank the auditor in chief and anonymous referees for their valuable comments and suggestions. This work is supported by PNBP Sebelas Maret University Grant for Fundamental Research 2017, No.623/UN27.21/PP/2017, Institute for Research and Community Service (LPPM), Sebelas Maret University, Surakarta, Indonesia.

IOP Conf. Series: Journal of Physics: Conf. Series **1022** (2018) 012007 doi:10.1088/1742-6596/1022/1/012007

Related Wheel Graphs and Its Locating Edge Domination Number

R. Adawiyah[1,2], I. H. Agustin[1,3], Dafik[1,2], Slamin [1,4] and E. R. Albirri[1,2]
[1]CGANT University of Jember Indonesia
[2]Mathematics Edu. Depart. University of Jember, Jember Indonesia
[3]Mathematics Depart. University of Jember, Jember, Indonesia
[4] Information System Depart. University of Jember, Jember, Indonesia

E-mail: d.dafik@unej.ac.id, ikahestiagustin@gmail.com, robiatul@unej.ac.id

Abstract. A subset D of $E(G)$ is called an edge dominating set of G if every edge not in D is adjacent to some edges in D. In this paper, we initiate to study a new concept in edge dominating set. It is locating edge dominating set. A set $D \subseteq E$ is a locating edge dominating set if every two edges $e_1, e_2 \in E(G) \setminus D$ satisfy that $\emptyset \neq N(e_1) \cap D \neq N(e_2) \cap D \neq \emptyset$. The locating edge domination number $\gamma'_L(G)$ is the minimum cardinality of locating edge dominating set. In this research, we analyze the locating edge domination number of some related wheel graphs. We also analyze the upper bound of locating edge domination number.

Keywords: Locating Edge Dominating set, locating edge domination number, related wheel graphs

1. Introduction

All graphs in this paper are simple, undirected an connected. For more details definition of graph see [2, 9, 4, 5, 6]. By related a wheel graphs, we mean some graphs developed in which wheel as a base graph. For example gear graph, helm graph and fan graph. The concept of edge domination number was introduced by Mitchell and Hedetniemi in [10]. A subset D of $E(G)$ is called an edge dominating set of G if every edge not in D is adjacent to some edges in H [1]. The edge domination number $\gamma'(G)$ of G is the minimum cardinality taken over all edge dominating sets of G [11].

The research about domination has been carried out extensively in numerous types of dominating set. Dominating set has various applications in real life such as in network topology, transportation system, facility location problems, and several other fields [3, 7, 8, 12]. The locating edge dominating set is one of a new study in dominating set topic. This concept is natural extension of edge dominating set. For further definition about locating edge domination number, we will describe in the following section. In this paper, we mainly discuss about the locating edge domination number of related wheel graph namely gear graph, helm graph and fan graph. We also analyze the upper bound of locating edge domination number of any graphs G.

2. Locating Edge Domination Number

An edge in graph G dominates it self and adjacent edges. Let D be a set of dominator, a set $D \subseteq E$ is a locating edge dominating set if every two edges $e_1, e_2 \in E(G) \setminus D$ satisfy that $\emptyset \neq N(e_1) \cap D \neq N(e_2) \cap D \neq \emptyset$. The locating edge domination number $\gamma'_L(G)$ is the minimum cardinality taken over all locating edge dominating sets of G. To emphasis the definition, we will show an example by taking H_6, see Figure 1. We can verify a locating edge dominating sets such as $D = \{cx_1, cx_2, cx_3, cx_4, cx_5, cx_6\}$. In order to prove D is locating edge dominating set, let us consider the neighborhood $E - D$. The set of edges which is not included in dominator is $E - D = \{x_ix_{i+1}; 1 \leq i \leq 5 \cup \{x_1x_6\} \cup x_iy_i; 1 \leq i \leq 6\}$. The intersection of D and the neighborhood of $E - D$ are as follows:

$N(x_ix_{i+1}; 1 \leq i \leq 5) \cap D = \{cx_i, cx_{i+1};\ 1 \leq i \leq 5\}$
$N(x_1x_6) \cap D = \{cx_1, cx_6\}$
$N(xiyi; 1 \leq i \leq 6) \cap D = \{cx_i; 1 \leq i \leq 6\}$

The example shows that the dominator dominates all edges in H_6 and also the neighborhoods are different. We conclude that D is locating dominating set. Since D is the minimum cardinality of locating dominating set, thus $\gamma'_L(H_6) = 6$.

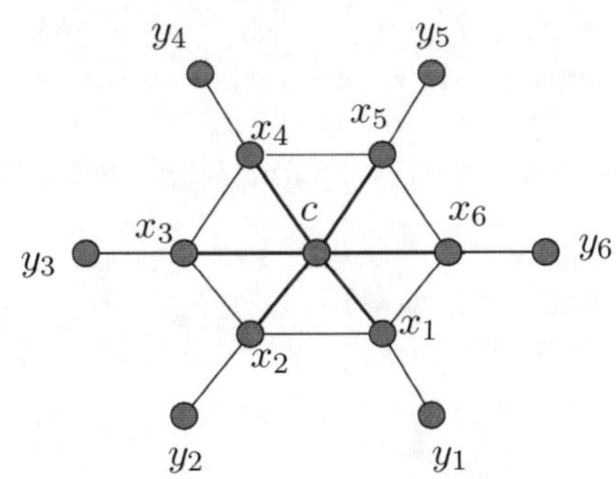

Figure 1. Locating Edge Dominating Set of Helm Graph Order 6

3. Main Results

The following theorems show the properties of the locating edge domination number on related wheels, namely gear graph, helm graph and fan graph. We denote them respectively by G_n, H_n and F_n. From now on, we will show our theorems as the new results of this paper.

Theorem 3.1 *Let G_n be a gear graph with order $2n+1$. For $n \geq 4$, the locating edge domination number of gear graph is $\gamma'_L(G_n) = n$.*

Proof. The gear graph G_n is a connected graph with vertex set $\{c\} \cup \{x_i; 1 \leq i \leq n\} \cup \{y_i; 1 \leq i \leq n\}$ and edge set $E(G_n) = \{cx_i; 1 \leq i \leq n\} \cup \{x_iy_i : 1 \leq i \leq n\} \cup \{y_ix_{i+1}; 1 \leq i \leq n-1\} \cup \{y_nx_1\}$. The cardinality of vertices $|V(G_n)| = 2n+1$ and the cardinality of edges $|E(G_n)| = 3n$.

In order to prove $\gamma'_L(G_n) = n$ with $n \geq 4$, we need to show $\gamma'_L(G_n) \leq n$ and $\gamma'_L(G_n) \geq n$. To show $\gamma'_L(G_n) \geq n$, we assume $\gamma'_L(G_n) < n$. Take the edge dominating set in G_n is $D = \{x_iy_i; 1 \leq i \leq n-1\}$. Thus, the cardinality of the edge dominating set is $|D| = n-1$. The edge set minus the element of edge dominator is $E - D = \{cx_i; 1 \leq i \leq n\} \cup \{x_iy_i; i = n\} \cup \{y_ix_{i+1}; 1 \leq i \leq n\} \cup \{y_nx_1\}$. The intersection between the edge neighborhood $E - D$ with D are as follows:

IOP Conf. Series: Journal of Physics: Conf. Series **1022** (2018) 012007 doi:10.1088/1742-6596/1022/1/012007

$N(cx_i; 1 \leq i \leq n) \cap D = \{x_iy_i; 1 \leq i \leq n-1\}$
$N(x_iy_i; i = n) \cap D = \emptyset$
$N(y_ix_{i+1}; 1 \leq i \leq n-2) \cap D = \{x_iy_i; 1 \leq i \leq n\,,\ x_{i+1}y_{i+1};\ 1 \leq i \leq n-1\}$
$N(y_ix_{i+1}; i = n-1) \cap D = \{x_iy_i; i = n-1\}$
$N(y_ix_1; i = n) \cap D = \emptyset$

It is easy to see that there are some empty sets from the intersection between the edge neighborhood and the dominator. It implies that dominator set D does not dominate all edges in G_n. We easily see that the neighborhood of edge $cx_i; 1 \leq i \leq n$ and edge $y_ix_{i+1}; i = n-1$. The intersection between neighborhood and the dominator $N(cx_i; 1 \leq i \leq n) \cap D = \{x_iy_i; 1 \leq i \leq n-1\}$ and $N(y_ix_{i+1}; i = n-1) \cap D = \{x_iy_i; i = n-1\}$. The edge $\{x_iy_i; i = n-1\}$ is one of the element of edge set $\{x_iy_i; 1 \leq i \leq n-1\}$. It implies that the intersection between edge cx_i and $y_ix_{i+1}; i = n-1$ with the dominator are the same. Based on the definition of locating edge dominating set, $\emptyset \neq N(e_1) \cap D \neq N(e_2) \cap D \neq \emptyset$, the intersection between edge neighborhood with the dominator may not be empty and must be different. Since $D = \{x_iy_i; 1 \leq i \leq n-1\}$ with $|D| = n-1$ does not comply the condition of a graph to be locating edge dominating set, it is a contradiction. It concludes that $\gamma'_L(G_n) \geq n$.

Furthermore, we will show that the upper bound of locating edge domination number of $\gamma'_L(G_n) \leq n$. Let us choose a locating edge dominating set in G_n as $D = \{x_iy_i; 1 \leq i \leq n\}$. The non-dominator set in G_n $E(G) - D = \{cx_i; 1 \leq i \leq n\} \cup \{y_ix_{i+1} : 1 \leq i \leq n-1\} \cup \{y_ix_1 : i = n\}$ We get the intersection between the neighborhood $N(E)$ with $E \in E(G) - D$ and dominator set D as follow:

$N(cx_i; 1 \leq i \leq n) \cap D = \{x_iy_i, 1 \leq i \leq n\}$
$N(y_ix_{i+1}; 1 \leq i \leq n-1) \cap D = \{x_iy_i; 1 \leq i \leq n,\ x_{i+1}y_{i+1}\ ; 1 \leq i \leq n-1\}$
$N(y_ix_1; i = n) \cap D = \{x_iy_i; i = 1,\ x_iy_i;\ i = n\}$

Based on the results above, it is easy to see the intersection between the edge neighborhood $E - D$ with D are all different and not empty set. Thus that the dominator D dominates all edges in G_n. It can be concluded that $\gamma'_L(G_n) \leq n$ attains the condition of locating edge dominating set. Hence, the lower bound and upper bound of locating edge domination number respectively are $\gamma'_L(G_n) \geq n$ and $\gamma'_L(G_n) \leq n$. It completes that the locating edge domination number of G_n is $\gamma'_L(G_n) = n$. □

Theorem 3.2 *Let H_n be a helm graph with order $2n+1$. For $n \geq 4$, the locating edge domination number of Helm graph is $\gamma'_L(H_n) = n$*

Proof. The helm graph H_n is a connected graph with vertex set $\{c\} \cup \{x_i; 1 \leq i \leq n\} \cup \{y_i; 1 \leq i \leq n\}$ and edge set $E(G_n) = \{cx_i; 1 \leq i \leq n\} \cup \{x_iy_i; 1 \leq i \leq n\} \cup \{x_ix_{i+1}; 1 \leq i \leq n-1\} \cup \{x_ix_1; i = n\}$ with the cardinality of vertices $|V(G_n)| = 2n+1$ and the cardinality of edges $|E(G_n)| = 3n$.

In order to prove that $\gamma'_L(H_n) = n$ with $n \geq 4$, we need to show $\gamma'_L(H_n) \leq n$ and $\gamma'_L(H_n) \geq n$. In order to show $\gamma'_L(H_n) \geq n$, we assume $\gamma'_L(H_n) < n$. Take the locating edge dominating set in H_n is $D = \{cx_i; 1 \leq i \leq n-1\}$. Thus, the cardinality of the edge dominating set is $|D| = n-1$. The edge set minus the element of edge dominator is $E - D = \{cx_i; i = n\} \cup \{x_iy_i; 1 \leq i \leq n\} \cup \{x_ix_{i+1}; 1 \leq i \leq n-1\} \cup \{x_ix_1; i = n\}$. The results of intersection between the edge neighborhood $E - D$ with D are as follow:

$N(cx_i; i = n) \cap D = \{cx_i, cx_{i+1}, cx_{i+2}, ..., cx_n; 1 \leq i \leq n\}$

$N(x_iy_i; 1 \leq i \leq n-1) \cap D = \{cx_i;\ 1 \leq i \leq n-1\}$
$N(x_iy_i; i = n) \cap D = \emptyset$
$N(x_ix_{i+1}; 1 \leq i \leq n-2) \cap D = \{cx_i; 1 \leq i \leq n-2,\ cx_{i+1}\ 1 \leq i \leq n-1\}$
$N(x_ix_{i+1}; i = n-1) \cap D = \{cx_i;\ i = n-1\}$
$N(x_ix_1; i = n) \cap D = \{cx_i;\ i = 1\}$

From the results above, it is easy to see the intersection between the edge neighborhood and the dominator is an empty set. It means that dominator set D does not dominate all edges in H_n. Furthermore, let us see the neighborhood of edge $\{x_iy_i; 1 \leq i \leq n-1\}$, edge $\{x_ix_{i+1}; i = n-1\}$ and edge $\{x_ix_1; i = n\}$. The intersection between neighborhood and the dominator $N(x_iy_i; 1 \leq i \leq n-1) \cap D = \{cx_i; 1 \leq i \leq n-1\}$,$N(x_ix_{i+1}; i = n-1) \cap D = \{cx_i; i = n-1\}$ and $N(x_ix_1; i = n) \cap D = \{cx_i; i = 1\}$. Edge $\{x_iy_i; i = n-1\}$ is one of the element of edge set $\{x_iy_i; 1 \leq i \leq n-1\}$. It implies that the intersection between edge c_i, $y_ix_{i+1}; i = n-1$ with the dominator are the same. Based on the definition of locating edge dominating set, $\emptyset \neq N(e_1) \cap D \neq N(e_2) \cap D \neq \emptyset$, the intersection between edge neighborhood with the dominator may not be empty and must be different. Since $D = \{x_iy_i; 1 \leq i \leq n-1\}$ with $|D| = n-1$ does not comply the condition of a graph to be locating edge dominating set, it is a contradiction. It concludes that $\gamma'_L(H_n) \geq n$.

Furthermore, we will show that the upper bound of locating edge domination number of $\gamma'_L(H_n) \leq n$. Let choose a locating edge dominating set in H_n $D = \{x_iy_i; 1 \leq i \leq n\}$. The non-dominator set in H_n $E(G) - D = \{cx_i; 1 \leq i \leq n\} \cup \{y_ix_{i+1}; 1 \leq i \leq n-1\} \cup \{y_ix_1; i = n\}$ We get intersection between the neighborhood $N(E)$ with $E \in E(G) - D$ and dominator set D as follow:

$N(cx_i; 1 \leq i \leq n) \cap D = \{x_iy_i;\ 1 \leq i \leq n\}$
$N(y_ix_{i+1}; 1 \leq i \leq n-1) \cap D = \{x_iy_i;\ 1 \leq i \leq n,\ x_{i+1}y_{i+1};\ 1 \leq i \leq n-1\}$
$N(y_ix_1; i = n) \cap D = \{x_iy_i;\ i = 1,\ x_iy_i\ i = n\}$

Based on the results above, we easily see the intersection between the neighborhood $E - D$ with D are all different and not empty set. Thus, the dominator D dominates all edges in H_n. It can be concluded that $\gamma'_L(H_n) \leq n$ attains the condition of locating edge dominating set. Hence, the lower bound and upper bound of locating edge domination number respectively, are $\gamma'_L(H_n) \geq n$ and $\gamma'_L(H_n) \leq n$. It completes that the locating edge domination number of H_n is $\gamma'_L(H_n) = n$. □

Theorem 3.3 *Let F_n respectively fan graph with order $n+1$. For $n \geq 2$, the locating edge domination number of Fan graph is:*

$$\gamma'_L(F_n) = \begin{cases} n & \text{for } n = 2 \\ n-1 & \text{for } 3 \leq n \leq 5 \\ \frac{2n}{3} & \text{if } n \equiv 0 \ (mod\ 3) \ \text{ and } n \leq 6 \\ \frac{2n+1}{3} & \text{if } \ n \equiv 1 \ (mod\ 3) \ \text{ and } n \leq 6 \\ \frac{2n+2}{3} & \text{if } \ n \equiv 2 \ (mod\ 3) \ \text{ and } n \leq 6 \end{cases}$$

Proof. The graph F_n is a Fan graph with vertex set $\{c\} \cup \{x_i; 1 \leq i \leq n\}$ and edge set $E(F_n) = \{cx_i; 1 \leq i \leq n\} \cup \{x_ix_{i+1}; 1 \leq i \leq n-1\}$ with the cardinality of vertices $|V(G_n)| = n+1$ and the cardinality of edges $|E(G_n)| = 2n-1$. In order to prove the locating edge domination number of fan graph, lets divide the the proof into five cases.
Case 1 ; $\gamma'_L(F_n) = n$ if $n = 2$
Graph F_2 consist of three edges $\{cx_i; 1 \leq i \leq 2\} \cup \{x_ix_{i+1}; i = 1\}$ and three vertices

$\{c\} \cup \{x_i; 1 \leq i \leq 2\}$. Assume we take only one edges as locating edge dominating set in F_2 such that $D = cx_1$. The intersection between the edge neighborhood $E - D$ with D are as follow:

$$N(cx_2) \cap D = \{cx_1\}$$
$$N(x_1x_2) \cap D = \{cx_1\}$$

Based on the results above, it is easy to see the intersections between the neighborhood $E - D$ with D and dominator D are the same. Based on the definition of locating edge dominating set, the intersection between the neighborhood $N(E)$ with $E \in E(G) - D$ and dominator D must be different and may not be empty. It is contradiction. It can be concluded that $\gamma'_L(F_2) = 2$.

Case 2 : $\gamma'_L(F_n) = n - 1$ for $3 \leq n \leq 5$

In order to prove $\gamma'_L(F_n) = n - 1$ with $3 \leq n \leq 5$, we need to show $\gamma'_L(F_n) \leq n - 1$ and $\gamma'_L(F_n) \geq n - 1$. In order to show $\gamma'_L(F_n) \geq n - 2$, we assume $\gamma'_L(F_n) < n - 1$. Suppose, the locating edge dominating set in F_n is $D = \{x_ix_{i+1}; 1 \leq i \leq n - 2\}$. Thus, the cardinality of the edge dominating set is $|D| = n - 2$. The edge set minus the element of edge dominator is $E - D = \{cx_i; 1 \leq i \leq n\} \cup \{x_ix_{i+1}; i = n - 1\}$. The results of intersection between the edge neighborhood $E - D$ with D are as follow:

$$N(cx_i : i = 1) \cap D = \{x_1x_2\}$$
$$N(cx_i : i = n) \cap D = \emptyset$$
$$N(cx_i : 2 \leq i \leq n - 3) \cap D = \{x_{i-1}x_i, x_ix_{i+1}; 2 \leq i \leq n - 2\}$$

Based on the result above, it is easy to see the intersection $E - D$ with D in $cx_i; i = n$ is an empty set. It means that dominator set D does not dominate all edges in F_n. Based on the definition of locating edge dominating set, $\emptyset \neq N(e_1) \cap D \neq N(e_2) \cap D \neq \emptyset$, the intersection between edge neighborhood with the dominator may not be empty and must be different. Since $D = \{x_ix_{i+1}; 1 \leq i \leq n - 2\}$ with $|D| = n - 2$ does not comply the condition of a graph to be locating edge dominating set, it is a contradiction. We can conclude that $\gamma'_L(F_n) \geq n - 1$. Furthermore, we will show that the upper bound of locating edge domination number of $\gamma'_L(F_n) \leq n - 1$. Let choose a locating edge dominating set in F_n $D = \{x_ix_{i+1}; 1 \leq i \leq n - 1\}$. The non-dominator set in F_n $E(G) - D = \{cx_i; 1 \leq i \leq n\} \cup \{x_ix_{i+1}; i \equiv 1 \ (mod\ 3)\}$ We get intersection between the neighborhood $N(E)$ with $E \in E(G) - D$ and dominator set D as follow:

$$N(cx_i : i = 1) \cap D = \{x_1x_2\}$$
$$N(cx_i : i = n) \cap D = \{x_{n-1}x_n\}$$
$$N(cx_i : 2 \leq i \leq n - 3) \cap D = \{x_{i-1}x_i, x_ix_{i+1}; 2 \leq i \leq n - 1\}$$

Based on the results above, we easily see the intersection between the neighborhood$E - D$ with D is unique and not empty set. Hence, the lower bound and upper bound of locating independent domination number respectively, are $\gamma'_L(F_n) \geq n - 1$ and $\gamma'_L(F_n) \leq n - 1$. It completes that that the locating edge domination number of F_n for $3 \leq n \leq 5$ is $\gamma'_L(F_n) = n - 1$.

Case 3 :$\gamma'_L(F_n) = \frac{2n}{3}$ if $n \equiv 0 \ (mod\ 3)$ and $n \leq 6$

In order to prove $\gamma'_L(F_n) = \frac{2n}{3}$ if $n \equiv 0 \ (mod\ 3)$ for $n \leq 6$, we need to show $\gamma'_L(F_n) \leq \frac{2n}{3}$ and $\gamma'_L(F_n) \geq \frac{2n}{3}$. To show $\gamma'_L(F_n) \geq \frac{2n}{3}$, we assume $\gamma'_L(F_n) < \frac{2n}{3}$. Take the edge dominating set in F_n is $D = \{x_ix_{i+1}; 1 \leq i \leq n - 2\ ; i = 1(mod3), 2(mod3)\}$. Thus, the cardinality of the edge dominating set is $|D| = \frac{2n}{3} - 1$. The intersection between the edge neighborhood $E - D$ with D are as follows:

$$N(cx_i : i = 1) \cap D = \{x_1x_2\}$$
$$N(cx_i : i = n) \cap D = \emptyset$$
$$N(cx_i : 2 \leq i \leq n - 3) \cap D = \{x_{i-1}x_i, x_ix_{i+1}; 2 \leq i \leq n - 2\}$$

Based on the result above, it is easy to see the intersection between the edge neighborhood and the dominator in $cx_i : i = n$ is an empty set. It implies that by assuming $D = \{x_i x_{i+1}; 1 \leq i \leq n-2\ ; i = 1(mod3), 2(mod3)\}$ with $|D| = \frac{2n}{3} - 1$ does not comply the condition to be locating edge dominating set. It is contradiction. It can be concluded that $\gamma'_L(F_n) \geq \frac{2n}{3}$. Furthermore, we will show that the upper bound of locating edge domination number of $\gamma'_L(F_n) \leq \frac{2n}{3}$. Let choose a locating edge dominating set in F_n $D = \{x_i x_{i+1}; 1 \leq i \leq \frac{2n}{3}\}$. The non-dominator set in F_n $E(G) - D = \{cx_i; 1 \leq i \leq n\} \cup \{x_i x_{i+1}; i = \frac{2n}{3}\}$ We get intersection between the neighborhood $N(E)$ with $E \in E(G) - D$ and dominator set D as follow:

$N(cx_i : i = 1) \cap D = \{x_1 x_2\}$
$N(cx_i : i = n) \cap D = \{x_{n-1} x_n\}$
$N(cx_i : 2 \leq i \leq n-3 : i \equiv 1\ (mod\ 3), 2\ (mod\ 3)) \cap D = \{x_{i-1}x_i, x_i x_{i+1}; 2 \leq i \leq n-2\ ; i \equiv 1\ (mod\ 3), 2\ (mod\ 3)\}$
$N(cx_i; 2 \leq i \leq n-3; i \equiv 0\ (mod\ 3)) \cap D = \{x_{i-1} x_1; i \equiv 0\ (mod\ 3)\}$
$N(x_i x_{i+1} : i \equiv 0\ (mod\ 3)) \cap D = \{x_{i-1}x_i, x_{i+1}x_{i+2}; i \equiv 0\ (mod\ 3)\}$

Based on the results above, it can be seen that the intersection between the neighborhood $N(E)$ with $E \in E(G) - D$ and dominator D is unique and not empty set, it means that dominator set D dominates all edges in F_n for $n \equiv 0\ (mod\ 3)$ and $n \leq 6$. It can be concluded that $\gamma'_L(F_n) \leq \frac{2n}{3}$ attains the condition of locating edge dominating set. Hence, the lower bound and upper bound of locating edge domination number respectively, are $\gamma'_L(F_n) \geq \frac{2n}{3}$ and $\gamma'_L(F_n) \leq \frac{2n}{3}$. It can be concluded that the locating edge domination number of F_n for or $n \equiv 0\ (mod\ 3)$ and $n \leq 6$ is $\gamma'_L(F_n) = \frac{2n}{3}$.

Case 4 :$\gamma'_L(F_n) = \frac{2n+1}{3}$ if $n \equiv 1\ (mod\ 3)$ and $n \leq 6$

In order to prove $\gamma'_L(F_n) = \frac{2n+1}{3}$ if $n \equiv 1\ (mod\ 3)$ for $n \leq 6$, we need to show $\gamma'_L(F_n) \leq \frac{2n+1}{3}$ and $\gamma'_L(F_n) \geq \frac{2n+1}{3}$. To show $\gamma'_L(F_n) \geq \frac{2n+1}{3}$, we assume $\gamma'_L(F_n) < \frac{2n+1}{3}$. Take the edge dominating set in F_n is $D = \{x_i x_{i+1}; 1 \leq i \leq n-1\ ; i = 1(mod3), 2(mod3)\}$. Thus, the cardinality of the edge dominating set is $|D| = \frac{2n+1}{3} - 1$. The edge set minus the element of edge dominator is $E - D = \{cx_i; 1 \leq i \leq n\} \cup \{x_i x_{i+1}; i = n-1\} \cup \{x_i x_{i+1}; i \equiv 0\ (mod\ 3)\}$. The results of intersection between the edge neighborhood $E - D$ with D are as follow:

$N(cx_i : i = 1) \cap D = \{x_1 x_2\}$
$N(cx_i : i = n) \cap D = \emptyset$
$N(cx_i : 2 \leq i \leq n-3) \cap D = \{x_{i-1}x_i, x_i x_{i+1}; 2 \leq i \leq n-2\}$

Based on the result above, we easily see the intersection between the edge neighborhood and the dominator in $cx_i; i = n$ is an empty set. Thus that the dominators D does not dominate all edges in F_n. Based on the definition of locating edge dominating set, $\emptyset \neq N(e_1) \cap D \neq N(e_2) \cap D \neq \emptyset$, the intersection between edge neighborhood with the dominator may not be empty and must be different. Since $D = \{x_i x_{i+1}; 1 \leq i \leq n-1\ ; i = 1(mod3), 2(mod3)\}$ with $|D| = \frac{2n+1}{3} - 1$ does not comply the condition of a graph to be locating edge dominating set, it is a contradiction. We can conclude that $\gamma'_L(F_n) \geq \frac{2n+1}{3}$.

Furthermore, we will show that the upper bound of locating edge domination number of $\gamma'_L(F_n) \leq \frac{2n+1}{3}$. Let choose a locating edge dominating set in F_n $D = \{x_i x_{i+1}; 1 \leq i \leq \frac{2n+1}{3}\}$. The non-dominator set in F_n $E(G) - D = \{cx_i; 1 \leq i \leq n\} \cup \{x_i x_{i+1}; i = \frac{2n+1}{3}\}$ We get intersection between the neighborhood $N(E)$ with $E \in E(G) - D$ and dominator set D as follow:

$N(cx_i : i = 1) \cap D = \{x_1 x_2\}$

IOP Conf. Series: Journal of Physics: Conf. Series **1022** (2018) 012007 doi:10.1088/1742-6596/1022/1/012007

$N(cx_i : i = n) \cap D = \{x_{n-1}x_n\}$
$N(cx_i : 2 \leq i \leq n-3 : i \equiv 1\ (mod\ 3), 2\ (mod\ 3)) \cap D = \{x_{i-1}x_i, x_ix_{i+1}; 2 \leq i \leq n-2\ ; i \equiv 1\ (mod\ 3), 2\ (mod\ 3)\}$
$N(cx_i; 2 \leq i \leq n-3; i \equiv 0\ (mod\ 3)) \cap D = \{x_{i-1}x_1; i \equiv 0\ (mod\ 3)\}$
$N(x_ix_{i+1} : i \equiv 0\ (mod\ 3)) \cap D = \{x_{i-1}x_i, x_{i+1}x_{i+2}; i \equiv 0\ (mod\ 3)\}$

Based on the results above, it is easy to see the intersection between the neighborhood $E - D$ with D are all different and not empty set. Thus that the dominator D dominates all edges in F_n for $n \equiv 1\ (mod\ 3)$ and $n \leq 6$.It can be concluded that $\gamma'_L(F_n) \leq \frac{2n+1}{3}$ attains the condition of locating edge dominating set. Hence, the lower bound and upper bound of locating edge domination number respectively are $\gamma'_L(F_n) \geq \frac{2n+1}{3}$ and $\gamma'_L(F_n) \leq \frac{2n+1}{3}$. It completes that the locating edge domination number of F_n for or $n \equiv 1\ (mod\ 3)$ and $n \leq 6$ is $\gamma'_L(F_n) = \frac{2n+1}{3}$.

Case 5 :$\gamma'_L(F_n) = \frac{2n+2}{3}$ if $n \equiv 2\ (mod\ 3)$ and $n \leq 6$

In order to prove $\gamma'_L(F_n) = \frac{2n+2}{3}$ if $n \equiv 1\ (mod\ 3)$ for $n \leq 6$, we should show $\gamma'_L(F_n) \leq \frac{2n+2}{3}$ and $\gamma'_L(F_n) \geq \frac{2n+2}{3}$. To show $\gamma'_L(F_n) \geq \frac{2n+2}{3}$, we assume $\gamma'_L(F_n) < \frac{2n+2}{3}$. Take the edge dominating set in F_n is $D = \{x_ix_{i+1}; 1 \leq i \leq n\ ; i = 1(mod3), 2(mod3)\}$. Thus, the cardinality of the edge dominating set is $|D| = \frac{2n+2}{3} - 1$. The edge set minus the element of edge dominator is $E - D = \{cx_i; 1 \leq i \leq n\} \cup \{x_ix_{i+1}; i = n-1\} \cup \{x_ix_{i+1}; i \equiv 0\ (mod\ 3)\}$. The results of intersection between the edge neighborhood $E - D$ with D are as follow:

$N(cx_i : i = 1) \cap D = \{x_1x_2\}$
$N(cx_i : i = n) \cap D = \emptyset$
$N(cx_i : 2 \leq i \leq n-3) \cap D = \{x_{i-1}x_i, x_ix_{i+1}; 2 \leq i \leq n-2\}$

It is easy to see the intersection between the edge neighborhood and the dominator in $cx_i : i = n$ is an empty set. Thus that the dominators D does not dominate all edges in F_n. Based on the definition of locating edge dominating set, $\emptyset \neq N(e_1) \cap D \neq N(e_2) \cap D \neq \emptyset$, the intersection between edge neighborhood with the dominator may not be empty and must be different. Since $D = \{x_ix_{i+1}; 1 \leq i \leq n-2\ ; i = 1(mod3), 2(mod3)\}$ with $|D| = \frac{2n+2}{3} - 1$ does not comply the condition of graph to be locating edge dominating set, it is contradiction. We can conclude that $\gamma'_L(F_n) \geq \frac{2n+2}{3}$. Furthermore, we will show that the upper bound of locating edge domination number of $\gamma'_L(F_n) \leq \frac{2n+2}{3}$. Let us choose a locating edge dominating set in F_n $D = \{x_ix_{i+1}; 1 \leq i \leq \frac{2n+2}{3}\}$. The non-dominator set in F_n $E(G) - D = \{cx_i; 1 \leq i \leq n\} \cup \{x_ix_{i+1}; i = \frac{2n+2}{3}\}$ We get intersection between the neighborhood $N(E)$ with $E \in E(G) - D$ and dominator set D as follow:

$N(cx_i : i = 1) \cap D = \{x_1x_2\}$
$N(cx_i : i = n) \cap D = \{x_{n-1}x_n\}$
$N(cx_i : 2 \leq i \leq n-3 : i \equiv 1\ (mod\ 3), 2\ (mod\ 3)) \cap D = \{x_{i-1}x_i, x_ix_{i+1}; 2 \leq i \leq n-2\ ; i \equiv 1\ (mod\ 3), 2\ (mod\ 3)\}$
$N(cx_i; 2 \leq i \leq n-3; i \equiv 0\ (mod\ 3)) \cap D = \{x_{i-1}x_1; i \equiv 0\ (mod\ 3)\}$
$N(x_ix_{i+1} : i \equiv 0\ (mod\ 3)) \cap D = \{x_{i-1}x_i, x_{i+1}x_{i+2}; i \equiv 0\ (mod\ 3)\}$

Based on the results above, it is easy to see the intersection between the neighborhood $E - D$ with D are all different and not empty set. Thus that the dominator D dominates all edges in F_n for $n \equiv 2\ (mod\ 3)$ and $n \leq 6$. It can be concluded that $\gamma'_L(F_n) \leq \frac{2n+2}{3}$ attains the condition of locating edge dominating set. Hence, the lower bound and upper bound of locating edge domination number respectively, are $\gamma'_L(F_n) \geq \frac{2n+2}{3}$ and $\gamma'_L(F_n) \leq \frac{2n+2}{3}$. It can be completes that the locating edge domination number of F_n for or $n \equiv 2\ (mod\ 3)$ and $n \leq 6$

IOP Conf. Series: Journal of Physics: Conf. Series **1022** (2018) 012007 doi:10.1088/1742-6596/1022/1/012007

is $\gamma'_L(F_n) = \frac{2n+2}{3}$. □

To finish our paper, we will show the upper bound of locating edge domination number of any graph. This bound is useful in locating edge domination number study. Since this research is a new type of dominating sets, thus finding a sharp upper bound is highly important. The result is as follow:

Lemma 1 *Let G be a graph with edge set $E(G)$ and vertex set $V(G)$. Let q be the number of edges in G and $\gamma'_L(G)$ be the locating edge domination number of graph G. The locating edge domination number of graph G satisfies*

$$\gamma'_L G \leq q - 1$$

Proof. Assume that we take $q-1$ edges as locating edge dominating set for any graph of q edges, there will be only one edge which is dominated by $q-1$ different edge. The intersection between the edge and its neighborhood will always be different. This condition satisfies the requirement of locating edge domination number. It completes the proof. □

4. Concluding Remarks

In this paper, we have determined the exact values of locating edge domination number on related wheel, namely gear graph (G_n), helm graph (H_n) and fan graph (F_n). We also analyze the upper bound of locating edge domination number of any graph G, since this research is a new topic in the edge dominating set study. Many problems related to this topic have not been discovered yet. Consequently, we propose the following open problem.

Open Problem 1 *Let G be any connected graph, determine the lower bound of locating edge domination number of any graph G.*

Open Problem 2 *Determine the exact locating edge domination number of special graphs and its operations.*

5. Acknowledgments

We gratefully acknowledge the support from CGANT - University of Jember of year 2017.

6. References

[1] Arumugam S and Velammal S 1998 Edge Domination in Graphs *Taiwanese Journal of Mathematics* **2(2)** 173-179.
[2] Chartrand G and Lesniak L 2000 *Graphs and digraphs 3rd* ed (London: Chapman and Hall).
[3] Colbourn C J, Slater P J and Stewart L K 1987 Locating-dominating sets in series-parallel networks *Congr. Numer.* **56** 135-162
[4] Dafik, Slamin, Tanna D, Semanicova-Fenovckova A and Baca M 2017 Constructions of H-antimagic graphs using smaller edge-antimagic graphs *Ars Combin.* **133** pp 233-245
[5] Dafik, Meganingtyas D E W, Purnomo K D, Tarmidzi M D and Agustin I H 2017 Several classes of graphs and their r-dynamic chromatic numbers *J. Phys.: Conf. Ser.* **855** 012011
[6] Dafik, Purnapraja A K and Hidayat R 2015 Cycle-Super Antimagicness of Connected and Disconnected Tensor Product of Graphs *Procedia Computer Science* **74** pp 93-99
[7] Foucaud F, Henning M A, Löwenstein C and Sasse T 2016 Locating dominating sets in twin-free graphs *Discrete Applied Mathematics* **200** 52-58
[8] Gembong A W, Slamin, Dafik, and Agustin I H 2017 Bound of Distance Domination Number of Graph and Edge Comb Product Graph *J. Phys.: Conf. Ser.* **855** 012014
[9] Gross J L, Yellen J and Zhang P 2014 *Handbook of graph Theory* (London : Second Edition CRC Press Taylor and Francis Group)
[10] Hedetniemi S T and Mitchell S 1977 Edge Domination in Graphs *Congr Numer* **19** 489-509
[11] Kulli V R 2013 The Semientire Edge Dominating Graph *Ultra Scientist* **25(3)A** 431-434
[12] Slater P J 1987 Dominating and location in acyclic graphs *Networks* **17** 55-64

Distance Domination Number of Graphs Resulting from Edge Comb Product

Slamin[1,2], **Dafik**[1,3], **Gembong Angger Waspodo**[1,4]
[1] CGANT Research Group, University of Jember, Indonesia
[2] Information System Study Program, University of Jember, Indonesia
[3] Mathematics Education Study Program, University of Jember, Indonesia
[4] Mathematics Department, University of Jember, Indonesia

Abstract. Let G be a simple, finite and connected graph with a vertex-set $V(G)$ and an edge-set $E(G)$. For an integer $1 \leq k \leq \text{diam }(G)$, a distance k-dominating set of a connected graph G is a set S of vertices of G such that every vertex of $V(G)\backslash S$ is at distance at most k from some vertex of S. The k-domination number of G, denoted by $\gamma_k(G)$, is the minimum cardinality of a k-dominating set of G. In this paper, we determine the exact value of k-domination number of graphs resulting from an edge comb product of two graphs G_1 and G_2, where G_1 is a wheel, a friendship graph, or a triangular book and G_2 is a cycle.

Keywords: distance domination, dominating set, diameter, edge comb product

1. Introduction

Let G be a simple, finite and connected graph with a vertex-set $V(G)$ and an edge-set $E(G)$. Let u and v be the vertices in $V(G)$. The distance from u and v in a connected graph G, denoted by $d(u, v)$, is the minimum of the length of the $u - v$ paths in G. The neighbour of the vertex $v \in V(G)$ is a vertex which is adjacent to v. The open neighbourhood of v, denoted by $N(v)$, is the set of all neighbours of $v \in V(G)$, while the closed neighbourhood of v is the set $N[v] = N(v) \cup \{v\}$. The closed k-neighbourhood of v, denoted by $N_k[v]$, is defined as the set of all vertices within distance at most k from $v \in V(G)$; that is, $N_k[v] = \{u \; d(u, v) \leq k\}$ [3]. When $k = 1$, $N_k[v] = N[v]$. The degree of a vertex $v \in G$, denoted by $d(v)$, is the number of neighbours of $v \in V(G)$, that is $|N(v)|$. The minimum and maximum degree among all the vertices of G are denoted by $\delta(G)$ and $\Delta(G)$, respectively. For detail definition of graph terminology, see for example [2].

Meir and Moon [9] introduced the notion of distance k-dominating set (or or k- covering) in a graph. A set S is a k-dominating set of G if every vertex is within distance at most k from some vertex of S; that is, for every vertex v of G, we have $d(v, S) \leq k$. The k-domination number of G, denoted by $\gamma_k(G)$, is the minimum cardinality of a k-dominating set of G. When $k = 1$, the 1-domination number of G is precisely the domination number of G, that is, $\gamma_1(G) = \gamma(G)$. Dominating set has a broad range of applications. One of the application is in the game of chess, where the goal is to cover or dominate various squares of a chessboard by certain chess pieces [2]. The distance domination is now widely studied, for example see [1, 3, 4, 7, 10, 11].

Many researchers have studied the exact value of distance domination number of some classes of graphs. Some results are related to the domination number of some product of graphs, for example see [5, 8, 12]. Most of the research on the distance k-domination number is still limited to $k = 1$, while for $k \geq 2$, the exact value of distance domination number of some graph classes is widely open. In this paper, we determine the exact value of k-domination number of graphs resulting from an edge comb product of two graphs G_1 and G_2, where G_1 is a wheel, a friendship graph, or a triangular book and G_2 is a cycle. The edge comb product of two graphs G_1 and G_2, denoted by $G_1 \trianglerighteq G_2$, is a graph obtained by taking one copy of G_1 and $|E(G_1)|$ copies of G_2 and grafting the i-th copy of G_2 at the i-th edges of G_1 [6].

2. Main Results

In this section we determine the exact value of distance k-dominating number of $G_1 \trianglerighteq C_m$ for $G_1 \in \{F_n, W_n, B_n\}$. For simplicity, let $(C_m)_e$ is subgraph of $G_1 \trianglerighteq C_m$ that isomorphic to C_m and $E((C_m)_e) \cap E(G_1 \trianglerighteq C_m) = \{e\}$. We start with the following theorem regarding to the distance k-dominating number of the graph obtained from edge comb product of a friendship graph and a cycle.

Theorem 1. *If F_n is a friendship graph on $n+1$ vertices for $n \geq 3$ and C_m is a cycle on $m \geq 3$ vertices, then the distance k-dominating number of $G = F_n \trianglerighteq C_m$ with grafting edge $c_1c_2 \in V(C_m)$ is*

$$\gamma_k(G) = \begin{cases} 1, & m \leq 2k \\ 2n(\gamma_k(C_m) - 1) + n\gamma_k(C_m) + 1, & m \geq 2k+1, m \equiv 0 \mod (2k+1) \\ 3n(\gamma_k(C_m) - 2) + 2n + 1, & m \geq 2k+1, m \equiv 1 \mod (2k+1) \\ 3n(\gamma_k(C_m) - 1) + 1, & \textit{other value of } m. \end{cases}$$

Proof. Let G be the graph obtained from edge comb product of F_n and C_m with grafting edge $c_1c_2 \in V(C_m)$. Then the vertex set of G is $V(G) = V(F_n) \cup ((V(C_m) \backslash \{c_1, c_2\}) \times E(F_n))$ and the edge set of G is $E(G) = E(F_n) \cup \{(c_i, e)(c_{i+1}, e) | 3 \leq i \leq m-1, e \in E(F_n)\} \cup \{f(c_3, e) | e \in E(F_n)\}$, where f is a vertex such that $e = ff', f' \in V(F_n)\} \cup \{f(c_n, e) | e \in E(F_n)$, and f is a vertex such that $e = ff', f' \in V(F_n)\}$. It is easy to verify that $|V(G)| = (2n+1) + (m-1)3n$ and $|E(G)| = 3mn$.

If $m \leq 2k$, then $\gamma_k(G) = 1$ by choosing $c \in S$. Every vertex $v \in V(\cup_{e \in E(F_n), e = cf_i}(C_m)_e), 1 \leq i \leq 2n$ has property $d(c, v) \leq k$ because $\gamma_k(\cup_{e \in E(F_n), e = cf_i}(C_m)_e) = 1$, for $k \leq 2m$ and $\forall v \in V(\cup_{e \in E(F_n), e = f_if_{i+1}}(C_m)_e), d(c, v) \leq k$ as $\left\lceil \frac{m-2k}{2k+1} \right\rceil = 0$, for $m \leq 2k$. Therefore, in order to reach vertices on $\cup_{e \in E(F_n), e = f_if_{i+1}}(C_m)_e$ the additional vertex for S is not necessary. Thus we can obtain $\gamma_k(G) \leq 1$. If the value of $|S| = \gamma_k(G) = 1$, then $|S| \geq 1$ because the set $S \neq \emptyset$.

For $m \geq 2k+1$, we divide into three cases that are concerned with the variation of $\gamma_k(C_m)$ when the cycles C_m are grafted into the edges of F_n.

Case 1: $m \equiv 0 \mod (2k+1)$.

First we prove that $\gamma_k(G) \leq 2n(\gamma_k(C_m) - 1) + n\gamma_k(C_m) + 1$. To reach all the vertices in G, we choose $c \in S$. Let $\cup_{e \in E(F_n), e = cf_i, 1 \leq i \leq 2n}(C_m)_e$ is a number of $2n$ of cycle such that c is its common vertex. Since $\gamma_k(C_m)$ is minimum k-domination number of C_m, then $\gamma_k(\cup_{e \in E(F_n), e = cf_i}(C_m)_e) \leq 2n(\gamma_k(C_m) - 1) + 1$. Next, we will count k-domination number of $\gamma_k(\cup_{e \in E(F_n), e = f_if_{i+1}}(C_m)_e)$ where $i = 2l-1, 1 \leq l \leq n$. Take arbitrary $(C_m)_{f_if_{i+1}}$ and let $m = p(2k+1)$, because $c \in S$, then

$$\gamma_k((C_m)_{f_if_{i+1}}) \leq \left\lceil \frac{m-2k}{2k+1} \right\rceil = \left\lceil p - \frac{2k}{2k+1} \right\rceil = p = \left\lceil \frac{m}{2k+1} \right\rceil = \gamma_k(C_m).$$

Therefore, $\gamma_k(\cup_{e \in E(F_n), e = f_if_{i+1}}(C_m)_e) \leq n(\gamma_k(C_m))$.
All together, we have

$$\begin{aligned} \gamma_k(G) &\leq \gamma_k(\cup_{e \in E(F_n), e = cf_i}(C_m)_e) + \gamma_k(\cup_{e \in E(F_n), e = f_if_{i+1}}(C_m)_e) \\ &\leq (2n(\gamma_k(C_m) - 1) + 1) + n(\gamma_k(C_m)). \end{aligned}$$

Next, it will be proved that $\gamma_k(G) \geq 2n(\gamma_k(C_m) - 1) + n\gamma_k(C_m) + 1$. By contradiction, we suppose that $\gamma_k(G) = 2n(\gamma_k(C_m) - 1) + n\gamma_k(C_m)$. The value of $|S| = \gamma_k(G)$ will be distributed to all $(C_m)_e, e \in E(F_n)$. There exists $2n$ subsets of S where it's elements are $\gamma_k(C_m) - 1$ elements and there exists n subsets of S where it's elements are $\gamma_k(C_m)$ element. Without loss generality, let $\forall (C_m)_e, e = f_if_{i+1}, \gamma_k((C_m)_e) = \gamma_k(C_m)$ and $\forall (C_m)_e, e = cf_i, 1 \leq i \leq 2n, \gamma_k((C_m)_e) = \gamma_k(C_m) - 1$. With this kind of distribution $(C_m)_{f_if_{i+1}}$ will be adjacent with $(C_m)_{cf_i}$ and $(C_m)_{cf_{i+1}}$. Since $\gamma_k((C_m)_{cf_i}) \leq \gamma_k(C_m) - 1$, then $f_i \in S$ and $\gamma_k((C_m)_{cf_i}) = \gamma_k((C_m)_{f_if_{i+1}}) = \gamma_k(C_m)$. However $f_{i+1} \notin S$, it is because if $m \neq 1 \mod (2k+1)$ and $s_1, s_2 \in S$, then $d(s_1, s_2) > 1$. Without common dominator vertex on $(C_m)_{cf_{i+1}}$, then $\gamma_k((C_m)_{cf_{i+1}}) = \gamma_k(C_m) - 1$ and $\exists v \in V((C_m)_{cf_{i+1}})$ such that $d(v, s) > k$ for some $s \in S$.

Case 2: $m \equiv 1 \mod (2k+1)$.

First we prove that $\gamma_k(G) \leq 3n(\gamma_k(C_m) - 2) + 2n + 1$. On cycle C_m where $m \equiv 1 \mod (2k+1)$ there exist $c_1, c_2 \in S$ such that $d(c_1, c_2) = 1$. By using c_1c_2 as edge contact of operation $F_n \trianglerighteq C_m$,

IOP Conf. Series: Journal of Physics: Conf. Series **1022** (2018) 012008 doi:10.1088/1742-6596/1022/1/012008

then each vertex of F_n are dominator elements. To reach all the vertices in G, we choose $c \in S$. Let $\cup_{e\in E(F_n),e=cf_i,1\leq i\leq 2n}(C_m)_e$ is a number of $2n$ of cycle such that c is a common vertex. Since $\gamma_k(C_m)$ is minimum k-domination number of C_m, then $\gamma_k(\cup_{e\in E(F_n),e=cf_i}(C_m)_e) \leq 2n(\gamma_k(C_m)-1)+1 = 2n(\gamma_k(C_m)-2)+2n+1$. Next, we will count k-domination number of $\gamma_k(\cup_{e\in E(F_n),e=f_if_{i+1}}(C_m)_e)$ where $i=2j-1, 1\leq j\leq n$. Take arbitrary $(C_m)_{f_if_{i+1}}$, because $f_i, f_{i+1}\in S$, then $\gamma_k(((C_m)_{f_if_{i+1}})\backslash\{f_i,f_{i+1}\})) \leq \gamma_k(C_m)-2$. Therefore,

$$\gamma_k(\cup_{e\in E(F_n),e=f_if_{i+1}}(C_m)_e) \leq n(\gamma_k(C_m)-2)+2n.$$

All together, we have

$$\begin{aligned}\gamma_k(G) &\leq \gamma_k(\cup_{e\in E(F_n),e=cf_i}(C_m)_e)+\gamma_k(\cup_{e\in E(F_n),e=f_if_{i+1}}(C_m)_e)-\\ &\quad |\cup_{e\in E(F_n),e=cf_i}(C_m)_e\cap\cup_{e\in E(F_n),e=f_if_{i+1}}(C_m)_e\cap S|\\ &\leq (2n(\gamma_k(C_m)-2)+2n+1)+(n(\gamma_k(C_m)-2)+2n)-2n\\ &= 3n(\gamma_k(C_m)-2))+2n+1.\end{aligned}$$

Next, it will be proved that $\gamma_k(G) \geq 3n(\gamma_k(C_m)-2))+2n+1$. By contradiction, we suppose that $\gamma_k(G) = 3n(\gamma_k(C_m)-2))+2n$. A number $3n(\gamma_k(C_m)-2))$ element of S will be distributed into $3n$ subgraph $(C_m)_e, e\in E(F_n)$. Therefore, $\forall(C_m)_e, \gamma_k(C_m)_e \geq \gamma_k(C_m)-2$. There are $2n$ elements of S will be used as an additional dominator element for $(C_m)_e, e\in E(F_n)$. Since $m\equiv 1 \mod (2k+1)$, then for all $(C_m)_e$ there exists $s_1,s_2\in S, d(s_1,s_2)=1$. Therefore, $\forall v\in V(F_n), v\in S$. Since $|V(F_n)|=2n+1>2n$, then $\exists v\in V(F_n), v\notin S$. Without loss generality, let $f_1\notin S$ and $f_2, c\in S$. Consider $(C_m)_{cf_1}, (C_m)_{cf_2}$ and $(C_m)_{f_1f_2}$, if $f_1\notin S$ and $f_2,c\in S$, then $\gamma_k((C_m)_{cf_1}) = \gamma_k((C_m)_{f_1f_2}) = \gamma_k(C_m)-1$. Therefore, $\exists v\in V((C_m)_{cf_1}))\cup V((C_m)_{f_1f_2}))$ such that $d(v,s)>k, s\in S$.

Case 3: Remaining value of m.

First we prove that $\gamma_k(G)\leq 3n(\gamma_k(C_m)-1)+1$. To reach all the vertices in G we first choose $c\in S$. Let $\cup_{e\in E(F_n),e=cf_i,1\leq i\leq 2n}(C_m)_e$ is a number of $2n$ of cycle such that c is a common vertex. Since $\gamma_k(C_m)$ is minimum k-domination number of C_m, then $\gamma_k(\cup_{e\in E(F_n),e=cf_i}(C_m)_e)\leq 2n(\gamma_k(C_m)-1)+1$. Next, we determine the distance k-domination number of $\gamma_k(\cup_{e\in E(F_n),e=f_if_{i+1}}(C_m)_e)$ where $i=2j-1, 1\leq j\leq n$. Take arbitrary $(C_m)_{f_if_{i+1}}$and let $m=p(2k+1)+j, 2\leq j\leq 2k$. Since $c\in S$, then

$$\gamma_k((C_m)_{f_if_{i+1}}) \leq \left\lceil\frac{m-2k}{2k+1}\right\rceil = \left\lceil p+\frac{j}{2k+1}-\frac{2k}{2k+1}\right\rceil = \left\lceil p+\frac{j-2k}{2k+1}\right\rceil = p = \frac{m-j}{2k+1}.$$

Let $\{x\}$ is a decimal part of $x=\frac{m}{2k+1}$, then

$$\frac{m-j}{2k+1} = x-\{x\} = x+(1-\{x\})-1 = \lceil x\rceil -1 = \left\lceil\frac{m}{2k+1}\right\rceil -1 = \gamma_k(C_m)-1.$$

Therefore, $\gamma_k(\cup_{e\in E(F_n),e=f_if_{i+1}}(C_m)_e)\leq n(\gamma_k(C_m)-1)$. All together, we have,

$$\begin{aligned}\gamma_k(G) &\leq \gamma_k(\cup_{e\in E(F_n),e=cf_i}(C_m)_e)+\gamma_k(\cup_{e\in E(F_n),e=f_if_{i+1}}(C_m)_e)\\ &\leq (2n(\gamma_k(C_m)-1)+1)+n(\gamma_k(C_m)-1)\\ &= 3n(\gamma_k(C_m)-1)+1.\end{aligned}$$

Next, it will be proved that $\gamma_k(G)\geq 3n(\gamma_k(C_m)-1)+1$. By contradiction, we suppose that $\gamma_k(G) = 3n(\gamma_k(C_m)-1)$. By distributing all elements S into $(C_m)_e, e\in E(F_n)$, then $\forall(C_m)_e, \gamma_k((C_m)_e)\geq \gamma_k(C_m)-1$. Consider $(C_m)_{cf_1}, (C_m)_{cf_2}$ and $(C_m)_{f_1f_2}$, because $(C_m)_e$ is a subgraph of G which is isomorphic with C_m, then $\exists(C_m)_e, \gamma_k((C_m)_e) = \gamma_k(C_m)$. Without loss generality let $\gamma_k((C_m)_{cf_1}) = \gamma_k(C_m)$, this can happen by setting $f_1\in S$. Thus, there exists $(C_m)_{f_1f_2}, \gamma_k(C_m)_{f_1f_2} = \gamma_k(C_m)-2$. To cover the lack of dominator on $(C_m)_{f_1f_2}$, then on $(C_m)_{cf_2}$ vertex $f_2\in S$. Now consider subgraph $(C_m)_{cf_2}$, without f_2 the value of $|S\backslash\{f_2\}\cap V((C_m)_{cf_2}| = \gamma_k(C_m)-2$. Whereas the number of dominators that needed to dominate $m-(2k+1)$ vertices on $(C_m)_{cf_2}$ are

$$\left\lceil\frac{m-(2k+1)}{2k+1}\right\rceil = \left\lceil\frac{m}{2k+1}-1\right\rceil = \left\lceil\frac{m}{2k+1}\right\rceil -1 = \gamma_k(C_m)-1 > |S\backslash\{f_2\}\cap V((C_m)_{cf_2}|.$$

IOP Conf. Series: Journal of Physics: Conf. Series **1022** (2018) 012008 doi:10.1088/1742-6596/1022/1/012008

Therefore, $\exists v \in (C_m)_{cf_2}, d(v,s) > k, s \in S$. □

The next theorem presents the distance k-domination number of the graph obtained from edge comb product of a wheel and a cycle as follows.

Theorem 2. *If W_n is a wheel on $n+1$ vertices for $n \geq 3$ and C_m is a cycle on $m \geq 3$ vertices, then the distance k-domination number of $G = W_n \trianglerighteq C_m$ with grafting edge $c_1c_2 \in V(C_m)$ is*

$$\gamma_k(G) = \begin{cases} 1, & m \leq 2k \\ 2n(\gamma_k(C_m) - 1) + \lceil \frac{n}{2} \rceil + 1, & m \geq 2k+1, m \equiv 0 \mod (2k+1) \\ 2n(\gamma_k(C_m) - 2) + n + 1, & m \geq 2k+1, m \equiv 1 \mod (2k+1) \\ 2n(\gamma_k(C_m) - 1) + 1, & \text{other value of } m. \end{cases}$$

Proof. Let G be the graph obtained from edge comb product of W_n and C_m with grafting edge $c_1c_2 \in V(C_m)$.. Then vertex set of G is $V(G) = V(W_n) \cup ((V(C_m) \backslash \{c_1, c_2\}) \times E(W_n))$ and edge set of G is $E(G) = E(W_n) \cup \{(c_i, e)(c_{i+1}, e) | 3 \leq i \leq m-1, e \in E(W_n)\} \cup \{w(c_3, e) | e \in E(W_n)$, where w is a vertex such that $e = ww', w' \in V(W_n)\} \cup \{w(c_n, e) | e \in E(W_n)$, w is a vertex such that $e = ww', w' \in V(W_n)\}$.. It is easy to verify that $|V(G)| = (n+1) + (m-2)(2n)$ and $|E(G)| = 2mn$.

If $m \leq 2k$, then $\gamma_k(G) = 1$ by using the same argument in the second paragraph in the proof of Theorem 1.

For $m \geq 2k+1$, the proof of theorem is divided into three cases which depend on the variation of $\gamma_k(C_m)$ when the cycles C_m are grafted into the edges of W_n.

Case 1: $m \equiv 0 \mod (2k+1)$.

First we prove that $\gamma_k(G) \leq 2n(\gamma_k(C_m) - 1) + \lceil \frac{n}{2} \rceil + 1$. To reach all the vertices in G, we choose $c \in S$. Let $\cup_{e \in E(W_n), e = cw_i, 1 \leq i \leq n}(C_m)_e$ be all n cycles such that c is a common vertex. Since $\gamma_k(C_m)$ is minimum k-domination number of C_m, then $\gamma_k(\cup_{e \in E(W_n), e = cw_i}(C_m)_e) \leq n(\gamma_k(C_m) - 1) + 1$. Next, we count k-domination number of $\gamma_k(\cup_{e \in \{w_iw_{i+1}, w_nw_1\}}(C_m)_e)$ where $1 \leq i \leq n-1$. Take arbitrary $(C_m)_{e \in \{w_iw_{i+1}, w_nw_1\}}$and let $m = p(2k+1)$, because $c \in S$, then

$$\gamma_k((C_m)_{w_iw_{i+1}}) \leq \left\lceil \frac{m-2k}{2k+1} \right\rceil = \left\lceil p - \frac{2k}{2k+1} \right\rceil = p = \left\lceil \frac{m}{2k+1} \right\rceil = \gamma_k(C_m).$$

Therefore, $\gamma_k(\cup_{e \in \{w_iw_{i+1}, w_nw_1\}}(C_m)_e) \leq n(\gamma_k(C_m))$. However, for all $(C_m)_e$ that are adjacent can share its dominator vertex. Thus, there exists $\lceil \frac{n}{2} \rceil$ subgraph $(C_m)_e$ where $\gamma_k((C_m)_e) = \gamma_k(C_m)$.
All together, we have

$$\begin{aligned} \gamma_k(G) &\leq \gamma_k(\cup_{e \in E(W_n), e = cw_i}(C_m)_e) + \gamma_k(\cup_{e \in \{w_iw_{i+1}, w_nw_1\}}(C_m)_e) \\ &\leq (n(\gamma_k(C_m) - 1) + 1) + (n(\gamma_k(C_m) - 1) + \left\lceil \frac{n}{2} \right\rceil) \\ &= (2n(\gamma_k(C_m) - 1) + 1) + \left\lceil \frac{n}{2} \right\rceil. \end{aligned}$$

Next, it will be proved that $\gamma_k(G) \geq (2n(\gamma_k(C_m) - 1) + 1) + \lceil \frac{n}{2} \rceil$. By contradiction, we suppose that $\gamma_k(G) = 2n(\gamma_k(C_m) - 1) + \lceil \frac{n}{2} \rceil$. The value of $|S| = \gamma_k(G)$ will be distributed to all $(C_m)_e, e \in E(W_n)$. First, a number of $2n(\gamma_k(C_m) - 1)$ element of S will be placed to a number of $2n$ subraph $(C_m)_e$ and a number $\lceil \frac{n}{2} \rceil$ elements of S will be placed to $\forall v \in V(W_n)$. Without loss generality, let $\forall e \in E(W_n), \gamma_k((C_m)_e) = \gamma_k(C_m) - 1$ and $|V(W_n) \cap S| = \lceil \frac{n}{2} \rceil$. If $c \in S$, then there exists $(C_m)_{w_iw_{i+1}}$ such that $\gamma_k((C_m)_{w_iw_{i+1}}) = \gamma_k(C_m) - 1$. Since $m \equiv 0 \mod (2k+1)$, then on $(C_m)_{w_iw_{i+1}}$ need at least $\gamma_k(C_m)$ element of S to dominate the remaining vertices on that subgraph. Thus, there exists $v \in V((C_m)_{w_iw_{i+1}})$ such that $d(v,s) > k, s \in S$. If $c \notin S$, then there exists $(C_m)_{cw_i}$ such that $\gamma_k((C_m)_{cw_i}) = \gamma_k(C_m) - 1$. With arguments similar to case $c \in S$ we can get the same contradiction. Consequently, with all the possibilities it has been shown that the value $\gamma_k(G) \geq 2n(\gamma_k(C_m) - 1) + \lceil \frac{n}{2} \rceil$ is imposible. Therefore, for $m \equiv 0 \mod (2k+1)$, $\gamma_k(G) \geq 2n(\gamma_k(C_m) - 1) + 1 + \lceil \frac{n}{2} \rceil$.

Case 2: $m \equiv 1 \mod (2k+1)$.

First we prove that $2n(\gamma_k(C_m) - 2) + n + 1$. On cycle C_m where $m \equiv 1 \mod (2k+1)$ there will be exist $c_1, c_2 \in S$ such that $d(c_1, c_2) = 1$. By using c_1c_2 as edge contact of operation $W_n \trianglerighteq C_m$, then each vertices of W_n are dominators element. To reach all the vertices in G, we choose $c \in S$. Let $\cup_{e\in E(W_n),e=cw_i,1\leq i\leq n}(C_m)_e$ be all n cycles such that c is a common vertex. Since $\gamma_k(C_m)$ is minimum k-domination number of C_m, then $\gamma_k(\cup_{e\in E(W_n),e=cw_i}(C_m)_e) \leq n(\gamma_k(C_m)-1)+1 = n(\gamma_k(C_m)-2)+n+1$. Next, we will count k-domination number of $\gamma_k(\cup_{e\in\{w_iw_{i+1},w_nw_1\}}(C_m)_e)$ where $1 \leq i \leq n-1$. Take arbitrary $(C_m)_{e\in\{w_iw_{i+1},w_nw_1\}}$, because $w_i, w_{i+1} \in S$, then $\gamma_k(((C_m)_{e\in\{w_iw_{i+1},w_nw_1\}})\backslash\{w_i,w_{i+1}\})) \leq \gamma_k(C_m)-2$.Therefore,

$$\gamma_k(\cup_{e\in\{w_iw_{i+1},w_nw_1\}}(C_m)_e) \leq n(\gamma_k(C_m)-2)+n.$$

All together, we have

$$\begin{aligned}\gamma_k(G) &\leq \gamma_k(\cup_{e\in E(W_n),e=cw_i}(C_m)_e) + \gamma_k(((C_m)_{e\in\{w_iw_{i+1},w_nw_1\}})\backslash\{w_i,w_{i+1}\})) \\ &\leq (n(\gamma_k(C_m)-2)+n+1) + (n(\gamma_k(C_m)-2)) \\ &= 2n(\gamma_k(C_m)-2))+n+1.\end{aligned}$$

Next, it will be proved that $\gamma_k(G) \geq 2n(\gamma_k(C_m)-2))+n+1$. By contradiction, we suppose that $\gamma_k(G) = 2n(\gamma_k(C_m)-2))+n$. A number $2n(\gamma_k(C_m)-2))$ element of S will be distributed into $2n$ subgraph $(C_m)_e, e \in E(W_n)$. Therefore, $\forall (C_m)_e, \gamma_k(C_m)_e \geq \gamma_k(C_m)-2$. A number n element of S will be used as an additional dominator element for $(C_m)_e, e \in E(W_n)$. Since $m \equiv 1 \mod (2k+1)$, then for all $(C_m)_e$ there exists $s_1, s_2 \in S, d(s_1,s_2)=1$. Therefore, $\forall v \in V(W_n), v \in S$. Since $|V(W_n)| = n+1 > n$, then $\exists v \in V(W_n), v \notin S$. Without loss generality, let $w_1 \notin S$ and $w_2, c \in S$. Consider $(C_m)_{cw_1}, (C_m)_{cw_2}$ and $(C_m)_{w_1w_2}$, if $w_1 \notin S$ and $w_2, c \in S$, then $\gamma_k((C_m)_{cw_1}) = \gamma_k((C_m)_{w_1w_2}) = \gamma_k(C_m)-1$. Therefore, $\exists v \in V((C_m)_{cw_1})) \cup V((C_m)_{w_1w_2}))$ such that $d(v,s) > k, s \in S$.

Case 3: Remaining value of m.

First we prove that $\gamma_k(G) \leq 2n(\gamma_k(C_m)-1)+1$. To reach all the vertices in G, we choose $c \in S$. Let $\cup_{e\in E(W_n),e=cw_i}(C_m)_e$ be all n cycles such that c is its common vertex. Since $\gamma_k(C_m)$ is minimum k-domination number of C_m, then $\gamma_k(\cup_{e\in E(W_n),e=cw_i}(C_m)_e) \leq n(\gamma_k(C_m)-1)+1$. Next, we will count k-domination number of $\gamma_k(\cup_{e\in\{w_iw_{i+1},w_nw_1\}}(C_m)_e)$ where $1 \leq i \leq n-1$. Take arbitrary $(C_m)_{w_iw_{i+1}}$ and let $m = p(2k+1)+j, 2 \leq j \leq 2k$. Since $c \in S$, then

$$\gamma_k((C_m)_{w_iw_{i+1}}) \leq \left\lceil \frac{m-2k}{2k+1} \right\rceil = \left\lceil p + \frac{j}{2k+1} - \frac{2k}{2k+1} \right\rceil = \left\lceil p + \frac{j-2k}{2k+1} \right\rceil = p = \frac{m-j}{2k+1}.$$

Let $\{x\}$ is a decimal part of $x = \frac{m}{2k+1}$, then

$$\frac{m-j}{2k+1} = x - \{x\} = x + (1-\{x\}) - 1 = \lceil x \rceil - 1 = \left\lceil \frac{m}{2k+1} \right\rceil - 1 = \gamma_k(C_m) - 1.$$

Therefore, $\gamma_k(\cup_{e\in\{w_iw_{i+1},w_nw_1\}}(C_m)_e) \leq n(\gamma_k(C_m)-1)$.
All together, we have

$$\begin{aligned}\gamma_k(G) &\leq \gamma_k(\cup_{e\in E(W_n),e=cw_i}(C_m)_e) + \gamma_k(\cup_{e\in\{w_iw_{i+1},w_nw_1\}}(C_m)_e) \\ &\leq (n(\gamma_k(C_m)-1)+1) + n(\gamma_k(C_m)-1) \\ &= 2n(\gamma_k(C_m)-1)+1.\end{aligned}$$

Next, it will be proved that $\gamma_k(G) \geq 2n(\gamma_k(C_m)-1)+1$. By contradiction, we suppose that $\gamma_k(G) = 2n(\gamma_k(C_m)-1)$. By distributing all elements S into $(C_m)_e, e \in E(W_n)$, then $\forall (C_m)_e, \gamma_k((C_m)_e) \geq \gamma_k(C_m)-1$. Consider $(C_m)_{cw_1}, (C_m)_{cw_2}$ and $(C_m)_{w_1w_2}$, because $(C_m)_e$ is a subgraph of G which is isomorphic with C_m, then $\exists (C_m)_e, \gamma_k((C_m)_e) = \gamma_k(C_m)$. Without loss generality let $\gamma_k((C_m)_{cw_1}) = \gamma_k(C_m)$, this can happen by setting $w_1 \in S$. Thus, there exists $(C_m)_{w_1w_2}, \gamma_k(C_m)_{w_1w_2} = \gamma_k(C_m)-2$. To cover the lack of dominator on $(C_m)_{w_1w_2}$, then on $(C_m)_{cw_2}$ vertex $w_2 \in S$. Now consider subgraph $(C_m)_{cw_2}$, without w_2 the value of $|S\backslash\{w_2\}\cap V((C_m)_{cw_2}| = \gamma_k(C_m)-2$. Whereas the number of dominators that needed to dominate $m-(2k+1)$ vertices on $(C_m)_{cw_2}$ are

$$\left\lceil \frac{m-(2k+1)}{2k+1} \right\rceil = \left\lceil \frac{m}{2k+1} - 1 \right\rceil = \left\lceil \frac{m}{2k+1} \right\rceil - 1 = \gamma_k(C_m) - 1 > |S\backslash\{w_2\} \cap V((C_m)_{cw_2}|.$$

Therefore, $\exists v \in (C_m)_{cw_2}, d(v,s) > k, s \in S.$ □

We close this section with the distance k-domination number of the graph obtained from edge comb product of a triangular book and a cycle.

Theorem 3. *If B_n is a triangular book on $n \geq 2$ sheets and C_m is a cycle on $m \geq 3$ vertices, then the distance k-domination number $G = B_n \trianglerighteq C_m$ with grafting edge $c_1c_2 \in V(C_m)$ is*

$$\gamma_k(G) = \begin{cases} 1, & m \leq 2k \\ (2n+1)(\gamma_k(C_m)-1)+2, & m \geq 2k+1, m \equiv 0 \mod (2k+1) \\ (2n+1)(\gamma_k(C_m)-2)+n+2, & m \geq 2k+1, m \equiv 1 \mod (2k+1) \\ (2n+1)(\gamma_k(C_m)-1)+1, & \textit{other value of } m. \end{cases}$$

Proof. Let G be the graph obtained from edge comb product of B_n and C_m with grafting edge $c_1c_2 \in V(C_m)$. Then the vertex set of G is $V(G) = V(B_n) \cup ((V(C_m)\backslash\{c_1,c_2\}) \times E(B_n))$ and the edge set of G is $E(G) = E(B_n) \cup \{(c_i,e)(c_{i+1},e) | 3 \leq i \leq m-1, e \in E(B_n)\} \cup \{b(c_3,e) | e \in E(B_n), b$ is a vertex such that $e = bb', b' \in V(B_n)\} \cup \{b(c_n,e) | e \in E(B_n),$
b is a vertex such that $e = bb', b' \in V(B_n)\}$. It is easy to verify that $|V(G)| = (n+2)+(m-2)(2n+1)$ and $|E(G)| = (2n+1)m$.

If $m \leq 2k$, then $\gamma_k(G) = 1$ again by using the same argument in the second paragraph in the proof of Theorem 1.

For $m \geq 2k+1$, the proof is divided into three cases which depend on the variation of $\gamma_k(C_m)$ when the cycles C_m are grafted into the edges of B_n.

Case 1: $m \equiv 0 \mod (2k+1)$.

First we prove that $\gamma_k(G) \leq (2n+1)(\gamma_k(C_m)-1)+2$. To reach all the vertices in G, we choose $t_1 \in S$. Let $\cup_{e\in E(B_n),e=t_1x}(C_m)_e$ is a number of $n+1$ cycle such that t_1 is its common vertex. Since $\gamma_k(C_m)$ is minimum k-domination number of C_m, then $\gamma_k(\cup_{e\in E(B_n),e=t_1x}(C_m)_e) \leq (n+1)(\gamma_k(C_m)-1)+1$. Next, we count k-domination number of $\gamma_k(\cup_{e\in\{t_2x,x\neq t_1\}}(C_m)_e)$. Take arbitrary $(C_m)_{e\in\{t_2x,x\neq t_1\}}$and let $m = p(2k+1)$, because $c \in S$, then

$$\gamma_k((C_m)_{b_iw_{i+1}}) \leq \left\lceil \frac{m-2k}{2k+1} \right\rceil = \left\lceil p - \frac{2k}{2k+1} \right\rceil = p = \left\lceil \frac{m}{2k+1} \right\rceil = \gamma_k(C_m).$$

By setting $t_2 \in S$, then $\gamma_k(\cup_{e\in\{t_2x,x\neq t_1\}}(C_m)_e) \leq n(\gamma_k(C_m)-1)+1$.
All together, we have

$$\begin{aligned}\gamma_k(G) &\leq \gamma_k(\cup_{e\in E(B_n),e=t_1x}(C_m)_e) + \gamma_k(\cup_{e\in\{t_2x,x\neq t_1\}}(C_m)_e) \\ &\leq ((n+1)(\gamma_k(C_m)-1)+1) + (n(\gamma_k(C_m)-1)+1) \\ &= (2n+1)(\gamma_k(C_m)-1)+2.\end{aligned}$$

Next, it will be proved that $\gamma_k(G) \geq (2n+1)(\gamma_k(C_m)-1)+2$. By contradiction, we suppose that $\gamma_k(G) = (2n+1)(\gamma_k(C_m)-1)+1$. The value of $|S| = \gamma_k(G)$ will be distributed to all $(C_m)_e, e \in E(B_n)$. First, a number of $(2n+1)(\gamma_k(C_m)-1)$ element of S will be placed to a number of $2n+1$ subraph $(C_m)_e$ and one element of S will be placed to $v \in V(B_n)$. Without loss generality, let $\forall e \in E(G), \gamma_k((C_m)_e) = \gamma_k(C_m)-1$ and $|V(B_n) \cap S| = 1$. If $t_1 \in S$ or $t_2 \in S$, then there exists $(C_m)_{b_it_j}, 1 \leq i \leq n, 1 \leq j \leq 2$ such that $\gamma_k((C_m)_{b_it_j}) = \gamma_k(C_m)-1$. Since $m \equiv 0 \mod (2k+1)$, then on $(C_m)_{b_it_j}$ need at least $\gamma_k(C_m)$ element of S to dominate the remaining vertices on that subgraph. Thus, there exists $v \in V((C_m)_{b_it_j})$ such that $d(v,s) > k, s \in S$. If $t_j \notin S$, then There exists $(C_m)_{t_1t_2}$ such that $\gamma_k((C_m)_{t_1t_2}) = \gamma_k(C_m)-1$. With arguments similar to case $t_j \in S$we can get the same contradiction. Consequently, with all the possibilities it has been shown that $\gamma_k(G) \geq (2n+1)(\gamma_k(C_m)-1)+1$ is impossible. Therefore, for $m \equiv 0 \mod (2k+1)$, $\gamma_k(G) \geq (2n+1)(\gamma_k(C_m)-1)+2$.

Case 2: $m \equiv 1 \mod (2k+1)$.

First we prove that $\gamma_k(C_m) \leq (2n+1)(\gamma_k(C_m)-2)+n+2$. On cycle C_m where $m \equiv 1 \mod (2k+1)$, there will be exist $c_1, c_2 \in S$ such that $d(c_1,c_2) = 1$. By using c_1c_2 as edge contact of operation $B_n \trianglerighteq C_m$, then each vertices of B_n are dominators element. To reach all the vertices in G, we choose $t_1 \in S$. Let $\cup_{e\in E(B_n),e=t_1x}(C_m)_e$ is a number of $n+1$ cycle such that t_1 is its vertex. Since $\gamma_k(C_m)$

is minimum k-domination number of C_m, then $\gamma_k(\cup_{e\in E(B_n),e=t_1x}(C_m)_e) \leq (n+1)(\gamma_k(C_m)-1)+1 = (n+1)(\gamma_k(C_m)-2)+n+1$. Next, we count k-domination number of $\gamma_k(\cup_{e\in\{t_2x,x\neq t_1\}}(C_m)_e)$. Take arbitrary $(C_m)_{e\in\{t_2x,x\neq t_1\}}$, because $\forall v \in V(B_n), v \in S$, then $\gamma_k(((C_m)_{e\in\{t_2x,x\neq t_1\}})\backslash\{t_2,b_i\})) \leq \gamma_k(C_m)-2$. Therefore,

$$\gamma_k(((C_m)_{e\in\{t_2x,x\neq t_1\}})\backslash\{t_2,b_i\})) \leq n(\gamma_k(C_m)-2)+1.$$

All together, we have

$$\begin{aligned}\gamma_k(G) &\leq \gamma_k(\cup_{e\in E(B_n),e=t_1x}(C_m)_e) + \gamma_k(((C_m)_{e\in\{t_2x,x\neq t_1\}})\backslash\{t_2,b_i\}))\\ &\leq ((n+1)(\gamma_k(C_m)-2)+n+1)+(n(\gamma_k(C_m)-2)+1)\\ &= (2n+1)(\gamma_k(C_m)-2))+n+2.\end{aligned}$$

Next, it will be proved that $\gamma_k(G) \geq (2n+1)(\gamma_k(C_m)-2))+n+2$. By contradiction, we suppose that $\gamma_k(G) = (2n+1)(\gamma_k(C_m)-2))+n+1$. A number $(2n+1)(\gamma_k(C_m)-2))$ element of S will be distributed into $2n+1$ subgraph $(C_m)_e, e \in E(B_n)$. Therefore, $\forall (C_m)_e, \gamma_k(C_m)_e \geq \gamma_k(C_m)-2$. A number $n+1$ element of S will be used as an additional dominator element for $(C_m)_e, e \in E(G)$. Since $m \equiv 1 \mod (2k+1)$, then for all $(C_m)_e$ there exists $s_1, s_2 \in S, d(s_1,s_2) = 1$. Therefore, $\forall v \in V(B_n), v \in S$. Since $|V(B_n)| = n+2 > n+1$, then $\exists v \in V(B_n), v \notin S$. Without loss generality, let $b_1 \notin S$ and $b_2, t_1 \in S$. Consider $(C_m)_{b_1t_1}, (C_m)_{b_1t_2}$ and $(C_m)_{t_1t_2}$, if $b_1 \notin S$ and $b_2, t_1 \in S$, then $\gamma_k((C_m)_{t_1b_1}) = \gamma_k((C_m)_{t_2b_1}) = \gamma_k(C_m)-1$. Therefore, $\exists v \in V((C_m)_{t_1b_1})) \cup V((C_m)_{t_1b_2}))$ such that $d(v,s) > k, s \in S$.

Case 3: Remaining value of m.

First we prove that $\gamma_k(G) \leq (2n+1)(\gamma_k(C_m)-1)+1$. To reach all the vertices in G, we choose $c \in S$. Let $\cup_{e\in E(B_n),e=t_1x}(C_m)_e$ is a number of $n+1$ cycle such that c is its common vertex. Since $\gamma_k(C_m)$ is minimum k-domination number of C_m, then $\gamma_k(\cup_{e\in E(B_n),e=t_1x}(C_m)_e) \leq (n+1)(\gamma_k(C_m)-1)+1$. Next, we count k-domination number of $\gamma_k(\cup_{e\in\{t_2x,x\neq t_1\}}(C_m)_e)$. Take arbitrary $(C_m)_{b_iw_{i+1}}$ and let $m = p(2k+1)+j, 2 \leq j \leq 2k$. Since $c \in S$, then

$$\gamma_k((C_m)_{b_iw_{i+1}}) \leq \left\lceil \frac{m-2k}{2k+1} \right\rceil = \left\lceil p + \frac{j}{2k+1} - \frac{2k}{2k+1} \right\rceil = \left\lceil p + \frac{j-2k}{2k+1} \right\rceil = p = \frac{m-j}{2k+1}.$$

Let $\{x\}$ is a decimal part of $x = \frac{m}{2k+1}$, then

$$\frac{m-j}{2k+1} = x - \{x\} = x + (1-\{x\}) - 1 = \lceil x \rceil - 1 = \left\lceil \frac{m}{2k+1} \right\rceil - 1 = \gamma_k(C_m) - 1.$$

Therefore, $\gamma_k(\cup_{e\in\{t_2x,x\neq t_1\}}(C_m)_e) \leq n(\gamma_k(C_m)-1)$. All together, we have

$$\begin{aligned}\gamma_k(G) &\leq \gamma_k(\cup_{e\in E(B_n),e=t_1x}(C_m)_e) + \gamma_k(\cup_{e\in\{t_2x,x\neq t_1\}}(C_m)_e)\\ &\leq ((n+1)(\gamma_k(C_m)-1)+1)+n(\gamma_k(C_m)-1)\\ &= (2n+1)(\gamma_k(C_m)-1)+1.\end{aligned}$$

Next, it will be proved that $\gamma_k(G) \geq (2n+1)(\gamma_k(C_m)-1)+1$. By contradiction, we suppose that $\gamma_k(G) = (2n+1)(\gamma_k(C_m)-1)$. By distributing all elements S into $(C_m)_e, e \in E(B_n)$, then $\forall (C_m)_e, \gamma_k((C_m)_e) \geq \gamma_k(C_m)-1$.

Consider $(C_m)_{b_1t_1}, (C_m)_{b_1t_2}$ and $(C_m)_{t_1t_2}$, because $(C_m)_e$ is a subgraph of G which is isomorphic with C_m, then $\exists (C_m)_e, \gamma_k((C_m)_e) = \gamma_k(C_m)$. Without loss generality let $\gamma_k((C_m)_{b_1t_1}) = \gamma_k(C_m)$, this can happen by setting $t_1 \in S$. Therefore, there exists $(C_m)_{t_1t_2}, \gamma_k(C_m)_{t_1t_2} = \gamma_k(C_m)-2$. To cover the lack of dominator on $(C_m)_{t_1t_2}$, then on $(C_m)_{b_1t_2}$ vertex $t_2 \in S$. Now consider subgraph $(C_m)_{b_1t_2}$, without t_2 the value of $|S\backslash\{t_2\} \cap V((C_m)_{b_1t_2}| = \gamma_k(C_m)-2$. Whereas the number of dominators that needed to dominate $m-(2k+1)$ vertices on $(C_m)_{b_1t_2}$ are

$$\left\lceil \frac{m-(2k+1)}{2k+1} \right\rceil = \left\lceil \frac{m}{2k+1} - 1 \right\rceil = \left\lceil \frac{m}{2k+1} \right\rceil - 1 = \gamma_k(C_m) - 1 > |S\backslash\{f_2\} \cap V((C_m)_{b_1t_2}|.$$

Therefore, $\exists v \in (C_m)_{b_1t_2}, d(v,s) > k, s \in S$. □

3. Conclusion

This paper has presented the exact value of distance k-domination number of $F_n \trianglerighteq C_m, W_n \trianglerighteq C_m$, and $B_n \trianglerighteq C_m$. The distance k-domination number of edge combs product of two other families of graphs are remaining open for future works.

Acknowledgement

This research is supported by DRPM Ditjen Penguatan Risbang Kemenristekdikti under Penelitian Tim Pascasarjana 2017 research grant.

Refferences

[1] Y. Caro and Y. Roditty, A Note on the k-domination number of a graph, *Internat. J. Math. & Math. Sci.* **13** (1990), 205-206.
[2] G. Chartrand, *Graph and Digraph*, New York: Taylor and Francis Group (2011).
[3] R. Davila, C. Fast, M.A. Henning and F. Kenter, Lower bounds on the distance domination number of a graph, arXiv:1507.08745v1 [math.CO] 31 jul 2015.
[4] W. Duckworth and B. Mans, *Randomised Greedy Algorithms for Finding Small k-Dominating Sets of Regular Graphs*, John Wiley & Sons. Inc, Australia (2005).
[5] M.H. El-Zahar, S.M. Khamis, and M. Nazzal, On the domination number of the cartesian product of the cycle of length and any graph, *Discrete Applied Mathematics* **155** (4) (2007), 515-522.
[6] A.W. Gembong, Slamin, Dafik and I.H. Agustin, Bound of distance domination number of graph and edge comb product graph, *Journal of Physics: Conference Series* **855** (2017).
[7] M. A. Henning *Domination in Graphs: Advanced Topics*, T. W. Haynes, S. T. Hedetniemi and P. J. Slater (eds), Marcel Dekker, Inc. New York (1998).
[8] S. Klavzar and N. Seifter, Dominating Cartesian products of cycles, *Discrete Applied Mathematics* **59** (2) (1995), 129-136.
[9] A. Meir and J. W. Moon, Relations between packing and covering number of a tree, *Pacific J. Math.* **61** (1975), 225-233.
[10] F. Tian and J. M. Xu, A note on distance domination numbers of graphs, *Australasian J. Combin.* **43** (2009), 181-190.
[11] S K Vaidya, Distance k-domination of some path related graphs, *International Journal of Mathematics and Soft Computing* **4** (1) (2014), 1 - 5.
[12] X. Zhang, J. Liu, X. Chen and J. Meng, Domination number of Cartesian products of directed cycles, *Information Processing Letters* **111** (1) (2010), 36-39.

IOP Conf. Series: Journal of Physics: Conf. Series **1022** (2018) 012009 doi:10.1088/1742-6596/1022/1/012009

The profile of students' self-regulated learning at vocational high school

Asih Ciptaningtyas[1*], Hasih Pratiwi[2] and Mardiyana[3]

[1,2,3]Mathematics Education Department of Graduate Program Sebelas Maret University, Jl. Ir. Sutami 36A Kentingan Jebres Surakarta 57126, Indonesia

Corresponding email: *[1]asihche25@gmail.com, [2]hasihpratiwi@gmail.com [3]mardiyana@staff.uns.ac.id

Abstract. Self-regulated learning is a power in the individual through the individualization process. Self-regulated learning will occur when the student is active to control himself from everything done, plan something, evaluate, and deeply reflect what he has experienced. This study aims to determine the profile of students' self-regulated learning in SMK Giripuro, Sumpiuh, Banyumas Regency. This study is a qualitative research with questionnaire and interview methods. This study used triangulation method technique to obtain from the questionnaire and interview to get valid data. The subjects in this study are three 10th Grade students who have different self-regulated learning in SMK Giripuro Sumpiuh. The results showed that the high self-regulated learning student has characteristics: 1) independent of others, 2) believe in their abilities, 3) awareness in learning, and 4) be able to reflect on their learning. Medium self-regulated learning student has characteristics: 1) independent of others, 2) believe in their abilities, 3) awareness in learning, and 4) do not reflect on learning. Low self-regulated learning student has characteristics: 1) dependent on others, 2) do not believe in their abilities, 3) lack awareness of learning, and 4) do not reflect on learning.

1. Introduction

Learning outcomes are influenced by several factors, these factors are divided into two, namely internal factors and external factors [1]. Internal factors are factors that come from within the self that includes motivation, interest in learning, and learning independence, while for learning facilities and learning environment is an external factor.

Self-regulated learning is an attitude that individuals have to learn with their own initiative in an effort to internalize knowledge without dependent or get direct guidance from others [2]. Students who have self-regulated learning will continue to build a willingness to understand what is learned without relying on the help of others even if what is learned is difficult material.

Learning mathematics is a process to understand problems to solve problems related to the material being studied. Students who have high self-regulated learning will strive to understand the material to the fullest by utilizing more than one learning resource that can be used so that what has not been understood well can be understood.

Self-regulated learning has emerged as an important new construct in education. The concept has been embraced by policy makers, teachers, educators, and parents [3]. Self-regulated learning can be seen in student learning habits in every day as well as how students plan and learn. Higher student self-regulated learning is needed in improving mathematics learning achievement because it gives effect to the creation of spirit in students to learn.

Self-regulated learning is moderately positive correlated with academic achievement [4]. Achievement is one of the important factors in determining students' graduation. Students with higher achievement certainly have higher self-regulated learning. Conversely, students with lower learning achievement have lower self-regulated learning.

Self-regulated learning is recognized as an important predictor of student academic motivation and achievement [5]. Student achievement will have an effect on success, students who have good learning achievement have big chance to succeed. Students with good learning achievement will have the ability to utilize learning resources independently. Thus students with a good learning independence will have a good learning achievement as well.

The revised 2013 curriculum expects students to have learning independence so that students can construct knowledge that has been learned, thus the learning material will be remembered for the longest time by each student. Proven that self-regulated learning has a central role in life, including in learning and learning achievement in school [6].

All high schools have used the 2013 curriculum so students can improve their skills without help from others. Self-regulated learning of senior high school students are different with vocational high school students, at the level of vocational high school students, assume that the math, in particular, is not much needed at work, but with learning mathematics, students will be able to think logically and systematically.

With the policy that the Vocational High School is required to use the curriculum 2013, the students' self-regulated learning is required to develop so that each student is able to develop his ability without the help of others. Therefore, this study will discuss the profile of students' self-regulated learning at Vocational High School.

2. Theoretical Background

Self-regulated learning is learning the process that has been done without any help [7]. Without any help from the other, the students can explore all their potential, so what is learned becomes more meaningful and remembered. Finding out the problem by yourself, it will make students more responsible for their decisions that have related to the learning process. In addition, the student will have the ability to execute the decisions that have been taken.

Self-regulated learning is an important attitude that every student must have to be independent. At school, independence is very important, this attitude is required by each student to make them discipline and blame for their decisions. Self-regulated is learning at your own way, own responsibility, and successful learning [8]. Student with self-regulated learning will affect in their achievements, whether or not successful in learning is determined by themselves.

Early theories about self-regulated learning used the real of learning process as a primary goal in self-regulation and focus on the description strategies of metacognitive that must be possessed by students, so they do not have abstraction in learning process [9]. In addition, students with good self-regulated do not depend on another student, so they must develop their metacognitive ability in learning process.

The goal of the students' self-sufficiency is that students can self-organize to enable them to achieve high levels, enjoy learning, and develop life-long learning skills [10]. The existence of student self-regulated learning will trigger more positive impact in students' life.

The step of the activities on student self-reliance are designing, monitoring, evaluating, and reflecting [11]. Designing student's learning step can be analyzing tasks, establishing learning goals, and designing learning strategies. Monitoring students step can be asking themselves questions about the strategies used are effective, students can focus on learning, and strategies used well. The step of evaluating, the student can check how the strategy used is well done, the learning outcomes achieved, and the strategies used are in accordance with the tasks obtained. Reflecting students step can be interpreting each stage that is from step one into the fourth step.

The students who can learn from their selves have special characteristics. The characteristics possessed by students who have self-regulated learning are 1) students are independent of others, 2) believe in their own ability, 3) willing to conduct their own learning activities, 3) responsible for what is done, 4) can plan to learn, 5) evaluate learning, and 6) reflect on learning activities [12].

The indicators of students having self-regulated learning in this study are 1) independent of others, 2) believe in their ability, 3) have awareness in learning, and 4) can reflect on learning.

3. Methods

This study is a qualitative research with strategy of study case that aims to find the profile of student's independence in accordance with the development of a new curriculum that requires students to have self-regulated learning, so the material that learned can be understood well and remembered.

3.1. Participants

The subjects of this study are three students who have different self-regulated learning in SMK Giripuro Sumpiuh. Sampling in this research used purposive sampling technique. The researchers choose three students who have different level of self-regulated learning they are student with high self-regulated, student with medium self-regulated, and students who have low self-regulated learning and each student has good communication.

3.2. Data Collection

The data collection techniques in this study are questionnaires and interviews. The indicators of self-regulated learning students' questioner are 1) independent of others, 2) believe in own ability, 3) have awareness in learning, and 4) can reflect on their learning.

Independent of others means students can choose learning strategies that are used independently, complete the tasks assigned to the teacher according to their abilities, and can add knowledge by using other references. Believing in the own ability means that the students dare to express opinions to others, have the confidence to achieve learning objectives, and can complete the task independently even though the task is quite difficult.

Students have learning awareness means spurring themselves to always be passionate about learning, able to focus attention in classroom learning activities, and change the time of delayed learning of mathematics. While, can reflect on their learning means that their learning activities will have a positive impact on them, to look at the increase or decrease in learning outcomes, and always try to improve their learning outcomes when there is decrease.

The questions that used an interview with students are 1) how the strategies learning is used, 2) how to do the tasks obtained, 3) how if there are differences of opinion with friends in completing the task, 4) how to reflect on learning. The questions are used in the interview to check whether the subjects have filled the questionnaire are in accordance with the actual situation so that obtained the valid data.

3.3. Data Analysis

The data analyzing technique in this study is done by describing or explaining the data that have been collected through interview and filling questionnaire. This study used triangulation method technique to obtain from the questionnaire and interview to get valid data.

4. Results

4.1. Subjects with High Self-Regulated Learning

Based on the data analysis of answering questionnaires and interviews obtained that the subjects belong to high self-regulated learning category have a good strategy in learning, that is by listening to the music at the same time in learning activities, completing task that obtained by using their ability possessed without asking for help to other people, learned by using various sources, for example from books or using other media such as internet.

The subjects who have high self-regulated learning have a target achievement of learning and what they get, it will be in accordance with the targets. It will also always motivate them to keep getting the best of what they were done, when there is a difference of opinion with another subject in high self-regulated learning will be able to defend his belief and be able to convey it to others.

The subjects with high self-regulated learning can replace the time of study that postpones in accordance with what has been agreed with their selves, the subject is always focused on learning and observe the increase and decrease in learning results then will follow up what is obtained, therefore the results of learning will always be good. So that, subjects with high learning attitudes are able to reflect on their learning.

4.2. Subjects with Medium Self-Regulated Learning

From the analyzing data of questionnaire answers and interviews, it was found that subjects with medium self-regulated learning can choose the learning strategies that used in doing the tasks that done without help from others and learn by using more than one references.

The subjects have confidence with the results to be obtained but sometimes the results that have got are not in accordance with the targeted, when there are differences of opinion with other subjects with the medium of self-regulated learning are able to express what their beliefs are, and complete the tasks have maximally got with their ability.

The subjects are able to focus themselves on learning, especially learning mathematics, the subject will replace the study time is delayed on another occasion based on their commitments, and they can motivate themselves in learning.

They do not depend on others, belief in their abilities, have a consciousness in learning, but the subject does not have the belief that learning at this time will have a positive impact on the change of life, the change in learning outcomes is not important to him, whether or not the results obtained do not matter because the most important for them the subject have been learned.

4.3. Subjects with Low Self-Regulated Learning

Based on the data analysis of answering questionnaires and interviews, subjects with low levels of self-regulated learning always rely on others in their learning activities, subjects unable to determine learning strategies that used, learn under the control of others, and have no other reference in learning, they got material from the notes it obtained when the teacher gave an explanation in class.

They rely on answers from others when they get a job because they do not have confidence in their abilities. Therefore, they always ask for help from another friend, and when there is a difference of opinion, they cannot express their opinion.

Subjects with low self-regulated learning at the time of learning often ask for permission to leave the class for many reasons, they do not try to do a good learning plan, not focusing on learning, and never changing learning time at a particular time.

The achievements are always bad, but it does not make them change his habits to learn well, they never pay close attention to his learning results, for him the learning activities do not demand and never change in their life.

5. Discussion and conclusion

Self-regulated learning is an important attitude that every student should have to be independent of others. At school, the self-regulation is also very important attitude is required by each student. Therefore, students able to discipline their selves and have a responsibility towards what they have chosen.

Students self-regulated are the power of individual through the process of individualization. The self-regulation learning will come true when the students are active to control him selves from everything doing, planning something, evaluating, and following up more deeply into what they have experienced.

The Profile of students' self-regulated learning at SMK Giripuro Sumpiuh as follows.

a. Students with high self-regulated learning have characteristics independent of others, believe in their abilities, have awareness in learning, and are able to reflect on their learning.
b. Students with medium self-regulated learning have characteristics that are independent of others, believe in their abilities, have awareness in learning, and are unable to reflect on their learning.
c. Students with low self-regulated learning have characteristics dependent on others, do not believe in their abilities, have no awareness of learning, and are unable to reflect on their learning.

6. References

[1] Susanto A 2013 *Teori Belajar dan Pembelajaran di Sekolah Dasar* (Jakarta: Kencana Predana Media Group) p 62
[2] Peraturan Menteri Pendidikan Nasional Nomor 41 Tahun 2007 tentang Standar Proses untuk Satuan Pendidikan Dasar dan Menengah
[3] Boekaerts M 1999 Self-Regulated Learning: Where We Are Today *Intl. J. Edu. Res.* **31** 445
[4] Banarjee P and Kumar K 2014 A Study on Self-Regulated Learning and Academic Achievment among the Science Graduate Students *Intl. J. Mult. Appr. Stud.* **1** 329
[5] Zumbrunn S, Tadlock J and Roberts E D 2011 *Encouranging Self-Regulated Learning in the Classroom: A Review of the Literature* (Virginia Commonwealt University: Metropolitan Educational Research Consortium) p 3
[6] Boekaerts M 2015 Motivation and Self-Regulation: Two Close Friends *The Decade Ahead: Application and Contexts of Motivation and Achievement* **16** 72
[7] Slameto 2010 *Belajar dan Faktor-faktor yang Mempengaruhinya* (Jakarta: Rineka Cipta) p 15
[8] Uno H B 2011 *Teori Motivasi dan Pengukurannya: Analisis di Bidang Pendidikan* (Jakarta: Bumi Aksara) p 3.
[9] Sumarmo 2010 *Kemandirian Belajar : Apa, Mengapa, dan Bagaimana Dikembangkan pada Peserta Didik* Tersedia di http://math.sps.upi.edu/?p=61
[10] Boekaerts M 2015 Motivation and Self-Regulation: Two Close Friends *The Decade Ahead: Application and Contexts of Motivation and Achievement* **16** 74
[11] Stoten D W 2013 The Extended Project Qualification: An Example of Self-Regulated Learning in Sixth *Form Colleges Intl. J. Less.Lear. Stud.* **3** 1
[12] Rahmah 2011 *Regulasi Diri* http://belajarpsikologi.com/macammacam-metode pembelajaran/

Construction of the mathematical concept of pseudo thinking students

D Anggraini[1*], T A Kusmayadi[2] and I Pramudya[3]

[123]Postgraduate School of Mathematics Education, Sebelas Maret University
Jl. Ir. Sutami 36A Kentingan Jebres Surakarta 57126, Indonesia

*Corresponding email: dellaramandha@gmail.com

Abstract. Thinking process is a process that begins with the acceptance of information, information processing and information calling in memory with structural changes that include concepts or knowledges. The concept or knowledge is individually constructed by each individual. While, students construct a mathematical concept, students may experience pseudo thinking. Pseudo thinking is a thinking process that results in an answer to a problem or construction to a concept "that is not true". Pseudo thinking can be classified into two forms there are true pseudo and false pseudo. The construction of mathematical concepts in students of pseudo thinking should be immediately known because the error will have an impact on the next construction of mathematical concepts and to correct the errors it requires knowledge of the source of the error. Therefore, in this article will be discussed thinking process in constructing of mathematical concepts in students who experience pseudo thinking.

1. Introduction

The very interesting thing to discuss when learning math is how students construct mathematical concepts and solve math problems. Constructing mathematical concepts mean building mathematical knowledge through the attribution of one concept to another. The process of building knowledge in mathematics is done continuously so that it becomes a complete knowledge for students.

There are two possible outcomes when students construct mathematical concepts namely success or failure. The failed construction indicates the presence of student's difficulty in constructing mathematical concepts. These difficulties are often reflected in the form of mistakes made by the students. Such errors might be a visible results of the construction process rather than an output of actual mental activity [1]. The study of students' mistakes in constructing mathematical concepts has been largely undertaken by researchers [2,3,4].

Among students' mistake in constructing mathematical concepts is their tendency to take the form of pseudo thinking, the true pseudo thinking and false pseudo thinking [4]. The true pseudo thinking occurs when students get correct answers without being able to justify the answer. In other words, after being explored deeply, it turns out that what students think does not match the substance of the concept (wrong reasoning). The false pseudo thinking occurs when students give incorrect answers, but after undergoing the process of reflection, students are able to correct the answer and to reason correctly [4]. Research related to pseudo thinking have been largely undertaken by researchers [5,6,7]. Based on the results known that students who experience pseudo thinking needs to know the process of thinking when solving math problems. This is due to pseudo thinking is a unreal thinking, so the correct answer may not necessarily result from the right thinking process and incorrect answers are not necessarily generated from the wrong thinking process [4]. Pseudo thinking is a thinking process that generates answers to problems or constructs of "incorrect" concepts [8]. The construction of concept does not represent the actual thought.

Pseudo thinking has been studied by many researchers using different terms and contexts. Pseudo analytic versus Pseudo conceptual in the context of routine mathematical problem solving [9]. Established Experience (EE) versus Plausible Reasoning (PR) in the context of problem solving non routine [10]. Direct Translation Approach (DTA) versus Meaning Based Approach (MBA) in the context of solving word problems [11]. Dual Process Theory of Kahneman (Process System 1 versus Process System 2) in the context of solving algebra problems [12].

The results of preliminary study conducted by researchers by looking at the results of students' task when completing the daily test of straight line equation material in grade VIII E at SMP N 16 Surakarta. When the researcher did the interview to ask how student got the answer, he or she still did not understand what he wrote correctly. It shows that the student has not understood what he or she wrote correctly. This is called the pseudo thinking process, a state in which students do not really use their minds to solve math problems. In other words, the construction process undertaken by students in solving the mathematical problems is not in accordance with the mathematical concepts requested on the matter.

The pseudo-thinking process of constructing mathematical concepts needs to be known, because the result of concept construction do not match the students 'mental activity (students' wrong reasoning). Students' mistakes in constructing mathematical concepts in students experiencing pseudo thinking need to get some attentions, because if it isn't resolved then the error will have an impact on the construction of subsequent mathematical concepts. This is due to the mathematical concept that one has relevance with other mathematical concepts. Therefore, in this research will be described pseudo thinking process on the equation of straight-line equation.

2. Research Method

This research is a qualitative research. In this research will be described pseudo thinking process in constructing mathematical concepts on straight-line equation. This study involved ten junior high school students of 8rd grade. The instruments used in this research are the task sheets and interviews. Data collection techniques in this study were conducted with *think aloud method*, where students were given a task sheet to be completed, students were asked to solve problems in front of the researcher, and in the process of solving the problem, the students revealed what they were thinking. To obtain the data validation techniques in this research, one of the steps used is triangulation method. Triangulation method is done by comparing the result of student task and interview result.

In this study, to obtain the subject of research, the researcher gives the task sheet to the students. Further, the researchers classified the students' answers into groups of correct answers and incorrect answers based on the final answers the students gave on the completion sheet. Students who involved in this study were students who experience true pseudo thinking and false pseudo thinking. To determine which students are experiencing true pseudo thinking, the students with correct answer are asked to justify the answer, if the student is unable to justify the answer, the student is chosen to be the research subject (true pseudo thinking). Meanwhile, to determine students who experience false pseudo thinking. The students with incorrect answer are given the opportunity to reflect or re-examine the work. If after the reflection, students with incorrect answer are able to correct the answer to be the correct answer, the students with incorrect answer are chosen to be the subject of research (false pseudo thinking).

3. Result and Discussion

In this research, there were ten students of 8rd grade in junior high school required to complete the task sheet of straight-line equation material. The results are categorized as presented in table 1.

Table 1. Categories of Students' Job Results

Category	Students
Pseudo thinking	7
Totally right	1
Totally wrong	2

Based on table 1, it appears that there were still many students who experience pseudo thinking, as many as 7 students out of 10 (70%). However, in this article the description of students' pseudo thinking process in constructing mathematical concepts was focused on student answers when determining the gradient and determining the equation of a straight line through two points.

Figure 1 is a picture of the students' work in determining the gradient that indicates that the student was in true pseudo thinking.

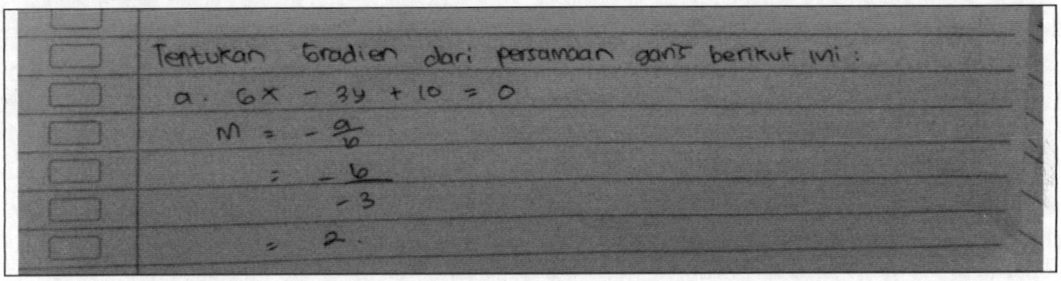

Figure 1. The Student's Answer Determines The Gradient

When determining the gradient, subjects illustrate the gradient formula which is $-\frac{a}{b}$. Furthermore, researcher conducted conversations with the students to explore and clarify the thinking process of students in determining the gradient. The dialogue between researcher and students are shown below::

Researcher	: Are you sure that the formula for determining the straight line gradient is $-\frac{a}{b}$?
Student	: Yes, miss ...
Researcher	: What is the reason?
Student	: yaaa, because usually illustrated like that by my teacher, miss ..
Researcher	: Hmm ..., *a* and *b* what does that mean?
Student	: *a* and *b* are variables, miss

Researcher continued the dialogue with students by asking questions that led to *cognitive conflict.*

Researcher	: Do you know the form of the straight line equation $y = mx + c$?
Student	: yes miss, it's a common form of straight line equation.
Researcher	: Then, what is the gradient of the equation $y = mx + c$?
Student	: Yes, it is clear that the gradient is '*m*'
Researcher	: What is '*m*'?
Student	: *m* is gradient, miss
Researcher	: I mean, what is '*m*'? is it variable or the coefficient of *x*?
Student	: (pause) '*m*' is a variable, miss..

IOP Conf. Series: Journal of Physics: Conf. Series **1022** (2018) 012010 doi:10.1088/1742-6596/1022/1/012010

Researcher continued the question

Researcher	: Is $ax + by + c = 0$ a straight-line equation?
Student	: Yes miss, it is a straight-line equation other than a general form.
Researcher	: Then what's the gradient?
Student	: (The student replied quickly) $-\frac{a}{b}$,miss
Researcher	: Why is it $-\frac{a}{b}$?
Student	: Yaaa.., because my teacher used to teach that the formula determines the gradient is $-\frac{a}{b}$ (student was unable to give a reason).

Based on the dialogue, it can be inferred that student experienced true pseudo thinking. Student was able to give a correct answer when asked to determine the gradient of the straight-line equation, but after undergoing further thinking, it turns out what the student thought was not in accordance with the substance of the concept in determining the gradient. The reason of the students in determining the gradient is to use the formula $-\frac{a}{b}$ by looking at a and b as variable. The constructs a and b as variables represent one form of error because a and b (in the context of a straight-line equation) not the type of variable but the coefficients of x and y. Furthermore, based on the results of the interview, the student's mistake occured also happened when the student declared m as variable and the student was unable to give the reason why the gradient of the straight-line equation was $ax + by + c = 0$ is $-\frac{a}{b}$. Therefore, the concept of construction in determining the gradient was not intact or what is commonly called as the construction hole. The construction hole is a hole in the student's thinking structure caused by an incomplete concept construction process [1]. The concept construction process that determines the gradient of students with true pseudo thinking is presented in figure 2.

Fault of true pseudo thinking process in constructing the concept occurs because students just "imitate" procedures done by the teacher often without knowing the reason why such procedures are used. As the result, students' reasoning in constructing mathematical concepts does not develop optimally as they assume that to solve the problem, they can simply choose the procedure of completion in accordance with the problems given previously.

Furthermore, false pseudo thinking occurs when students were asked to determine the equation of a straight-line through two known points. The following results of the student's work in solving the problem of determining the equation of the straight-line through the points A (3, -2) and point B (6.0) are presented in figure 2.

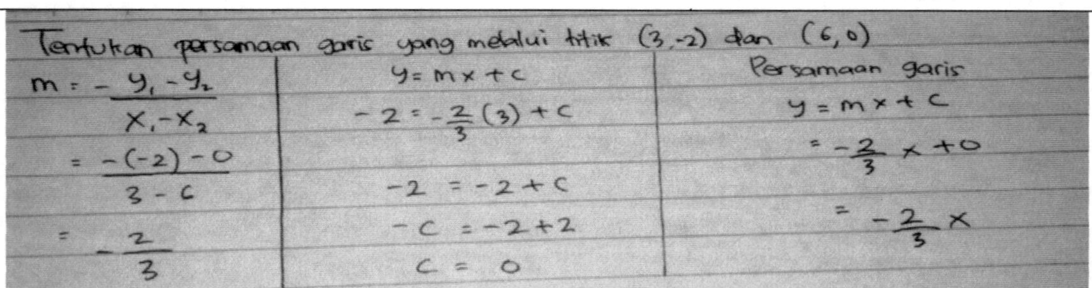

Figure 2. The Student's Answer Determines The Equation of A Straight-Line Through Two Points

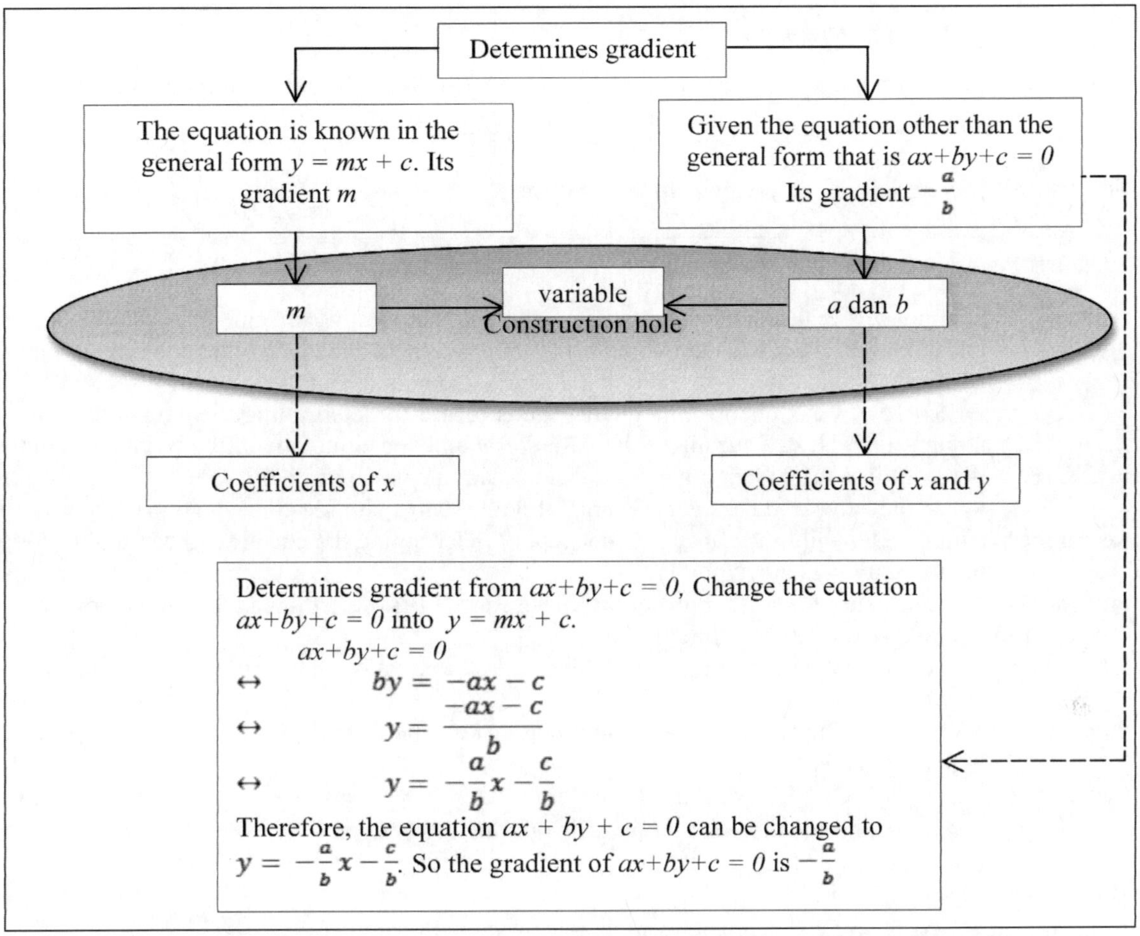

Figure 3. Construction Process with The Concept of True Pseudo Thinking

Based on the results of the answers given by students, students are experiencing the process of System 1 (S1) which is fast, automatic, without effort, unconscious, and inflexible [12]. This can be seen from the student's answer when determining the gradient through two points until it finally determines the equation of a straight-line through two points. Student also experienced the process of thinking "fuzzy memory" or remembering vaguely [9]. Student remembered that determining the equation of a straight-line through two points can be done by determining the gradient of the line first, but the student's memory is only vague. Therefore, the answers produced were incorrect. After the reflection student was able to improve the answer to be the correct answer. In other words, student experienced a false pseudo thinking. A dialogue showing that student experienced false pseudo thinking is presented below.

Researcher : Why do you define a gradient through two points using the formula $-\frac{y_1-y_2}{x_1-x_2}$?
Student : The formula determines the gradient through two points is $-\frac{y_1-y_2}{x_1-x_2}$
Researcher : Are you sure of your answer ?
Student : (pause) may I try again, miss?
Researcher : Please ...
Student : Determining the gradient through two points means $m = \frac{\text{selisih ordinat A dan B}}{\text{selisih absis A dan B}}$ or

IOP Conf. Series: Journal of Physics: Conf. Series **1022** (2018) 012010 doi:10.1088/1742-6596/1022/1/012010

$m = \frac{\Delta y}{\Delta x}$ (student paused for a moment) means $m = \frac{y_2 - y_1}{x_2 - x_1}$.
(student fixed the answer to $m = \frac{0-(-2)}{6-3} = \frac{2}{3}$). (Student was able to improve the answer in determining the gradient throught two points).

The researcher continued the interview with the student

Researcher : Why, in determining the equation of the straight-line, do you should use the equation of line $y = mx + c$? Can you explain?
Student : Ehmmm .... because it is the equation of the line that I know, ma'am... because straight-line equation material mmmm... it is already... I was looking and on the matter which shows that the line through two points namely point A (3, -2) and B (6,0). So, I can choose one of the points to use in determining the equation of the straight-line. Next, I substitute the value of m and substitute one of the points of point A (3, -2) into the equation $y = mx + c$, to obtain the value of c.
When the value of c has been obtained it will obtain a straight-line equation.
Researcher : Hmm..., Did you forget about the concept of determining the equation of a straight line through two known points?
Student : (With a smile)Yes miss, I forgot that. Therefore, I first determined the gradient and then used the equation of line $y = mx + c$.
Researcher : Do you remember the concept of determining the equation of a straight-line through a known point?
Student : (While recalling the formula) If I'm not mistaken, the formula is $y - y_1 = m(x - x_1)$. Is that right miss?
Researcher : That is right. Do you think the formula $y - y_1 = m(x - x_1)$ can be used to determine the equation of a straight-line through two points?
Student : (Pause for a moment) Seems to be like that, ma'am ... May I try it again?
Researcher : Please...
Student : $y - y_1 = m(x - x_1)$, The formula to determine the gradient is $m = \frac{y_2 - y_1}{x_2 - x_1}$.
Substitute the value of m to the equation $y - y_1 = m(x - x_1)$.
Therefore, we obtain $y - y_1 = \frac{y_2 - y_1}{x_2 - x_1}(x - x_1)$. As on the matter, note that the line through two points, which is point A (3, -2) and B (6, 0). So, we assume that point A (x_1, y_1) and point B (x_2, y_2). Subtitute point A and point B into the equation $y - y_1 = \frac{y_2 - y_1}{x_2 - x_1}(x - x_1)$. (students fix the answer to $y - (-2) = \frac{0-(-2)}{6-3}(x - 3)$).
So, The equation of straight-line that obtained is $y + 2 = \frac{2}{3}(x - 3)$ atau $y = \frac{2}{3}x - 4$.
Is it correct, miss?
Researcher : Yes, it is.

Based on the dialogue it is known that the student was actually able to correctly answer the problem, but because the student's thinking process was vague in constructing the concept of determining the equation of the straight-line, the answer was wrong. This is the student's answer after do reflection presented in figure 4.

False pseudo thinking process was spontaneous, without any reflection, and it was used to infer the answer also known as Direct Translated Approach thinking process [11] and thinking processes of students was also fast and unconscious [12]. In fact, when reflection was done, the student was able to correct the answer, in other words the students experienced a false pseudo thinking.

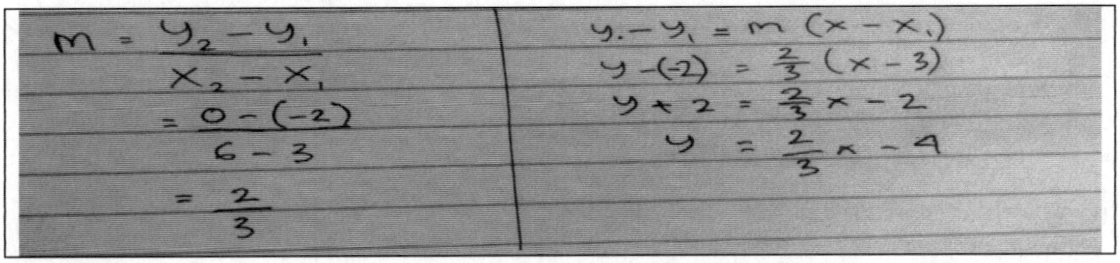

Figure 4. The Student's Answer After Reflection

The concept of construction process of determining the equation of a straight-line through two points done by the student with false pesudo thinking is presented in figure 5.

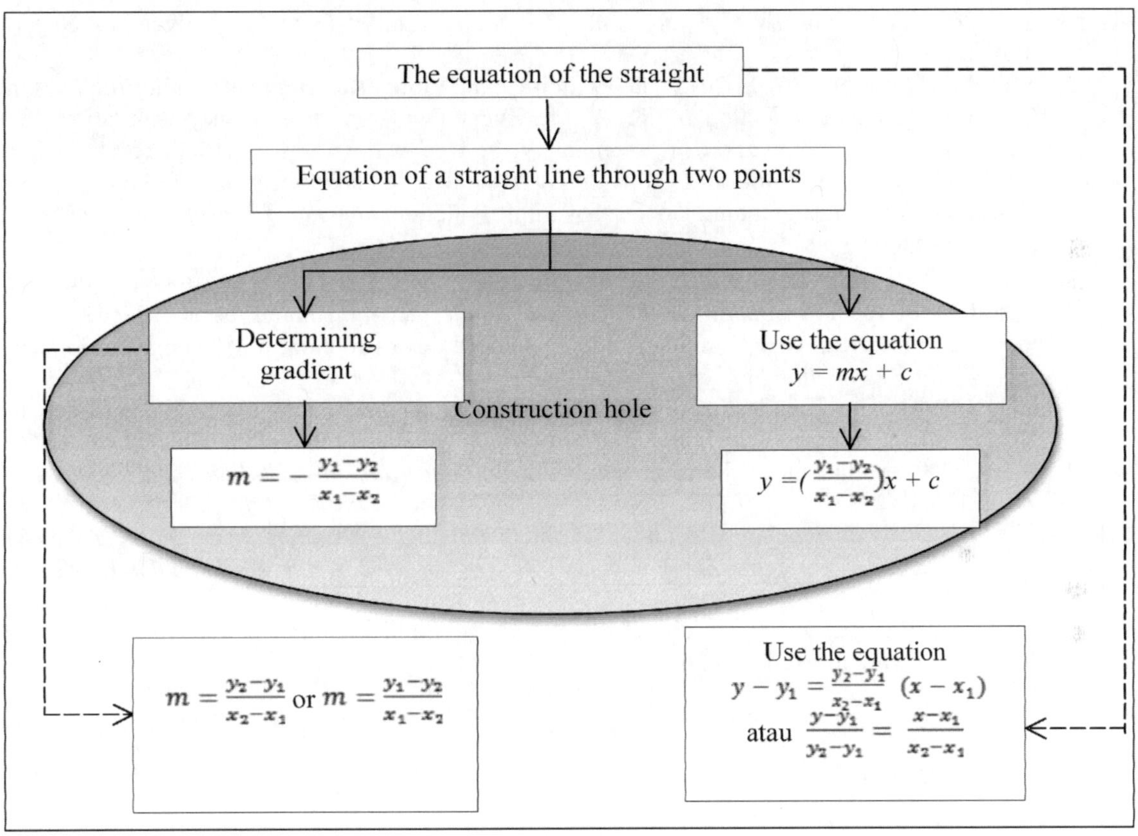

Figure 5. Construction Process with The Concept of False Pseudo Thinking

The use of the pseudo-thinking process in constructing a mathematical concept was wrong because student's thinking processes were spontaneous, losing control on what he thought or did, and such process gave the vague procedure. Therefore, the student's answer was incorrect. However, after the reflection, student was able to correct the answer.

4. Conclusion

Based on the results of the research, students 'mistakes in constructing mathematical concepts of students experiencing pseudo thinking occur because the students' thought processes are memorizing, spontaneous, not controlling what they think or doing, and remembering the procedures that occur vaguely. Consequently, mathematical concepts constructed by students are not intact or are called construction holes. Students' mistakes in constructing mathematical concepts in students experiencing pseudo thinking need attention, because if not addressed then the error will have an impact on the construction of later mathematical concepts. This is because the mathematical concept that one has a relationship with other mathematical concepts.

5. References

[1] Subanji 2015 *Teori Kesalahan Konstruksi Konsep dan Pemecahan Masalah Matematika* (Malang: Universitas Negeri Malang)reference
[2] A Brodie K 2010 *Teaching Mathematical Reasoning in Secondary School Classrooms* (New York: Springer Dordrecht Heidelberg London)
[3] Subanji and Nusantara 2013 Karakteristik Kesalahan Berpikir Siswa dalam Mengonstruksi Konsep Matematika. *Jurnal Ilmu Pendidikan (JIP)* **19** pp 208-217
[4] Subanji 2011 *Teori Berpikir Pseudo Penalaran Kovariasional* (Malang: Universitas Negeri Malang)
[5] Subanji R and Supratman A M 2015 The Pseudo-Covariational Reasoning Thought Processes in Constructing Graph Function of Reversible Event Dynamics Based on Assimilation and Accommodation Frameworks *J. Korean Society of Mathematical Education* **19** pp 61-79
[6] Herna Nusantara T Subanji and Mulyati S 2016 The Characterization of True Pseudo Construction in Understanding Concept of Limit Function *Journal of Research & Method in Educations* **6** pp 77-87
[7] Caglayan and Olive 2010 Eighth grade students' representations of linear equations based on a cups and tiles model *Educational Studies in Mathematics* **74** pp 143-162
[8] Subanji and Nusantara 2016 Thinking Process of Pseudo Construction in Mathematics Concept *International Education Studies* **9** 1913-9039
[9] Vinner S 1997 The Pseudo-Conceptual and The Pseudo-Analytical Thought Processes in Mathematics Learning *Educational Studies in Mathematics* **34** pp 97-129
[10] Lithner J 2000 Mathematical Reasoning in Task Solving *Educational Studies in Mathematics* **41** pp 165-190
[11] Pape S J 2004 Middle School Children's Problem-Solving Behaviour: A Cognitive Analysis from A Reading Comprehension Perspective *Journal of Research in Mathematics Education* **35** pp 187-219
[12] Leron U and Hazzan O 2009 Intuitive vs Analytical Thinking Four Perspective *Educational Studies in Mathematics* **71** pp 263-278

The implementation of portfolio assessment by the educators on the mathematics learning process in senior high school

Ida Lestariani[1*], Imam Sujadi[2] and Ikrar Pramudya[3]

[123]Universitas Sebelas Maret, Jl. Ir. Sutami 36 A, Surakarta, Indonesia

*Corresponding email: idha.laney@gmail.com

Abstract. Portfolio assessment can shows the development of the ability of learners in a period through the work so that can be seen progress monitored learning of each learner. The purpose of research to describe and know the implementation of portfolio assessment on the mathematics learning process with the Senior High school math teacher class X as the subject because of the importance of applying the assessment for the progress of learning outcomes of learners. This research includes descriptive qualitative research type. Techniques of data collecting is done by observation method, interview and documentation. Data collection then validated using triangulation technique that is observation technique, interview and documentation. Data analysis technique is done by data reduction, data presentation and conclusion. The results showed that the steps taken by teachers in applying portfolio assessment obtained focused on learning outcomes. Student learning outcomes include homework and daily tests. Based on the results of research can be concluded that the implementation of portfolio assessment is the form of learning results are scored. Teachers have not yet implemented other portfolio assessment techniques such as student work.

1. Introduction

Assessment is the process of information collecting and processing to measure the achievement of learning outcomes of learners. Assessment is an important part of education [1]. For educators, are collecting information about learners learning and information about the learning process. For learner, the assessment is the process of information them about the progress of learning. Assessment has become an important key to the improvement of the quality of education [2]. The quality of education can be reached by improving the quality of learning and assessment system [3].

Assessment by the educator is an assessment conducted by educators/teacher of a subject that aims to measure the comprehension of the learners. Assessment can be done by educators by making assessment instruments including test and assessment formats. *Permendikbud No 23 Tahun 2016* has set the provisions made by educators in assessing the learning outcomes of learners. The provisions of the educator in conducting the assessment of the learners on the knowledge and skills aspects of the form and usefulness of the assessment, assessment mechanism, assessment procedures, assessment instruments performed [4].

Assessment in the curriculum 2013 requires educators to conduct authentic assessments that include attitudinal assessment, knowledge and skills [4]. In mathematics subjects, the authentic assessment used is the assessment of knowledge and skills [5]. Assessment as a form of implementation of Basic Competence 3 and Competence 4 in the curriculum. In this study, focused on portfolio assessment. Portfolio assessment is one of the techniques in the execution of skill assessments contained in authentic assessment [6].

Portfolio assessment is an assessment of a collection of work of learners in a semester or a year that shows the effort, development and achievement of learning with certain criteria that are considered good work [7]. Pandey asserts that portfolio assessment is the collection and evaluation of a carefully chosen selection of learners' work [8] The number and types of selections included in a portfolio may vary, but are typically agreed upon by the educator and learners for the purpose of representing what that learners has learned.

The assessments done by the educators are in accordance with the assessment standards set by the government. It is contained in assessment standard of *Kurikulum 2013*. The procedure of assessing the skills aspect is done through the stages: a) planning the assessment; c) developing assessment instruments; c) executing the assessment; d) utilizing the assessment results; and e) reporting the assessment results in the form of numbers with a scale of 0-100 and description [4]

The purpose of this research is to describe the application of portfolio assessment on the learning process of mathematics. Portfolio assessment is one of the authentic assessments that not only to measure what the learners know, but rather to emphasize what the learners do [6]. While in school, educators rarely do the assessment with non-test [9]. The results of research conducted by Sukmawati showed that there are differences in the assessment technique of mathematics learning with portfolio techniques with written techniques to learners' learning outcomes. The better the application of portfolio assessment techniques conducted by educators will be the better results of learning mathematics learners [10]. On the other hand educators have done portfolio assessment on the learning process of mathematics but not fully implemented because educators have not clearly understood the portfolio assessment [11]. In relation to the valuation function that is very important for educators and learners it has become a teacher's skill to be able to do a good assessment according to what assessment standards on the *Kurikulum 2013*.

2. Method

This study is descriptive qualitative research used to determine the application of portfolio assessment conducted by educators [12]. Data were obtained by observation, interview and documentation. The subject of this research is the teacher of mathematics of class X of SMAN 1 Ngawi, East Java, Indonesia because this school is pilot projek *Kurikum 2013* in the regency. Determination of the subject done interviews to educator's class X, XI and XII on the implementation of portfolio assessment on the mathematic learning process. Interview results show that educators more often perform skills assessment using portfolio techniques than with other techniques. Educators who more often do skills assessment with portfolio techniques is a educators of mathematics class X. Based on the results of interviews, researchers set the subject on this research is the educators of mathematics class X.

The results of interview, researchers conducted observations into the classroom on the process of learning mathematics took place using the help of digital cameras to record teaching activities. Researchers observe the learning process from educators prepare learners to follow the learning, educators to assess the learning outcomes of learners until the learning process ends. Based on the observation, the subjects were re-interviewed to validate the observed data in the class. Through interviews, researchers also ask about the constraints experienced by educators during the assessment of the portfolio. Then the researcher did method of instrument documentation and portfolio assessment result done by educator on learners learning outcome. Based on the data obtained through observation, interview and documentation then performed data analysis techniques with data reduction, data presentation and conclusion [9].

3. Results and Discussion

During the learning process, digital camera was being used to record all learning activities in the class. Stages of assessment conducted by educators are a) planning the assessment; B) developing

IOP Conf. Series: Journal of Physics: Conf. Series **1022** (2018) 012011 doi:10.1088/1742-6596/1022/1/012011

assessment instruments; C) executing the assessment; D) reporting the results of scoring in the form of numbers with a scale of 0-100 [4].

The result of observations at the stage of planning, educators set the purpose of assessment, made question appropriate with basic materials and competences tested and developing of the assessment instrument like a figure 1 and made assessment rubric like figure 2. Figure 1 shows the compulsory mathematics subjects that have been made by the educator in accordance with the material to be tested after the material of one Competency is completed. Educators conducted portfolio assessments by providing questions such as quiz/written test. In the stage of developing of the assessment instrument, educators tested the questions that had been made before they were tested to the learners. In addition to making a questions, the educators made a assessment rubric for the question tested.

Soal Uji Kompetensi Trigonometri Matematika Kelas X

Waktu 75 Menit

Kerjakan soal berikut!

1. Segitiga ABC siku – siku di B. panjang sisi AB = 4 cm dan $\angle A = 60^\circ$ panjang BC adalah ...cm

2. Diketahui segitiga PQR siku – siku di Q. jika $\sin P = \frac{5}{13}$, maka $\sin R - \cos R = \cdots$

3. Jika $\sin 20^\circ = p$ maka $\cos 1190^\circ = \cdots$

4. Nilai dari $2\sin(-30^\circ)\cos(-135^\circ) = \cdots$

Figure 1. First Quiz

Rubrik Penilaian

NO	Soal	Jawaban	Skor
1	Segitiga ABC siku-siku di B. Panjang AB=4 cm dan $\angle A = 60^\circ$ panjang BC adalah	 $\tan A = \frac{BC}{AB}$	5
		$\leftrightarrow \tan 60^\circ = \frac{BC}{4}$	10
		$\leftrightarrow BC = 4 \cdot \tan 60^\circ$	5
		$\leftrightarrow BC = 4\sqrt{3}\ cm$ ■	5
			Total = 25

Figure 2. Assessment Rubric

Based on the results of interview, educators made quiz questions in accordance with the subject that can be made for portfolio assessment. Before tested to the students, educators tried to test the questions by finishing it first. When it can be solved by the students and have been in accordance with the material and the lattice, then the questions are tested to the learners. The educators made a

IOP Conf. Series: Journal of Physics: Conf. Series **1022** (2018) 012011 doi:10.1088/1742-6596/1022/1/012011

assessment rubric for each question. This rubric became a reference for educators in providing value objectively.

The educators executed a portfolio assessment by giving a written test of quiz and then collected back and scored. Student's work Sheets that had been scored became the assessment of the educator's portfolio like Figure 3. After the material of one Basic Competency has been completed, then the educator provides questions/quizzes to check the ability of learners.

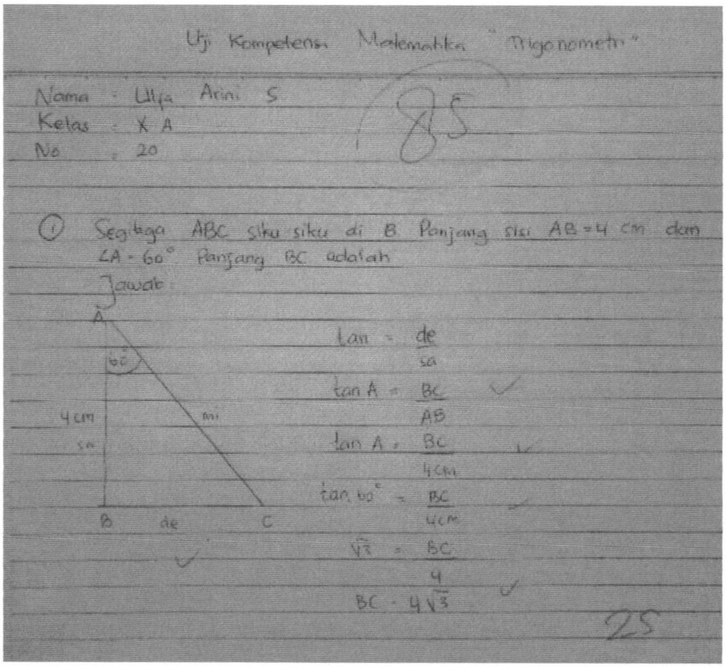

Figure 3. Work Sheet of Student

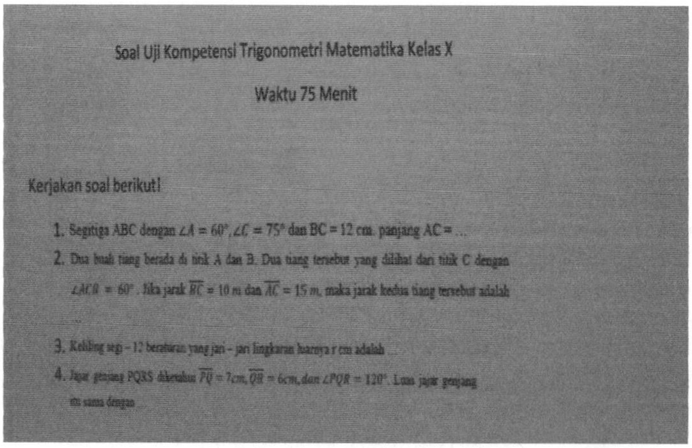

Soal Uji Kompetensi Trigonometri Matematika Kelas X

Waktu 75 Menit

Kerjakan soal berikut!

1. Segitiga ABC dengan $\angle A = 60^\circ, \angle C = 75^\circ$ dan BC = 12 cm. panjang AC = ...
2. Dua buah tiang berada di titik A dan B. Dua tiang tersebut yang dilihat dari titik C dengan $\angle ACB = 60^\circ$. Jika jarak $\overline{BC} = 10\ m$ dan $\overline{AC} = 15\ m$, maka jarak kedua tiang tersebut adalah ...
3. Keliling segi - 12 beraturan yang jari - jari lingkaran luarnya r cm adalah ...
4. Jajar genjang PQRS diketahui $\overline{PQ} = 7cm, \overline{QR} = 6cm, dan\ \angle PQR = 120^\circ$. Luas jajar genjang itu sama dengan

Figure 4. Second Quiz

IOP Conf. Series: Journal of Physics: Conf. Series **1022** (2018) 012011 doi:10.1088/1742-6596/1022/1/012011

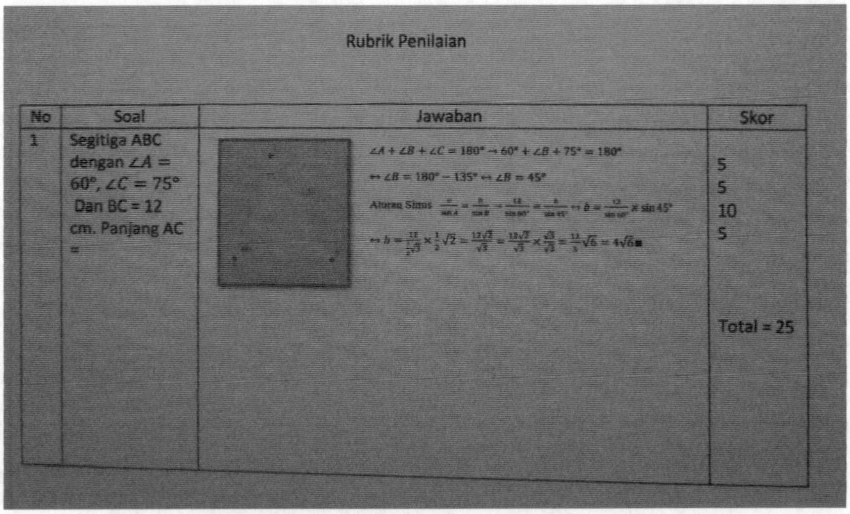
Rubrik Penilaian

No	Soal	Jawaban	Skor
1	Segitiga ABC dengan $\angle A = 60°$, $\angle C = 75°$ Dan BC = 12 cm. Panjang AC =	$\angle A + \angle B + \angle C = 180° \rightarrow 60° + \angle B + 75° = 180°$	5
		$\leftrightarrow \angle B = 180° - 135° \leftrightarrow \angle B = 45°$	5
		Aturan Sinus	10
		$\leftrightarrow b = 4\sqrt{6}$	5
			Total = 25

Figure 5. Assessment Rubric

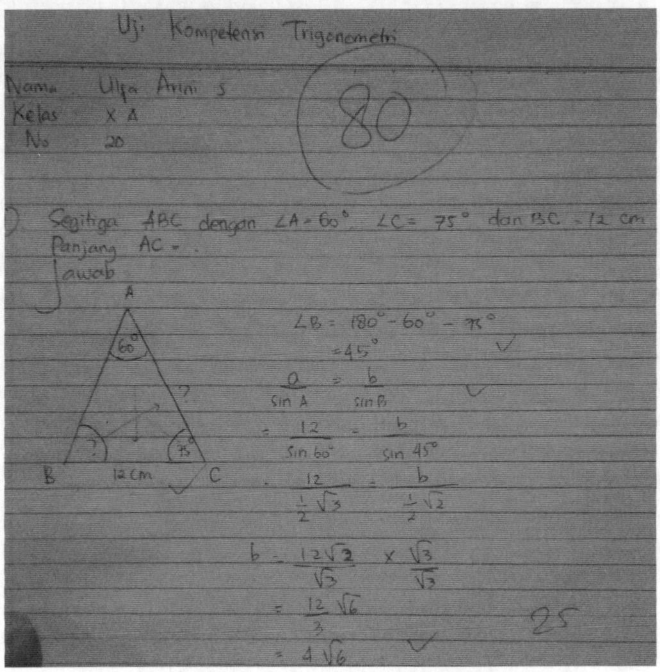

Figure 6. Learners Worksheet

Figure 4 shows the subject of the mathematics of choice given by the educator to measure the ability of the learner. In addition to making a problem, the educator also makes a rubric assessment for references providing score. This is shown in Figure 5. In the process of learning mathematics in the classroom, educators implementation of the assessment by providing questions to learners. Figure 6 is the result of the work of the learner who has been given score by the educator. Figures 3 and 6 are the work of learners gathered by educators. The worksheets of these learners as evidence that the educator has done a portfolio assessment.

This is certainly different from the definition of portfolio assessment in accordance with assessment standard of *Kurikulum 2013*. Portfolio assessment included in the authentic assessment is not only to measure what is known by the learners but to emphasize measurements that have been made by them. Not just measuring knowledge but measuring the skills of learners in terms of work. The work intended in the *Kurikulum 2013* is the performance/practice. Interview result showed is educators have difficulty in selecting material that can be used as portfolio assessment, long time need to did portfolio assessment and educator not understanding about kind of portfolio with other techniques besides of written test/quiz.

The results of the researcher's documentation showed the same rubric assessment as the rubric assessment on skill aspects. This is because the form of tests on portfolio assessment conducted by educators is in the form of written test. Rubric is created like a column with a value of 1-100 scale in each correct answer. Rubric created a question with a total score of 25, so with 4 questions if combined to a total value of 100 [4].

4. References

[1] Dimas A Cari Suparmi dan Handhika J 2016 The effect of assessment form tp the ability of student to answer the problem correctly *Intl. J. of Sci. and Appl. Sci.* **1** 40

[2] Hariadi 2017 Instrument Development of Autentic Assessment for Manipulatif Fundamental Motor Skill at Elementary School *Annual Applied Acience and Engineering Conference*

[3] Mardapi D 2012 *Pengukuran, penilaian dan evaluasi pendidikan* (Yogyakarta: Nuha Litera)

[4] Peraturan Menteri Pendidikan dan Kebudayaan No 23 Tahun 2016 Tentang Standar Penilaian. Indonesia

[5] Septiani A dan Rustaman 2017 Implementation of Performance Assessment in STEM (Science, Technology, Engineering, Mathematics) Education to Detect Science Process Skill. *Journal of Physics* 812

[6] Alimudin 2014 Penilaian dalam kurikulum 2013. *Prosiding Seminar Nasional* **01** 23

[7] Arifin Z 2009 *Evaluasi Pembelajaran* (Bandung: Remaja Karya)

[8] Ebrahimzadeh M and Khodareza M R 2016 *Porfolio Assessment as a Window into Reading Development.* **4** 13

[9] Salmayzury, Ruslan dan Pristiwaluyo 2015 Evaluasi Program Pembelajaran Matematika di SMA Negeri Watansoppeng. Riset Assesmen *J. Pen. dan Eval. Pend.* **1**

[10] Sukmawati R 2016 Pengaruh Teknik Penilaian Portofolio Terhadap Hasil Belajar Matematika Siswa. *J. Prim.* **5** 41

[11] Sumardi and Maudyla C V 2016 Implementasi Model Penilaian Portofolio dalam Pembelajaran Matematika Berbasis Lesson Study. *Prosiding Seminar Matematika dan Pendidikan Matematika* (Surakarta: UNS)

[12] Lawrence J and Tar U 2004 The use of grounded theory technique as a practical tool for qualitative data collection and analysis. *J. of Trav. and Tour. Mark.* **31** 417

IOP Conf. Series: Journal of Physics: Conf. Series **1022** (2018) 012012 doi:10.1088/1742-6596/1022/1/012012

Ethnomathematics elements in Batik Bali using backpropagation method

Mei Lestari[1], Ari Irawan[2], Wanti Rahayu[3] and Ni Wayan Parwati[4]

[1234]University Indraprasta PGRI, Jl. Raya tengah No.80, Jakarta Timur 13760, Indonesia

Corresponding email: *[1]mei_lestari@unindra.ac.id, [2]mascan_89@yahoo.com, [3]wanti_reiku@gmail.com, [4]ni_wayanparwati@unindra.ac.id

Abstract. Batik is one of traditional arts that has been established by the UNESCO as Indonesia's cultural heritage. Batik has varieties and motifs, and each motifs has its own uniqueness but seems similar, that makes it difficult to identify. This study aims to develop an application that can identify typical batik Bali with etnomatematics elements on it. Etnomatematics is a study that shows relation between culture and mathematics concepts. Etnomatematics in Batik Bali is more to geometrical concept in line of strong Balinese culture element. The identification process is use backpropagation method. Steps of backpropagation methods are image processing (including scalling and tresholding image process). Next step is insert the processed image to an artificial neural network. This study resulted an accuracy of identification of batik Bali that has Etnomatematics elements on it.

1. Introduction

Indonesia has varied cultural diversity; one of them is Batik which known as an icon of Indonesia. Almost every region in Indonesia has Batik culture and they have unique characteristics that vary in each region. As a form of International appreciation of Batik, UNESCO (United Nations Educational, Scientific and Cultural Organization) officially crowned Indonesian Batik as World Culture Heritage on October 2nd 2009 in France. World culture heritage is only specializing on Batik Tulis which became an authentic characteristic of Indonesian Batik and not printed Batik. Is mentioned more than 181 batik motifs. The amount does not include a variety of local batik motif continues to grow [1]. Batik is a cultural heritage not because of the batik itself but because of the art in making the batik [2].

Marvin and Wahyuni et.al. said that culture is defined as a whole aspects of human life in society, acquired by way of learning, include thoughts and behavior [3]. Wahyuni et.al., states that culture is something that will affect the level of knowledge including the system of ideas or ideas contained in human mind, so that in everyday life, the culture is abstract.

Meanwhile the embodiment of culture is the objects created by human as cultured creatures, such as behaviour and objects that are real, for example; patterns of behaviour, language, life tools, social organization, religion, art, and et cetera, which intended to help human in social life.

From the understanding above, it can be concluded, culture is as a manifestation of habits, objects and tradition which is in heritage from generation to generation which still used by the inheritors. Batik is a traditional fabric that became one of the nation's cultural richness of Indonesia[1]. Batik has a long history, which began before the era of Raden Wijaya (1294 -1309), the first King of Majapahit. During its development, style, motif, and colour of batik is also influenced by the culture from outside, such as Hinduism, Islam, Netherlands, China, and Japan In the book "Spirits of Batik Indonesia" is mentioned more than 181 batik motifs. Ethnomatematics research is suitable to do in Indonesia which has a lot of culture. Despite the progress of the times happening, many ethnic in Indonesia who survived by clinging to the doctrine of his ancestors[4].

Batik is pictorial fabric (patterned or fermented) which is made in certain way; first written and scratched with candles, followed by colouring with *tarum* and *soga.* Bali is a province in Indonesia [5]. Capital district in Bali is Denpasar. Bali is also the name of the island in this area. Wastika states in Balinese cultural which most population is Hinduism, has their cultural character built since in the fetus until death[6]. Their lives are fulfilling with religious and traditional ceremonies to be done. From the understanding above it can be conclude that Balinese batik is tangible artworks or fabric that have special motif which is influence by Hindu doctrine which exist in Bali.

Ethno-mathematics is a study about math that calculate cultural considerations where math arise by understanding the system of the math that they use[3]. Ethno-mathematics is math which is grow and evolve in a certain culture[7]. Etnomatematics is a concept, knowledge, study, or approaches that associate mathematics with culture [8].The presence of math that have cultural nuances will give big contribution for school mathematics, because school is social institution which is different with others, it increase the possibilities of some cultural socialization[9]. Ethno-mathematics defined as cultural anthropology of mathematics and mathematics education. From the understanding above, it can be concluded that ethno-mathematics is one of math-elements which is exist inside of the surrounding society. Consciously or not, that cultural have mathematical elements which applied on daily life. The ethnomathematician mediator plays a major role in this approach. It is his or her reflections on the things that has been said and done, and the perceptual shifts that he or she undergoes during the investigation, that keep the dialogue going[10].

Based on those things, need a deeper identification in order to increase the devotion of batik culture especially for Indonesia itself and worldwide. In this case for some region which hasn't exposed yet from the batik culture is Bali. Worldwide only knew that Bali is a place for tourist destination for domestic or international tourist. Besides, there are thing stored well and maintained; Balinese Batik which had vary variety and motif that is different with others regions especially Java. Balinese batik have pattern that used ethno-mathematic, that unconsciously use mathematical elements.

Subjects that matter on ethno-mathematic are:

a. Emblems, concepts, principals, and mathematical skills which exist on national group, tribe, or other communities.
b. The differences and similarity on mathematically things between each group and also factors behind every differences and similarity.
c. Interesting or specific thing between one group and several groups, for example: the way to think, behave, speak, and et cetera that related with math.
d. Every aspect on social life is related with mathematic, for example: financial literacy and economic awareness, social justice, cultural awareness, and political awareness.

There are many people who doesn't know mathematical elements on batik's motif therefore it needs deeper study about those elements. In addition to overcome the ignorance of the Balinese people in particular about the motive and variety of batik in Bali it is important to have a method that can be used to identify the motive of Balinese Batik. This is what underlies the use of back propagation algorithms in recognizing Balinese motives.

Back-propagation algorithm itself is one of the most common methods from artificial neural networks for the pattern identification and had high accuracy. Back-propagation trainee the networks to balance between network capability to identify the pattern used and also the capability to give the accurate response for similar pattern (but different) with pattern that use on training period[11]. On training period of back-propagation algorithm there are three phases: forward propagation, backward propagation and weight changes. First, propagate the input signal to the hidden screen with activation functions which determined. Second, back-propagation phase, every output unit which already done on the feed forward phase will receive pattern that targeted that related with input pattern to calculate the mistakes numbers. Third, weight modifications in order to decrease the numbers of mistake.

2. Methods

This research is use qualitative methods with explorative approach to reveal ethno-mathematical elements in Batik's motif. Research development of the application maker of Batik motif identifies and contains ethno-mathematic elements. Sample which used in this research is five motifs that contain ethno-mathematics element inside.

3. Result and Discussion

Balinese batik had unique characteristics; vary of the motifs, colours and also fabrics. The name of the Balinese batik also gained because of the varies of the motifs, for instance; *Balinese batik buketan, merak abyorhokokai, singabarong, ulamsari mas, jagatan pisang,* et cetera. Each motif drawn in order to accentuate a characteristic that reminds us to Balinese culture which is really sacred, mythology, tradition, culture, and the artworks. In Bali the use of Balinese Batik still preserve by still use Batik in every traditional and religious ceremonies.

Figure 1. Motif Abyorhokokai

On *abyorhokoki*i motif seen only on the rectangular frame that surrounds the image of the peacock. There is only a reflective element, however it's created with different colour between each flowers. The most predominate on this motif is there are two peacock which is face-to-face. The meaning is where the destination of human soul afterlife. From this illustration can be concluded that death is only body damage, meanwhile the soul still alive run into the creator (God).

Figure 2. Motif Buketan

Buketan mean a bunch of flowers, and this name adopted by French language. The motif of this Balinese batik is flowers pattern or small plants arranged along the fabric with an additional decoration such as butterfly, hong, stork, and also vines that increase the values of the art from this batik. Ethno-mathematic elements from this batik can be seen from the reflection of the geometry element; black fabric and gold coloured flowers.

Figure 3. Motif Jagatan Pisang

Some people called this motif as Balinese banana, this motif describes Balinese banana which has been distilled. There are called as Balinese banana which mean "back again/retro". This batik usually given from someone to their loves one who will travel far away in order to make the loves one will return. Based from the picture above, there are mathematical elements; curved lines with variety arch and every arch are connected. Besides, there are many reflection elements and relatable between each motifs

Figure 4. Motif Singa Barong

Barong lion motifs has a very few ethno-mathematical elements, because this batik is the two-pieces image of *Barong* which different and also facing each other. It is only can be seen by geometry element; circles and curved lines that facing each other. The meaning and philosophy of this batik is one of mythological or magical animals, because in Balinese or Javanese culture, "*barong*" means "magical". The miracle of it is being can be seen by the existence of these elements: integration between lion and tiger (body, foot, and eye), elephants (trunk), garuda (wings), dragon (grinning mouth with tongue sticking out). The embodied animals in Barong lion not just because of it is physical strengthen, but also because of the spiritual symbols.

Figure 5. Motif UlamSari Mas

Batik *ulam sari mas* motif is a very simple batik motif if viewed from the ethno-mathematical element which exists on the batik. In this batik the patterns mostly fulfil with reflection of fish and shrimp contradict each other when played. Ulamsari mas are symbol from the society who is livelihood as a fisherman; fish also represented prosperity from people who live in coastal areas.

Data of the batik motif that has been analysed by ethno-mathematical elements will be process with image processing; scaling, grayscale and tresholding stages. Those stages are done to obtain input with binary image followed with learning artificial neural network back-propagation.

IOP Conf. Series: Journal of Physics: Conf. Series **1022** (2018) 012012 doi:10.1088/1742-6596/1022/1/012012

3.1. Scaling

Image of Abyorhokokai batik motif measure 350 x 225 pixel, minimize into 50 x 50 pixel

Figure 6. Minimize Batik Abyorhokokai Motif Image

After the pixel size changes, then find the value of RGB every pixel. On the following table will showed RGB value of Abyorhokokai batik motif only to the size 5 x 5 pixel.

Table1. Value of RGB Abyorhokokai Motif

Pixel	1	2	3	4	5
1	252,255,255	252,255,255	250,252,249	254,255,251	254,254,255
2	245,253,248	255,255,246	255,255,241	252,251,233	255,255,243
3	255,255,230	145,136, 105	156,147,108	165,156,113	148,136,96
4	255,255,233	157,145,107	157,145,97	155,142,89	253,137,85
5	255,255,228	147,138,105	85,74,29	116,104,54	170,154,105

3.2. Grayscale

After scaling, image will perform grayscale stage, in this stage image will turn into gray.

Figure 7. Citra Grayscale

Thereafter find the grayscale value of each pixel with formulation: *Grayscale* = (R + G + B) / 3
With the result that obtains grayscale value up to size 5 x 5 pixel as follows:

Table 2. Value of Grayscale Buketan Motif

Pixel	1	2	3	4	5
1	254	254	250,3	253,3	254,3
2	248,7	252	250,3	245,3	251
3	246,7	128,7	137	144,7	126,7
4	247,7	136,3	133	128,7	158,3
5	246	130	62,7	91,3	143

IOP Conf. Series: Journal of Physics: Conf. Series **1022** (2018) 012012 doi:10.1088/1742-6596/1022/1/012012

3.3. Tresholding
Tresholding is process that convert image into black and white image or binary image.

Figure 8. Thresholding

Calculation into binary form with this following rule:
a. value of threshold range between 0-255 limit T = 128.
b. if the value of the pixel more or same with 128, then changes into 1, meanwhile if the value of the pixel is less than 128, then changes into 0.

Table 3. Binary Value of Abyorhokokai Motif

Pixel	1	2	3	4	5
1	1	1	1	1	1
2	1	1	1	1	1
3	1	1	1	1	1
4	1	1	1	1	1
5	1	1	0	0	1

The output as shown in table 3 will be used as input in the learning of algorithm back-propagation.

3.4. Training
Input on training process is the result of treshloding process of batik abyorhokokai. Then, the motif transformed into 1 dimension array.
Stages of learning process are:
a. Set the value of epoch, rate of understanding, minimum error. For example: in this calculation limit that used are: epoch = 10 rate of understanding = 0.2, *range* 0- 1, in this case used value 0.2 as learning ratio with minimum error = 0
b. Initialize the initial weights with random values with intervals -0.5 – 0.5
c. Loop is performed as long as loop value is less than the maximum of the iteration and the square of error is greater than the minimum error value. Each data is looped counting step 4 and 5.
d. Feed-forward
 First loop
 First data
e. Back-propagation

On the second data do the same operation using the final result from the first data. This process repeated until the maximum iteration.

3.5. Identification
Identification is process to recognize the testing image identification. For example, there are image of Abyorhokokai batik motif scale 50 x 50 pixels which have been pre-processed and transformed from matrix into one array.

Identificationsteps are:

a. Do the learning process with feed-propagation and back-propagation. The learning process has been done and obtained the weight of learning outcomes and the value of Y.
b. After the obtained the weight of the learning outcomes, do the identification process with feed-forward-propagation, it will get the value of Y from the output of identification process.
c. After obtained the value of Y from the identification, the value of Y from identification compared with value of Y which on the learning target, then the data will be obtained. Y from identification is Y from calculation using the most optimal weight of W and V that obtained from the learning process that is done repeatedly until the epoch is met, therefore Y from identification will approach Y from learning process, so that the value of this Y is the matching data of learning and training.

3.6. Testing

Testing are aims to know the accuracy of the image recognition of Balinese batik motif from the back-propagation methods which applied on the simulation to identify the batik motif.
The performance testing plan that will run on this simulator is a test using various combinations of learning parameters; it is shown in table 4.

Table 4. Testing Parameter

Parameter	Value
Epoch	10
	50
Understanding rate	0,1
	0,2
Error Minimum	0,01

Table 5. Parameter combination

No	Epoch	Learning rate	Minimum Error
1	10	0,1	0,01
2	10	0,2	0,01
3	50	0,1	0,01
4	50	0,2	0,01

Table 6. Testing Result of Parameter 1

Motives name	Target	Result	Label
Abyorhokokai	1,0000	1,0028	Identified
Buketan	2,0000	2,0331	Identified
Jagatan Pisang	3,0000	3,0011	Identified
Singa Barong	4,0000	4,0001	Identified
Ulam Sari Mas	5,0000	4,3501	Un Identified
Akurasi=4/5*100%=80%			

Table 7. Testing Result of Parameter 2

Motives name	Target	Result	Label
Abyorhokokai	1,0000	1,0028	Identified
Buketan	2,0000	2,0031	Identified
Jagatan Pisang	3,0000	3,0011	Identified
Singa Barong	4,0000	4,0003	Identified
Ulam Sari Mas	5,0000	4,3501	Un Identified
Akurasi=4/5*100%=80%			

Table 8. Testing Result of Parameter 3

Motives name	Target	Result	Label
Abyorhokokai	1,0000	1,3596	Identified
Buketan	2,0000	2,2063	Identified
Jagatan Pisang	3,0000	2,3949	Un Identified
Singa Barong	4,0000	4,0024	Identified
Ulam Sari Mas	5,0000	4,9951	Identified
Akurasi=4/5*100%=80%			

Tabel 9. Testing Result Parameter 4

Motives name	Target	Result	Label
Abyorhokokai	1,0000	2,6637	Un Identified
Buketan	2,0000	2,0144	Identified
Jagatan Pisang	3,0000	3,0773	Identified
Singa Barong	4,0000	3,9109	Identified
Ulam Sari Mas	5,0000	5,0780	Identified
Akurasi=3/5*100%=60%			

IOP Conf. Series: Journal of Physics: Conf. Series **1022** (2018) 012012 doi:10.1088/1742-6596/1022/1/012012

Accuracy Rate

	1st Testing	2nd Testing	3rd Testing	4th Testing
Accuracy	80%	80%	80%	80%
Accuracy rates = (80%+80%+80%+80%)/4= 80%				

Application Design

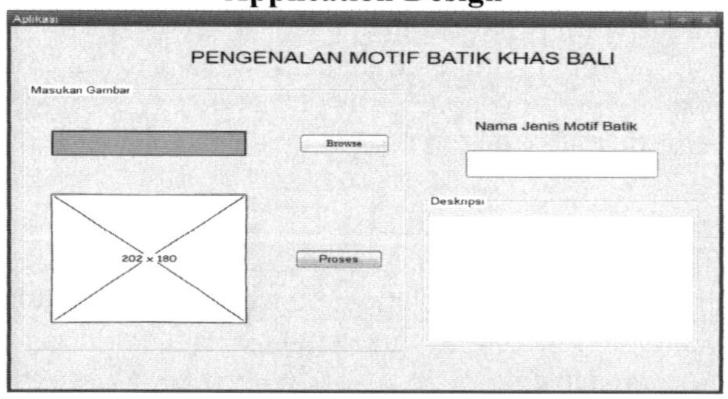

Figure 9. Application Design

4. Conclusion and Suggestion

4.1 Conclusion

Based on the result that obtained from the research and adjusted with the purpose, then obtained the conclusion, average accuracy rate is 80%, which mean the accuracy rate to identify the batik motif that contained ethno-mathematical elements are really good.

4.2 *Suggestion*

a. in order to increase the performance of this system, the sampling of the batik images need to be multipLied for the trainee of neural network system, to increase the accuracy of this system.
b. For the future work it is sugested to built an application to Identify bati's motif that has ethnomathematic element.

5. References

[1] R E Putra N, Suciati and A Y Wijaya 2011 Implementing Content Based Image Retrieval For Batik Using Rotated Wavelet Transform And Canberra Distance in *Bali International Seminar on Science and Technology* pp 1–5

[2] A H Rangkuti, Z E Rasjid and D J Santoso 2015 Batik image classification using treeval and treefit as decision tree function in optimizing content based batik image retrieval *Procedia Comput. Sci* **59** no Iccsci pp 577–83

[3] A Wahyuni, A A W Tias and B Sani 2013 Peran etnomatematika dalam membangun karakter bangsa in *Prosiding Jurusan Pendidikan Matematika FMIPA UNY. Yoyakarya: Universitas Negeri Yogyakarta* pp 113–18

[4] T Septianawati, Turmudi and E Puspita 2016 International Conference on Recent Trends in Physics 2016 (ICRTP2016) in *Journal of Physics: Conference Series* **755** pp 1–7

[5] R U Amalia 2010 Motif batik Pekalongan : studi dokumen koleksi museum batik Pekalongan *Seni dan Budaya* **6**(2) pp 125–41

[6] D N Wastika 2005 Penerapan konsep tri hita karana dalam perencanaan perumahan Di Bali *J. Pemukim. Natah* **3**(2) pp 72–7

[7] K R Puspadewi and I G N N Putra 2014 Etnomatematika di balik kerajinan anyaman Bali *J. Mat.* **4**(2) pp 80–9

[8] D A Kusuma, S P Dewanto and B Nurani 2016 The role of ethnomathematics in West Java (a preliminary analysis of case study in Cipatujah) in *Journal of Physics: Conference Series* pp 1–8

[9] F S Sirate 2012 Implementasi etnomatematika dalam pembelajaran matematika pada jenjang pendidikan sekolah dasar *Lentera Pendidik.* **15**(1) pp 41–54
[10] N A Adam 2010 Mutual interrogation: A methodological process in ethnomathematical research *Procedia - Soc. Behav. Sci.* **8** pp 700–7
[11] J J Siang 2009 *Jaringan Syaraf Tiruan & Pemrogramannya Menggunakan Matlab* Yogyakarta: Andi

An analysis of teacher's preparation in implementing 2013 revision curriculum on the mathematics learning as a specialization subject

T Susilo[1] and A Suryawan[2]

[1]Department of Mathematics Education, Sebelas Maret University, Indonesia.
[2]lecturer of Muhammadiyah Magelang University, Indonesia

Corresponding email: [1]trisusilo.math@student.uns.ac.id,[2]ari.surya_88@ummgl.ac.id

Abstract - This study aimed to determine the pedagogical competence of teachers, the readiness of planning and implementation of learning related to the implementation 2013 revised edition curriculum on mathematics specialization learning for senior high schools Wonogiri. Informants in this study there are 6 high school mathematics teachers X and XI class who teach in the school district Wonogiri. Data were collected using questionnaire method, interview, observation and documentation. Qualitative data analysis is done interactively through 4 paths: data collection, data reduction, data display, drawing conclusion. The results showed that high school mathematics teacher class X and XI in school district of Wonogiri City. The results show that most high school mathematics teachers in grade X and XI are ready to implement the 2013 revised edition curriculum and a few have not been able to implement due to internal or external factors. High school math teachers at Wonogiri district who are ready to face the 2013 revised edition curriculum have applied 10 teacher pedagogic competency indicators according to Regulation of the national education ministry Number 16 Year 2007 in learning. The readiness and implementation of mathematics learning is in line with the demands of the 2013 revised edition curriculum. Based on the teachers who are not ready, data on issues that arise in the implementation of the 2013 revised edition curriculum. Especially the problems in learning, namely mismatch of Core Competence (KI) and Basic Competence (KD) in teacher manual, material disregard in student handbook and lack of examples of problems that exist in teacher manual.

Key words : 2013 revision edition curriculum, teachers preparation, mathematics specialization learning

1. Introduction

Over time this has evolved the demand for a changing educational curriculum that prioritizes the need to build a nation's character and also guides students to be positive about everything for their own future good. This is based on the facts and perceptions of society about the decline in the quality of the attitude and morals of children or young people. What is required now is a curriculum of character education in the sense that the curriculum itself has a character, as well as oriented to the formation of the learner's character [1].

IOP Conf. Series: Journal of Physics: Conf. Series **1022** (2018) 012013 doi:10.1088/1742-6596/1022/1/012013

The curriculum reform is largely welcomed by teachers, undergraduate education and the general public. However, the reforms are not free from critics largely focused on implementing reforms. That research conducted so far shows that the lack of material support and training opportunities for teachers, resources and scarce materials, poor technology infrastructure and physical facilities are major obstacles before achieving the goals outlined in the reforms [2]. Critics also argue that the participation and involvement of civil society actors and universities in the preparation of curriculum programs and pilot cases has been quite limited [3].

The existing educational problems, Ministry of Education and Culture in Indonesia assess the need to develop a new curriculum, the Curriculum 2013 [4].The result of PISA analysis in 2015 which has represented all teachers in Indonesia after the use of the 2013 curriculum demonstrates the high educational role of teachers' learning. (OECD, PISA Database 2015, Tables II.5.9, II.5.10 and II.5.11.)

Revisions to the 2013 Curriculum document have been conducted in 2016. From the revision of the 2013 curriculum, it is expected that teachers are easier to implement. The results of this revision will further facilitate the work of teachers, especially in the assessment of their students. "If previous teachers of mathematics should assess the spiritual attitudes and social attitudes of all students, now not.These tasks are done teachers of religion and character and teachers of Education Pancasila and Citizenship (PPKn) directly.But the teacher of mathematics remains obliged to develop children's character, Children are not deprived of the roots of our nation's culture (Head of Curriculum and Book Center (Puskurbuk) Tjipto Sumadi, 2016).

Especially in mathematics subjects that in fact mathematics teachers are expected to enter the world of students' imagination to invite students to love and enjoy learning mathematics, which mathematics subjects are often considered difficult by students. The math teacher should also be the ideal teacher. This means that teachers who have a variety of competencies and intelligence that clearly evident from the character and behavior of everyday, both when as educators, in the professional community, as well as members of the community.

Curriculum 2013 has been tested in some elementary and middle schools, where in fact it shows that the 2013 curriculum has obstacles in its implementation in various regions because most teachers are not ready, the teacher's unpreparedness is not only related to his competence, but related to his creativity problem, which is also caused by The formulation of a slowly socialized curriculum by the Government [5]. In line with that opinion, teacher preparation for the implementation of the 2013 Curriculum of the hasty and less up-to-date revision edition. It can be seen from the very short time of teacher training and close to implementation time. Teachers in duty in the region will find it difficult to follow new things in a short time, especially with an integrative thematic approach that takes time to understand them. The government is determined to prepare the master book for teachers and students' handbooks, but teachers should first comprehend comprehensively the learning materials, not just read the handbook. From the explanation is likely to also affect the subjects of mathematics, especially the mathematics of interest, because it is difficult to implement the mathematics of interest with thematic approach.

To know the determinants of the success of the Curriculum 2013, especially the readiness of teachers in implementing the curriculum of 2013 revised edition of math in education, it is necessary to carry out the analysis of teacher readiness in implementing the 2013 revised edition curriculum on learning mathematics in the city of Wonogiri.

2. Theoretical Framework

2.1. Pedagogic Competency of Teachers

In Law No. 14 of 2005 mentioned that the competence is a set of knowledge, skills, and behaviors that must be owned, lived, and mastered by teachers or lecturers in performing professional duties. Law number 14 Year 2005 on National Education Standards defines teacher competency covering pedagogic competence, personality competence, social competence, and professional competence gained from professional education.

Regulation of the national education ministry number 16 year 2007 about academic qualification standard and teacher potency, mention in detail pedagogic competence include:

a. Understanding the characteristics of learners from the physical, moral, spiritual, social, cultural, emotional, and intellectual aspects.
b. Mastering learning theories and learning principles that educate.
c. Develop a curriculum related to the subjects being taught.
d. Managing educational learning.
e. Utilizing information and communication technology to learn.
f. Facilitate the development of potential learners to actualize potentials.
g. Communicate effectively, empathy, and courteous with learners.
h. Conduct assessment and evaluation of learning outcomes.
i. Utilizing assessment and evaluation results for learning benefits.
j. Conduct a reflective action to improve the quality of learning.

2.2. 2013 Revised edition Curriculum

Teacher pedagogical skills are indispensable when implementing the revised 2013 Curriculum edition. In this curriculum the learning method becomes one of the things that concern in the improvement of K-13. Some teachers consider the method of learning with the 5M thinking process (observing, asking, gathering information or trying, associating, communicating) procedural and mechanistic so as to shackle the creative space. During this time they view the method as the only approach in learning in all subjects.

The provision of creative space in the teacher as a supporter of pedagogic ability includes several things. The Government prepared syllabus is one of the models to inspire. Teachers can develop according to the relevant context. In thematic lessons (especially elementary level), teachers can develop themes and sub themes according to the relevant context. 5M is the ability of thought processes that need to be trained continuously through learning so that students are accustomed to thinking scientifically. 5M is not a procedure or steps or approach to learning.

2.3. Learning Planning in 2013 Revised edition Curriculum

The lesson plan is a necessary tool for providing direction on field work, and is often used as a basis for assessing learning. The lesson plans connect learning, practice, evidence, and time into a coherent plan [7]. Learning planning can also be interpreted as the process of preparing the lesson material, the use of instructional media, the use of approach, the use of approach or method of pursuit, in a time allocation that will be implemented in one semester to come to achieve the specified goal.

2.4. Implementation 2013 Revision Edition Curriculum

The improvement of K-13 is done by evaluative formative, one of them is by improving the point of competence and Basic Competence, syllabus and textbooks. The improvements were made based on inputs provided by the community such as teachers, education activists, practitioners, educational observers, and the general public.

Based on the results of the evaluation, it was found that a less precise understanding by the community caused by the presentation format and nomenclature in K-13, among them KD on KI which is considered less logical is associated with the characteristics of the maple. It also found an indication of inconsistency between KD with syllabus and textbook.One of the principles of syllabus improvement to make it easier for teachers to understand it so it is easy to implement.

Improvements to the syllabus include, among others, writing and formatting so that teachers can easily understand, efficient presentation (from more than 100 pages to an average of 20 pages per maple), without reducing the substance and consistently observing the scope and order of their knowledge , As well as providing clearer explanation of the characteristics of the maple, the scope of competence and learning materials.

IOP Conf. Series: Journal of Physics: Conf. Series **1022** (2018) 012013 doi:10.1088/1742-6596/1022/1/012013

3. Methods

3.1. Types of Research

This study uses a qualitative descriptive approach, with case studies, it is research focused on a single phenomenon chosen and wanted to be understood in depth, regardless of other phenomena. Qualitative descriptive research is a study aimed at expressing empirical facts objectively based on scientific logic, procedure and supported by strong methodology and theoretical according to the discipline of study [8].

3.2. Participants

In this research, the subject of research is the mathematics teacher in the interest of class X and XI in Wonogiri District Senior High School, while the object of the research is the readiness of the teacher in implementing the revised edition 2013 curriculum on mathematics specialization learning.

3.3. Source of Research Data

In this study the main data in the form of information about the problematic the Teacher's Preparation In Implementing 2013 Revision Edition Curriculum on the Mathematics Specialization Learning. Information on the misuse of the the Teacher's Preparation In Implementing 2013 Revision Edition Curriculum through documentation of supporting documents and math teacher interviews. Information about the problem of the Teacher's Preparation In Implementing 2013 Revision Edition Curriculum obtained through the observation of the implementation of learning activities on mathematics teachers.

3.4. Data Validity

Validation of data on problematic teacher's preparation in implementing 2013 revision edition curriculum on mathematics specialization learning using triangulation theory, that is the researcher use various theories, hence the researcher will get a complete and depth interpretation. These theories can succeed from the same discipline, as well as from different disciplines. Further triangulation techniques, namely testing the credibility of data by checking the data to the same source with different techniques. For examples data obtained by interview, then checked with observation, documentation, or questionnaire. Finally, the triangulation of methods, ie by using the same data source to search for specific data but using different data collection techniques.

3.5. Data Collection Technique

Data collection techniques in this study are interviews, observation, and documentation. Interviews in this study were conducted with unstructured interviews with the aim of obtaining more in-depth information about the problematic of teacher's preparation in implementing 2013 revision edition curriculum on mathematics specialization learning.

3.6. Data Analysis Technique

Data analysis is an activity undertaken by researchers in order to arrange the systematic data obtained from the observation. The arrangement is done by organizing data into categories, describing into units, synthesizing, composing into patterns, choosing what is important and what will be learned, and making conclusions so easily understood by yourself and others. This research includes three stages of qualitative data analysis that is data reduction, data presentation, and conclusion. In reducing the data of field records in transcripts the researchers complete the reduction by categorizing the data in preparing the learning scenario. The presentation of dataclassifies data based on steps in the process of composing the learning scenario. A conclusion based on the presentation of data on the preparation of learning scenario.

Table 1. Data Analysis from Six Teachers

No	Statements	Very Often	Often	Sometimes	Never	Category
1	Understand the student carracter	-	6	-	-	Good
2	Teach according to basic competence	-	6	-	-	Good
3	Creative in learning	2	4	-	-	Very Good
4	Giving material systematically	-	6	-	-	Good
5	Develop a lesson plan	-	3	3	-	Enough
6	Utilizing information technology and communication in learning mathematics	-	2	4	-	Enough
7	Develop student potential	4	2	-	-	Very Good
8	Communicate effectively, empathically, and courteous with student	6		-	-	Very Good
9	Determine the assessment procedure and evaluation	5	1	-	-	Very Good
10	Make a use of the assessment results and evaluation for learning purpose	1	5	-	-	Very Good

3.7. Research Procedure

Preparatory stage, at this stage of the instrument preparation activities. Implementation and analysis phase, documentation of supporting documents, conducting interview problematic teacher's preparation in implementing 2013 revision edition curriculum on mathematics specialization learning, then doing analysis of learning process planning data, observation of learning implementation in class, doing data analysis result of observation of process implementation Learning and ending interview about the implementation of learning. The conclusion withdrawal stage, at this stage the activities draw conclusions from the research.

4. Results

The results of this research is to know how the teacher's preparation in implementing 2013 revision edition curriculum on mathematics specialization learning, supporting factors and obstacles, and obstacles in implementing the 2013 revision edition curriculum in senior high school of class X and XI and how to overcome them.

The number of teachers who became respondents in this study were 3 teachers who teaches in grades X and XI, with the status of permanent teachers with education level Strata S1. The data obtained in this study comes from interviews with teachers based on the questions that have been prepared that include implementation, the supporting and inhibiting factors, and the constraints faced with in Implementation of the 2013 curriculum on mathematics subjects.

Table 2. Pedagogical Teachers Aspect to 2013 Revision Edition Curriculum

No.	*Statement List*	*Information*
1	Every teacher tries to understand individual differences of learners, especially differences in ability and attitude	Understanding of differences Individual students, especially differences in ability and attitude From three teachers began to develop
2	Each teacher teaches attention to Competence related to student attitudes	Teachers attention to competencies associated with students' attitudes of Teachers began to develop
3	Teachers apply various approaches, strategies, methods and learning techniques teach creatively in learning mathematics	Application of a scientific approach to moment learning strategy, methods and learning techniques educate creatively with in learning mathematics from all three teachers began to develop.
4	Teachers organize learning materials according to the correct approach chosen and characteristics of learners	Structuring learning materials on a exactly according to that approach Selected and characteristics of learners of the three teachers began to develop.
5	The teacher develops components learning design	development of components the learning component the teacher starts to look
6	Teachers utilize information technology and communication in learning mathematics	Utilization of information technology and communication in learning the math of the three new teachers Start to look. This is because in adequate facilities, such as LCD and projector.
7	Teachers facilitate the development of Students potential to actualize their potentials	Facilities giving in the development of potential learners to actualize various potentials owned from the three teachers began to develop
8	Teachers communicate effectively, empathically, and courteous with learners	Communication between teachers and students begin to culture and consistent. This is done at the time of learning, many students does not feel awkward in.
9	The teacher determines the assessment procedure and evaluation of the learning process and outcomes	Determination the process and learning outcomes of three teachers have begun to cultivate and consistent.
10	The teacher utilizes the results of the assessment and evaluation for learning purposes	Utilization of assessment results and evaluation for the benefit of Teachers from three teachers started developing.

The result of the questionnaire and observation above is also clarified by interviewing the teacher of class X and XI on 2013 revised edition curriculum on learning mathematics in the city of Wonogiri. The following is the result of interview about pedagogic competence related to the implementation of 2013 revised edition curriculum which applied by teacher of math class X and XI senior high school of class X and XI in Wonogiri regency.

By the interviewing:

Mr. Yustinus Murdoko, MM, M.Pd ask "according to the definition of the 2013 revised edition curriculum in learning, teachers are expected to use a scientific or scientific approach. Scientific approach that starts from the emergence, understand, ask, try and conclude.

Mr. Krisdiyasmo, S.Pd, M.Pd ask "The 2013 curriculum is a student-based curriculum". Students should try to understand themselves with teacher guidance. This means that the 2013 curriculum motivates students to, first the students have to search first by looking at the environment, see the teaching materials, then try what is observed, after that critical thinking, and then draw a result / communicate what was tried and seen earlier".

Mrs. Neny Diyatuti, S.Pd ask "This is an important curriculum for building knowledge, with this revised edition curriculum the teacher becomes easier to teach without bribing many judgments"

From the questionnaire it shows the teacher of class X and XI is ready in the planning of learning related to the implementation of Curriculum 2013 revision edition. This is held by teachers using the syllabus that has been provided by the government, teachers understand the whole thing related to integrative thematic syllabus before done. Learning, teachers understand beginner teachers and teachers, teachers and teachers who will be taught, teachers and understand the theories that will be taught, and teachers understand the utilization of facilities and learning resources that have been provided.

5. Discussion

5.1 Pedagogic Competency

Generally the teacher of math class X and XI has implemented the pedagogic competence. The pedagogic competencies that teachers have undertaken include: (a) every teachers seek to understand the difference of learners, disorders and attitudes, (b) subject teachers related to students' attitudes, (c) teachers apply various approaches, strategies, methods and learning techniques Mathematics learning, (d) teachers organize learning materials very well with the chosen approach and characteristics of learners, (e) teacher learning potential learners to actualize their potentials, (f) teachers communicate effectively, empathetically and courteously With participants (f) the teacher determines how to evaluate and evaluate the learning outcomes, and (h) the teacher utilizes the results of the assessment and evaluation for the learning purposes, (i) the teacher development of the learning components and (j) the utilization of information and communication technology in the lesson.

The first pedagogic competence that teachers must master is to understand the characteristics of learners. Teachers must understand the principles of personality development and learners' differences in order to actualize the potential of the students. This is supported by research conducted by Sanaky H.A.H (2005) entitled "*Certification and Teacher Professionalism in the Era of Educational Reform*" which definitely has an understanding of the nature, characteristics of students and its development, understanding some useful educational concepts to help students, teachers Some [9].

5.2. Teacher Readiness in Learning Planning

Teacher as the respondent has compiled a complete learning plan, with attitude, skill, and attitude. That is certainly in accordance with the demands in the Curriculum 2013.

5.3. Readiness Teachers in Learning Implementation

In the process of learning in the classroom running in accordance with the stage of learning activities of the initial activities, core activities and end activities

6. Conclusion

Based on the results of research analysis and discussion obtained can be concluded the readiness of teachers in implementing Curriculum 2013 revision edition on mathematics specialization learning in senior high school in Wonogiri regency.

IOP Conf. Series: Journal of Physics: Conf. Series **1022** (2018) 012013 doi:10.1088/1742-6596/1022/1/012013

a. Pedagogic competence of mathematics teacher in senior high school in Wonogiri regency is in accordance with the demands of Curriculum 2013 revision edition on mathematics specialization learning. Mathematics teachers who teach in senior high school in Wonogiri regency have applied 10 indicators of pedagogical competence according to Regulation of the national education ministry Number 16 of 2007. Pedagogic competencies that have been implemented by teachers include: a) each teaching teacher seeks to understand individual differences of learners, especially differences in abilities and attitudes; (b) teachers pay attention to competencies related to students' attitudes; (c) teachers apply creative approaches, strategies, methods and techniques in (e) teachers facilitate the development of learners' potential to actualize their potentials, (f) teachers communicate effectively, empathically and politely With learners, (g) the teacher determines the pen procedure Assessment and evaluation of learning process and outcomes, and (h) teachers utilizing assessment and evaluation results for learning purposes,(i) Teachers develop components of learning and (j) utilization of information and communication technology in learning.
b. Mathematics teachers at senior high school in Wonogiri regency and indicate their readiness in learning planning related to the implementation of Curriculum 2013 revised edition.
c. Mathematics teachers at senior high school in Wonogiri regency indicate their readiness in the implementation of learning related to the implementation of Curriculum 2013

7. References

[1] Rohman Muhammad 2012 *Kurikulum Berkarakter (Refleksi dan Proposal Solusi Terhadap KBK dan KTSP)* Jakarta: Prestasi Pustaka Jakarta

[2] Bulut M 2007 Curriculum reform in Turkey: *A case of primary school mathematics curriculum. Eurasia Journal of Mathematics, Science & Technology Education*, (Turkey University) chapter 3(3) p 203-212

[3] Akşit N 2007 Educational reform in Turkey *International Journal of Educational Development* (Turkey Univercity) chapter **27** p 129–137

[4] Kementrian Pendidikan dan Kebudayaan 2016 *Media Komunikasi dan Inspirasi, Jendela Pendidikan dan Kebudayaan ed.3: Empat Perbaikan Kurikulum 2013* Jakarta: Kementrian Pendidikan dan Kebudayaan

[6] Mulyasa 2013 Pengembangan dan Implementasi Kurikulum 2013 Bandung: Remaja Rosdakarya

[7] Hodgson D and Walford H 2007 Planning for learning and learning about planning in social work fieldwork Journal of Practice Teaching & Learning chapter **7(1)** pp 50-66 Whiting & Birch Ltd

[8] Mukhtar 2013 Metode Praktis Penelitian Deskriptif Kualitatif Jakarta: GP Press Group. Terbitan **1** p 10

[9] Sanaky HAH 2005 *Certification and Teacher Professionalism in the Era of Educational Reform* (Indonesia: Journal of islam education) chapter **1** pp 1-13

[10] Kementrian Pendidikan dan Kebudayaan **2013b** *Pedoman Pelatihan Implementasi Kurikulum 2013*. Jakarta: Kementrian Pendidikan dan Kebudayaan.

[11] Kementrian Pendidikan dan Kebudayaan **2013c** *Pedoman Pemberian bantuan Implementasi Kurikulum Tahun 2013* Jakarta : Kementrian Pendidikan dan Kebudayaan

Acknowledgement
In this opportunity, we would like to thank all those who support us that we cannot be mentioned here.

IOP Conf. Series: Journal of Physics: Conf. Series **1022** (2018) 012014 doi:10.1088/1742-6596/1022/1/012014

The effect of brain based learning with contextual approach viewed from adversity quotient

V Kartikaningtyas[1*], T A Kusmayadi[2] and R Riyadi[3]

[123)]Postgraduate School of Mathematics Education, Sebelas Maret University, Jl. Ir. Sutami 36A, Jebres, Surakarta, Central Java 57126, Indonesia

*Corresponding email: valeriakartika@gmail.com

Abstract. The aim of this research was to find out the effect of Brain Based Learning (BBL) with contextual approach viewed from adversity quotient (AQ) on mathematics achievement. BBL-contextual is the model to optimize the brain in the new concept learning and real life problem solving by making the good environment. Adversity Quotient is the ability to response and faces the problems. In addition, it is also about how to turn the difficulties into chances. This AQ classified into quitters, campers, and climbers. The research method used in this research was quasi experiment by using 2x3 factorial designs. The sample was chosen by using stratified cluster random sampling. The instruments were test and questionnaire for the data of AQ. The results showed that (1) BBL-contextual is better than direct learning on mathematics achievement, (2) there is no significant difference between each types of AQ on mathematics achievement, and (3) there is no interaction between learning model and AQ on mathematics achievement.

1. Introduction

Mathematics is about thinking, organizing, and also a logic proofing [1]. Students do a problem solving in mathematics learning. There are some issues in mathematics learning. The difficulty to understand the concept and solve some mathematics problem, teacher centered, and passive students are the issues that found in the implementation. Commonly, teacher gives the properties and formulas directly to the students in the geometry learning. Students have not experience in the solving contextual problem in geometry. Figure 1 shows the example of the contextual problem.

Solve this problem below.

A student would like to make a box. She prepares 12 × 8 cm two cardboards and the other two are 8 × 9 cm. How many types of cardboard that he needs to make the box? What is the size of cardboards?

Figure 1. Contextual Problem of The 3D Net

Some students could not solve this problem easily. Basically, this problem is about the nets of the box. Students only remember the properties of the cuboid to find cardboards that they need. Then, they arrange the different cardboards to be a box. So, the active learning methods need in learning process to support and optimize the students' ability in geometry.

The constructivism theory states that the students' achievement is an active activity where students construct their knowledge and find the meaning of what they learnt [2]. The students construct their knowledge through some activity. Then, they conclude what they have learnt. Based on the definition of student's achievement and mathematics, student's mathematics achievement in this research is a result of all active activities that have done during the mathematics learning process and also do an interaction with their social environment and mathematical objects.

Brain Based Learning (BBL) facilitates students to active and enjoy the learning. BBL is a learning that related to how our brain works naturally in the learning process [3]. The teacher's roles are important. The teacher should motivate, support, and guide the students in the process. So, it builds the positive environment for the students.

Student is the main subject in learning process who does some activities to achieve the best achievement. One of the learning approaches is contextual approach. This approach facilitates the students to construct their knowledge. It involves students actively to solve the real problem. The students could share their idea through the group discussion. There are seven principles of contextual approach in the instruction. They are constructivism, inquiry, questioning, learning community, modeling, reflection, and authentic assessment [4].

The definition of BBL-contextual is the model to optimize the brain in the learning of new concept and the real life problem solving by making a good environment. The implementation of BBL-contextual model in the geometry learning, specifically in the learning of 3D nets, shows in Figure 2. Generally, there are seven steps of BBL-contextual model, such as pre-learning, preparation, initiation and acquisition, elaboration, incubation and formation of memory, verification, and integration [5].

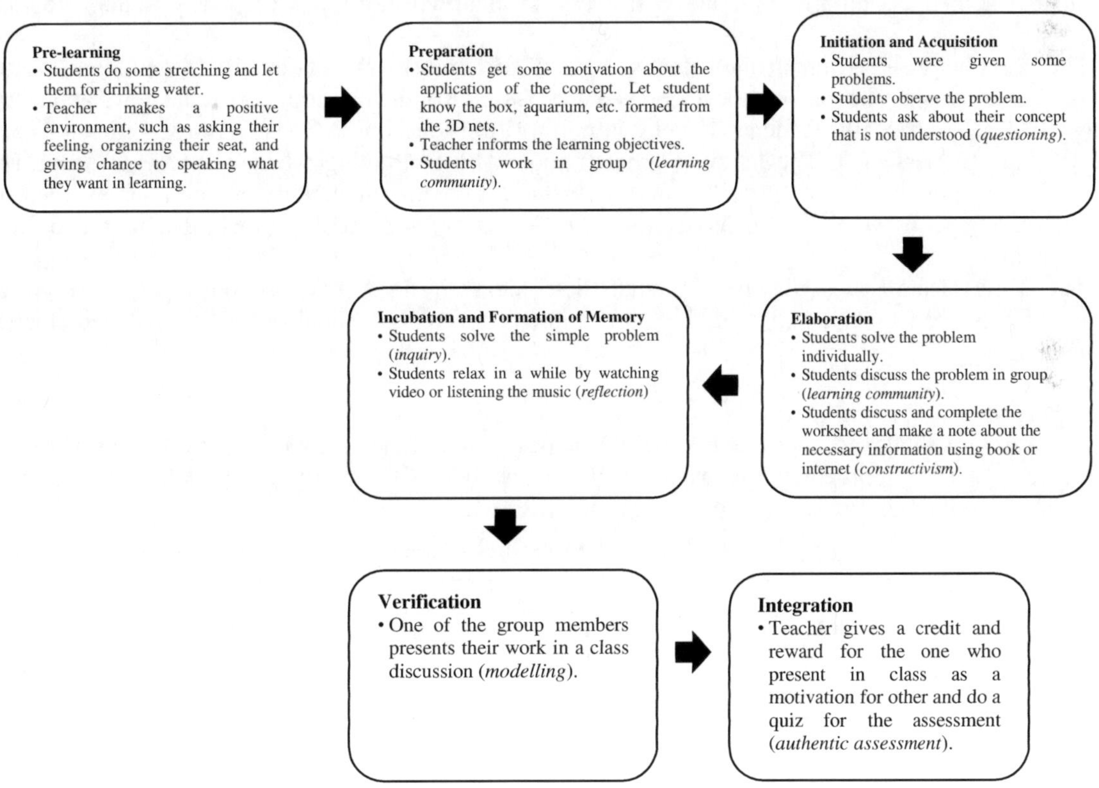

Figure 2. The Steps of BBL-Contextual Model

In the second step (preparation), students get motivation about the application of the 3D nets in their life, such as the box, aquarium, etc. formed by the net of cuboid or cube. In the third step (initiation and acquisition), students bring the some boxes, and then they cut the box in many ways to get the nets of the boxes. The net 1, net 2, net 3, etc. in Figure 3 are the nets of cube that students would get. In elaboration, students complete their worksheet by drawing the nets of the boxes that they got as shown in Figure 3.

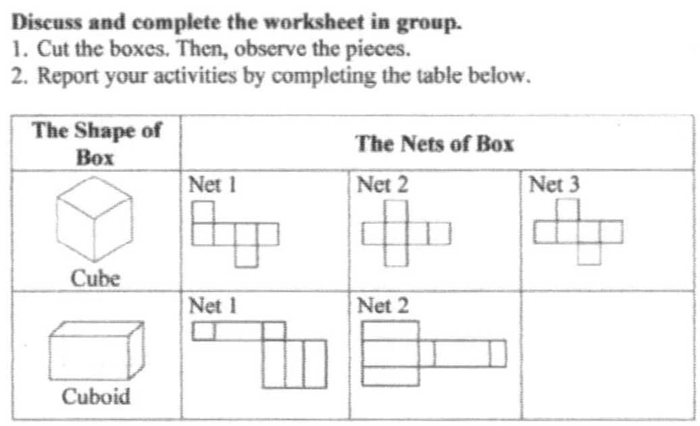

Figure 3. Students' worksheet of 3D net

In the learning process, there are many factors that influence students in their study. The factors could be come from external or internal. One of the factors is Adversity Quotient (AQ). AQ is the ability to response and face the problems, also how to turn the difficulties into chances. The types of AQ divided into three types. They are quitters, campers, and climbers [6]. Quitter is someone who is no effort or spirit to achieve their goal. Camper is someone who satisfies to the result that he got. Climber is someone who always thinks that every obstacle will not block their way to achieve their success.

The previous research about the effectiveness of BBL model shows that BBL was significantly positive in increasing the mathematics achievement [7,8]. Also, the previous research shows about AQ shows that AQ impacts the students in their learning [9]. Besides that, AQ also has a significant role in the learning process [10]. The learning process should be facilitating what the students need. The learning also should be more interactive and meaningful to helps students in the good achievement in geometry. BBL-contextual gives a chance for the different AQ of students by making a good condition for learning.

According to the issues, the study about BBL-contextual and AQ is essential to do to know the effect of this model, AQ, and also the interaction between the models and AQ on mathematics achievement.

2. Experimental Method

This research method was a quasi-experiment. The research design was 2x3 factorial designs as shown in Table 1. The experiment group was the group that use BBL-contextual model in the learning process. The control group was the group that use direct learning model in the learning process.

Table 1. The Research Design

Group	Adversity Quotient (AQ)		
	Quitters (b_1)	Campers (b_2)	Climbers (b_3)
Experiment (a_1)	a_1b_1	a_1b_2	a_1b_3
Control (a_2)	a_2b_1	a_2b_2	a_2b_3

where

a_1b_1: mathematics achievement of quitters' students by using BBL-contextual learning
a_1b_2: mathematics achievement of campers' students by using BBL-contextual learning
a_1b_3: mathematics achievement of climbers' students by using BBL-contextual learning
a_2b_1: mathematics achievement of quitters' students by using direct learning
a_2b_2: mathematics achievement of campers' students by using direct learning
a_2b_3: mathematics achievement of climbers' students by using direct learning

IOP Conf. Series: Journal of Physics: Conf. Series **1022** (2018) 012014 doi:10.1088/1742-6596/1022/1/012014

The population was the students of grade 8th of public junior high school. This research did in Surakarta. The sample was chosen by using stratified cluster random sampling. The sample was 109 students. The students divided into two groups; 49 students as the experiment group and 60 students as the control group.

The independent variables were the learning models and AQ. The dependent variable was student's mathematics achievement. The instruments were test and questionnaire for the data of AQ. The test and questionnaire were given to the two groups after the treatment. The test for mathematics' achievement was 25 items of multiple choice test. The test validated by the masters of mathematics education and the questionnaire of AQ validated by the masters of psychology. Then, the instruments tried out to the other 76 students who did not the subject of this research.

The normality and homogeneity test did before the data analysed. The data was analysed by using two ways ANOVA to know the interaction of two independent variables to the dependent variable. The hypothesis in this study were

2.1. H_{0A}: $\alpha_i = 0$, for every $i = 1,2$
H_{1A}: at least one α_i was not zero
2.2. H_{0B}: $\beta_j = 0$, for every $j = 1,2,3$
H_{1B}: at least one β_j was not zero
2.3. H_{0AB}: $(\alpha\beta)_{ij} = 0$, for every $i = 1,2$ and $j = 1,2,3$
H_{1AB}: at least one $(\alpha\beta)_{ij}$ was not zero

where
α_i : the effect of row i^{th} to dependent variable
i = 1,2
1 = BBL-contextual model
2 = direct learning model
β_j : the effect of column j^{th} to dependent variable
j = 1,2,3
1 = quitters
2 = campers
3 = climbers
$(\alpha\beta)_{ij}$: The interaction of row i^{t} in column j^{th} to the dependent variable

The level of significant is $\alpha = 0.01$. The criteria of H_0 is rejected if $F > F_{table}$ [10].

3. Result and Discussion

The data of the mean score of mathematics achievement shows in Table 2. The data is about the mean score of each AQ by using BBL-contextual or direct learning model. Table 2 shows that the mean score of mathematics achievement of each AQ in experiment group was higher than control group. The mean score of quitters, campers, and climbers students by using BBL-contextual were higher than the quitters, campers, and climbers' students by using direct learning.

Table 2. The mean score of mathematics achievement

Model	AQ			Mean Score
	Quitters	Campers	Climbers	
BBL-contextual	61.91	55.69	65.54	61.05
Direct Learning	46.43	54.81	58.00	53.08
Mean Score	54.17	55.25	61.77	

Normality and homogeneity test did before the hypothesis testing. Normality test was done to know the data normally distributed or not. If the significant was less than 0.01 or $L_{obs} > L_{table}$, then H_0 is rejected. The result of normality test by using Lilliefors method shows in Table 3 and Table 4. The data that used in normality test was the data of mathematics achievement test of each group. Table 3 shows that the L_{obs} score of mathematics achievement in experiment group was less than L_{table} = 0.15 and also L_{obs} mathematics achievement in control group was less than L_{table} = 0.13. So the population normally distributed for both groups. Table 4 shows that the L_{obs} score of mathematics achievement of quitters' students was less than L_{table} = 0.17. Also, the L_{obs} score of mathematics achievement of campers' students was less than L_{table} = 0.15 and L_{obs} mathematics achievement of climbers' students was less than L_{table} = 0.19. So the population normally distributed for both groups. Then, it continued to the homogeneity test.

Table 3. Normality test of mathematics achievement in each group

Group	Normality Test (Lilliefors Method)			Conclusion
	L_{obs}	L_{table}	Interpretation	
Experiment	0.09	0.15	H_0 is not rejected	Normal
Control	0.10	0.13	H_0 is not rejected	Normal

Table 4. Normality test of mathematics achievement in each type of AQ

AQ	Normality Test (Lilliefors Method)			Conclusion
	L_{obs}	L_{table}	Interpretation	
Quitters	0.12	0.17	H_0 is not rejected	Normal
Campers	0.10	0.15	H_0 is not rejected	Normal
Climbers	0.13	0.19	H_0 is not rejected	Normal

Homogeneity test used Bartlett testing by using chi-square test. Homogeneity test was done to know the classes had the same variance or not. If $\chi^2 > \chi^2_{table}$, then H_0 is rejected. The data of mathematics achievement test of each type of AQ was compared. The results wereχ^2_{obs} of each group is less than χ^2_{table}= 9.21. This result shows in Table 5. Table 5 shows that the variances of score of students' mathematics achievement of quitters, campers, and climbers students in experiment group were homogen. Also, the variances of score of students' mathematics achievement of quitters, campers, and climbers students in experiment group were homogen.

Table 5. Homogeneity test

Group	χ^2_{obs}	χ^2_{table}	Interpretation	Conclusion
Experiment	0.30	9.21	H_0 not rejected	Homogen
Control	0.63	9.21	H_0 not rejected	Homogen

The results by using two ways ANOVA shown that F = 9.91 > F_{table} = 6.89 as shown in Table 6. It concluded that there was significant different of learning by using BBL-contextual or direct learning on mathematics achievement. Table 6also shows there was significant different between each type of AQ since F = 3.20 < F_{table} = 4.82. It means that there was no difference between quitters, campers, and climbers students on mathematics achievement. In the other hand, there was no interaction between learning model and AQ since F = 3.08 < F_{table} = 4.82.

Table 6. Two ways ANOVA

Source	JK	dk	RK	F_{obs}	F_{table}	Interpretation
Model	1538.21	1	1538.21	9.91	6.89	H_0 is rejected
AQ	992.79	2	496.40	3.20	4.82	H_0 is rejected
Interaction	957.02	2	478.51	3.08	4.82	H_0 is not rejected
Error	15988.13	103	155.23			
Total	19244.92	108				

As shown as Table 2, the mean score of experiment group by using BBL-contextual model is 61.05 which higher than the mean score of direct learning is 53.08. It means BBL-contextual model is better than direct learning model in mathematics instruction.

Based on the result, there is significant different between learning mathematics by using BBL-contextual and direct learning. BBL-contextual model facilitates the students to optimize their brain on mathematis learning. The students need a good or positive environment to study well. The motivation and support from the teacher are important for them. The learning process would be smooth if the students do well in their study. In the other hand, the teacher is only explain the material to the students. The students drill by solving many problems, but less contextual. The students get bored in the learning. They are only sit and hear the explanation. It would impact to their achievement if they are not focus in study. The research about BBL supported by the previous research that shows BBL was significant effected the students' achievement [5,7,8]. BBL also gave the positive contribution to the achievement [12]. So, BBL is better than traditional method. Then, BBL-contextual is better than direct learning in the mathematics learning.

AQ is the ability to response and face the problems, also how to turn the difficulties into chances. Quitters have no effort or spirit to achieve their goal. So, the students that have the quitters type would be give up easily in the difficult problem. Camper is someone who satisfies to the result that he got. The students that have the type are unwilling to increase their level even though they have the ability to develop. Climber is someone who always thinks that every obstacle will not block their way to achieve their success. The students that have the type would be full of spirit to solve every problem even the difficult case. Besides that, there are also another factors, such as students' intelligent that did not research in this study. If the climber that have a motivation to study and face the problem has the low intelligent, then they might not get the best achievement because the geometry problems need a good analysing. The intelligent or AQ only could not guarantee to get the high achievement, but the intelligent, AQ, and emotional related each other [13]. So, there is no difference between quitter, camper, and climber on mathematics achievement.

The result also shows that there is no interaction between learning model and AQ. The learning model implemented in the learning process to facilitate students to achieve the better achievement. For the each type of AQ, BBL-contextual as the active learning is better than direct learning in mathematics learning. In fact, there are also many factors that influence students in the process. Teacher, as a facilitator, could not always handle all of the factors. Students distracted by another factors during the learning process. So, the mathematics achievements of quitter, camper, and climber students were same for each learning models (BBL-contextual and direct learning).

4. Conclusion

According to the result, it concluded the learning by using BBL-contextual is better than direct learning. It was shown by the result of two way ANOVA with different cell testing that the score of $F = 9.91 > F_{table} = 6.89$. Also, the mean score of BBL-contextual group is higher than control group which learning by using direct learning. So, BBL-contextual could be implemented in geometry learning. The model facilitates the students to learn actively and effectively. It also concludes there was no different on the achievements of quitter, camper, and climber students. It was shown by the result of two way ANOVA testing that the score of $F = 3.20 < F_{table} = 4.82$. The research also concludes that there is no interaction between learning model and AQ. BBL-contextual supports the good condition for study and solving the contextual problem. The students learnt by their own way and they were not bored then participate actively in the discussion. The teacher also gives an attention to the students' difficulty. So, the teacher knows their weakness and develop the learning media to minimize the students' mistake in problem solving.

Finally, it would be better if mathematics' teacher along could implement BBL-contextual in geometry learning and develop this model with another learning model or learning approach to increase the student's mathematics achievement. One more note, the students have different ability, intelligent, personality, etc. Therefore, in the further research, this factor could be attention. For other researcher, they could research about another variables or comparing BBL-contextual and the other learning model or approach.

5. References

[1] Suherman E, Turmudi, Suryadi D, Herman T, Suhendra and Prabawanto S 2003 *Strategi Pembelajaran Matematika Kontemporer* (Bandung: FMIPA Universitas Pendidikan Indonesia JICA) p 19

[2] Markaban 2006 *Model Pembelajaran Matematika dengan Penemuan Terbimbing* (Yogyakarta: Pusat Pengembangan dan Penataran Guru Matematika) p 7

[3] Jensen E 2008 *Brain-Based Learning* Terj. Narulita Yusron (Yogyakarta: Pustaka Belajar) p 12

[4] Rusman 2012 *Model-model Pembelajaran: Mengembangkan Profesionalisme Guru* (Jakarta: Rajawali Press) p 193-7

[5] Kartikaningtyas V, Kusmayadi T A and Riyadi 2017 Brain Based Learning with Contextual Approach to Mathematics Achievement *Accepted on International Journal of Science and Applied Science*

[6] Stoltz P G 2003 *Adversity Quotient: Mengubah Hambatan Menjadi Peluang* Terj Hermaya (Jakarta: PT Grasindo) p 9

[7] Gozuyesila E and Dikicib A 2014 The effect of brain based learning on academic achievement: a meta-analytical study *Educational Sciences: Theory & Practice* **14** 642–8

[8] Ozden M and Gultekin M 2008 The effects of brain-based learning on academic achievement and retention of knowledge in science course *Electronic Journal of Science Education* **12** 1–17

[9] Hema and Gupta S M 2015 Adversity quotient for prospective higher education*The International Journal of Indian Psychology* **4** 49–64

[10] Devakumar M 2012 A study of adversity quotient of secondary school students in relation to their academic self concept and achievement motivation Mumbai: University of Mumbai

[11] Budiyono 2016 *Statistika untuk Penelitian Edisi Ke-2* Surakarta: UNS Press pp 229-230

[12] Eggen P and Kauchak D 2012 *Strategi dan Model Pembelajaran: Mengajarkan Konten dan Keterampilan Berpikir* Terj Satrio Wahono Jakarta Barat: PT Indeks

[13] Maiquez R, Preolco A, Sausa L and Tagatagod K 2015 Predictive Ability of Emotional Intelligence and Adversity Quotient on Academic Performance of USC College Students *Thesis* p 33

Acknowledgments

The authors express a great thanks to some institutions. Especially, SMP Negeri 3 Surakarta, SMP Negeri 23 Surakarta, and SMP Negeri 26 Surakarta that supported us during data collection, so we can accomplish this research.

IOP Conf. Series: Journal of Physics: Conf. Series **1022** (2018) 012015 doi:10.1088/1742-6596/1022/1/012015

Natural science modules with SETS approach to improve students' critical thinking ability

A P S Budi[1*], W Sunarno[2] and Sugiyarto[3]

[1,2] Science Education Department of Post Graduate Programme, Faculty of Teacher Training and Education, Sebelas Maret University, Jl. Ir. Sutami 36A, Surakarta 57126, Indonesia

[3]Department of Biology, Faculty of Mathematics and Science, Sebelas Maret University, Jl. Ir. Sutami 36A, Surakarta 57126, Indonesia

*Corresponding email: ardianpuguh219@gmail.com

Abstract. SETS (Science, Environment, Technology and Society) approach for learning is important to be developed for middle school, since it can improve students' critical thinking ability. This research aimed to determine feasibility and the effectiveness of Natural Science Module with SETS approach to increase their critical thinking ability. The module development was done by invitation, exploration, explanation, concept fortifying, and assessment. Questionnaire and test performed including pretest and posttest with control group design were used as data collection technique in this research. Two classes were selected randomly as samples and consisted of 32 students in each group. Descriptive data analysis was used to analyze the module feasibility and t-test was used to analyze their critical thinking ability. The results showed that the feasibility of the module development has a very good results based on assessment of the experts, practitioners and peers. Based on the t-test results, there was significant difference between control class and experiment class (0.004), with n-gain score of control and the experiment class respectively 0.270 (low) and 0.470 (medium). It showed that the module was more effective than the textbook. It was able to improve students' critical thinking ability and appropriate to be used in learning process.

1. Introduction

The development of technology will bring various informations, knowledge, and the emergence of various problems. Therefore, it is necessary to manage the knowledge and information, and to solve the problems faced [1]. Several studies have shown that, the ability to use technology and information that involves self-thinking and problem solving independently, greatly helps students in terms of developing their critical thinking competencies [2]. Critical thinking is a method or way of thinking that maximizes results. In this way, students can enjoy their lessons and the teacher can stimulate them to get fun outcomes and analyze information [3].

Critical thinkers are people who are able to analyze and evaluate information, create abstract ideas, opened and directional thinking, and communicate effectively with others [4]. Critical thinking is an effectively needed cognitive skill and intellectual disposition to identify, analyze and evaluate, arguments and truths to discover and overcome personal prejudices and biases and to formulate and provide compelling reasons to support conclusions and make sensible decisions about what must believe and what to do [5]. Critical thinking skills are important because it enable students to handle social, scientific and practical problems effectively [6].

With this in mind, educators must help students so that they become successful in future world of work to be ready to achieve success in life by focusing education in developing students' critical thinking skills. With this skill, students are ready to work together successfully, think critically and analytically, communicate effectively and solve problems efficiently. Such activities require students to engage in active learning, engage in high level problem solving skills and to be able to participate in activities. Thus, students will develop strong leadership, communication, teamwork, cross cultural and cross country skills, and most importantly a belief in their ability to contribute in science and engineering field in society [7].

Through critical thinking, students can be taught to critically examine different perspectives on issues regarding the impact of science and technology on everyday life, and to evaluate these issues from a social and environmental perspective. The encouragement of science by educators for students in critical thinking about science and technology will also help to develop their analysis, skills, and abilities to make informed choices in everyday life [8]. Critical thinking as a combination of knowledge, attitudes and performance in each individual that includes ability in understanding skills, identification of hypotheses, conclusions, analysis and evaluation of rational arguments [9].

Science education plays a role in the effort to produce quality learners in accordance with the curriculum and the existing situation. The learning process of science should not only emphasize the understanding of science concepts but must connect with other elements of environmental, technological and community elements incorporated in SETS (Science, Environment, Technology and Society), so that in the learning process students are able to link its four elements in a unity of learning. In the process of learning in middle school, teachers are more likely to provide materials in the form of science concepts only, without linking science with the environment, technology and society. In addition, in the learning process is only oriented in the completion of all learning materials without associating with everyday life. As a result, students lack the ability to view science learning as a whole unit that related to the environment, technology and society [10].

Based on the observation of the researcher, the learning process that occurs in the class does not improve the critical thinking ability and less directly related to daily life. It is characterized by more learning characteristics referring to short-term goals (passing school exams), focusing on procedural capabilities, one-way communication, more dominant routine and low-level questions. Appropriate observations in the learning process and evaluation questions in the learning activities have not been oriented towards the development of students' critical thinking skills, so students have not been trained in developing their critical thinking yet. Thus, it can lead to low students' critical thinking ability. Students tend to be difficult to analyze information obtained and tend to accept whether there is any information in the book, students are less active in learning, students tend to be passive in expressing questions or in answering, and not yet able to convey ideas or ideas in solving problems.

This illustrates that students' critical thinking skills are still low. Science learning should be not just about memorizing material and material understanding, but students need to understand exactly the meaning and value contained in the materials. Through criticizing an event, will train students to think critically also on what is experiencing in Indonesia right now. In addition, critical approach to science teaching can lead to critical dialogue between teachers and students as well as students themselves on the issues they are facing and finding solutions to their solutions. By conducting this critical dialogue, students can discuss and brainstorm what they know about the profound meaning of the science matter [11].

Development of learning is needed to overcome these problems which will be poured in the module that can be used to train critical thinking skills and improve student learning outcomes. The module is a form of teaching material that is packed intact and semantic, containing a set of planned learning experiences designed to help learners understand specific learning objectives. The minimum module contains learning objectives, materials or learning substance and evaluation. The module served as an independent learning tool, so that learners can learn independently at their own pace [12].

The developed module is an natural science module design with the SETS approach. Students are not only invited to think about the use of the concept of science in the form of related technology, but also possible consequences that occur in the learning process of science which being studied into the form of technology to society and its environment. The core goal of SETS education is that this education can make students understand the key elements of SETS and the interrelationship among these elements when studying science. In other words, critical thinking is needed to learn every SETS element [13].

SETS starts from the belief that the relationship between students and the real world must be established. This process will lead the students to recognize the positive problems they have. Environments are created, where students can gather data for their problem solving, consider alternative solutions, determine the best way to solve problems and practice them [14].

The main reason for choosing SETS is because SETS learning is oriented towards students' active participation. Students are more guided to have the ability to think critically on environmental issues, technological developments and the community and also actively seek the problem [15]. The SETS approach is an effective learning approach to improve academic and science achievement [16]. Using the SETS approach can improve students' understanding of the nature of science and attitudes toward science significantly compared to students using traditional learning [17].

Teachers and learners like to have a decisive role in achieving learning goals in SETS learning. The teacher's role creates a thinking pattern that sees the future with its implications, bringing learners to always have integrative thin pattern, invites learners to think critically in facing of something with reference to SETS. Quality learning has a significant influence in producing qualified graduates. To produce a quality learning process, there are many aspects that influence it. These aspects include: qualified teachers and lecturers with qualifications mandated by Teachers and Lecturers from Law field, use of interesting and varied teaching methods, positive learning behavior of learners, and use of appropriate learning media in support the learning process itself [18].

The results of previous research indicated that 95% of students think if the SETS concept is incorporated into the learning process, it provides an opportunity for them to gain knowledge and enhance their understanding among the branches of science, so that hopefully through learning activities that are insightful to each other will get ideas about the results of technology from transformation of science, without destroying or harming the environment and society [19].

2. Method

This type of research is Research and Development (R & D). The development model of this research was Four-D (4-D), which includes define, design, development and disseminate. This research was conducted at State Junior High School 2 Jatisrono. True experimental design method with pre-test and post-test of control group was applied in this research was a research design. The subjects of the study for large-scale trials were students of class VII C as many as 32 students as control class (using BSE package book) and class VII D number of 32 students as experimental class (using IPA module approach SETS), while students of class VII E total 9 students as a limited trial class. The sampling technique which used in this research was simple random sampling.

The instrument of data collection in this research was done by questionnaire and test method. Questionnaires to obtain the data about student's response and the data about teachers' response about effectiveness of modules developed, while the test was used to determine the results of students' critical thinking skills. The tests used in this study were pre-test and post-test in the control and experimental class. The obtained data were analysed descriptively based on the expert validation results, education practitioners and peers as well as the results of questionnaire responses of students and teachers. The test results of critical thinking ability were analysed by calculating the number of relevant to students' answers. Then the test results were tested by the independent sample, t-test to determined whether there are differences before and after treatment was given between the control class and the experimental class. The students 'critical thinking skills were also analysed with an n-gain score to determine the improvement of students' critical thinking skills during the learning process.

3. Result and Discussion

3.1 Research Result

In the module development process, it started with the validation of draft module which consisted on materials, language, and media components by a lecturer who was the expert of teaching materials, a linguist lecturer, a lecturer who was also the expert of media, one educational practitioner and 2 colleagues. The lecturer who was as the expert of material evaluated the content, the linguits checked the language and the picture, the lecturer that was a media expert evaluated the graduation, one practitioner of education (peers review) evaluated the feasibility of presentation in the module. In table 1 below, it can be seen the results of expert validator test which obtained, it showed the average value which was 95.31%. It was concluded that the IPA module with the SETS approach belongs to the very feasible category.

Table 1. The Product Validation Results of The SETS Approach Module

Validation	N	Score	Average	Max Score	Percentage (%)
Material Expert (Content Feasibility)	1	60	60	64	93.75
Language Expert (Language Eligibility)	1	40	40	40	100
Media Expert (Feasibility of Graficity)	1	35	35	36	97.22
Educational Practitioner (Feasibility of Presentation)	1	189	189	208	90.86
Friends (Feasibility of Presentation)	2	394	197	208	94.71
Average value					95.31
Description					Very Decent

The limited trial stage which involved 9 students of class VII E State Junior High School 2 Jatisrono to find out the feasibility and effectiveness of natural science module with SETS approach on ecosystem theme. In this limited trial, students worked on the module and then the students' work results were analyzed to determine the feasibility and effectiveness of the module. Student participated in the trial which was limited to a different level, it was intended that the results obtained could represent the entire students in the field who had various capabilities. The results of module-limited trials could be seen from the student's work on the natural science module with SETS Approach in Table 2. The eligibility and effectiveness of the module was measured from the students' achievement in form of their critical thinking skill.

In each of the learning activity involved three students to work on modules on a limited trial. The average of students' critical thinking ability in learning activity obtained the value of 79.16 with good criteria, learning activity II got value 81.94 with good criteria and learning activity III had value 83.33 with good criteria, so that in the average limited trial the overall was 81.47 with good criteria. Thus, it can be concluded that the module feasibility in the limited trial was good or feasible to be used for large-scale trials and students 'critical thinking skills increased because of the natural science module with SETS approach and the module used in learning was effective because the level of students' understanding of the module increased with high performance.

Table 2. The Recapitulation of Trial Results Which Was Limited to Students' Critical Thinking Skills

Learning Activity	Respondents	Fascione's critical thinking indicators						Average	Average per learning Activity
		Interpretation (%)	Analysis (%)	Explanation (%)	Evaluation (%)	Inferention (%)	Self regulation (%)		
I	Student A	75	50	75	75	50	75	66.67	79.16
	Student B	75	75	75	75	100	100	83.33	
	Student C	75	100	100	100	75	75	87.50	
II	Student D	75	75	75	100	75	75	79.17	81.94
	Student E	100	75	100	100	75	75	87.50	
	Student F	75	100	50	75	75	100	79.17	
III	Student G	75	75	75	100	75	75	79.17	83.33
	Student H	100	75	100	100	75	100	91.67	
	Student I	75	75	75	100	75	75	79.17	
			Average						**81.47**

To determine the effect of using the SETS learning approach module to improve students 'critical thinking skills, the scores of students' critical thinking test in the control class were compared with those of experimental class based on the size of critical thinking developed [20]. The mean and standard deviations of pre-test value of their critical thinking were 69.66 and 10.38 for the control class and 69.40 and 8.96 for the experimental group. The t test results showed that there was no significant difference in the pre-test value of the critical thinking of the two groups with p-value = 0.915 ($p > 0.05$), implying that both classes of students had equal critical thinking skills before participating in the learning activities.

After the learning activity, the final value or post-test of critical thinking in both classes is taken. Table 3 shows the t-test results of the post-test value of the two classes. It was found that group post-test values differ significantly with p-value = 0.004 ($p < 0.05$). In addition, the mean post-test grade of the experimental grade (84.11) was higher than the control class (77.86), indicating that the students in the experimental class demonstrated a much better critical thinking ability than the control group. Consequently, it was concluded that the SETS approach module had helped students in improving their critical thinking skills.

Table 3. The Differences in The Ability to Think Critically on Posttest Values of Both Groups

Groups	N	*Mean*	SD	*p-value*	*n-gain*
Control class	32	77.86	9.19	0.004*	0.27
Experimental class	32	84.11	7.36		0.47

*Independent sample t-test

An analysis to know the effectiveness of learning module by using gain score for control class post-test and experiment class. Gain score was good indicator to show the effectiveness in learning. Based on the calculation of n-gain scores, the control class obtained a value of 0.27 (low), while for the experimental class had a value of 0.47 (medium).

The data of students' responses were used to obtain students' opinions about the feasibility of the ecosystem theme module with the SETS approach applied in the learning process. Questionnaire given to 32 students in class VII D (experimental class) which covered two statements that were positive and negative statements. The number of positive statements there were 11 items and negative statements were 10 items. The response of the students in class VII D answered strongly agree on the positive statement with the average gain of 3.47 which means the module was very good and feasible to use, whereas in the negative statements, the students answered strongly disagree with the average gain of 1.58 which means the module was also very good and decent Used. In addition to students, teachers also provided an assessment of the development of the natural science module with SETS approach. The results of the assessment and response of the average science teachers amounted to 3.73 included in the category of very good and feasible to be used.

3.2 Discussion

The first stage was the definition through field observation by spreading the questionnaire to analyze the needs of teachers and students conducted at schools. The results showed that the learning process to train critical thinking skills had never been taught, the students still needed direction from the teacher, the learning process was only centered on the teacher. In addition, teachers and students lacked of learning resources to optimize their critical thinking skills. From the results of observation, it was also known that SETS approach in learning had never been done at school so that students' critical thinking skills did not maximized yet. The lessons that have been applied using team teaching but in reality, in the classroom learning process was still done alternately according to the area of expertise of each teacher.

The second stage was the design stage by drafting the draft module which consisted of the initial module design containing cover, francis page, introduction, table of contents, list of drawings, table list, introduction, learning activities (I, II and III), assessment, key answers, assessment rubrics, glossary and bibliography. The module compilation refers to the learning path of SETS, a module that contains the linkage of SETS elements and modules used in classroom learning and self-study at home. The material was presented using SETS steps, so that finally found the correlation between one concept with other concept especially related to daily life. It was done so that students could think critically, so that it could relate between the concept that had been accepted with daily life, made wrong in making decisions and made learning more meaningful.

In the third stage was the development stage which included the initial product validation results in the form of draft module, it was a module that had been assessed by experts, education practitioners and peers. Validation is a process to test the suitability of the module with the competencies that are the target of learning. If the content of the module is appropriate, the meaning is effectively used to study the competencies which are the target of learning, then the module is said to be valid. However if the validation result states invalid, then it needs to be repaired until the module has valid category [12]. Based on the validation result, it can be concluded that the natural science module with SETS approach was feasible to be used in learning. After the validation of the product, further revisions were made, based on suggestions and inputs from reviewers, the weaknesses that found in the module design were then improved, so that the module that had been developed, had paid attention to the ease, and clarity for students in studying the material independently.

In accordance with the purpose of making modules is to train the independence of learners, reduce the role of teachers in learning activities, train students honesty, help learners to measure the mastery of learning materials, and the last is to accommodate the various levels and speed of learning learners. The preparation of this module had also paid attention to the purpose of composing the module itself, but there was one objective that could not be achieved with the preparation of this module. To determine whether this module could be said as a teaching material that was able to accommodate the various levels and speed of learners, it is necessary to test the use of this module in the actual class [21].

Learning modules could help learners to build what they learn and understand and facilitated their active participation in the learning process. Modules could be studied independently and also have a specific theme. It could give students the information necessary to acquire and assess the knowledge and skills specified. The modules require students to actively interact with learning materials, not just passively reading the material only. Students were asked to do various things in the learning activities and got feedback about what they do. Some types of evaluation strategies presented in the module told the students whether they achieved full mastery of the material and what to do if they could not

IOP Conf. Series: Journal of Physics: Conf. Series **1022** (2018) 012015 doi:10.1088/1742-6596/1022/1/012015

achieve the required mastery [22]. It is known that one of the functions of the module was as a student guide in self-study, so that basic skills matched the needs and abilities of students in elementary education stage [23].

Implementation of Natural Science module with SETS Approach was conducted to determine the feasibility and effectiveness of the module itself. Module eligibility could be seen from the responses or responses of students and teachers after using the module. The effectiveness of the module could be seen from the improvement of students' critical thinking skills after using the module. In a large-scale trial of data analysis used to determine differences in critical thinking skills in the control class and experimental class using prerequisite test that was normality test, homogeneity test and t-test.

Based on the prerequisite test results, the hypothesis test using parametric statistical test that is t-test Independent Sample t-Test. In the p-value of 0.004, the significance level of $\alpha = 0.05$ ($p < 0.05$), the result of the analysis of critical control thinking skill score and experiment class found that there was a significant difference between the control and experiment class scores, Which was significant between students' critical thinking skills between classes that used modules with classes that did not use modules. Significant differences in the use of the SETS approach natural science module with the ecosystem theme in the product testing class (experimental class) presented positive results because learners were interested in learning to use natural science module. In addition, the use of modules also has a positive effect on students' critical thinking skills.

The effectiveness of this module could be seen from the n-gain results for the post-test of control class and the experimental class, based on the n-gain results in table 3 it could be explained that the effectiveness of the experimental class (0.47) was higher than the control class (0.27). This proved that learning using natural science module with SETS approach was more effective to improve students' critical thinking ability compared to conventional learning using borrowed books from school (Electronic Books package from school).

Through critical thinking, students could be taught to critically examine different perspectives on issues regarding the impact of science and technology on everyday life, and to evaluate these issues from a social and environmental perspective. The encouragement of science by educators for students in critical thinking about science and technology, would also help to develop their analysis, skills, and abilities to make informed choices in everyday life [8].

According to the literature review, most research emphasizes that thinking could be taught. In other research findings indicated that life skill training could emphasize positive effectiveness and increase critical thinking of students in primary school. This finding also indicated an increase in the critical thinking subscale of the experimental group [24]. The students' critical thinking skills will be developed by working together in small groups. It was indeed necessary to attempt to teach the thinking skills and the contents of integrated thinking and curriculum strategies [25].

After the learning activity which done using the module had been completed, students were given a questionnaire that contained the responses, feedback and assessment of the module. The result of the overall students' response was excellent and the module was feasible to use. Learning using modules with the SETS approach was able to cultivate students' curiosity, provided new information, and encouraged students to seek additional information that is further evidenced by student responses. During the learning process students actively answer questions from the teacher. Discussion process was also done by students by actively searched information through library or internet sources.

The fourth stage was disseminate, at this stage was done by spreading of modules in the 5 junior high schools in the District Jatisrono, Wonogiri regency. The distribution was done to the science teacher and then given a questionnaire containing the teacher's response to the developed module. Teacher response results were used to elicit teachers' opinions on the feasibility of ecosystem modules with the SETS approach developed by using teacher response questionnaires. The overall teacher response outcomes were excellent and the modules were eligible to use. Modules with communicative language and examples or images contained in the module, could not be denied to provide a very large message for student learning outcomes, the number of images clarified with the description of the material would make students interested to learn the material ecosystem. With the interest of students on the material that had been taught, then the students would be more motivated to get a better value. Teachers also provided a positive response that the module could be studied independently and anywhere.

4. Conclusions

Based on the result of research, it can be concluded that the quality of module based on the module validation result Natural science with SETS approach on the theme of developed ecosystem included in very good category, while based on the responses of students and teachers, it showed that the module was very good and feasible to be used. Natural Science module with SETS approach on ecosystem theme was effective to improve students' critical thinking ability based on N-gain score of 0.470 showing medium category.

5. References

[1] Chu H C, Hwang G J, Tsai C C and Chen N S 2009 An innovative approach for promoting information exchanges and sharing in a web 2.0-based learning environment *J. Interact. Learn. Environ* **17**(3)11-23
[2] Johnson T E, Archibald T N and Tenenbaum G 2010 Individual and team annotation effects on students' reading comprehension, critical thinking, and meta-cognitive skills *Computers in Human Behavior* **26** 1496-507
[3] Chee Choy S, and Kin Cheah P 2009 Teacher perceptions of critical thinking among students and its influence on higher education *Int. J. Teach. Learn. High. Edu* **20** 198-206
[4] Duron R, Limbach B, and Waugh W 2006 Critical thinking framework for any discipline *J. Teach. Learn. High. Edu* **17** 160-66
[5] Bassham G, William J, Henry N and James M W 2002 *Critical Thinking A Students Introduction* (USA: McGraw-Hill Companies, Inc)
[6] Shakirova D M 2007 Technology for the shaping of college students' and upper-grade students' critical thinking *Russian. J. Educ. Soc* **49** 42-52
[7] Kalonji G 2005 Capturing the imagination: High priority reforms for engineering educators. In National Academy of Engineering (Ed) *Educating the engineer of 2020: Adapting Engineering Education to the New Century* Washington, DC: National Academies Press pp 146-50
[8] Mapeala R 2015 The development and validation of a test of science critical thinking for fifth graders *J. Springer Plus* **4** 741
[9] Watson G and Glaser E M 2000 Measure the ability to think critically and Problem solve Harvard University Press
[10] Binadja, Sri W dan Sigit N 2008 Keberkesanan pembelajaran kimia materi ikatan kimia bervisi sets pada hasil belajar siswa *J. Inov. Pend. Kim* **2** 256-62
[11] Supriatna N 2007 *Konstruksi Pembelajaran Pembelajaran Sejarah Kritis* Bandung: Historia Utama Press
[12] Daryanto 2013 *Menyusun Modul* Yogyakarta: Vega Media
[13] Binadja A 1999 Pendidikan bervisi sets dan master plan percepatan peningkatan mutu pendidikan dasar dan menengah di propinsi riau, Universitas Islam Riau 10-15 Pekanbaru
[14] Yager R E 1990 The science/technology/society movement in the United States Its Origin Evolution and Rationale *Social Education* **54** 198-200
[15] Sutarno N 2009 *Materi dan Pembelajaran IPA SD* (Jakarta: Universitas Terbuka)
[16] Rosario B I D 2009 Science and technology section science, technology, society and environment (STSE) approach in environmental science for nonscience students in a local culture *CHED Accredited Research Journal, Category B Liceo J. High. Edu. Res* **6** 2094-1064 Davao Doctors College

[17] Akcay 2015 Effectiveness of science-technology-society (sts) instruction on student understanding of the nature of science and attitudes toward science *J. Edu. Math. Sci. Technol* **3** 37-45

[18] Colburn 2010 An inquiry primer, science scope *J. Sci. Edu* **22** 42-44

[19] Frank M and Barzilai A 2006 Project-Based Technology: Instructional Strategy for Developing Technological Literacy *J. Technol. Edu* **18** 39-53

[20] Yeh Y C 2009 Integrating e-learning into the direct-instruction model to enhance the effectiveness of critical-thinking instruction *Instructional Science* **37** 185-203

[21] Prastowo A 2012 *Panduan Kreatif Membuat Bahan Ajar Inovatif: Menciptakan Metode yang Menarik dan Menyenangkan* Jogjakarta: Diva Press

[22] Dick W, Carey L and Carey J O 2001 *The Systematic Design of Intruction* Fifth Education New York: Longman

[23] Akinyemi O 2010 Analysis of Science Process skills in west african senior secondary school certificate physics practical examination in Nigeria *Am. eur. J. Sci. Res* **5** 234-40

[24] Scott G, Leritz L E and Mumford M D 2004 The effectiveness of creativity training: A quantitative review *Cre. Res J* **16** 361-88

[25] Abrams Z I 2005 Asynchronous cmc, collaboration and the development of critical thinking in a graduate seminar in applied linguistics *Can. J. Learn. Technol* **31** 23-47

IOP Conf. Series: Journal of Physics: Conf. Series **1022** (2018) 012016 doi:10.1088/1742-6596/1022/1/012016

Student's profile about science literacy in Surakarta

D Nur'aini[1], S B Rahardjo[2] and Elfi Susanti V H[3]

[123]Postgraduate, Science Education, Sebelas Maret University, Surakarta, Indonesia

Corresponding email: [1]diahnuraini_05@student.uns.ac.id; [2]sentotbr@yahoo.com; [3]bunda3f@gmail.com

Abstract. This research was conducted to find out student's initial profile of science literacy. The method used was descriptive with 46 students as subjects. The instrument used is science literacy question referring to PISA 2015. Data processing technique used are scoring on each question, changing the score values, grouping the level subjects obtain based on the value and conclusion. Competencies measure in this test are explaining scientific phenomena, interpretation of data and evidence scientifically, and evaluate and design scientific inquiry. The results of the three competencies are 30,87%, 40,20% and 24,90%. Achievement level of science literacy achieved by students in level 1 47,82%, level 2 33,82%, level 3 42,93%, level 4 26,50%, level 5 21,73%. Based on the result of research, it can be concluded that the ability of science literacy students in Surakarta relatively low.

1. Introduction

People's lives are evolving along science and technology advances. Therefore, need a society that knows and understands the problem in society. In this case, education is becoming the determinant of society's ability to face the progress. The educational process is expected to form students who are science and technology literate. Science literacy is considered important in development that occurred because science literacy can train critical thinking ability and science process skills [1]. Understanding of science and technology is essential for a person's readiness to live in a modern society [2]

Definition of science literacy according to PISA is "An individual's scientific knowledge and use of that knowledge to identify questions, to acquire new knowledge, to explain scientific phenomena, and to draw evidence-based conclusions about science-related issues, understanding of the characteristic features of science as a form of human knowledge and enquiry, awareness of how science and technology shape our material, intellectual, and cultural environments, and willingness to engage on science-related issues, and with the ideas of science, as a reflective citizen" [2]. Miller (2008) defines science literacy as the level of understanding of science and technology which is needed in people's lives [1]. Dani (2009) defines science literacy as an understanding of concept and process of science necessary for decision making [3]. Generally, science literacy is defined as the ability to use scientific knowledge, identify questions, make factual inferences and apply in daily life.

Science literacy is an important an necessary ability, especially to face up the problems in the 21st century. Science literacy make enables one to make conclusions and make informed decisions [3]. In the era of globalisation, a people would need science literacy. Someone who literate science will live well in this era of globalisation [4]. According to Sadler (2009), someone who literate in science is able to make decisions regarding the issues in force, able to face situations where scientific ideas and process are involved [5] [6]. Thus, it can be concluded that science literacy is an important ability to be possessed by a person. Science literacy helps one to solve problems and make scientific decisions.

PISA (Programme for International Student Assessment) is a literacy study that aims to examine periodically the ability of learners aged 15 years. The research is done every three years. Research by PISA aims to know the readiness of students in facing the challenges of the 21st century [6].

The PISA test result indicates that student's literacy skills in Indonesia are still low. Indonesia is ranked 64 out of 65 countries. The scores obtained for science literacy are 382 [7]. PISA test results in 2015, Indonesia is ranked 62 out of 70 countries. The scores obtained for science literacy are 402 [8]. The scores obtained by Indonesia on PISA test 2012 and 2015 are at level 1. According to PISA 2015, there are 6 levels used to measure science literacy. The levels are level 1a&b with minimum score 334,9, level 2 with minimum score 409,5, level 3 with minimum score 484,1, level 4 with minimum score 558,7, level 5 with minimum score 633,3, and level 6 with minimum score 707,9 [9].

1st scientific competence measured in science literacy is identifying scientific problems (recognising the issues that are likely to scientifically investigate, identifying keyword to search scientific information, recognising the key features of scientific inquiry). The second competence measured in science literacy is explaining phenomena scientifically (applying knowledge of science in a given situation, describing or interpreting phenomena scientifically and predicting changes, identifying appropriate descriptions, explanation, and predictions). The last competence measured in science literacy is using scientific evidence (interpreting scientific evidence and making and communicating conclusions, identifying the assumptions, evidence and reasoning behind conclusions, reflecting on the societal implications of science and technological development) [10].

2. Method

This study was conducted to find out the initial profile of science literacy students in Surakarta. The method used in this research is a descriptive method. Descriptive method is a research method used to collect information on the subject of research [11].

The subjects in this study are students of class XI MIA, amounting 46 people. Subjects come from two different schools. The instrument used in this research is science literacy test that adapted from PISA 2015. The questions of test amount 13 questions. The competencies assessed in this test are interpreted data and evidence scientifically, evaluate and design scientific inquiry, and explain phenomena scientifically [8]. The competencies are divided into levels 1-5. In the implementation of the test, each student has distributed 1 sets of test questions. The time given to answer the questions is 30 minutes.

Data processing technique performed are:

1. Scoring for each question. The score for the perfect answer is 1. The maximum score if answer the questions perfectly is 13.
2. Converting score into value. Written in equation 1 [12].

$$\text{value} = \frac{\text{scores obtained}}{\text{maximum scores}} x\ 100 \text{..........} \quad (1)$$

3. Grouping of research subject based on a value obtained. Subject are grouped into science literacy levels.
4. Calculate the competence achievement science literacy
5. Calculate the average ability of students science literacy
6. Analyse the data obtained and describe the data.

IOP Conf. Series: Journal of Physics: Conf. Series **1022** (2018) 012016 doi:10.1088/1742-6596/1022/1/012016

3. Result and Discussion

3.1. *Result*

Questions of test and categorization of science literacy levels adapted from PISA 2015. The result of science literacy test according to levels of science literacy shows in Table 1.

Table 1. Result of Science Literacy Test According Levels of Science Literacy

Levels of Science Literacy	Percentage	Students Who Can Answered Questions on Level
Level 1	47,82%	22
Level 2	33,82%	15
Level 3	42,93%	20
Level 4	26,50%	13
Level 5	21,73%	10
Level 6	-	-

If viewed from the achievement of science literacy competence from PISA, the result of science literacy test can be seen in Table 2.

Table 2. Result science literacy test based on science competencies measured

Competencies	Percentage
explain phenomena scientifically	30,87 %
interpret data and evidence scientifically	40,42 %
evaluate and design scientific inquiry	24,90 %

3.2. *Discussion*

Based on the results of the science literacy tests shown in Table 1, Know that as many as 47.82% or 22 students can answer test questions that are at level 1. That's mean, as much as 52,18% student cannot answer questions level 1 and considered unable to perform the competencies contained at level 1. There are 33.82% or 15 students can answer the test questions that are at level 2. That's mean, as much as 66,18% can't answer questions level 2 and considered unable to perform competencies contained at level 2. There are 42.93% of students or 20 students can answer the test questions that are at level 3. That's mean, as much as 57,07% can't answer questions level 3 and considered unable to perform competencies contained at level 3. There are 26.5% or 13 students can answer the question at level 4. That's mean, as much as 73,5% students can't answer questions level 4 and considered unable to perform competencies contained at level 4. 21,73% or 10 students can answer the questions that are at level 5. That's mean, as much as 78,27% students can't answer questions level 5 and considered unable to perform competencies contained at level 5. The competencies levels are shown in Table 3 [2].

IOP Conf. Series: Journal of Physics: Conf. Series **1022** (2018) 012016 doi:10.1088/1742-6596/1022/1/012016

Table 3. Science Literacy Levels

Levels of science literacy	Lower score limit	What students can typically do at each level
Level 6	707,9	At Level 6, students can consistently identify, explain and apply scientific knowledge and knowledge about science in a variety of complex life situations. They can link different information sources and explanations and use evidence from those sources to justify decisions. They clearly and consistently demonstrate advanced scientific thinking and reasoning, and they use their scientific understanding in support of solutions to unfamiliar scientific and technological situations. Students at this level can use scientific knowledge and develop arguments in support of recommendations and decisions that centre on personal, social or global situations.
Level 5	633,3	At Level 5, students can identify the scientific components of many complex life situations, apply both scientific concepts and knowledge about science to these situations, and can compare, select and evaluate appropriate scientific evidence for responding to life situations. Students at this level can use well-developed inquiry
Level 5	633,3	abilities, link knowledge appropriately and bring critical insights to situations. They can construct explanations based on evidence and arguments based on their critical analysis.
Level 4	558,7	At Level 4, students can work effectively with situations and issues that may involve explicit phenomena requiring them to make inferences about the role of science or technology. They can select and integrate explanations from different disciplines of science or technology and link those explanations directly to aspects of life situations. Students at this level can reflect on their actions and they can communicate decisions using scientific knowledge and evidence.
Level 3	484,1	At Level 3, students can identify clearly described scientific issues in a range of contexts. They can select facts and knowledge to explain phenomena and apply simple models or inquiry strategies. Students at this level can interpret and use scientific concepts from different disciplines and can apply them directly. They can develop short statements using facts and make decisions based on scientific knowledge.
Level 2	409,5	At Level 2, students have the adequate scientific knowledge to provide possible explanations in familiar contexts or draw conclusions based on simple investigations. They are capable of direct reasoning and making literal interpretations of the results of scientific inquiry or technological problem-solving.
Level 1		At Level 1, students have such a limited scientific knowledge that it can only be applied to a few, familiar situations. They can present scientific explanations that are obvious and follow explicitly from given evidence.

Based on table 2, the achievement of science literacy for each competency is still low. A total of 30.87% of students can meet the competence of explaining scientific phenomena. In this competence, students can remember the knowledge appropriate to a particular situation. Students can use knowledge to interpret and provide explanations related to phenomena. Such knowledge can also be used to generate hypotheses when there is a lack of data.

A total of 40.42% of students can meet the competence of data interpretation and scientific evidence. In this competence, students can interpret and understand the basic forms of scientific data and evidence used to make claims and make conclusions. Students can identify assumptions, points and reasoning, students can different arguments based on scientific evidence and theories based on considerations, And evaluate scientific arguments and scientific evidence from different sources (internet, newspapers, journals, etc).

A total of 24.90% of students can meet the competence of evaluating scientific design and inquiry. In this competence, students can identify the questions explored in a given scientific study. Students can distinguish possible questions to investigate scientifically. Students can evaluate how to explore the questions given scientifically. Students can explain and evaluate the various ways that scientists use to ensure data's truth and objectivity.

From the data that has been presented, can be described the ability of science literacy students in Surakarta. Based on the result of the test, the ability of science literacy in Surakarta is still quite low. This is seen by a number of students who have not been able to answer the test question, even a matter of tests at levels 1 and 2 are easily classified. If viewed from the achievement of competence, students' literacy skills are also low, This can be seen from the achievement of the three competencies that are below 50%, That is, students still can not dominate the competencies tested by PISA 2015.

4. Conclusions

Based on the result of research and data analysis, obtained the following conclusions the result of science literacy test according to levels of science literacy is 47,82% students can answer the level 1 questions, 33,80% students can answer level 2 question, 42,90% students can answer the level 3 questions, 26,50% can answer the level 4 questions, and 21,73% can answer the level 4 questions. The result of science literacy test according to competencies science literacy measured are 30,87% students achieving competence explain scientific phenomena, 40,20% students achieving competence interpret data and evidence scientifically, and 24,90% students achieving competence evaluate and design scientific inquiry. The ability of science literacy students in Surakarta relatively low. This can be seen from the achievement of low competence, and low-level science literacy achievement.

5. References

[1] Hobson A 2008 The surprising effectiveness of college scientific literacy courses *Phys. Teach.* **46** 404
[2] Anon 2009 PISA 2009 Assessment Framework Key competencies in reading , mathematics and science
[3] Martinez-hernandez K and Kimaru I 2015 Perspectives on science literacy : a comparative study of United States and Kenya perspectives on science literacy : a comparative study of United States 25–34
[4] Holbrook J and Rannikmae M 2007 Enhancing scientific literacy the nature of science education for 37–41
[5] Sadler T D, Zeidler D L, Hall N, Box P O, Education S, Florida S and Ave E F 2009 Scientific Literacy , PISA , and Socioscientific discourse : assessment for progressive aims of science education **46** 909–21
[6] Soobard R and Rannikmäe M 2011 Assessing student ' s lèvel of scientific literacy using interdisciplinary scenarios **22** 133–44
[7] Fensham P J 2009 Real world contexts in pisa science : implications for context-based science education **46** 884–96
[8] Anon 2012 PISA 2012 Results in Focus

[9] Anon 2015 Pisa 2015
[10] Bybee R, Mccrae B and Laurie R 2009 PISA 2006 : an assessment of scientific literacy **46** 865–83
[11] Mukhtar 2013 *Metode Penelitian Deskriptif Kualitatif* (Jakarta: Gaung Persada Pers Group)
[12] Arikunto S 2012 *Prosedur Penelitian : Suatu Pendekatan Praktek* (Jakarta: Rineka Cipta)

IOP Conf. Series: Journal of Physics: Conf. Series **1022** (2018) 012017 doi:10.1088/1742-6596/1022/1/012017

The need analysis of chemistry module based on REACT (relating, experiencing, applying, cooperating and transferring) to improve critical thinking ability

D M Tyffani[1*], S B Utomo[2] and S B Rahardjo[3]

[123]Magister of Science Education Program, Faculty of Teacher Training and Education, Sebelas Maret University

*Corresponding email: choidynda@gmail.com

Abstract. This research was aimed to find out how students' need of chemistry module based REACT (Relating, Experiencing, Applying, Cooperating and Transferring) to improve students' critical thinking ability. The subjects of this research was the studentsof XI grade in three school in even semester of academic year 2016-2017 that contained of 48 students of Senior High School 2 Bandar Lampung, 38 students of Senior High School 3 Bandar Lampung and 46 students of Senior High School 12 Bandar Lampung. The data was gathering used non-test method by using open questionnaire with 13 questions. The results showed that 84,84% of students stated that the development of chemistry module based REACT on colloid material is needed.The analysis of hand's book was used aspects of critical thinking proposed by Facione (2011) are interpretation, analysis, evaluation, conclusion, and explanation. Based on the result of the analysis of hand's book at Senior High School 12 Bandar Lampung for critical thinking in colloid material that indicate 50% indicator is appropriate, while for indicator of inference and explanation only 16,67% appropriate, then for indicator analysis and evaluation doesn't have conformity. Based on the results of the analysis shows that the hand's book used have not empowered critical thinking ability with maximum. The development of chemistry module on colloid material is needed to overcome the problem of hand's book that hasn't maximized critical thinking ability, then the development of module oriented to REACT learning model (Relating, Experiencing, Applying, Cooperating, and Transferring).

Key words: *critical thinking ability, module, REACT*

1. Introduction

Education is a process in influencing students to adjust to the environment, so that students are expected to function in the life of the community. The government seeks to improve the quality of education through the development of the education system, including in the 2013 curriculum and the Law on the education system. One of the objectives to be achieved through education is the formation of an Indonesian human who has the spiritual, religious and life skills and intelligence needed for himself, the people, the nation and the state, in accordance.

The curriculum of 2013 has one of the characteristics of developing core competencies. The core competencies emphasize to students to have religious competence, social competence, cognitive competence and good psychomotor competence. Ideal learning process can involve students actively and not only emphasizes the cognitive aspect but also on the psychomotor and affective aspects. The expected learning is innovative learning, relevant to the needs and the active role of the students in learning.

One of the abilities that must be developed in chemistry learning is critical thinking skills. Critical thinking skills of students is the ability of high-level thinking is important to have students because of the ability to think critically can equip students in facing problems in the future not only in classroom learning. Critical thinking skills is one of the characteristics of scientific attitude which is one of the goals of chemistry education in high school. The guidance of critical thinking at all levels of education is felt to be very important, so how to teach students to think critically is an important issue in education.

Critical thinking can't be taught through lecture methods, because critical thinking is an active process. The intellectual skills of critical thinking include thinking of analysis, thinking synthesis, reflective thinking and should be learned through actualization of appearance. Critical thinking can be taught through laboratory activities, inquiry, term paper, homework that presents opportunities for evocative thinking, and exams designed to promote critical thinking skills [1].

Modules can potentially trick critical thinking skills because critical thinking trainings can be done independently, flexibly and repeately. Learning using a module allows students to enhance optimal learning activities in accordance with the level of ability and progress students gain during the learning process [2]. The module can potentially improve students' critical thinking skills if the contents of the module can spur students to solve and analyze problems and evaluate information [3]. Teaching method and approach will determine the level of proficiency of students. Teachers should use student-centered approach to stimulate the students mind thus help students acquire and master the knowledge [4].

Based on observations in three state high schools in Bandar Lampung indicate that the learning process that is carried out by teachers in general is still a teacher center and students are given less opportunity to develop the thinking skills (student center). Student participation to study chemistry is very low, it is proven when the researcher participate in teaching and learning activity seen the student just listen and record explanation of teacher in front of class. Students in Senior High School 3 Bandar Lampung during the learning process only hold on to worksheets as learning resources, while for Senior High School 12 Bandar Lampung, student learning has been given facility in the form of printed book which is loan from school.Based on the results of the analysis of teaching materials using aspects of critical thinking proposed by[5] indicates that the teaching materials used have not empowered critical thinking skills with the maximum.

To overcome the problem of teaching materials that have not maximized critical thinking ability, the development of module oriented to REACT learning model (Relating, Experiencing, Applying, Cooperating, and Transferring). Activities based on REACT strategy increase students' interest and affect their learning positively. According to physics students, learning process gained more meaning because they could enjoy the learning process by directly apply it in real activities and know the implementations of what they learned. Learning process remained as a real experience [6]. Based set of problems above, this study makes early analysis about students' need for REACT based chemistry module.

2. Theoretical Framework

2.1. Theory of Learning

In the view of constructivism, learning is an active process that students build their own knowledge [7]. Therefore,students must be actively involved and students become centre of learning activities in the classroom. The constructivism approach in teaching applies cooperative learning on the theory that students will more easily find and understand difficult concepts when discussing such issues with their peers [8].

Vygotsky's theoretical implications of learning using this module apart from being used as independent teaching materials, this module is also integrated in learning through discussions and experiments conducted in small groups. Providing assistance in the form of guidance, warning, encouragement by the teacher during the early stages of learning is done so that the more students can take over responsibility independently. Knowledge was shaped by the student through a problem or case given by the teacher associated with the application in the daily life of the environment, so that students are expected to gain meaningful learning [9].

2.2. REACT Learning Model

REACT learning model according to [10] has stages are as follows: (1) relating, (2) experiencing, (3) applying, (4) cooperation, and (5) transferring. Here is a process of implementation of learning with REACT learning model which is a cycle of activities, meaning the process is never interrupted.

REACT learning model has several advantages compared with conventional learning, which are: (1) The relating stages in this model can help teachers in relating the material / concepts that are taught to real situations, (2) The experiencing stage can encourage students in connecting the knowledge they already process with its application in daily life, (3) The applying stage can increase students' understanding of the material given by the teacher by linking it to everyday life, so the material can be easily understood without having to memorize it, (4) The cooperating stages can increase student activity in the learning process where students are asked to be active in cooperating with friends of one group, and (5) The transferring stages can improve students' ability in conveying or transferring a new concept into a more complex problem [11].

REACT methodology can be applied in context based teaching successfully (1) when the projects and activities are chosen that are related to students‘ daily life, (2) when students are motivated to get necessary data, (3) when students get the chance of applying of knowledge, (4) when students study cooperatively, (5) and when students are helped to discover connections that enable them to transfer knowledge from one context to another[12].

Teachers can maximize the outcomes of this type of learning by successfully applying the REACT strategies in their everyday lessons: relating what is being taught into the context of the real world, experiencing the new knowledge, applying new concepts to the real world situations, solving problems by communicating with each other, and transferring that knowledge to an experience that they will have in the future[13]. In the contextual learning process, learning occurs when the content of the knowledge creates meaning in learners' mind, making connection between the context to be studied and real world situation. This is because students could learn a lot of physics terms in the learning process [6]. Learning based on REACT methodology may train students' critical thinking ability.

2.3. Critical Thinking Ability

Critical thinking is thinking to systematically investigate the thought process itself. It means not only deliberate thinking, but also exploring how we and others use evidence and logic [14]. Critical thinking in terms of these skills, identifying assumptions, checking out the validity of them, examining different perspectives, and then making informed decisions [15]. Critical thinking aims to achieve a critical assessment of what we will receive or what we will do for logical reasons [8]. Critical thinking was focuses on skills as involving solving problems, formulating inferences, calculating likelihoods, and making decisions [16].According to definitions above, critical thinking means one's ability to process consideration about some information with their cognitive ability. Critical thinking aspects are interpretation, analysis, evaluation, inference, explanation, and self-regulation [5].

2.4. Module

Module as an independent learning package that includes a series of experience learning experience that has been planned and systematically designed to facilitate students in achieving learning objectives [17]. Module is one of learning material which is intact and systematic. It includes set of planned learning experiences and is designed to help students master the goal of the learning [18].

3. Method

The research method used is qualitative method. The data were collected by four stages, that were observation, interview, questionnaire distribution and material analysis. The study was conducted in February 2017 at three public schools in Bandar Lampung. The subjects of the study were students of class XI IPA of 134 students that contained of 48 students of Senior High School 2 Bandar Lampung, 40 students of Senior High School 3 Bandar Lampung, and 46 students of Senior High School 12 Bandar Lampung. The interview stage was conducted to 3 chemistry teachers, to know the condition of the school and the process of learning activity in the classroom directly by the speaker ie the chemistry subject teacher. Observations were made by collecting data on the state of the school and the ongoing learning process. The questionnaire distribution stage is given to the students. Questionnaire given in the form of an open questionnaire, to find out the reason students answer yes or no on each question. Calculated percentage of students answered yes and no for each question, then analyzed to know the needs of students of chemical modules Then the last stage of the analysis of teaching materials done by assessing the suitability of the content, delivery and use of books and modules using aspects of critical thinking skills by [5]. Result of analysis of teaching materials used to know the extent of fulfillment of each aspect of critical thinking in teaching materials used in Senior High School 2 Bandar Lampung, and Senior High School 12 Bandar Lampung.

4. Result

Observation has done by collecting data about the state of the school and the learning process that took place in Senior High School 2 Bandar Lampung, Senior High School 3 Bandar Lampung and Senior High School 12 Bandar Lampung directly. Based on the observations in Senior High School 3 Bandar Lampung shows that the learning process that is implemented by teachers in general is still the teacher centre and students are given less opportunity to develop the thinking skills (student centre). During the learning process, students only stick to the student worksheet as a learning resource. Student participation to study chemistry is very low, it is proven when the researcher participate in teaching and learning activity seen the student just listen and record explanation of teacher in front of class. However when students are confirmed about the material presented by the teacher, students can not answer and the chemistry concepts obtained are less precise. Similarly, the observation results at Senior High School 12 Bandar Lampung, that the study of chemistry is still using the lecture method. Through this method, teachers play an active role so that students are not given space to develop their potential. During the learning process students have been given facilities in the form of a printed book that is a loan from the school.

Questionnaires are given to students to find out the students' needs on chemical modules. Questionnaires pertain to chemical books used by students, systematic modules, modules that help develop critical thinking skills, and modules equipped with practicum activities. From the result of questionnaire, it is known that students do practicum in chemistry learning (35,82%), laboratory and library availability can help to understand chemical concept (41,79%), using chemistry book as study guide (65,67%), chemistry book (54.47%), students use more than one chemistry book for additional literature (47.76%), teachers use special teaching materials to teaching chemical concept (29,8%), teacher give chemical module to study chemical concepts (45.98%), students enthusiastic in following chemistry lessons (81.34%), easy to understand colloidal material (55.22%), students find difficult to understand colloidal material with chemical books they use and methods applied by teacher (44,02%), teacher used contextual learning method in chemistry learning process (35,07%), student need alternative teaching material to learn colloid concept with more easy and interesting (79,1%), students agree that module based REACT help them to understand colloid material(84.84%).

The analysis of teaching materials conducted using aspects of critical thinking proposed by [5] are interpretation, analysis, evaluation, conclusion, and explanation. Based on the results of the analysis of teaching materials at Senior High School 2 Bandar Lampung indicates that 50% of interpretation indicators are appropriate, while for the analysis and explanatory indicator is only 16.67% appropriate, and for the conclusion indicator is only 33.33% appropriate. Then for the evaluation indicator there has been no compliance. Similarly, the result of the analysis of teaching materials that have been done in Senior High School 12 Bandar Lampung obtained for critical thinking ability in Colloidal material is showing 50% indicator of interpretation is appropriate, while for indicator of inference and explanation only 16,67% accordingly, then for indicator analysis and evaluation there is no match. Based on the results of the analysis shows that the teaching materials used have not empowered critical thinking skills with the maximum. So it is necessary to develop modules on colloidal material.

5. Discussion and Conclusion

The students' need for modules based REACT is reinforced by the results of the questionnaire given to the students. Students of Senior High School 2 Bandar Lampung and Senior High School 12 Bandar Lampung have used the book as a source of reference learning, while students in Senior High School 3 Bandar Lampung only use worksheet as a learning resource. Books used in schools have not empowered critical thinking skills to the maximum.

Another finding is that the teaching experience given by the teacher is more emphasized on lecture, exercise and laboratory is not optimal. These activities seem monotonous and have not emphasized the students' active activities in building the concept. One way to tackle the problem is to use an innovative model of learning that is appropriate in its application in the classroom.

Modules that have the potential to improve students' critical thinking skills are activity-based modules one of them is a module oriented to the REACT learning model. Modules based REACT are modules that are characterized by the syntax of REACT learning proposed by [10]relating, experiencing, applying, cooperating and transferring.This module based REACT is used by students as a guide in studying chemistry as in colloid material.Based on the above description it can be concluded that students need modules based REACT to improve critical thinking skills.

6. References

[1] Sadya I W 2009 Model Pembelajaran yang Efektif untuk Meningkatkan Keterampilan Berpikir Kritis (Denpasar: Program Pascasarjana, Universitas Pendidikan Ganesha)
[2] Lunenburg F C 2011 *Critical Thinking Constructive. Tech. Improve. Student Achievment* **21** 1–9
[3] Yanti F A 2015 Pengembangan Modul Pembelajaran Fisika SMA/MA Berbasis Masalah untuk Meningkatkan Keterampilan Berpikir Kritis Siswa. *Jurnal Inkuiri UNS.4(3)* 96-103
[4] Matanluk O, Mohammad B, Kiflee D N A and Imbug M 2013 The Effectiveness of Using Teaching Module Based on Radical Constructivism Toward Students Learning Process *Procedia - Soc. Behav. Sci.***90** 607–15
[5] Facione P A 1991 Using the California Critical Thinking Skills Test in Research, Evaluation, and Assessment *Calif. Acad. Press* 1–20
[6] Ozbay S A 2015 The Use of REACT Strategy for the Incorporation of the Context of Physics into the Teaching English to the Physics English Prep **4** 91–117
[7] Suparno P 2007 *Metodologi Pembelajaran Fisika.* (Yogyakarta: FKIP UNY)
[8] Slavin R E 2008 *Cooperative learning: Teori, Riset dan Praktik* (Bandung: Nusa Media)
[9] Kozulin A, Gindis B, Ageyev V S and Miller S M 2003 Vygotsky's educational theory in cultural context *Educ. Res.*
[10] Crawford M L 2001 Teaching Contextually in Mathematics and Science (Texas: CCI Publishing)
[11] Dewi novi a 2015 No Title *Pengembanganan Modul Fis. Berbas. Relating, Exp. Applying, Coop. Tranferring pada Mater. Alat Opt. untuk Meningkat. Kecerdasan Majemuk dan Kreat. Siswa. J. Inkuiri UNS* **4** 47–56
[12] Ultay E 2012 Implementing REACT Strategy in a Context-Based Physics Class : Impulse and Momentum Example **4** 233–40
[13] Davtyan R 2014 Contextual learning *ASEE 2014 Zo. 1 Conf.* 3–4
[14] Johnson E B 2014 *Contextual Teaching and Learning; Menjadikan Kegiatan Belajar Mengajar*

Mengasyikkan dan Bermakna (Bandung: Kaifa)
[15] Brookfield S D 2012 *Teaching for Critical Thinking: tools and techniques to help students questions their assumptions* (San Francisco)
[16] Halpern D 1998 Teaching Critical Thinking for Transfer Across Domains: Dispositions, Skills, Structure Training, and Metacognitive Monitoring *Am. Psychol.* **53(4)** 449–455
[17] Mulyasa E 2004 *Kurikulum Berbasis Kompetensi : Konsep, Karakteristik, dan Implementasi* Bandung: Remaja Rosdakarya
[18] Daryanto 2013 *Menyusun Modul: Bahan Ajar untuk Persiapan Guru dalam Mengajar* Yogyakarta: Gava Media

IOP Conf. Series: Journal of Physics: Conf. Series **1022** (2018) 012018 doi:10.1088/1742-6596/1022/1/012018

The development of thematic materials using project based learning for elementary school

M Yuliana[1*]**, S A Wiryawan**[2] **and Riyadi**[3]

[1,2,3]Post Graduate, Sebelas Maret University, Jl. Ir. Sutami 36A, Surakarta 57126, Indonesia

*Corresponding email: mitayuliana234@yahoo.co.id

Abstract. Teaching materials is one of the important factors in supporting on learning process. This paper discussed about developing thematic materials using project based learning. Thematic materials are designed to make students to be active, creative, cooperative, easy in thinking to solve the problem. The purpose of the research was to develop thematic material using project based learning which used valid variables. The method of research which used in this research was four stages of research and development proposed by Thiagarajan consisting of 4 stages, namely: (1) definition stage, (2) design stage, (3) development stage, and (4) stage of dissemination. The first stage was research and information collection, it was in form of need analysis with questionnaire, observation, interview, and document analysis. Design stage was based on the competencies and indicator. The third was development stage, this stage was used to product validation from expert. The validity of research development involved media validator, material validator, and linguistic validator. The result from the validation of thematic material by expert showed that the overall result had a very good rating which ranged from 1 to 5 likert scale, media validation showed a mean score 4,83, the material validation showed mean score 4,68, and the mean of linguistic validation was e 4,74. It showed that the thematic material using project based learning was valid and feasible to be implemented in the context thematic learning.

1. Introduction

The improvement of quality of education in Indonesia belongs to one of the educational systems that has function to increase the quality of human resources. Schools are considered as formal institutions and a part of plans which made by the government in educational area with the curriculum as the grown operation. Curriculum as one of important components in education and became the main attention of the government. The curriculum in Indonesia had some changes. The latest change of curriculum was the change from KTSP to the 2013 curriculum that contained of thematic learning. Based on the Constitution No. 20 Year 2016 about national educational system, the education in Indonesian would develop educated human who and pious to the Almighty, had noble characteristic, healthy, learnt, capable, creative, autonomous, and became democratic citizen with their responsibility. Thus, national education should have function in the optimum state as the first mode in building the nation and characters. It refers to the order of the 2013 curriculum that put forward the establishment of students' characteristics.

The 2013 curriculum which implemented in elementary schools used thematic learning with scientific approach. Thematic learning as one of the curriculum implementation models that was being

IOP Conf. Series: Journal of Physics: Conf. Series **1022** (2018) 012018 doi:10.1088/1742-6596/1022/1/012018

recommended to be implemented in the elementary grade. Having the same opinion thematic approach was one of the teaching strategy that used themes towards creating an active, interesting and meaningful learning [1]. Thematic learning is one integrated learning model which is part of learning system that enables students, individually or in groups, to be active and find scientific principles in a holistic, accented and authentic [2]. Thematic learning accommodate students to be able to comprehend the main concepts and to be able to connect between concepts in thematic learning.

The implementation of the 2013 curriculum in Indonesia also had some challenges. Abidin (2016:24) explained that one of the challenges faced by the teachers in implementing the 2013 curriculum was learning sources [4]. Learning sources considered as important components for the teachers and students in the learning process. Learning sources could be defined as information, tools and/or lesson needed by the teachers to plan and review the learning implementation [5]. The 2013 curriculumas learning sources used by the teachers nowadays, it refers to the teacher's book and students' textbook provided by the government.

In practice, the government's textbooks used by the teacher and students in the learning process seem to have some weaknesses both intheir content and materials. The current teacher's and students' textbooks used in elementary schools only contain few students' worksheets in which the materials coverage were not complete and less wide. Thus, it makes both teacher and students should find other learning sources by themselves or additional materials that can be used in the teaching and learning process. The problematic about the minimum materials of learning sources used by the teacher and students were proved from the results of interview conducted to 14 elementary school teachers who had been approved to implement 2013 curriculum in Wirosari City. Based on the results of conducted interview, 10 of 14 teachers complained that the lack of leaning sources materials for them. They also felt the difficulty in implementing the 2013 curriculum if they just relied on the learning sources provided by the government. Teachers needed thematic learning sources which were more complete in term of content and had wider material coverage in order teacher and students could use them as materials or learning sources in the teaching and learning process.

Besides that, based on the result of observation on direct observation at schools and deep interview to the teachers of Wirosari City elementary schools who taught fourth grade students in seven elementary schools in Wirosari City, it was found some facts including: 1) there was no sufficient learning sources which supported the 2013 curriculum, thus it was needed for learning sources as learning materials; 2) the schools needed the provided learning sources which appropriate with the 2013 curriculum sufficiently in the form of print out, since it was known that the material in current learning sources was not complete; 3) in the learning sources, it was needed to be inserted various and innovative learning models, in expecting the students to be motivated and learning process can be considered as successful process.

Its implementation in the elementary schools, the learning sources going to be developed bad on the 2013 curriculum and the learning activity was based on scientific approach. It is in line with Permendikbud No. 65 about Standard Process which stated that in order to strengthen scientific approach and integrated theme, the discovery learning model needed to be implemented. In the effort to encourage students to produce contextual works both individually and in group, it was advised that the learning process should use project based learning [3]. Thus, one of the solutions conducted in completing the lack of current learning sources is the development of learning sources in the form of Project Based Learning. The reason why the researcher choose Project Based Learning model because it can help students to be more active in thinking, solving problems, be more creative, and be able to work together with peers. It is in line with Abidin (2016:167) who says that Project Based Learning model is a learning model that directly involve the students in the learning process by observation activity to do and finish a certain learning project [4]. One of the strengths of this model is Project Based Learning considered as one of the learning model which was very good in developing various basic skills that should be mastered by the students including thinking skill, decision making skill, creativity skill, problem solving skill, and it also was viewed as effective method in developing students' confidence and self-management.

Some researchers have proved the effectiveness of project based learning model. The research result conducted by Aziz, et.al (2013) showed that project based learning can attract the students to involve well in their learning processes [6]. Project based learning also help them to improve their reasoning skill in learning. In addition, the research results of Woro Sumarni (2015) showed that the strength of project based learning implementation as a learning model by using scientific approach would accustom the students to conduct research process in which it was believed it could improve the quality of education [7]. In this research, it emphasized that implementing the project based learning from constructive approach, it will activate students centre learning, and thus it will ease them to transfer their knowledge to the real world situation innovatively and creatively.

Based on the result of the observation about learning sources used by the teacher and students, it was found out that there was no inserted learning models that triggering students' activity in the learning process. Thus, learning books based on Project Based Learning that will be developed as a learning source potentially toward cognitive aspect. With the use of arranged to be learning book that use learning model based on daily life problems can sharpen students' knowledge towards something new. The connection with the real life problem faced by the students can help them to ease in absorbing the knowledge. Learning book as learning source potentially effect towards affective aspect. With the use of the learning book, students' behaviour in daily activities can be implemented through practice and group discussion. Learning book as learning source potentially effect towards psychomotoric aspect. With the use learning book can train students' skill that focuses on their work in doing practice or observation, through problem based learning, the connection between problem and learning teach them to find the solution through practice or observation. Based on the explanation above, this research and development proposes the title of "The Development of Thematic Materials Using Project Based Learning for Elementary School".

2. Methodology

This research belongs to research and development method (R & D research). The model used as the basic to develop the thematic learning source in the term of Project Based Learning, and it related with the result of the adaptation of Four-D model [8]. The development of thematic learning source in the term of Project Based Learning was conducted using Four-D modification models that consisted of 4 stages, namely: define, design, develop, and disseminate.

Dealing with this R & D research, one of the step in research and development is development stage, this stage to product validation from expert toward thematic materials using project based learning for fourth grade students of elementary school that will be validated by 9 experts who consisted of 3media experts, 3 material experts, and 3 linguists experts before the initial field trial. The data collecting technique was experts' validation. The judgment scale employed here was Likert scale with 5 point scale, namely 5=very relevant, 4=relevant, 3=less relevant, 2=not relevant, 1= very not relevant. After getting the results validation of thematic materials using project based learning, then it is interpreted based on the table below:

Table 1. The Criteria of Each Component Average Score [9]

Category	Score	Criteria
A	4.20-5.00	Very Good
B	3.40-4.19	Good
C	2.60-3.39	Enough
D	1.80-2.59	Less Good
E	1.00-1.79	Very Less Good

It applied the value of feasibility of thematic materials using project based learning at least get "B" with the criteria of "Good". Thus, if the results of the expert of learning media, learning material, and linguist were "B" or "Good" then the developed product was feasible to be e used as instructional material of learning.

3. Result and Discussion

3.1 Results

The results of this study was to produce a thematic materials using project based learning for grade IV elementary school. The results of media, material experts and linguists' validation obtained as follow:

Table 2. The Validation Results

Aspects	Average Score	Criteria	Judgement
Media experts	4,83	Very good	Worthy
Material experts	4,68	Very good	Worthy
Linguist expert	4,74	Very good	Worthy

Based on the experts judgments and it was viewed from media aspect, the score was 4,83. The conversion data which belongs to the category of "very good". Based on the data conversion above viewed from media aspect, the thematic materials based on the project based learning, it was feasible to be used. Based on the experts judgments' and viewed from the material aspects, the score was 4,68. The conversion data belong to the category of "very good". Based on the data conversion above viewed from the point of material aspect, the thematic material which based on project based learning sentence was feasible to be used. Based on the experts judgments' viewed from linguistic aspect, the score was 4,74. The conversion data belong to the category of "very good". Based on the data conversion above viewed from the linguistic aspect, the thematic materials which based on project based learning, the sentences were feasible to be used.

3.2 Discussion

Based on the figure, it can be concluded that the average media validation score was 4,83 included in category A which means that it has a very good predicate. In developing this learning source, the media experts gave some suggestions such as: (1) There should be given instruction for each activity; (2) The decrease of writing activity and ask the students to be more active; (3) The choice of colors should be appropriate with color gradation; (4) The removing of upper header since it was too full. Besides that, another revision toward the media aspect, it was the pictures in the learning material and they were not big enough. Pictures were considered as visual media which used in delivering information. Shabiralyani, et.al (2015) explained that visual aids must be visible to the whole class [10]. All calligraphy and illustration must be large adequate to be seen easily by the students farthest from the aids. As for the elementary school students, pictures were really needed since they tend to be more attracted if they saw many pictures. In the process of learning material, good pictures' compositions would give effect to attract the students in learning.

The material experts provided some suggestions among the following: 1) the addition to some parts of the materials as complementary to related material, thus the material became broader and more complete; 2) to combine scientific approach with project based learning; 3) the material or activity which triggers the students to be active in bringing up unseen questions; 4) the project of based learning activities were less challenging and less motivating the students to be more active; 5) some difficult words were hard to be understood by the students and it would be better to simplify those words.

One of the revisions from the material aspect was material or activity that encouraged the students to be active in bringing up the unseen questions. Bringing up the questions was considered as one of the activities in scientific approach process. Trilling and Fadel (2009: 91) stated that the scientists observed the world through questions, while the discoverers were motivated by challenging problems [11]. In the step of bringing up the questions, the students conducted observation toward certain object. Based on their observation, they built a question which should be answered through doing research activity. After sending the question, they verified the question, made hypothesis, did the experiment, analyzed the data and made the conclusion.

The revision from the aspect of materials, it was needed to be done for the learning material based on project based learning in the learning activities. Learning activities in the learning materials which based on the project based learning needed to be combined with scientific approach. In'am and Hajar (2017) stated that the steps of scientific approach were as follow observing, asking question, reasoning, attempting, and presenting [12]. The steps of implementing the project-based learning included six stages as follows: starting with the essential question, designing a plan for the project, creating a schedule, monitoring the student and the progress of the project, assessing the outcome, evaluating the experience (National Science Foundation, 2012; Thomas, 2002):Six steps of the project based learning and five steps of scientific approach which were being integrated could produce learning activities [13]. The following table showed the combination of project based learning and scientific approach in learning material based on project based learning which has been developed.

Table 3. The Combination of Scientific Approach and Project Based Learning

Steps of a scientific approach	Steps of project based learning	Scientific steps approach combined with project based learning
1. Observing	a. Start With the Essential Question	1a Let's observe
2. Asking question	b. Design a Plan for the Project	2b Let's plan
3. Reasoning	c. Create a Schedule	3c Let's practice
4. Attempting	d. Monitor the Student and the Progress of the Project	4d Let's monitor
5. Presenting	e. Assess the Outcome	4e Let's judge
	f. Evaluate the Experience	5f Let'sreflect

Based on the revision of the aspect of materials about the combination between scientific approach and project based learning, it could be said that in the process of project based learning there was an activity where the students solved certain problem in a group. It is in line with the main point of scientific approach that it was directed to build students' ability in solving the problem, in which it is in line with the objective of project based learning. Drew (Lowe, 2007: 4) explained that research is a process conducted to solve the problem and to extend the knowledge [13]. In simply, Lowe (2007: 4-5) says that research is a third way for us to understand the world around us [14]. Based on the scientific approach process, if it is related and combined with project based learning, it will be very suitable to be implemented in elementary school.

Besides that, in the learning process with scientific approach combination and project based learning, it was needed an evaluation that could be used to measure the weakness and the effectiveness of the learning method and learning material used, to give feedback to the students, and to revise learning process. Gronlund and Gullo (2005) said that evaluation as a process of making decision about achievement, score, the success of educational program, the success of project, material quality, or the advantage of certain technique [15]. Thus, the position of evaluation was very important to draw the decision in learning process.

The arranged learning materials should be suited with the learning objectives. The learning objectives considered as a foundation for dialogue, curriculum planning, and assessment [16]. The learning material in this research was learning material based on project based learning. In order to acquire the learning objectives in the maximum state, thus the revision of learning material being conducted should follow the learning steps which are the result from the combination of scientific approach and project based learning.

The evaluation by the linguistics experts tend to advice some points, namely: 1) improving the effectiveness of grammar and vocabulary since elementary school students need simple and short linguistic ; 2) some errors in EYD, punctuation, and letters needed to be paid attention; 3) the use of linguistics in the project making activity needed to use simple linguistic. One of the revisions in the linguistics aspects was improving the effectiveness of grammar and vocabulary since elementary school students needed simple and short linguistics pattern. Reading is considered a cognitive enterprise that entails three components including reader, text and activity [17].

One of the components in reading is text. A good and easily understood text is a text which uses vocabulary which is appropriate with students' development growth. Cooper (1984) described vocabulary as being the key ingredient to reach successful reading while other researchers argued that "no text comprehension is possible, either in one's native linguistic or in a foreign linguistic , without understanding the text's vocabulary" [18]. In comprehending a text, it was needed vocabulary that appropriate with students' development. As for elementary school students, it was needed vocabulary which easily understood by them. If the vocabulary was easily understood by the students, of course the material and knowledge would be received well. The importance of vocabulary in reading comprehension was needed in order that the students got no difficulty in comprehending the text. Yorio (1971) surveyed second type of students, who stated that vocabulary was their most important problem in reading comprehension [19].

Based on the three experts' validation, it could be obtained results that the average validation score of the development of thematic materials using project based learning for elementary school as a whole was 4.75 which included in the category of Very Good and declared eligible to be used in grade IV of elementary school in learning and teaching process.

4. Conclusions

A thematic materials using project based learning for elementary school in grade IV primary school learning had received a proper statement from the media experts, material experts, and linguists that with a minimum worth of "B" with the "Good" criteria. Based on the results and discussion, it was known that the thematic book had been declared eligible by the experts where the media experts showed the score 4,83 of scale 5 included in the category of very good, the material experts gave 4,68 from scale 5 so that it categorized very well, and also it had been declared feasible with a score of 4,74 from a scale of 5 that it also belong to very good category. The researcher gave suggestion such as for grade 4 primary school teachers and students. For the teachers, they were expected to be able to use thematic materials using project based learning in learning to support the content of thematic learning. For students, they could use thematic materials using project based learning according to the directions and in order to make them enjoy the learning process well.

5. References

[1] Chon Min, Rashi and Nazri 2012 Teachers' Understanding and Practice towards Thematic Approach in Teaching Integrated Living Skills (ILS) in Malaysia. *International Journal of Humanities and Social Science p* 273

[2] Rusman 2013 *Model-Model Pembelajaran* Jakarta: PT Raja grafindo Persada

[3] Kemendikbud 2013 *Permendikbud No. 65 tentang Standar Proses* Jakarta: Kementrian Pendidikandan Kebudaayaan.

[4] Yunus Abidin 2016 *Desain Sistem Pembelajaran Dalam Konteks Kurikulum 201* Bandung: PT RefikaAditama

[5] Hamdani 2011 *StrategiBelajarMengajar* Bandung: PustakaSetia

[6] Aziz Z, Shamsuri S M and Damayanti L 2013 Project Based Learning to Pose Reasoning Skills for Year 1 Pupil Journal of Primary Education *Canadian And Center Science Of Education* **5**(4) 82-87

[7] Sumarni Woro 2015The Strengths and Weaknesses of the Implementation of Project Based Learning: A Review *Intl. J. of Sci. and Res. (IJSR)* **4**(3) 478-84

[8] Thiagarajan Sivasailam, dkk 1974 *Instructional Development for Training Teachers of Exceptional Children* Washinton DC: National Center for Improvement Educational System
[9] Sudjana Nana 2005 *Dasar-Dasar Proses Belajar Mengajar*. Bandung: Sinar Baru Algesindo
[10] Shabiralyani, Ghulam et al 2015 Impact of Visual Aids in Enhancing the Learning Process Case Research: District Dera Ghazi Khan *J. of Edu. and Pract* **6**(19)
[11] Trilling B and Fadel C 2009 *Problems as Possibilities: Problem Based Learning for K-16 Education* Alexandria: ASCD
[12] In'am, Akhsanul and SitiHajar 2017 Learning Geometry through Project based learning Using a Scientific Approach *International Journal of Instruction* **10** (1) pp 55-70
[13] National Science Foundation 2012 *Project-based Learning and Universal Design for Learning Retrieved June 21 2014 from National Center for Supply Chain Technology Education*
[14] Lowe M 2007 *Beginning Research: a Guide for Foundantion Degree Students* London: Routledge
[15] Gronlund N E 1993 *How to Make Achievement Test and Assessment* Boston: Allyn and Bacon
[16] Austin Zubin 2016 How to design and use learning objectives in clinical learning *The Pharmaceutical Journal* **296** 7885
[17] Snow C 2002 *Reading for Understanding: Toward an R&D Program in Reading Comprehension* Santa Monica: The RAND Corporation
[18] Cooper P 1984 *The Assessment of Writing Ability: A Review of Research*. Princeton. NJ: Educational Testing Services
[19] Yorio C 1971 *Some Sources of Reading Problems for Foreign Linguistic Learners. Linguistic Learning* **21** 107-115
[20] Depdiknas 2006 *Kurikulum Tingkat Satuan Pendidikan* Jakarta: Depdiknas
[21] Kemendikbud 2013 *Kurikulum 2013* Jakarta Kementerian Pendidikan dan Kebudayaan
[22] Thomas J W 2000 *A Review of Research on Project-based Learning* San Rafael California: The Autodesk Foundation

Acknowledgments
The acknowledgment is special dedicated to Prof. Dr. Sri Anitah Wiryawan, M. Pd and Dr. Riyadi, M. Si as the researcher's supervisor who had provided guidance and review. The researcher also would like to say thanks to some experts, namely Dr. Suharno, M.Pd and Dr.Roemintoyo, M.Pd, and Dr. Rukayah, M. Hum, furthermore thanks a lot to ICOSMEE which had organized an international seminar and gave me opportunity to publish this article.

IOP Conf. Series: Journal of Physics: Conf. Series **1022** (2018) 012019 doi:10.1088/1742-6596/1022/1/012019

Reading for tracing evidence: developing scientific knowledge through science text

R M Probosari[1*], F Widyastuti[2], S Sajidan[3], S Suranto[4] and B A Prayitno[5]

[12345]Science Education Doctoral Program, Faculty of Teacher Training and Education, Sebelas Maret University, Jl. Ir. Sutami No. 36A Surakarta, Indonesia

*Corresponding email :riezkymaya@fkip.uns.ac.id

Abstract. The purposes of this study were to investigate students' learning progression on reading activity, science concept comprehension and how they imply it in scientific communication in the classroom. Fifty-nine biology education students participated in this study. This classroom research was developed to portray students' reading activity, factors affecting reading comprehension, and the development of reading motivation. Qualitative analysis was used to describe the whole activities, involve the instruction, process and the product of reading activity. The result concluded that each student has their own way in interpreting the information from scientific text, but generally, they can filter and apply it in their argument as a part of reasoning and evidence. The findings can be used to direct reading activity to the goal of inquiry in order to support the nature of reading as evidence.

1. Introduction

Science is a structured activity to gain knowledge in the real world systematically, so it requires a scientific method to observe, identify, explain and investigate problems. On the other hand, science also can not detach from the discourse includes the use of language, especially written language to formulate scientific discoveries and build a theory that explains natural phenomena, communicates knowledge, principles, ways of thinking and how to make arguments to others [1]–[3].

1.1 The essence of reading science text

Reading is basically an activity that can be cultivated, but the essence of reading is not merely putting together words, sentences or paragraphs, but looking for the essence or main idea contained in the text[4]. Cognition can be built and developed through appropriate strategies reading because the main purpose of reading literature critically is to bring innovation and move students as critical agents and subjects in making decisions [5]–[7].

However, reading scientific literature is more difficult than reading a popular article or non-scientific reading. People who able to understand the reading of science can interpret the text into semantic information and knowledge into the brain so that they get the equal understanding with the author, in the special degree.[8]. Learning to read science texts not only leads to how to understand the substantive content of science but also how does the epistemology [9], [10].

1.2 Reading as inquiry
The cumulative nature of science makes a researcher or potential researcher need to know everything that happens before they learn what's new. This can be done through reading science, especially research articles published from reputable journals. However, real cases or problems are often different from those from previous studies for various reasons, so there is a need for compromise in adapting the scientific literature used [10]–[12].

Reading scientific texts requires several strategies to help students think about what they are reading. Students do not only think of questions arise after reading but at the same time should be able to predict and anticipate what kind of material they should read after that to get what they want. Students view general reading content by skimming through titles, abstracts, content, and illustrations, then marking problems encountered. Next, the students read more detail to find the answer. After reading, they make some citation or reading paraphrasing and then answer the question with the conclusion of the citation and complete it with another source needed [13, 14].

Reading is a multidimensional activity involves complex cognitive processes [15]. Reading action by a scientist is categorized into inquiry activities so that reading as part of inquiry should be included as part of science learning, but unfortunately, science curriculum and science learning have not supported this yet[9, 16, 17]. The lecturer should understand the "message" present in the literature to support the students understanding of the reading used. Reputable science readings play an important role in introducing authentic examples, the complexities and actual sections of scientific arguments[18], [19]. A proper reading of science provides a strong foundation for students in arguing activity, especially in terms of formulating claims, showing evidence, reasoning, and positioning, whether they support a statement or not.

The purposes of this study were to investigate students' learning progression on reading activity, science concept comprehension and how they imply it in scientific communication in the classroom.

2. Methods
2.1 Research Design
This study used a qualitative methods research design [20], [21]. Qualitative analysis data were used to explore students' reading activity, science concept comprehension, factors affecting reading comprehension, and the development of reading motivation.

2.2 Participants
This research took place in a 2nd-grade biology education classroom in a state university in Central Java and taught by team members. The partners were 59 students (50 females and 6 males) who had taken plant embryology class. They are not experienced with related communication, especially on how to build scientific arguments to convince others. Based on the previous observation, the class discussion has been dominated by only a few students, and most of them rely on lecturer as their major learning resource.

2.3 Data collection and analysis
Data were taken during the eight-week period from many sources including students' writing assignments, structured interviews and lecturer's notes. Data used to capture students' reading activities and habits, and how they use the information to support their statement as a part of evidence beyond analysing the difference between the before and after the intervention. Intervention by lectures was done in two weeks to develop students' skill of science reading includes selecting necessary points, utilizing information and locating material in their library. To portray students' reading activity and the use of information, data were analysed focusing on amounts of reading, process of reading, and developing skill in using information.

3. Results and Discussion
The lessons focused on three topics, i.e. self-infertility, apomixes and embryo rescue. Observation was done to examine the importance of reading in science education. Findings showed several facts from the data: (1) students' reading habit, included amount of references and citation management, (2)

process of reading, (3) science concept comprehension, and (4) developing skill in using information as an evidence. Writing assignments analysis showed students' reading behaviour as follows:

Table 1. Students' Reference Resources and Reading Activities Before The Interventions

Amounts of citations	Type of references	Use of Citation
< 3 references (11 students)	Textbook, primary literature, research journal (8 students)	Support a claim (27 students) As data (59 students) Agreement (59 students)
3 – 5 references (26 sudents)		
6 – 8 references (14 students)	Textbook, research journal (14 students)	
> 8 references (8 students)	Textbook, research journal and web/blog (37 students)	

Table 1 described that based on the number of references, most students refer less than 8 sources to support the paper he wrote. Based on the reference type, only 8 students complete a complete three genre of reading including textbook literature, primary literature, and research journal. These three genres have been recommended as an ideal reference source for scientists and science students [17]. Popular magazines, web or blogs that are referred to by students though can be used as reference sources, are actually addressed to the general public, not for the science community. Students who refer more web pages or popular news shows low motivation and reading habit because they take information instantly from internet browsing or refer to the conclusions of previous research without seeing scientific process happened in it [22], [23], [12]. It has an impact on the students' lack of understanding of science comprehensively, especially in appreciating how scientists think and act, how a study leads to new findings and how to use the results of research as empirical foundation of further research[24], [25].

In connection to how students locate the information and use it, it is found that more students use information from references to express approval of a statement or use data taken to indicate support for a particular claim. However, many students do not yet understand how to distinguish raw data from evidence, whereas raw data can not be regarded as an evidence without a strong theoretical foundation [26], [27]. This is apparent with the writings of the students using the words "I support the research ... which shows that ...", "I agree with the results of the research which shows that, ...", "I agree with the research ... which results ...". Some examples of the use of information from science reading are shown in Figure 1.

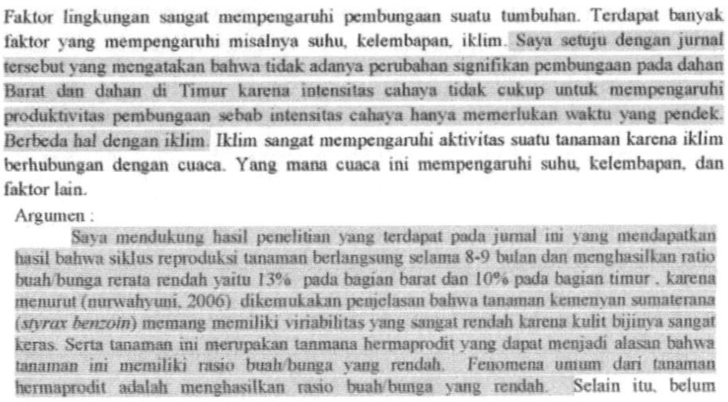
Faktor lingkungan sangat mempengaruhi pembungaan suatu tumbuhan. Terdapat banyak faktor yang mempengaruhi misalnya suhu, kelembapan, iklim. Saya setuju dengan jurnal tersebut yang mengatakan bahwa tidak adanya perubahan signifikan pembungaan pada dahan Barat dan dahan di Timur karena intensitas cahaya tidak cukup untuk mempengaruhi produktivitas pembungaan sebab intensitas cahaya hanya memerlukan waktu yang pendek. Berbeda hal dengan iklim. Iklim sangat mempengaruhi aktivitas suatu tanaman karena iklim berhubungan dengan cuaca. Yang mana cuaca ini mempengaruhi suhu, kelembapan, dan faktor lain.

Argumen :

Saya mendukung hasil penelitian yang terdapat pada jurnal ini yang mendapatkan hasil bahwa siklus reproduksi tanaman berlangsung selama 8-9 bulan dan menghasilkan ratio buah/bunga rerata rendah yaitu 13% pada bagian barat dan 10% pada bagian timur . karena menurut (nurwahyuni, 2006) dikemukakan penjelasan bahwa tanaman kemenyan sumaterana (*styrax benzoin*) memang memiliki viriabilitas yang sangat rendah karena kulit bijinya sangat keras. Serta tanaman ini merupakan tanmana hermaprodit yang dapat menjadi alasan bahwa tanaman ini memiliki rasio buah/bunga yang rendah. Fenomena umum dari tanaman hermaprodit adalah menghasilkan rasio buah/bunga yang rendah. Selain itu, belum

Figure 1. The Way Students Locating The Information and Using It Before The Interventions

The figure showed students' weakness in taking scientific information from science reading as a component of evidence, as they review the results but do not integrate it with the discussion and use it to support claims. There are still many students who cannot distinguish between "data" and "evidence".

Intervention is done through tutoring science reading by teacher, including selection of up-to-date reference sources and suitable for scientific community, use of citation management, reading strategy to develop comprehension and use the information as a part of evidence[22], [23], [28]. After the intervention, the data obtained as follows
:

Table 2. Students' Reference Resources and Reading Activities After The Interventions

Amounts of citations	Type of reference	Use of Citation
< 3 references (0 students)	Textbook, primary literature, research journal (32 students)	Support a claim (59 students) As evidence (59 students) Positioning (59 students)
3 – 5 references (0 students)	Textbook, research journal (27 students)	
6 – 8 references (21 students)	Textbook, research journal and web/blog (0 students)	
> 8 references (38 students)		

After the intervention was done, the students' behaviour changes especially in choosing the scientific reading source. Sources from non-reputable articles and blogspot are no longer used, and more and more reference counts are used, especially from the Adapted Primary Literature (APL) category, i.e. research articles. This source contains elements of evidence and reasons inside to reinforce claim and arguments [17], [25], [29]. The observation results showed that students who make citations of accurate science text genre are more skilled in expressing their ideas and able to provide clear scientific explanations, a strong theoretical foundation is accompanied by accurate evidence. Some student activities in the discussion activities are described as follows:

Perbedaan dapat diidentifikasi dari beberapa aspek, yaitu stomata daun, bentuk spora, pollen, buah, biji dan embrionya. Perbedaan tingkat ploidy menunjukkan perbedaan ukuran sel dan stomata. Sebagaimana penelitian Griffiths *et al.* (1996) pada tanaman tembakau yang menunjukkan bahwa jika tingkat ploidy semakin tinggi, maka semakin besar ukuran sel dan stomata (Damayanti, 2007). Penelitian yang dilakukan oleh Cohen (2015) menggunakan 4 jenis kondisi *Hylocereus*, yaitu *H. monacanthus* diploid, autotetraploid line D-27, allotriploid S-75, dan autohexaploid line D-2.3. Perbandingan dari keempat kondisi tersebut akan difokuskan pada *H. monacanthus* autotetraploid line D-27 dan allotriploid S-75. Berikut adalah gambar stomata daun, bentuk spora, dan pollen dari *H. monacanthus* diploid (A), autotetraploid line D-27 (B), allotriploid S-75 (C), dan autohexaploid line D-2.3 (D) (Cohen, Fait, & Tel-Zur, 2013)

Figure 2. Students' Explanation and Evidence

The synergy between science and literacy is undeniable in science learning, proved by a better understanding of students' science concepts as they become more familiar with reading appropriate scientific references. In the discussion activities, it appears that students who are actively answering are students who have more references. The claim they propose is clearer with apparent empirical and theoretical evidence. Some sample student dialogs are summarized in Figure 3 below:

Pertanyaan: apa faktor penyebab autopoliploid dan allopolyploid? Karena tanaman yang digunakan oleh presentator ada yang memiliki jalur poliploidi yang autopoliploid dan ada yang allopoliploidi?
Jawaban: factor gen atau DNA dan kadar kolkisin
Alasan dan bukti: penyebab munculnya kedua jalus tersebut dalam satu genus tanaman adalah perbedaan gen atau DNA penyusun kedua spesies dan kadar kolkisin yang kemungkinan ada pada kedua spesies tersebut ditemukan dalam bentuk dan kadar serta keaktifan yang berbeda-beda. Perbedaan kadar dan keaktifan masing-masing spesies nantinya berdampak pada jalur pemebentukan poliploidinya. Maka munculnlah dua kelas poliploidi yaitu autoploiploidi dan allopoliploidi (Wang et al, 2006)

1. Fajar : Bagaimana perbedaan konsentrasi efektif pada perlakuan kolkisin antara tanaman bunga dengan buah dan bagaimana pengaruh kolkisin tersebut ?
Jawaban awal : Pemberian kolkisin pada tanaman bunga dan buah itu berbeda karena pada tanaman bunga membutuhkan sedikit lebih banyak kolkisin agar bunga dapat bertahan lama (mekar) sedangkan tanaman buah membutuhkan lebih sedikit kolkisin agar bunga cepat menghasilkan buah.
Jawaban akhir : Respon terhadap pemberian kolkisin pada masing-masing tumbuhan berbeda, tergantung pada tingkat ploid individu, struktur genom, modus reproduksi, perenniality, dan organ tanaman yang dibudidayakan (Sattler et al., 2016). Sehingga jika tanaman bunga membutuhkan konsentrasi sedikit lebih banyak daripada tanaman buah. Pemberian kolkisin berpengaruh pada proses pembungaan yang relative lama sehingga konsentrasi yang dibtuhkan tanaman bunga dan buah berbeda (Yulianti et al., 2014). Pengaruh kolkisin yaitu semakin tinggi konsentrasi kolkisin yang diinduksikan maka ukuran tanaman lebih kecil dibanding tanaman kontrol. Terhambatnya pertumbuhan tunas jeruk siam simadu karena proses pembelahan sel yang abnormal dikarenakan sel yang mengganda. Pemberian kolkisin pada tanaman dapat mengakibatkan penundaan pertumbuhan akibat jaringan yang rusak dan memerlukan waktu yang relative lama untuk tumbuh. Selain itu pemberian kolkisin yang berlebihan dapat menyebabkan kematian pada tanaman (Yulianti et al., 2014).

Figure 3. Students' Dialogue in Classroom Discussion

The facts clearly illustrate the synergy between science and literature. Students are no longer accustomed using the phrase "in my opinion" or "I think, .." or other expressions of doubt, but using "previous research shows that ...", "from the experiment ... .." which shows evidence and arguments clear. This shows that students' understanding of the concept of science increases as they become more experienced with scientific reading. Students sportively want to revise their answers when they get a more accurate reference. In relation to the search for evidence to support the assertion, students should be able to link textual content with previously owned knowledge structures, build meaning, represent knowledge through different perspectives, and retain their interests and motivations[17], [29]

4. Conclusions

The result concluded that each student has their own way in interpreting the information from scientific text, but generally they can filter and apply it in their argument as a part of reasoning and evidence. The findings can be used to direct reading activity to the goal of inquiry in order to support the nature of reading as evidence.

5. References

[1] D S McNamara and P Kendeou 2011 Translating advances in reading comprehension research to educational practice *Int. Electron. J. Elem. Educ* **4**(1) pp 33–46

[2] M Ness 2016 When readers ask questions: inquiry-based reading instruction *Read. Teach* **70** (2) pp 189–96

[3] A Styhre 2016 Scholarly reading as professional practice: A reappraisal *Scand. J. Manag* **32**(3) pp 121–6

[4] P Deane *Writing Assessment and Cognition* New Jersey 2011

[5] P Kendeou, P Van Den Broek, A Helder and J Karlsson 2014 A cognitive view of reading comprehension: Implications for reading difficulties *Learn. Disabil. Res. Pract.* **29**(1) pp 10–6

[6] P. Van Den Broek and C. A. Espin 2012 Connecting cognitive theory and assessment : measuring individual differences in reading comprehension *School Psych. Rev.* **41**(3) pp 315–25,.

[7] J Zhang, X Luo, L Lu and W Liu 2012 An acquisition model of deep textual semantics based on human reading cognitive process *Int. J. Cogn. Informatics* **6**(2) pp 82–103

[8] M Cao, J Tian, D Cheng, J Liu and X Sun 2016 What makes it difficult to understand a scientific literature? in *Proceedings - 2015 11th International Conference on Semantics, Knowledge and Grids, SKG 2015* pp 89–96

[9] S P Norris, H Falk and M Federico-Agraso 2009 Reading science texts — epistemology, inquiry, authenticity — a rejoinder to jonathan osborne *Res. Sci. Educ.*, **39** 3 pp 405–10

[10] N Bryce 2011 Meeting the reading challenges of science textbooks in the primary grades *Read. Teach.* **64**(7) pp 474–85
[11] M Enfield 2014 Reading scientifically: practices supporting intertextual reading using science knowledge *J. Sci. Teacher Educ.* **25** 4 pp 395–412
[12] D J Ford 2009 Promises and challenges for the use of adapted primary literature in science curricula: Commentary *Res. Sci. Educ* **39**(3) pp 385–90
[13] D Jairam, K A Kiewra and M P K. *Marxhausen SOAR versus SQ3R : a test of two study systems* 2013
[14] G T Martin 2002 Reading, writing, and comprehending : encouraging active reading in the science classroom *Sci. Teach* **69**(7) pp 56–9
[15] E H Jeon and J Yamashita 2014 L2 reading comprehension and its correlates: A meta-analysis," *Lang. Learn* **64**(1) pp 160–212
[16] L M Phillips and S P Norris 2009 Bridging the gap between the language of science and the language of school science through the use of adapted primary literatur *Res. Sci. Educ.* **39**(3) pp 313–19
[17] A Yarden, S P Norris and L M Phillips 2015 Adapted Primary Literature The Use of Authentic Scientific Texts in Secondary Schools **22**. *Dordrecht: Springer Science+Business Media Dordrecht*
[18] C Poon 2014 Five decades of science education in singapore in inquiry into the singapore science classroom Singapore: *Springer* pp 1–25
[19] U Wingate 2012 Argument!' helping students understand what essay writing is about *J. English Acad. Purp* **11**(2) pp 145–154
[20] J W Creswell *Research Design: Qualitative, Quantitative, and Mixed Methods Approaches*. California: SAGE Publication 2014
[21] J W Creswell *Qualitaive Inquiry and Reseach Desgin: Choosing Among Five Approaches*. London: SAGE Publication 2007
[22] E T Troscianko 2013 Reading imaginatively: The imagination in cognitive science and cognitive literary studies *J. Lit. Semant* **42**(2) pp 181–98
[23] D L Jordan *Teaching Reading in Science A Supplement to Teaching Reading in the Content Areas Teacher ' s Manual (2nd Edition)* Alexandria: Association for Supervision and Curriculum Development 2001
[24] G Cervetti and P D Pearson 2012 Reading, Writing and Thinking Like a Scientist," *J. Adolesc. Adult Lit.* **55** April pp 580–86
[25] H Falk and A Yarden 2009 Here the scientists explain what i said.' Coordination practices elicited during the enactment of the results and discussion sections of adapted primary literature *Res. Sci. Educ* **39**(3)pp 349–83
[26] J Leach 1999 Students' understanding of the co-ordination of theory and evidence in science," *Int. J. Sci. Educ* **21**8 pp 789–806
[27] W A Sandoval and K A Millwood 2010 Cognition and instruction the quality of students' use of evidence in written scientific explanations the quality of students' use of evidence in written scientific explanations **23**(1) pp 23–55
[28] M Sailors and L R Price Development That Supports the Teaching of Cognitive Reading 1971
[29] S P Norris, H Falk, M Federico-Agraso, M P Jiménez-Aleixandre, L M Phillips, and A Yarden 2009 Reading science texts - Epistemology, inquiry, authenticity - A rejoinder to Jonathan Osborne *Res. Sci. Educ* **39**(3) pp 405–10

An analysis of logical thinking using mind mapping

S Swestyani[1*], M Masykuri[2], B A Prayitno[3], Y Rinanto[4] and S Widoretno[5]

[1,2,3]Science Education Department of Post Graduate Programme, Faculty of Teacher Training and Education, Sebelas Maret University, Jl. Ir. Sutami 36A, Surakarta 57126, Indonesia

[4,5]Biology Education Department, Faculty ofTeacher Training and Education, Sebelas Maret University, Jl. Ir. Sutami 36A, Surakarta 57126, Indonesia

*Corresponding email: swestyani@yahoo.co.id

Abstract. Brains can remember information in different forms, i.e images, symbols, sounds, and senses, and the information is connected by logical gate. This information needs imagination and association to construct new meaningful images. The purpose of this research was to describe a method of teaching which based on Tony Buzan's mind mapping technique. This research showed how mind mapping could be used to measure students' logical thinking and how mind mapping could promote students' understanding in meaningful way. The test of mind mapping that involved 31 students of XI grade in SMA Batik 2 Surakarta was used as the data collecting method in this research. Then, the Ohassta's mind mapping rubric was used to analyze the structure and content of mind mapping. The rubric includes four aspects, i.e knowledge, communication, thinking, and application. A qualitative analysis Miles and Hubberman's was used to assess the obtained data. The result showed that the percentage of knowledge aspect was 53,23 %, communication aspect was 28,33 %, thinking aspect was 28,33 %, and knowledge aspect was 41,53 %. Mind mapping makes logical thinking visible so that the quality of learning that has occurred can be seen and explored. Using mind mapping in the course of teaching means that learning is no longer a complex and intractable process, measurable is not only by proxy but also by an observable phenomenon.

1. Introduction

Education in the 21st century is needed to practice students' problem-solving skills and thinking skills. A problem must have a solution, but this requires a series of processes. One's mental operation activities in solving a problem is considered as a logical thinking [1], but it is also said that a logical thinking is one's ability to solve a problem scientifically [2]. Logical thinking is a s a rational process of brain by which people find correct conclusions, but it is one of the learning concepts being the most difficult to achieve [3]. It is achieved through the development of the logic capability and a rational relation between the different factors involved in every particular situation. Therefore logic, from this perspective, can be considered like the standardization of human thinking [4]. Logic reasoning, like a cognitive central component, depends on the theories of understanding, memory, learning, visual perception, planning, problem solving and decision-making [5].

Problem solving logically is a search process through known data, information is added to data in order to complement the basic file of this particular subject [6]. For example, to solve a math problem the brain applies logical reasoning: first it searches for the file how to apply mathematics, until it founds the information related with the problem, but if it does not find the information the requirement itself guides brain to make science in order to find and discovering information, thus it will fill the blanks it has in its files. This process is possible because the person has been educated in logic and he has developed a reasoning that allow him following or constructing a road, through steps carefully structured and taking care that all of these steps is supported firmly on previous knowledge. This problem-solving step requires a logical path of thinking visible from the hierarchy, a system that organizes information in a class, starting from the most general to the specific [7]. The hierarchy is poured in lines, symbols, words and pictures summarized in the mind map [8]. The mind mapping technique developed by Buzan [9] is a technique which presents associations among concepts, thoughts and information through a network or a non-linear diagram, using verbal and symbolic elements [10]. The main principles of making mind maps are as follows: 1) first the central theme (i.e. the main idea) defined and put in the center, 2) key ideas radiate out from the center, like the branches of a tree, 3) the branches contain key drawing or words written in capitals over the line. Each basic idea sprouts a further set of ideas, connected by arrows, like twigs at the end of a branch, 4) the branches for a network of nodes. At this point, words rather than sentences have to be used for an effective mind map and colors can be beneficial for visual impact [11]. In this way you can construct visual and meaningful relationships between ideas, which will assist you in recall, review and understanding [8].

The assumption of mind mapping is based on the associative memory model. The human brain is equipped with structures enabling the storage of some information [12]. Each piece of information reaching the brain is connected to other information by means of associations, forming a network of interconnected elements. This model reflects a system of connections between neurons observed during information processing.It is possible to add words, colors and visual appearances in order to help the conceptualization and contextualization of the brain's ideas with other ideas for the most effective mind mapping. Therefore, the emerging figure can be said to be like a colorful tree, as it binds all ideas together. These maps with a tree-like or hierarchical appearance are constructed in such a way as to account for a number of issues, such as a story's flow, the central topic, cause and effect relationships or the relationships of ideas to each other in a clear way[13].

As well as a tool for measuring logical thinking mind map can also be used to improve students' conceptual understanding. Generally, words are used as tool of thinking by human and thus neglect to combine non-verbal pictures, chart, concept map and semantic network. Mind mappingg is helpful for problem-solving process and deduction. Under visually and brainstorming provided by pictures and characters, one is able to use both hands and brain. Thoughts are evoked while drawing. Messages obtained from the process can be attached to the concepts and trigger learning motivation[14]. If students are taught to use these non-verbal symbols to think (so-called "Abstract Thinking") and being equipped with this ability, creativity, comprehension and problem-solving ability will be enhanced [15].

Mind map as an effective note-taking method used to look at the profile of logical thinking ability of students [8]. This paper shows how to know the profile of students' logical thinking ability through mind mapping analysis using rubric of ohassta. Mind map has four aspects: knowledge, communication, thinking, and application [16].

2. Methodology

In this research, a qualitative approach with the type of a descriptive qualitative was employed. The subjects were the 31 second semester students of XI grade in SMA Batik 2 Surakarta, Indonesia. The data were obtained trough observations during the test of mind mapping in order to understand the student logical thinking. Then, the structure and the content of the mind mapping are analysed using ohassta's mind mapping rubric. The rubric includes four aspects, i.e knowledge, communication, thinking, and application. Each aspect has four levels: level 1, level2, level 3, and level 4. A qualitative analysis Miles and Hubberman's [17] was used to assess the obtained data, performed in three components include data reduction, data presentation, and conclusion. Data reduction starts from the collection of students' mind map results and calculates the percentage of logical thinking from the quality of mind map. The quality of mind map is calculated from the score of each aspect divided by the total score of each aspect of mind map then multiplied by 100%. Presentation of data is described in the form of calculation of students' mind mapping scores and logical thinking ability diagrams. Verification is taken from the conclusion of logical thinking data viewed from the quality of students mind map. An analysis of the document of the results of the tests was also made to analyse the series of stages in the student logical thinking. The percentage obtained is then interpreted according to criteria according to Arikunto (2009) which is excellent (81% -100%), good (61% -80%), enough (41% -60%), less (21% -40%), and very less (<21%).

CRITERIA	PERFORMANCE INDICATORS (Observable descriptors indicating extent to which a criterion is met.)			
	Level 1	Level 2	Level 3	Level 4
Depth of coverage (Knowledge)	• Bare minimum of content covered • No extension of ideas evident	• Shows a basic level of coverage of key ideas only • Attempts extension of a few ideas	• Shows a solid grasp of most of the content • Shows extensions of most key ideas	• Shows a solid grasp of all the content covered • Extensions of the key ideas show a deep understanding of the content
Central Image (Communication)	• Present but difficult to separate from other information	• Present but not clearly related to key idea	• Clear use of picture or image that relates to key idea	• Stands out meaningfully and grasps the key idea through metaphor or humour
Ideas have key images (Communication)	• A little evidence of key images. Has only a few keywords	• Images and keywords are evident, but either too few or some are imprecise	• Images and key words clearly show an understanding of the content	• Images and key words clearly and dynamically show an understanding of the content. (One or more of: use of metaphor, humour, cut-outs from magazines, clipart, illustrations.)
Colour or codes or links used to illustrate connections between ideas (Thinking)	• A little use of colour, codes or links to illustrate connections between ideas	• Obvious attempt is made to use colour, codes or links to enhance clarity and memory. Still some inconsistency of application	• Clearly uses colour, codes, or links to clarify connections and to assist with memory for most aspects of Mind Map	• Effectively uses colour, codes, or links to meaningfully clarify connections for all aspects of Mind Map
Ideas radiate out from central image and from most to least complex (Application)	• Some ideas are connected to and radiate out from centre • Some confusion in moving from most to least complex	• All ideas radiate out from centre • Still some confusion in moving from most to least complex	• Ideas clearly connect to central image and ideas • Generally moves from most to least complex	• Ideas clearly connect to central image and ideas • Consistently and accurately shift from most to least complex

Figure 1. The Mind-Map Rubric [16]

3. Result and Discussion

Observation to know the logical thinking of students is done by analysing the mind map on human reproduction system material. The results of the later observations are accumulated into achievements of the students' logical thinking percentage. Figure 2 shows the mind-map results of students with low achievement in thinking and communication but have high achievement in knowledge and application aspect. It seen that the mind map has a little evidence of key images, has only a few keywords, a little use of colour, and codes or links to illustrate connections between ideas. Figure 3 shows the mind-map results of students with high achievement in thinking, communication, knowledge, and application aspect.

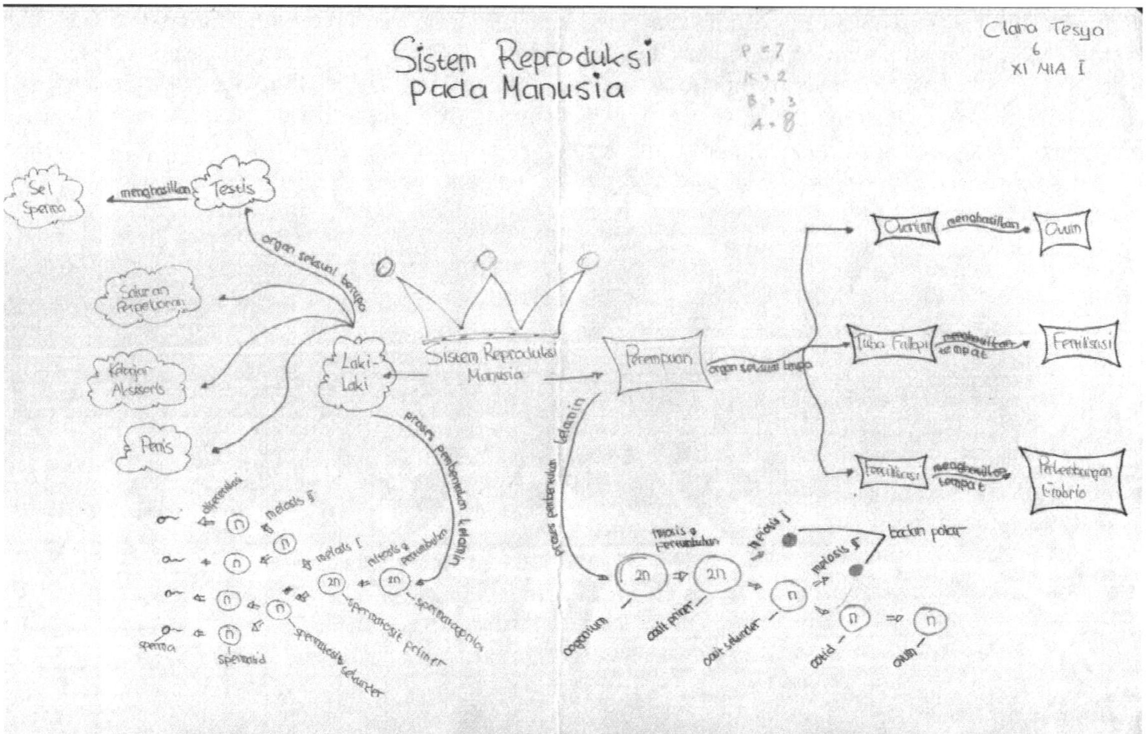

Figure 2. The Mind-Map Results of Students With Low Achievement in Thinking And Communication

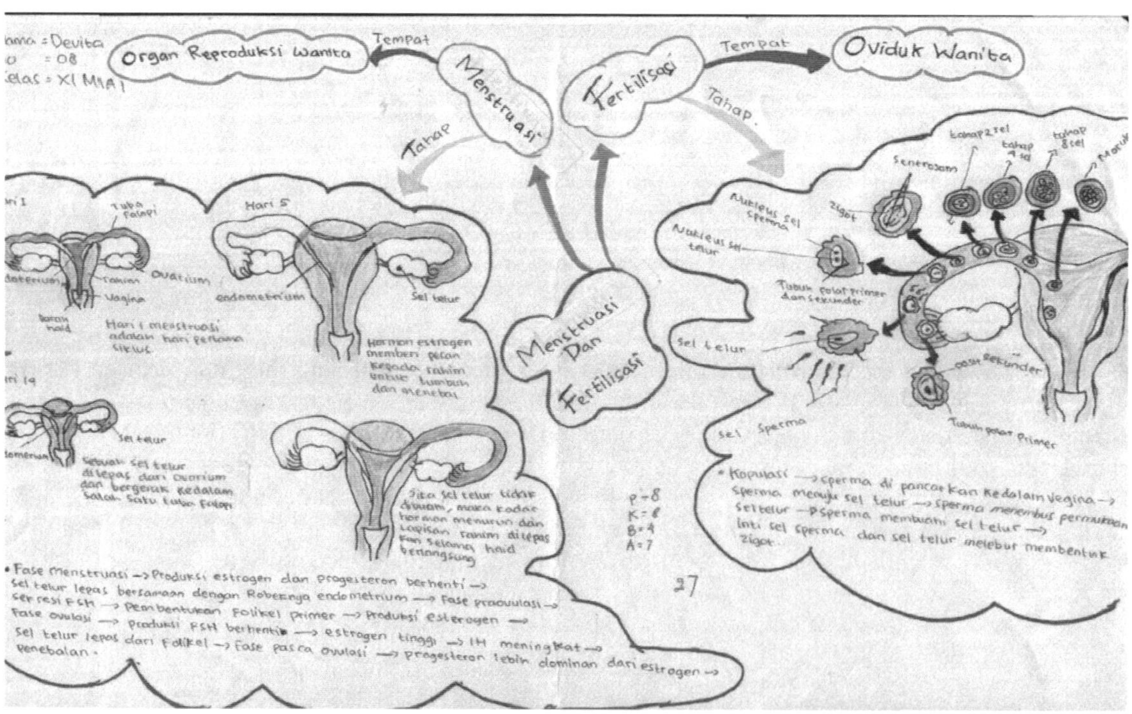

Figure 3. The Mind-Map Results of Students with High Achievement in Thinking and Communication

Achievement percentage of logical thinking aspect from 31 students can be known from table 1. Based on table 1 can be seen that highest achievement is knowledge aspect that is 55%, second rank is application 42,92%, and at bottom rank is communication and thinking aspect that is 28,33%.

Table 1. The Percentage of Achievement of Each Aspect of Logical Thinking

No.	Logical Thinking Aspects	Percentage of achievements (%)	Criteria
1.	Knowledge	55%	Enough
2.	Communication	28,33%	Less
3.	Thinking	28,33%	Less
4.	Application	42,92%	Enough

The percentage of achievement of each aspect of logical thinking is formed from the achievement of diverse scores of students. The logical thinking score of each student is described in figure 3. The scores on each aspect range from 0-8. Achievement of the highest value of knowledge aspect is 6 and lowest 3. Achievement of highest communication aspect value is 6 and lowest 2. Achievement of highest thinking aspect value is 2 and lowest 1, while achievement of highest application aspect value is 5 and lowest 2.

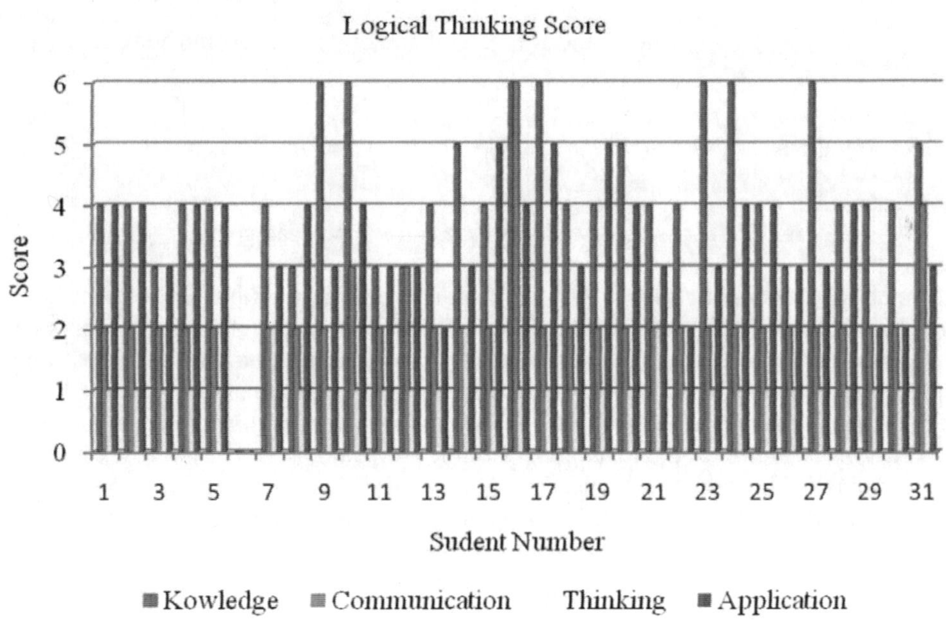

Figure 4. Logical Thinking Score

The percentage of the number of students by value category of each aspect can be described in Figure 4. Based on Figure 4, percentage of the number of students who categorized less still dominate in aspects of communication, thinking, and application. In the dominant category knowledge aspect is sufficient. The percentage of the number of students who categorized enough on the knowledge aspect of 48.39%. The percentage of the number of students who categorized less on the communication aspect of 90.23%, the percentage of students who categorized less on the aspect of thinking of 83.87%, and the percentage of the number of students who categorized less on the aspect of the application of 51.61. There are no students who categorized excellent on the aspect of knowledge and there are no students who categorized both on aspects of communication, thinking, and applications. There is a percentage of the number of students categorized very less because there is a student who does not make a mind map.

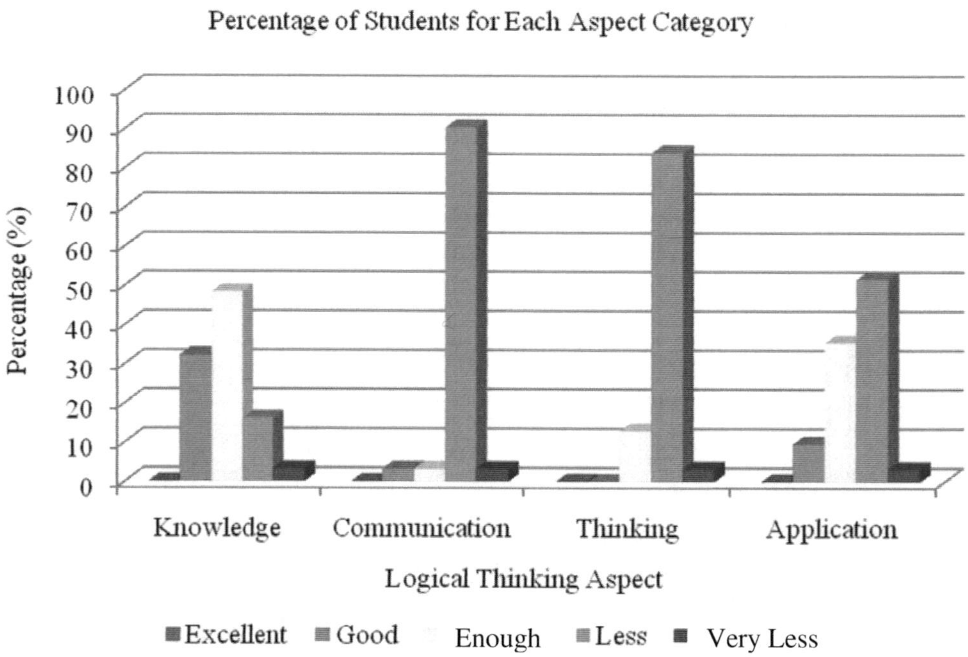

Figure 5.Percentage of Students for Each Aspect Category

Mind map is used as a measuring tool to see the profile of logical thinking ability because mind map is a record that contains the demands of thinking and students learn information. The students' learning information includes the depth of the content of the material learned during the learning and translation of the content. Student learns information can be seen from the aspect of knowledge and application. The thinking demands of the students include the demands of thinking from the common to the special and the visualization of the material in the form of symbols and keywords. The thinking demands of students can be seen from the aspects of communication and thinking. The knowledge aspect includes the depth of the content. Achievement of the knowledge aspect affects the performance aspect of the application. This is because both aspects include understanding the concept of students in the form of depth of material content and the elaboration of material concepts. Aspects of communication include visualizing the information that students gain when learning into symbols. The achievement of communication aspect affects the achievement of thinking aspect. This is because the two aspects include the visualization of knowledge into symbols and keywords that represent the components of matter and the relationship between words.

Low achievement of communication and thinking aspect affect to the students understanding of learning concepts. Achieve [3]. It is achieved through the development of the logic capability and a rational relation between the different factors involved in every particular situation. Logical reasoning depends mainly on the skill for structuring and formulating logical procedures and applying inference processes using a precise language [19]. There is a relation between the concepts of reasoning and logical thinking. He considers the reasoning like a common definition of thinking, which sometimes is used like a synonym in the broad sense of the word thinking [20].

4. Conclusions

This research result shows that students under the low category possesses a low ability in logical reasoning, meaning that they cannot make use of logical thinking stages viewed from the mind mapping result. Those under the medium category may be said that their logical thinking pattern is relatively logic, though some of their thinking is less logic, while those under the high ability category may said to have high logical thinking. Mind mapping is a thinking method which stimulates thinking and helps integrate thoughts and information, and it is also a strategy of thinking visualizing concepts.

Mind mapping is very important to be applied in learning because with mind mapping students will be able of widening the range of things they know and they understand, of promoting the self-knowledge, of understanding problems and presenting efficient and effective to daily problems. Therefore, educational systems must give the required importance to these fields and including them relationally in the course plans.

5. References

[1] Piaget J 1969 *The origins of intelligence in children* New York: International University Press
[2] Abdulkadir, Cagru A and Lutfi 2013 An analysis of mathematcis teacher candidates logical hinking level: Case of Turkey *J. Edu. Instruct. Stud. World.* **38**
[3] R Haygood and L Bourne 1965 Attribute and rule-learning aspects of conceptual behavior Psychological Review **7**(2) 175-95
[4] E Dahlin, L Nyberg, L Bäckman and A Neely 2008 Plasticity of executive functioning in young and older adults: Immediate training gains, transfer, and long-term maintenance Psychology and Aging **23** 720-30
[5] A Damasio 2003 *Looking for Spinoza: Joy, Sorrow and the Feeling Brain* Harcourt Brace & Co
[6] Kıncal R Y and Deniz Yazgan A 2010 Investigating the formal operational thinking skills of 7th and 8th grade primary school students according to some variables *Elementary Education Online* **9** 723-33.
[7] Adodo S O 2013 Effect of mind-mapping as a self-regulated learning strategy on students' achievement in basic science and technology *Mediterr. J. Soc. Sci.***4** 6
[8] Buzan Tony 2007 *Buku Pintar Mind Map* Jakarta: PT GramediaPustakautama
[9] Fah L Y 2009 Logical thinking abilities among form 4 students in the ınterior division of Sabah, Malaysia *J. Sci. Math. Edu. South. Asia* **32** 161-87
[10] DePorter, Bobbi and Mike Hernacki 2001 *Quantum Learning: MembiasakanBelajarNyaman danMenyenangkan* Bandung: Kaifa
[11] Riswanto and Putra P P 2012 The use of mind mapping strategy in the teaching of writing at SMAN 3 Bengkulu *Ina.Int. J. Hum.Soc.Sci.* **2**(21)
[12] Edward Caroline 2009 *Mind Mapping untuk Anak Sehat dan Cerdas* Yogyakarta: Sakti.
[13] Evrekli E, İnel D and Balım A G 2011 A research on the effects of using concept cartoons and mind maps in science education *Necatibey. Fac. Edu. Elect. J. Sci. Math. Edu.***5** 58-85.
[14] Jones B D, Ruff C, Snyder J D, Petriche B and Koonce C 2012 The effects of mind mapping activities on students' motivation *Inter. J. Sch. Teach. Learn. Georgia. Sout. Univ* **6**(1)
[15] C DeWall, R Baumeister and E Masicampo 2008 *Evidence that logical reasoning depends on conscious processing Consciousness and Cognition* **17** 628-45
[16] Ohassta 2004 *Mind Map Rubric Ontario History and Social Sciences Teachers' Association*
[17] Miles, Mattew B and A Michael Huberman 2007 *Analisis Data Kualitatif* Jakarta: Universitas Indonesia Press
[18] Arikunto Suharsimi 2009 *Dasar-dasarEvaluasi Pendidikan* Jakarta: Aneka Cipta

[19] Serna M, Edgar and Serna A Alexei 2015 Knowledge in engineering: a view from the logical reasoning *Int. J.Compu. Theo.Eng.* **7**(4)

[20] B. Nigel Dovetailing language and context: teaching balanced argument in legal problem answer writing *English for Specific Purposes* **21** 321-45

Acknowledgments

Authors wishing to acknowledge assistance or encouragement from colleagues, special work by technical staff or financial support from organizations should do so in an unnumbered Acknowledgments section immediately following the last numbered section of the paper.

IOP Conf. Series: Journal of Physics: Conf. Series **1022** (2018) 012021 doi:10.1088/1742-6596/1022/1/012021

Energy efficiency effect on the public street lighting by using LED light replacement and kwh-meter installation at DKI Jakarta Province, Indonesia

Panggih Sudarmono[1]*, Deendarlianto[2] and Adhika Widyaparaga[2]

1Master Program of Systems Engineering, Faculty of Engineering, Universitas Gadjah Mada
[2]Department of Mechanical and Industrial Engineering, Faculty of Engineering, Universitas Gadjah Mada

*Corresponding email: panggih.s@gmail.com

Abstract. Public street lighting consumes large energy for the public interest, but many street lights still do not use energy-saving technologies. In 2014, Provincial Government of DKI Jakarta operated 179,305 units of street lights. Of the number of installed armature, 92 % of them or 166,441 units are HPS (High-Pressure Sodium) armatures which are inefficient.
In 2016, the Provincial Government of DKI Jakarta cut down the energy used for street lighting, by implementing the programs of kWh-meter installation in every street lighting panel and use energy-saving lamps equipped with the smart system. The Provincial Government of DKI Jakarta is registered with 6,399 customer IDs in PLN (State Owned Electric Company), and gradually carried out the kWh Meter installation and changes to the contract. The program to use energy-saving lights done by replacing the HPS armature that is not energy efficient to LED armature which is known to be energy efficient. Until the end of 2016, the number of armatures that has been replaced was 89,417 units.
The research results on 25 samples of PLN customer IDs and the replacement of 2,162 units armature, showed that the energy efficiency through kWh meter installation and armature replacement reduce the power consumption from 330,414 kWh to 71,278 kWh or by 78.43%. Generally, there was a decrease in the value of electricity bill compared to the before the replacement. The program of kWh-meter installations and replacement of the armature has a payback period of 2.66 years.

1. Introduction

Public street lighting is street lighting was installed for public interest / shared/is public (www.pln.co.id) Electrical energy sold to public street lighting nationally in 2015 of 3,448.11 GWh with of its installed capacity 1,003.91 MVA (1). Most of the public street lighting lamps installed still use HPS lamps that are not energy efficient (2).

The research of energy saving on street lighting has been done in various places. The various scenario of energy saving in the street lighting has been developed, such us the use of energy saving lamps and the use of the smart system (2,3). In the scale of the pilot project was conducted to determine out the capabilities of LED lamps to replace HPS lamps as street lighting lamps (4–7). In the other studies has been considering financing scheme that can be done by local government, such us the scheme of ESCO (Energy Service Company)(2,8,9).

This paper, to evaluate the energy efficiency on the street lighting systems by using energy saving lamps and kWh-meter installation at Jakarta. This study based on actual condition of the streets lighting at Jakarta that are currently implemented the program.

Moreover, The Jakarta government follow up the energy-saving program by issuing the Governor Regulation of Jakarta no. 156 the year 2012 about energy saving and water, as for accompanying the implementation of the energy saving and water in Jakarta administration. The regulation it arranged also about the step of electric energy saving in the field of street lighting in accordance with the Minister Energy and Mineral Resources Regulation no. 13 years 2012 about Saving Discharging Electric power namely reduce the power used for street lighting of 50 % in 24 hours up to 5.30 am(10,11).

Provincial Government of DKI Jakarta currently listed as customer category public street lighting in PLN Jakarta Raya and Tangerang. Based on the data PLN 2015, PLN Jakarta Raya and Tangerang recorded customers category street lighting as many as 10,028 connection with a capacity attached 130.59 MVA and energy sold 466.90 GWh worth USD 50,860,328.68[1](1). If Provincial Government of DKI Jakarta able to perform of electric energy saving in street lighting then there will be the very significant saving, both at the energy consumption and the costs of an electricity bill.

Energy saving programs in street lighting can be done with various strategies, but there are three main strategies that can be implemented namely kWh-meter installation, use energy-saving lamps are equipped with smart system (12,13), which is currently the three step was implemented by the Provincial Government of DKI Jakarta gradually.

1.1. kWh-meter Installation

The installation of kWh-meter to be done at every panel of street lighting and so that the power consumption will be measured. At this time, for the payment of street lighting accounts much of local Government that enforces contract subscription or lump sum without using the kWh-meters. Electricity bill based on the number of lamps installed without kWh meter or un-metered.

In the contract, the lights assumed on 12 hours from hours 6 pm to 6 am for 30 days and assumed as 375 hours. So such as fitted mercury lamps with a capacity of 70 W will be calculated with a capacity of 200 VA and use in 375 an hour a month regardless the condition of the field. So with these contracts, local governments are reluctant to undertake energy efficiency, because even by using the energy-saving light bulb, the cost of electricity bills paid will remain, or absence of tangible benefits for local government.

IOP Conf. Series: Journal of Physics: Conf. Series **1022** (2018) 012021 doi:10.1088/1742-6596/1022/1/012021

Table 1. The Classification of The Power Lights on The Street Lighting In Unmetered Tariff

No.	Lamps type	Power load (W)	Power load at Bill (VA)
1.	Incandescent lamps	25 – 50	50
		51 – 100	100
		101 – 200	200
		201 – 300	300
		301 – 400	400
		401 – 500	500
		501 – 600	600
		601- 700	700
		701 – 800	800
		801 – 900	900
		901 – 1000	1000
2.	Gas discharge lamps	10 – 50	100
		51 – 100	200
		101 – 250	500
		251 – 500	1000

[1] USD stands for United States Dollars. This research uses exchange rate assumption 1 USD for IDR 13,600.

Therefore, necessary to change the contract and kWh-meter installation, so the number of used measurable and the fee for according to the power usage. The process kWh-meter installation, local governments are required to provide network, panel, time switches, and contactor.

1.2. Use of Energy Saving Lamps

Currently, street lighting lamps commonly use HID (High-intensity discharge) technology type of HPS (High-pressure sodium). This type of lamps produces a yellow light colour, long lifetime 12,000 – 24,000 hours or 2.5 – 5 years, with efficacy at 45 – 130 lm/watt. Currently, HPS lamp technology was mature, so likely to develop no longer significant.

Technology that is currently developing is led light, this type of lights start widely adopted for street lighting. Led lamps have high energy efficiency and the high quality of light. Led light having longer lifetime, 50,000 - 100,000 hours and high efficacy 70 - 150 lm / w.

From a comparison between the HPS lamps and LED lamps for street lighting, HPS lamps can be replaced with LED lamps and can meet the standard set forth (5,6). Jakarta has done the installation of LED lamps in some location, the lighting in the road surface have met SNI (6).

1.3. Smart System

Without the smart system, then the local government will turn off the 1 light between 2 lights, as happened in Jakarta Outer Ring Road (JOR) (12). To be able to turn off the light, then local governments must prepare two networks in a single line, so it requires considerable investment.

The smart system itself is a combination of several types of equipment such as a sensor, lighting fixture, transceiver, measurement and power meter, and some other equipment so that it allows automatic control (14). The lighting of the city which features smart system will be able to control the lighting system at the time of peak hours or times of weather disturbances occur, so will eventually accrue to the energy savings.

IOP Conf. Series: Journal of Physics: Conf. Series **1022** (2018) 012021 doi:10.1088/1742-6596/1022/1/012021

2. Research Methodology

The data used in this research is the data on street lighting electricity bills 2016 and data armature replacement 2016.

2.1. Data Analysis

Energy efficiency obtained by comparing data the electric bill before and after the electricity saving programs done. So that would be needed two data electricity bills that in pairs, before and after the electricity saving programs done. With the availability of the data, it can be done statistical tests are t-test: paired two sample for means.

T-test: Paired two sample for means can be used for observing the two samples collected in pairs. Each pair is observed under the homogeneous condition, but with different conditions from one pair to the other pair (15). The equation used is as follows:

$$T_0 = \frac{\bar{D}-\Delta_0}{S_D/\sqrt{n}} \quad (1)$$

Where:

$\bar{D}$: The average of the difference between the sample of observed

Δ_0 : The average difference in the population under *null hypothesis*

S_D : Standard deviation

n : Number of samples

2.2. Economic analysis

Economic analysis is used to determine a project feasible or not. The feasibility an investment one of which can be seen from payback period. Payback period is calculated by dividing the total cost of the investment by the savings that occur every year, so it will be known how long will start to get profits from this program. The expected payback period of time equal to or less than the time of product warranty (4,5).

$$Payback\ period = \frac{C}{R} \quad (2)$$

Where:

C (Cost) : Total cost

R (Revenue) : Revenue or annual saving

3. Research Result

In 2014, Provincial Government of DKI Jakarta, through the Industry and Energy Agency of the DKI Jakarta operated 179,305 units street lighting lamp in 5 areas of the city administration. The lights still dominated the HPS lamp, which reaches 92% of total the installed lamps, whereas the LED lamp, known as energy-saving lamps only 1.5% of the total the installed lamps.

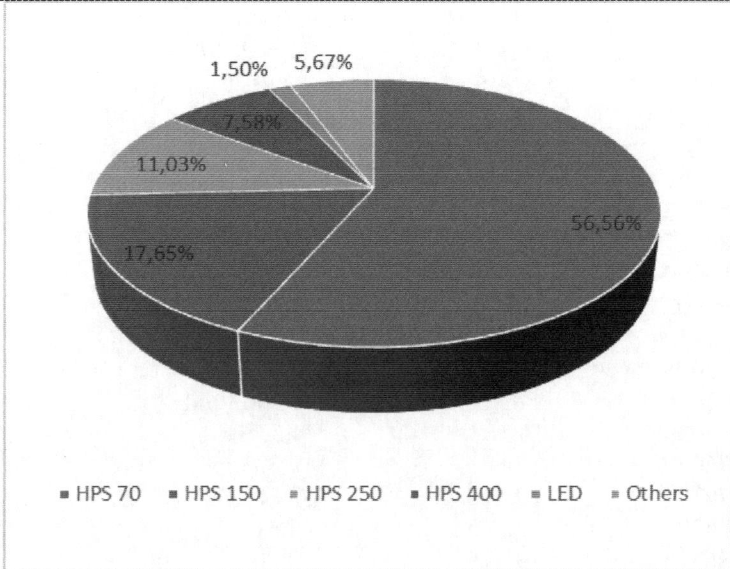

Figure 1. Composition of the installed street lighting lamps in DKI Jakarta of the year 2014

The programs of kWh meter installation and the lamps replacement with energy saving lamps are done simultaneously, so hopefully is a significant energy saving. HPS lamps are inefficient replaced with LED lamps that are efficient, E.g. 400 W HPS lamps replaced with LED lamps at maximum 200 W, without reducing the quality of street lighting. Comparison of the specification of the new lamps and old lamps used by Government of DKI Jakarta can be seen Table 2.

In 2016, the Provincial Government of DKI Jakarta carries out kWh meter installation and armature replacement from HPS to LED and until the end of 2016 was installed 89,417 LED lamps. Electricity consumption between un-metered with HPS lamps and metered with LED lamps can be seen in Table 3.

Table 2. Comparison of Old Armature and New Armature According to The Road Class

Parameter	Road class							
	Arterial road		Collector road		Local Road		Narrow street	
	Old	New	Old	New	Old	New	Old	New
Type	HPS	LED	HPS	LED	HPS	LED	HPS	LED
Power (W)	400	200	250	120	150	90	70	40
Lumen (Lm)	48,000	20,600	25,000	13,100	15,000	9,300	5,880	4,800
Color Temperature (K)	2,000	4,000	2,000	4,000	2,000	4,000	2,000	4,000
Price (USD[a])	220.59	713.38	202.21	558.82	154.41	481.18	71.91	441.18

Table 3. Data Sample Electricity Consumption Between Un-Metered with HPS Lamps and Metered with LED Lamps

No.	Customer ID	Power consumption (kWh)	
		January 2016	January 2017
1	541102453623	19,875	2,193
2	541300481235	19,875	4,731
3	543104591133	12,375	4,950
4	543600109454	8,625	2,395
5	543600564608	12,375	1,364

No.	Customer ID	Power consumption (kWh)	
		January 2016	January 2017
6	544102526433	3,975	745
7	544103336751	8,625	5,460
8	544103348141	12,375	2,304
9	547102992067	12,375	3,726
10	547103252763	8,625	3,186
11	547103289924	15,563	3,119
12	547103456090	12,375	2,392
13	547103494902	12,375	1,392
14	547400139110	12,375	1,889
15	547401075608	24,750	2,893
16	547401075616	8,625	1,260
17	547401076029	15,563	4,535
18	547401076060	15,563	2,333
19	547401076181	19,875	3,507
20	547401076307	12,375	3,311
21	547401076315	19,875	2,960
22	547500071757	12,375	3,582
23	547500387589	8,625	928
24	547500419787	12,375	4,266
25	547500504311	8,625	1,857

Table 3 shows the data customer ID who experienced activities kWh meter installation and lamp replacement. From 6,399 customer ID listed in PLN, there are 25 customer ID who experienced activities kWh meter installation and lamp replacement, sample available 25 samples, the amount would be adequate as a sample.

Further analysis conducted an analysis of t-test: Paired two sample for means by using Microsoft Excel 2013. The result of t-test: Paired two sample for means with 25 sample and the α value = 0.5, retrieved the $t_0 = 11.40727$, and the p-value 1.77×10^{-11}. With the value of $t_{0.05,\,25} = 1.708141$, then the $t_0 = 11.40727 > 1.708141$, can be concluded that by kWh meter installation and lamp replacement occurs of difference value of electric power consumption. Electricity consumption on average before kWh meter installation and lamp replacement higher than after kWh meter installation and lamp replacement.

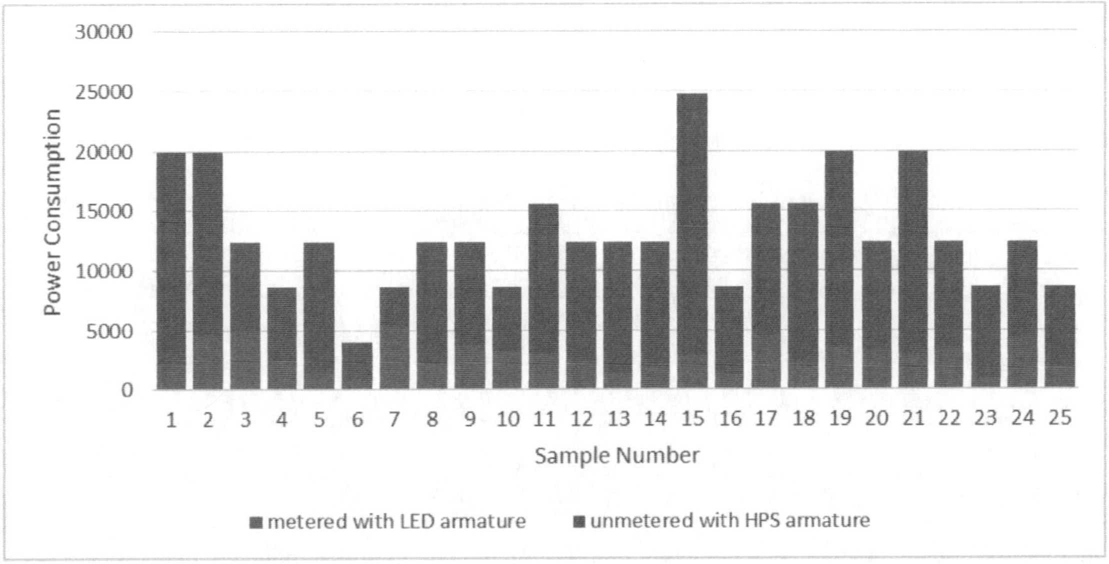

Figure 2. Comparison Of Electric Consumption Without Kwh Meter by Using HPS Lamps and After Kwh Meter Installation by Using LED Lamps

Figure 2 shows a comparison of electric consumption in P3 tariff or subscriptions or without kWh meter by using HPS lamps with a P31 tariff or after kWh meter installation by using LED lamps. Figure 2, can be seen a happening trend of decline in the electrical power consumption on the entire sample. Decline overall power consumption of 330,414 kWh in January 2016 be 71,278 kWh in January 2017 or saving of 78.43 %. The saving is combined saving between kWh-meter installation and lamp replacement.

Cost efficiency obtained from efficiency electricity bill and maintenance cost before and after the energy saving programs. LED armature besides reducing their electricity bill also reduce maintenance cost. LED armature different with HPS armature, where HPS armature consisting of many components that should be replaced periodically.

Table 4. Kwh Meter Installation and Armature Replacement

No.	Description	Volume	Unit price (USD)	cost (USD)
1.	KWh meter installation			
	Cable (25 location x @20m)	500	183.82	4,595.59
2.	LED armature installment			
	LED 40 Watt	1,120	441.18	494,117.65
	LED 90 Watt	670	481.18	322,388.24
	LED 120 Watt	64	558.82	35,764.71
	LED 200 Watt	308	713.38	219,721.76
3.	SS installment			
	Controller	25	177.94	4,448.53
	Total			1.081.036.48

The cost to the Provincial Government of DKI Jakarta of kWh meter installation is a cable from substation heading the street lighting panel, because the whole customer's ID registered have installed street lighting panel. While the cost of replacing the armature in form of the cost of purchasing the armature units. In full can be seen in Table 4.

From Table 5 can be seen the value of the cost-saving electricity bill must be paid by the government. That value is monthly saving on 25 samples so that annual saving obtained is USD 322,355.66. The value of the saving comes from 25 samples from total 6,399 customers ID.

Table 5. Monthly Saving Of Electricity Bill

No	Area	Power consumption (kWh)		Power efficiency (kWh)	Cost efficiency[a] (USD)
		January 2016	January 2017		
1	Central Jakarta	39,750	6,924	32,826	3,402.86
2	South Jakarta	290,664	64,354	226,310	23,460.11
Total		330,414	71,278	259,136	26,862.97

[a]) Electricity rates on the average year 2016: USD 0,10 / kWh

In addition, from Figure 3, can be seen the value of saving from the reduced cost of replacing a component of an HPS armature USD 81,815.52 per year, because LED armature unit replacement will be performed in the case of damage. This is one of the weaknesses of the LED armature, because the replacement units needing high cost, while in HPS armature to be done component replacement at a low cost.

Cost saving obtained from HPS armature component replacement USD 81,815.52, and reduced from electricity bill USD 322,355.66, with investment USD 1,076,440.88 obtained the payback period 2.66 year.

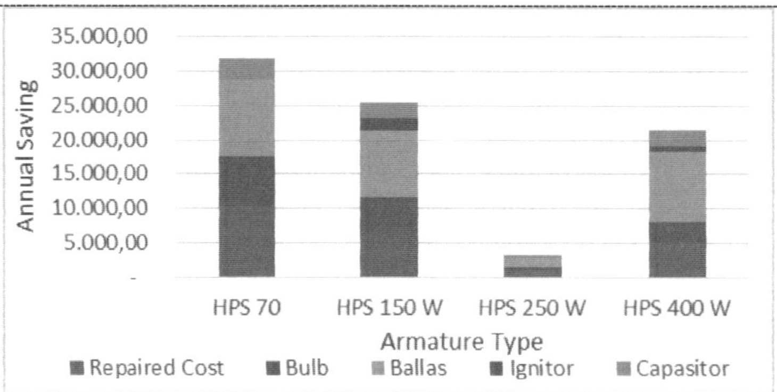

Figure 3. Annual Saving from Maintenance Cost.

4. Discussion

The program kWh meter installation and armature replacement conducted by Provincial Government of DKI Jakarta is able to save 78.43 % power used. Currently, there still 5,102 customer ID without kWh meter or unmetered, with potential saving 3,790,192 kWh.

Payback periods of this programs are 2.66 years when compared with warranty offered by the provider of 5 years, then it can be said this program very reasonable to run.

Currently, in South Jakarta and Central Jakarta, the HPS armature have been replaced with LED armature, but not all connected with metered panels, that has not been registered saving electric power. So should be driven the acceleration of the kWh meter installation program so that power saving and costs saving will be can be perceived.

While energy efficiency through dimming using a smart system until now not been made by Provincial Government of DKI Jakarta. This is due to the implementation still need further study, the associated location, so it doesn't reduce the security and convenience of the society in the work during the night.

5. Conclusions

Energy efficiency of street lighting through kWh meter installation and armature replacement saving 78.43 % power consumption.

Generally, occurs a reduction in the value electricity bills than before energy saving programs. If compared to the investment value, payback period obtained kWh meter installation activities and armature replacement with a payback period of 2.66 years.

6. References

[1] PLN. Statistik PLN 2015. 2016;

[2] Al Irsyad M I and Nepal R A 2016 Survey based approach to estimating the benefits of energy efficiency improvements in street lighting systems in Indonesia Renew Sustain Energy Rev [Internet] 58 1569–77

[3] Beccali M, Bonomolo M, Ciulla G, Galatioto A and Lo Brano V 2015 Improvement of energy efficiency and quality of street lighting in South Italy as an action of Sustainable Energy Action Plans *The case study of Comiso (RG) Energy* [Internet] **92**(3) 394–408

[4] Ramli R M, Arief Y Z and Aziz P D A 2015 Application of LED technology into public road lighting in Malaysia for replacing the high pressure sodium vapour lighting 2015 *Int Conf Sustain Energy Eng Appl [Internet]* 76–81

[5] Nogueira F J, Vitoi L A, Gouveia L H, Casagrande CG, Pinto DP and Braga HAC 2014 Street lighting LED luminaires replacing high pressure sodium lamps: Study of case 2014 11th *IEEE/IAS.Int.Conf. Ind. Appl IEEE INDUSCON* Electron Proc.

[6] Setyaningsih E, Fat J, Wardhani L and Zureidar I 2014 Performance of led lights installed on dki jakarta streets electr power electron commun control informatics semin 45–50

[7] Erol E, Kalkan E, and Ath D 2015 New Approaches in street lightings in turkey a brief history of street lighting in the world *4th Int Conf Renew Energy Res Appl* 22–5

[8] Huang S-C, Lee L, Jeng M-S and Hsieh Y-C 2011 Assessment of energy-efficient LED street lighting through large-scale demonstration

[9] Burgos-payán M, Correa-Morenoel F J and Riquelme-Santos J M 2012 Improving the Energy Efficiency of Street Lighting *A Case in the South of Spain* 1–8

[10] Peraturan Menteri ESDM No.13 Tahun 2012 tentang Pengehmatan Pemakaian Tenaga Listrik

[11] Peraturan Gubernur Provinsi DKI Jakarta No 156 Tahun 2012 tentang Penghematan Energi dan Air

[12] Irsyad MI al and Rivai A 2014 Peningkatan Ketahanan Energi Melalui Efisiensi Penerangan Jalan Umum *M&E* **12** 312(3) 66–77

[13] Direktorat Jenderal Energi Baru Terbarukan dan Konservasi Energi. Buku Pedoman Efisiensi Energi Pencahyaan Jalan Umum Buku II : Perencanaan Sistem PJU Efisien Energi. Kementerian Energi Dan Sumber Daya Mineral 2014

[14] Pinto M F, Mendonca T R F, Coelho F and Braga H A C 2015 Economic analysis of a controllable device with smart grid features applied to LED street lighting system *IEEE Int Symp Ind Electron* **11** 84–9

[15] Montgomery DC and Runger GC Applied Statistics and Probability for Engineers Third Edition 3rd ed New York: John Wiley & Sons Inc 2003

Performance investigation of an innovative Vertical Axis Hydrokinetic Turbine – Straight Blade Cascaded (VAHT-SBC) for low current speed

R Hantoro[1], J Prananda[2], A W Mahmashani[3], E Septyaningru[4] and F Imanuddin[5]

[1,3,4,5]Faculty of Industrial Technology, Department of Engineering Physics, Institut Teknologi Sepuluh Nopember Surabaya, Indonesia
[2] Faculty of Marine Technology, Department of Marine Engineering, Institut Teknologi Sepuluh Nopember Surabaya, Indonesia

Corresponding email: [1]hantoro@ep.its.ac.id, [2]juniarko@ne.its.ac.id, [3]mahmashany@gmail.com, [4]erna.septya@gmail.com, [5]fahmi13oct@gmail.com

Abstract. Research on the development and innovation of Vertical Axis Hydrokinetic Turbine (VAHT) to improve performance has been done. One of the important indicator that affects VAHT's performance is Coefficient of Performance (Cp). Theoretical Cp value for the VAT (Darrieus) turbine is 0.45. This paper presents the results of a performance investigation for an innovative Vertical Axis Hydrokinetic Turbine – Straight Blade Cascaded (VAHT-SBC) by modifying the number and the arrangement of blades using CFD simulation. Symmetrical NACA 0018 is used for this study, each model is simulated with current speed variation (U - m/s) of 0.5, 1 and 1.5. An increase in Cp value is shown in variation of 9 blades (3 blades cascaded in each arm) with Cp value of 0.396 at TSR of 2.27 which is reach 88% of the theoretical value. Furthermore, the streamline velocity of the pressure contour, velocity streamline and torque fluctuations are also presented in this paper to gain in deep information.

Key words: hydrokinetic turbine; darrieus straight blade; turbine performance; vertical axis turbine; computational fluid dynamics.

1. Introduction

One of the largest and cheapest renewable energy is hydropower. There are two mainly methods for energy harvesting from water, called hydrostatic and hydrokinetic methods. Hydrostatic method is the conventional technique of producing electricity using head and extracts the potential energy of water using turbo-machinery. While hydrokinetic method is a modern technique to extract the kinetic energy containing in a flow water with zero head [1].

A feasibility study of ocean current in Alas Straits [2] and Larantuka Straits [3] of Indonesia shows that the current speed is relatively high, i.e. could reach up to 2 m/s and 4 m/s respectively. A preliminary study for ocean current in several small islands in Indonesia shows that ocean current speed in Berhala, Anambas, and Biawak island have their mean on 0.135 m/s, 0.055 m/s, 0.272 m/s respectively, and Miangas island has the highest speed of 0.835 m/s, means that Indonesia has varying low current speed potential for hydrokinetic energy harvesting. Hence, hydrokinetic technology which having good capability to operate at low current speed is essential. The minimum workable current speed for hydrokinetic is about 0.8–1m/s, but the preferably current speed is 1.3–1.5m/s and goes up to a maximum of 3m/s [4]. The approximate fluid densities are 1000 kg/m3 and 1.223 kg/m3 for water and wind, respectively.

Darrieus is the most attractive type of turbine to develop because of its simplicity and because the turbine permits the high rotating turbine to develop in low current speed. Various researches have been conducted to gain the best performance of darrieus turbine. Kobold turbine is originally vertical axis turbine that modified the blade by adding the rough sand strips along the span to postponing the stall condition [5]. A research of vertical axis turbine straight blade in the effect of number of blade is done by adding arm to keep the blade as the turbine rotates. As a result, it is clarified that the power coefficient decreases with the increase of numbers of blades [6].

Innovative approaches are required to meet the needs for electricity production especially for isolated islands in Indonesia which has lower current speed compare to the straits. Although these ocean currents have suitable characteristics, their costs of energy are considered to be higher than that of wind power in Research and Development. To alleviate this problem, Computational Fluid Dynamics (CFD) is used to reduce the future experimental cost and optimizing the design of Vertical Axis Hydrokinetic Turbine Straight Blade Cascaded (VAHT-SBC) which is an advance development of vertical axis hydrokinetic turbine that has a favourable self-starting proposed by the author [7].

This paper describes the main factor that affect the performance of the turbine by modifying the number and the arrangement of blades without adding the number of arm. Furthermore, the flow profile of the pressure contour, velocity streamline and torque fluctuations are also presented in this paper to gain in deep information.

2. Description of VAHT-SBC

2.1. The Origin of VAHT-SBC

Hydrokinetic turbines could be categorized into two types. Firstly, the vertical-axis turbine, which the turning axis is perpendicular to the stream flow. Secondly, the horizontal turbine, which the rotational axis follows the direction of flow. Due to the omni-directional characteristic, vertical axis turbines are preferable in situations where flow direction changes, such as in tidal systems [4].

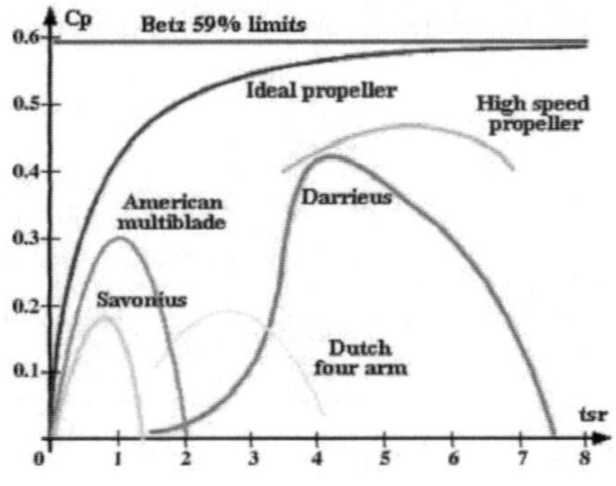

Figure 1. Coefficient Performance of Different Turbines

Theoretical coefficients performance of different free flow turbines for wind application are depicted in Figure 2 and the coefficient performance for darrieus turbine will not exceed 0.45 [8]. The conventional of darrieus straight-blade turbines with fixed pitch blades typically suffer from poor starting torque, low efficiency and shaking due to large fluctuations in both radial and tangential force with change of azimuth angle. The use of passive variable pitch can improve the aspect of performance of the three straight-blade vertical axis turbine [7].

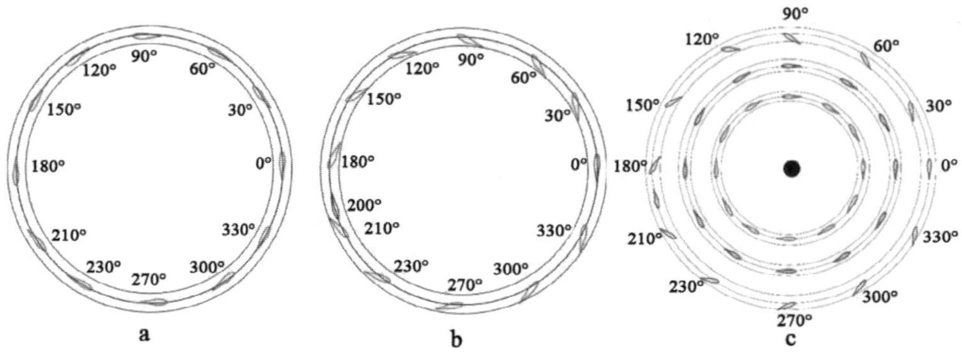

Figure 2. The Change of Pitch Angle In Azimuth For (a) Fixed-Blade, (b) Passive-Variable Pitch Blade And (c) Prediction in The Changes of Azimuth Angle Of VAHT-SBC

VAHT-SBC is an adaptation from darrieus straight-blade turbine that designed to rotate at low current speed. The result of previous research with current speed variations of 0.6 – 1 m/s, provides data on capability of self-start rotating at speed 0.8 m/s for fixed-pitch condition, and at speed <0.6 m/s for passive variable-pitch which confirms better ability to start rotating at intermediate Tip Speed Ratio (TSR) of 2.3 - 2.4 [7]. The pitch angel change regarding to fixed-blade and passive-variable pitch blade is given in figure 2 (a). Both type of blades in two different turbines are combine as one turbine without enlarge the dimension of the turbine depict in figure 2 (c). Henceforth, this combination is called as VAHT-SBC (Vertical Axis Hydrokinetic Turbine – Straight Blade Cascaded).

2.2. VAHT-SBC Configuration

NACA 0018 has been used as the airfoil of the blades for VAHT-SBC. Initial analysis of 4 symmetrical NACA 4-digit foils suggested that NACA 0018 has the best characteristics for vertical axis tidal current turbine [7]. The three model turbine configurations analyzed in this work are consist of passive pitch blade (red color) and fixed blade (blue color) and the number of blades for model 1, 2 and 3 respectively are 3, 6 and 9 as depict in figure 3. The turbine has 10 cm of cord, 80 cm of span, 80 cm of diameter, 3 cm shaft of dimeter, with the aspect ratio of 8 and H/D is 1.

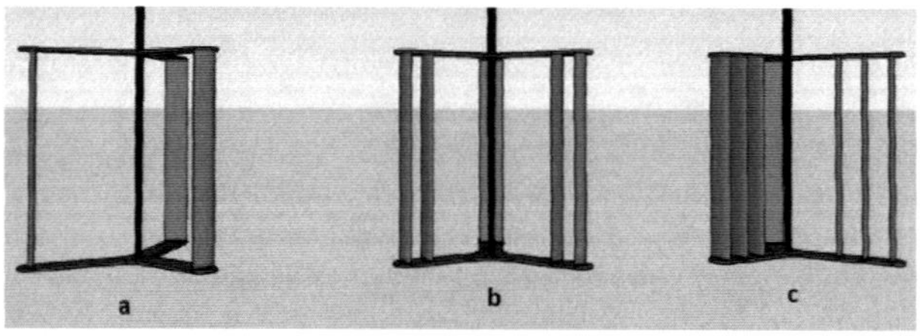

Figure 3. (a) Model 1, (b) Model 2 And (c) Model 3.

3. Numerical Setup

CFD simulation was performed as preliminary study to gain deep information regarding to the effect of blade number to the performance of VAHT-SBC. The SST (Shear Stress Transport) k-ω turbulence model was used. Some applications have shown that the k-ϵ model has limitations in boundary layer flows with adverse pressure gradients [9]. In contrast, SST k-ω turbulence model gives accurate predictions of the amount flow separation under adverse pressure gradients, and has been successfully used in the CFD simulation of wind or water turbines [12-13].

3.1. Computation Domain and Boundary Condition

In this simulation, the turbine's arms are neglected for simplicity and the economy of computational time. There are three domains used in this simulation, the external domain has dimension of $4.5D \times 2.1D \times 2D$, the height of dimension of rotating domain is 2D and the diameter is 1.5D.

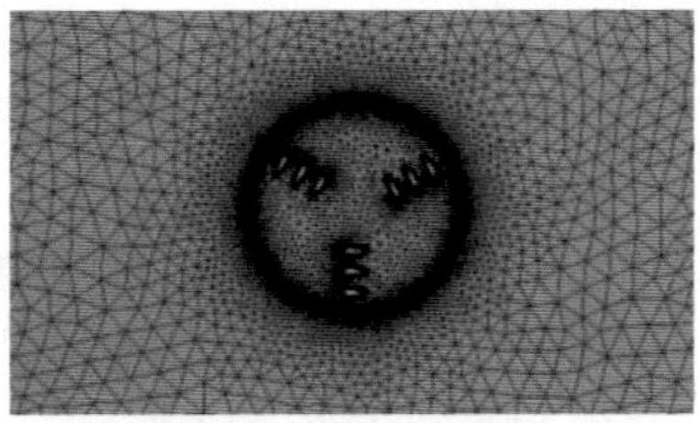

Figure 4. Mesh Generation

The meshing size are divided according to its domain. Turbine and shaft are ranging between 0.0005 m to 0.1m, external and rotating domain are between 0.0095 m to 0.3m, while for the inlet and outlet the meshing size is ranging from 0.01 m to 0.3 m with the same angular resolution of 18°. Three different grids amount that produced for meshing in this study are $\mathbf{8.7 \times 10^6}$, $\mathbf{12.5 \times 10^6}$, $\mathbf{17.3 \times 10^6}$ for model 1,2 and 3 respectively. The grid is fully tetrahedral unstructured in both external and rotating domains as shown in figure 4. The unstructured tetrahedral grids are chosen due the capabilities to discretize complex geometries with fast and minimum user intervention. The velocity condition for inlet boundaries are set into 0.5, 1, 1.5 m/s and the turbulence intensity is set for 5% with the outlet, right and left wall are set to opening condition.

4. Result and Discussion

In this research, CFD feasibility analysis can be used as a reference in preliminary design of a turbine. The used of CFD is done to minimize errors for further turbine design. In addition, CFD operation is zero cost and provide results that cannot be achieved from experimental studies such as torque fluctuation, streamline velocity and fluid pressure around the blade with the intention of improvisation design can be done before the fabrication.

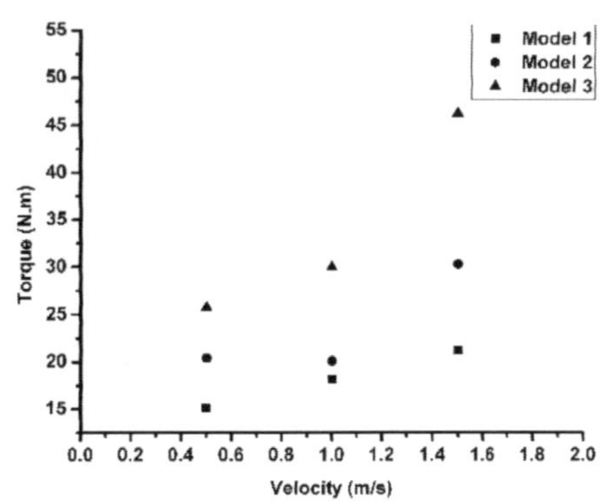

Figure 5. Number of Blade Effect To Average Torque Comparison in Freestream Velocity Variation

4.1. Torque Fluctuation

In this study, each model is simulated with three current speed variations of 0.5 m/s, 1 m/s, and 1.5m/s. Figure 8 depicts the average torque for each model in current speed that mentioned above. The magnitude of the torque is affected by freestream velocity, the higher the freestream velocity, the higher the torque can be generated by the turbine. The average torque (N.m) generated by the turbine model is 18.13, 23.59, 33.97 N.m for model 1,2,3 respectively with the highest torque of 46.18 N.m achieved by turbine model 3 at speed of 1.5 m/s. Further description of the torque per azimuth is given in Figure 6.

Simulations are performed 24 times for each model to get results at every 15° of azimuth in a full rotation (360°). Rotation of the turbine is an effect from kinetic energy that converted into different pressure and shear stress along the span. The sinusoidal pattern is formed by the change of angle of attack, as depicted in Figure 6. The pattern tends to be the same for every 120° due to three arm that also rotates in a full rotation. In all variation of current speed, the highest torque for model 1 are achieved in every 105°, 225°, 345° of azimuth, while for model 2 and 3 the highest toque are shifts and achieved in every 30°, 150°, and $270^{\circ}$of azimuth. The lowest torque for model 1 are also shifts compared to model 2 and 3, where the minimum torque for model 1 are in the azimuth of 60°, 180° and 300°, the lowest torque for model 2 are in the azimuth of 75°, 195°, 315°, and the lowest torque for model 3 are achieved in azimuth of 0°, 120°, and 240°. Further information of minimum and maximum torque will be discussed in figure 7 and figure 8.

The combination between passive-pitch and fixed pitch can shift both the highest and the lowest torque fluctuation pattern. The passive-pitch blade affects the Reynolds number which tend to be higher, the increment of Re number leads the fixed-pitch blade to get turbulence flow from passive-pitch blade and postpone the dynamic stall condition.

As the freestream velocity increases, the rises and falls of torque fluctuation for turbine model 1 tend to be more stable compared to model 2 and 3 in higher freestream velocity and vice versa for the lower current speed. In the real implementation, mostly turbines that have high magnitude of gap between maximum and minimum torque per-azimuth, also have higher intensity of vibrations, this phenomenon could directly impact on the turbine structure strength and generate poor power quality for electricity [12].

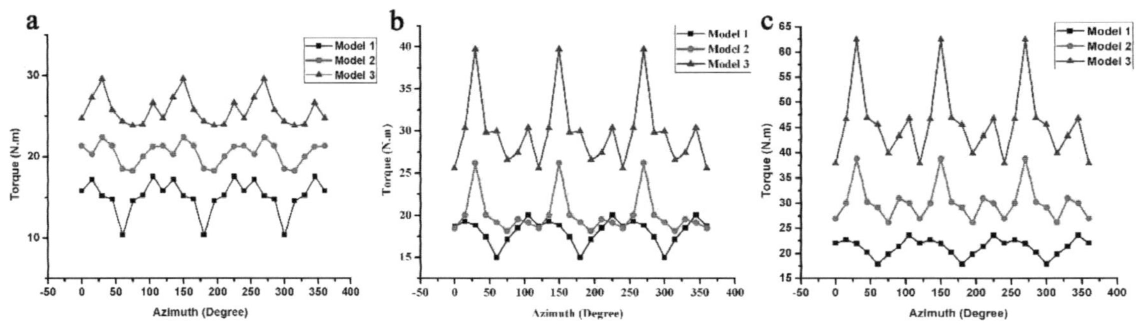

Figure 6. Torque Comparison Per Azimuth in Velocity of (a) 0.5 m/s, (b) 1 m/s and (c) 1.5 m/s

4.2. Pressure Distribution and Streamline Velocity

Since the freestream velocity and the fluid pressure relates to the aerodynamic force, the faster the freestream velocity, the greater the energy converted to the pressure that could be extracted by turbine. To gain more deep information for simulation result of VAHT-SBC performance, it is necessary to study detailed flow profile around the blade. Figure 7 and figure 8 present flow profile of pressure contour distribution and streamline velocity at minimum and maximum torque regarding to fluctuation torque mentioned in figure 6.

IOP Conf. Series: Journal of Physics: Conf. Series **1022** (2018) 012022 doi:10.1088/1742-6596/1022/1/012022

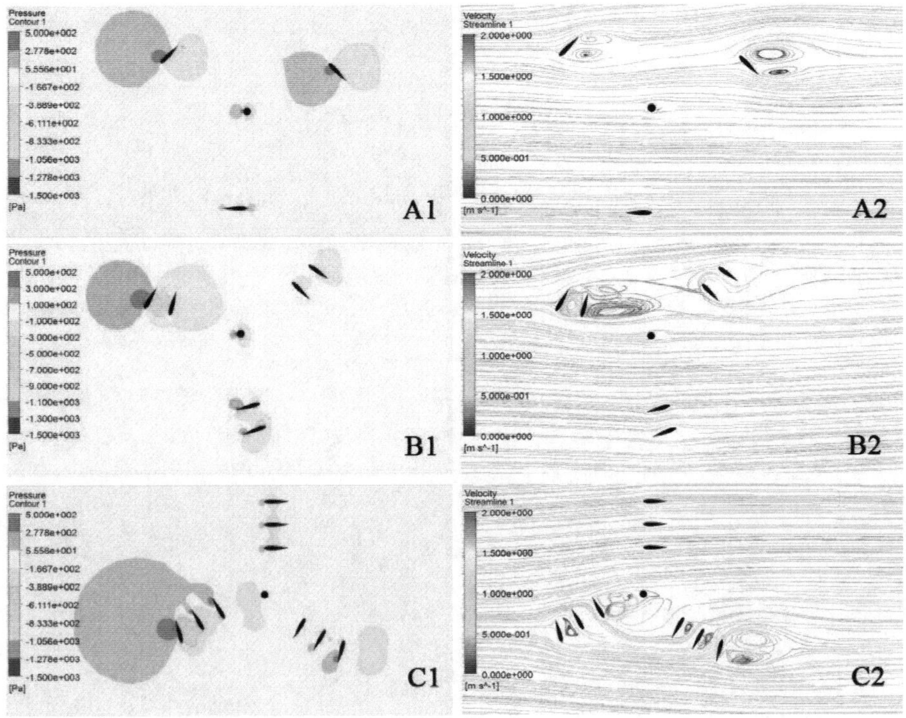

Figure 7. Pressure Distribution (1) and Streamline Velocity (2) of The Minimum Torque, Model 1(a) 60°, 180° and 300°, model 2 (b) 75°, 195°, 315°, model 3 (c) 0°, 120°, and 240°

In order simplify the analyses of the local flow pressure variations near the blade, for further descriptions one full rotation have been divide into 4 quadrants, i.e. 1st quadrant (0°-90°), 2nd quadrant (90°-180°), 3rd quadrant (180°-270°) and 4th quadrant (270°-360°).

The position of blades in the minimum torque achieved by model 1is in the azimuth of 60°, 180° and 300°. The blades in 2nd quadrant did not achieve a sufficient flow pressure (figure 7-A1) around it, while the blade in 4th quadrant obtain a low velocity due the vortex create by the blade in 1st quadrant (figure 7-A2).

Due the effect of adding a single blade cascaded into inner arm, greater vortex created by model 2 in the blade of 1st quadrant causes a minimum velocity obtained by the blade in 4th quadrant (figure 7-B2) as the consequences of minimum torque fluctuation achieved by model 2. Apart from the low pressure and velocity contour as seen in figure 7 B1 & B2, position of blade in 3rd quadrant is on the opposite direction against the freestream which affects the wrong position of the blade against the angle of attack.

By adding single more blade into the arm, model 3 obtains a great pressure in 2nd quadrant (figure 7-C1). As the consequences, a greater vortex also creates by the combination of the cascaded blade and impinging the blades in 3rd quadrant as seen in figure 7-C2.

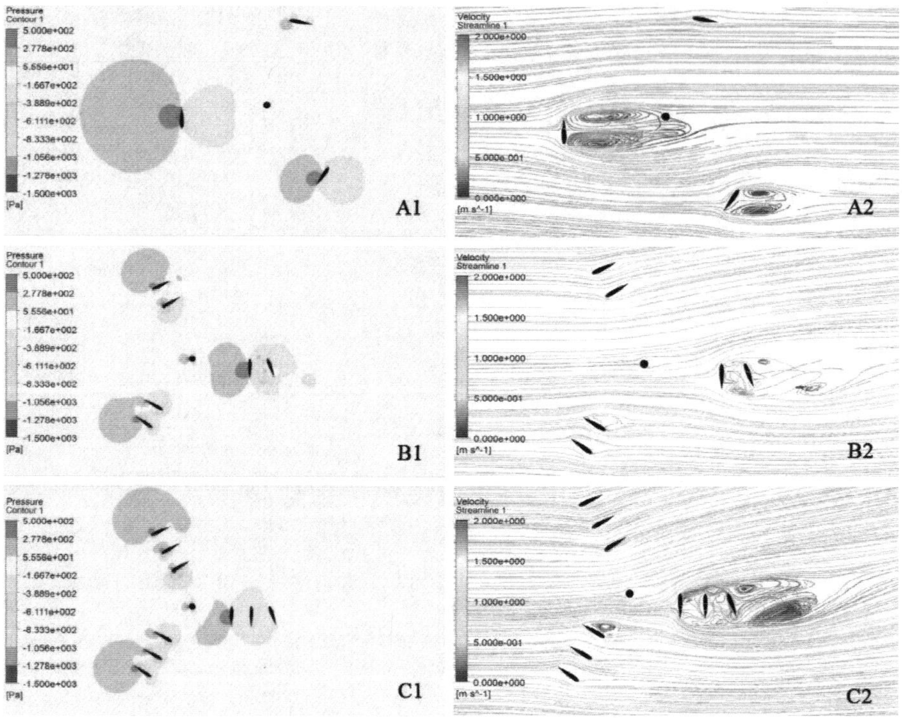

Figure 8. Pressure Distribution (1) and Streamline Velocity (2) of the Maximum Torque, Model 1(a) 105°, 225°, 345°, model 2 (b) and 3 (c) 30°, 150°, and 270° of Azimuth

The application of passive-variable pitch as the outmost blade, leads the angle of attack to adjust according to the azimuthal position and postpones the stall condition. Passive-variable pitch blades that employed in model 1 provide a greater performance as the increment of TSR. The maximum Cp for model 1 is 0.31 at TSR 3.2, as shown in figure 9. The less of number blade in this turbine, the less vortex impinging another blade, as seen on figure 8-A2 for blade in 3rd and 4th quadrant for turbine model1. In addition of the vorticity, the higher flow pressure (figure 8-A1) are well-distributed to another blade compared to figure 7-A1.

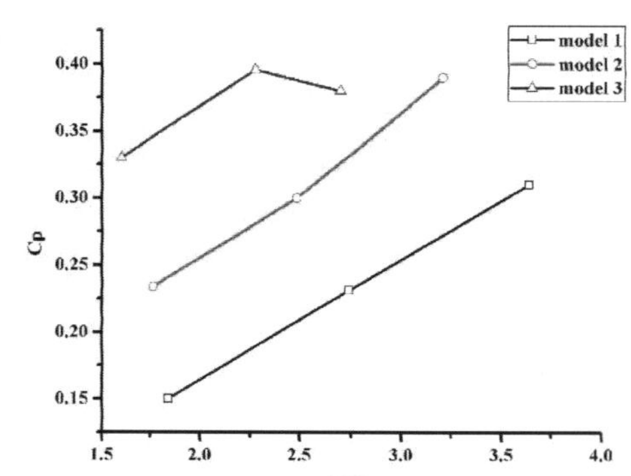

Figure 9. Number of Blade Effect on Turbine Performance Characteristic

Adding a single fixed-pitch blade into inner arm as model 2, gives the enhancement of Cp in lower TSR compared to model 1 (figure 8). The contribution of adding fixed-pitch blade can be seen on figure 8-B1. In the 3rd quadrant, the inner fixed-pitch blade tends to gain fresh current perpendicularly without interference from the movement of passive-variable pitch blade in the same quadrant as depict in figure 7-B2. The highest torque for model 3 is obtained in exactly same azimuth as model 2.

The highest of Cp is obtained by turbine model 3 (figure 9) at 0.396 with TSR of 2.27 which is reach 88% of the theoretical Cp value of darrieus turbine (figure 1). As the freestream velocity is 0.5 m/s, the VAHT-SBC provides the lowest TSR at 1.6, but in higher Cp compared to mode 1 and 2. Thus, this phenomenon confirms that VAHT-SBC is suitable for low current speed application which appropriate with ocean current condition in Indonesia.

5. Conclusions

Based on Computational Fluid Dynamics simulation investigation, it has been concluded that increase in the number of blades increases the torque output from VAHT-SBC. Three models of VAHT-SBC were studied in this research, which provides preliminary data for future experimental research.

As the results, the effectiveness of combining the passive-variable pitch blade and fixed-pitch blade is obtained. The interaction of the passive and fixed pitch blade to the incoming flow generate a complex flow profile that investigate into quadrants. The comparison of models, show that maximum Cp is achieved by model 3 of 0.396 at TSR of 2.27 which reach 88% of the darrieus turbine theoretical performance. The capability of model 3 to obtain a higher performance in lower TSR can make a significant contribution in the kinetic energy conversion for low speed ocean current application.

6. References

[1] M J Ã Khan, M T Iqbal and J E 2008 Quaicoe River current energy conversion systems: progress , prospects and challenges **12** pp 2177–193
[2] N Aziz 2009 Tidal Energy Resources Assessment In Indonesia A Case Study In Alas Strait University of Southampton
[3] K Orhan, R Mayerle and W W Pandoe 2015 Assesment of energy production potential from tidal stream currents in Indonesia *Energy Procedia* **76** pp 7–16
[4] M S Guney 2011 Evaluation and measures to increase performance coefficient of hydrokinetic turbines *Renew Sustain Energy Rev* **15** 8 pp 3669–675
[5] G Calcagno, F Salvatore and L Greco 2006 An experimental investigation and a theoretical and computational methodology to study an innovative technology for marine current exploitation : the kobold turbine pp 14–24
[6] Q Li, T Maeda, Y Kamada, J Murata, K Furukawa and M Yamamoto 2015 Effect of number of blades on aerodynamic forces on a straight-bladed vertical axis wind turbine *Energy* **90** pp 784–95
[7] R Hantoro, I K A P Utama and A Sulisetyono 2011 An experimental investigation of passive variable-pitch vertical-axis ocean current turbine **43** 1 pp 27–40
[8] B K Kirke and L Lazauskas 2011 Limitations of fixed pitch darrieus hydrokinetic turbines and the challenge of variable pitch *Renew. Energy* **36** 3 pp 893–97
[9] B Yang and C Lawn 2011 Fluid dynamic performance of a vertical axis turbine for tidal currents *Renew. Energy* **36** 12 pp 3355–3366
[10] W Tian, B Song, J H Van Zwieten, P Pyakurel, and Y Li 2016 Numerical simulations of a horizontal axis water turbine designed for underwater mooring platforms *Int. J. Nav. Archit. Ocean Eng.* **8** 1 pp 73–82
[11] A Pizzoferratoa, I M Kalkman and B Blocken 2015 Preliminary CFD simulations for the torque wind turbine *Trends Challenges Wind Energy Harvest* pp 87–96
[12] Y Li and S M Calisal 2011 Modeling of twin-turbine systems with vertical axis tidal current turbines: Part II — torque fluctuation **38** pp 550–58

IOP Conf. Series: Journal of Physics: Conf. Series **1022** (2018) 012023 doi:10.1088/1742-6596/1022/1/012023

Innovation in Vertical Axis Hydrokinetic Turbine – Straight Blade Cascaded (VAHT-SBC) design and testing for low current speed power generation

R Hantoro[1], I K A P Utama[2], I S Arief[3], A Ismail[2] and S W Manggala[1]

[1] Faculty of Industrial Technology, Department of Engineering Physics, Institut Teknologi Sepuluh Nopember Surabaya, Indonesia
[2] Faculty of Marine Technology, Department of Naval Architecture and Shipbuilding Engineering, Institut Teknologi Sepuluh Nopember Surabaya, Indonesia
[3] Faculty of Marine Technology, Department of Marine Engineering, Institut Teknologi Sepuluh Nopember Surabaya, Indonesia

Corresponding email: [1]hantoro@ep.its.ac.id, [2]kutama@na.its.ac.id, [3]irfansya@its.ac.id, [4]abdiismail1993@gmail.com, [5]seno.manggala@gmail.com

Abstract. This study examines an innovative turbine with the addition of the number and arrangement of straight blade cascaded (SBC). SBC is a combination of passive variable-pitch and fixed pitch of each turbine arm. This study was conducted in an open channel flow that has a current velocity (V-m/s) of 1.1, 1.2, and 1.3. RPM and torque ware measured for coefficient of performance (Cp) and tip speed ratio (TSR) calculation. Without changing the turbine dimension, the employment of cascaded blade (three blades in each arm) contributes to improve energy extraction significantly. A significant increase in Cp value is seen when 9 blades (3 cascaded blades per arm) are used with a Cp 0.42 value at TSR 2.19 . This value has reached 93% of the maximum theoritical Cp value.

Key words : vertical axis, hydrokinetic turbine, straight blade cascaded, passive variable pitch, fixed pitch, Cp

1. Introduction

Hydrokinetic turbines extract mechanical power from flowing water without dam. They operate essentially identical principles as wind turbines at low Froude numbers. Hydrokinetic turbines used for harnessing renewable energy sources such as rivers, drainage flows, tidal flows, and ocean currents [1]. The mechanical energy that has been generated from hydrokinetic turbine, then converted into electrical using a generator.

Straight-bladed or Darrieus turbines has number of blades that rotate around an axis that perpendicular to the flow. These turbines can receive flow from any direction, as long as the flow perpendicular to the axis of rotation. These turbines do not require yaw control [1]. Straight-bladed or Darrieus turbines is one type of Vertical Axis Hydrokinetic Turbine (VAHT).

IOP Conf. Series: Journal of Physics: Conf. Series **1022** (2018) 012023 doi:10.1088/1742-6596/1022/1/012023

VAHT should have high coefficient of perfomance (Cp) in order to extract current energy to mechanical energy as much as possible. Several methods to obtain optimal Cp have been developed by researchers. Aspects regarding these research are to optimize some design paramaters VAHT such as, number of blades [2], height to radius ratio [3], type of airfoil [4], and solidity [5].

Theoritical model was used for the design and performance simulation of Darrieus-type vertical axis turbine, such as number of blades affects, coefficient of performance (Cp), and structural stability. As number of blades decreases, range of operating tip speed ratio increases. It should be noted there is no significant difference of Cp_{max} when the turbines consist with one blade, two blades or three blades. The lower number of blades used, the wider Cp curve. Implementing one bladed turbine is not possible due to unbalanced centrifugal forces. Three bladed turbines gives more stability (Cp_{max} =0.1) than two bladed turbines. Generally, some research use 3 blades on VAHT [1] [3] [6] [7] [8].

Other design parameter is height to radius ratio (H/R). Newly developed vortex method was used for modelling three-dimensional effects vertical axis tidal current. The numerical results show that if the height is more than six times radius, the Cp_{max} increases slower correspond to the turbine height. The Cp_{max} reached more than 0.43 [3]. However, this study uses numerical study, not using experimental study.

Type of airfoil affects the value of Cp. Reference [4] investigated the effect of several airfoils toward Cp_{max} by using Computational Fluid Dynamics (CFD). These airfoils were NACA 00XX, NACA 63XXX, S-series, A-series and FX-series. Airfoil S-1046 had highest Cp_{max} among others. NACA 0018 had highest Cp_{max} among other NACA series. As a note, turbines which consist of symmetrical airfoils have better performance than asymmetric airfoil turbines. As a note, turbines consisting of symmetrical airfoils have higher performance than asymmetric arifoil turbines. NACA 0018 is the best airfoil for NACA series and usually used for Darrieus turbines [4]. Furthermore, NACA 0018 has the best characteristics for VAHT. Due to the high value of thickness to chord ratio so that have enough power to withstand bending, NACA 0018 is usually used for Darrieus turbines. [6]. Thus, the airfoil with the highest max Cp is NACA 0018 (Cp_{max} = 0.2964) and S-1046 (Cp_{max} = 0.4051) [4].

Solidity is another parameter affecting the performance of turbines. Reference [5] investigated the effect of solidity towards Cp_{max} by using numerical study. In this study, the solidity was variated of 0.3, 0.4, 0.5, and 0.6. This study confirmed that higher Cp_{max} was obtained in lower solidity, highest Cp_{max} is 0.43 obtained for solidity of 0.3. In addition, the field test was conducted for solidity of 0.84 and showed that Cp_{max} was 0.25.

Cp higher than 0.4 obtained on [3] [4] [5]. However, these researches are numerical study. To validate numerical calculation results, experimental study was used. Experimental study also give support for the design of the prototype which is investigate three types of variable pitch VAHT. Considering the control rules, the turbine can classified as cycloid type controllable-pitch, spring-control variable-pitch, and passive variable-pitch vertical axis turbine. Between those control rules, highest Cp_{max} is 0.33 obtained from spring-control variable-pitch [8].

Cp_{max} or Cp_{peak} which can be achieved by Darrieus type VAHT is 0.45 in the theoretical value [9]. So that still needed further research, to get Cp_{max} approaching the theoretical value. This experimental study proposes an innovative turbine with the addition of the number and arrangement of straight blade cascaded (SBC). SBC is a combination of variable passive pitch and fixed pitch of each turbine arm. This study was conducted in an open channel flow that has a current velocity (V-m/s) of 1.1, 1.2, and 1.3. RPM and torque ware measured for coefficient of performance (Cp) and tip speed ratio (TSR) calculation. Considering the number of blade, the experiments can classified as VAHT 3 blades, VAHT-SBC 6 blades, and VAHT-SBC 9 blades.

IOP Conf. Series: Journal of Physics: Conf. Series **1022** (2018) 012023 doi:10.1088/1742-6596/1022/1/012023

2. Experimental Methods

2.1. VAHT-SBC Specifications

Fiberglass was used for blade materials, while shaft, turbine arm, and frame used steel as material. Width and depth of the open channel was used to determine the design parameter of the turbine. Table 1 shows design parameter that was used in this research.

The position of passive variable-pitch gives the freedom to move within the range -20^0 to 20^0, as shown in figure 1 (a) & (b). There are airfoil plate on the top of variable pitch blade. The airfoil plate is used as a sacrifice that crash into the inner stopper and the outer stopper, as shown in figure 1 (c).

Passive variable-pitch located at turbine radius (R), while fixed pitch located at 0.75 R and 0.5 R. Considering the number of blade and the arrangement of blade, the experiments can classified as VAHT 3 blades, VAHT-SBC 6 blades, and VAHT-SBC 9 blades, as shown in figure 2. The VAHT 3 blades represents conventional turbines, while VAHT-SBC 6 blades and VAHT-SBC 9 blades represents the innovative turbines proposed in this study.

Tabel 1. Design of VAHT & VAHT-SBC

Design Parameter	Dimension
Turbine diameter (cm)	80
Span of blade (cm)	80
Number of *arm*	3
Shaft diameter (cm)	3.2
Airfoil of blade	NACA 0018
Length of chord (cm)	10
Pitch angle range (^{o})	-20 to 20
The placement of blade 1	R
The placement of blade 2	0,75 R
The placement of blade 3	0,5 R
Frame dimension (cm)	90 x 90 x 134

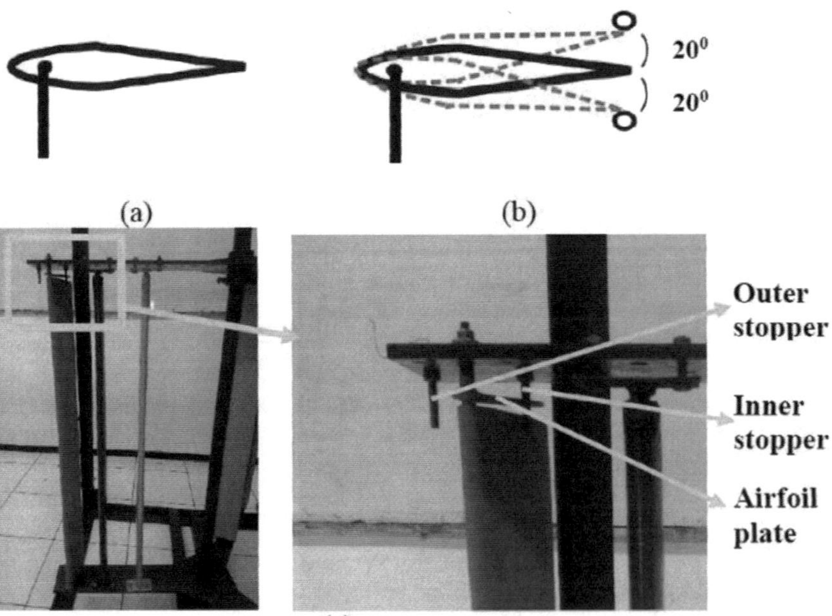

Figure 1. (a) Fixed Pitch, (b) Passive Variable-Pitch and The Pitch Angle, (c) Stopper Mechanism

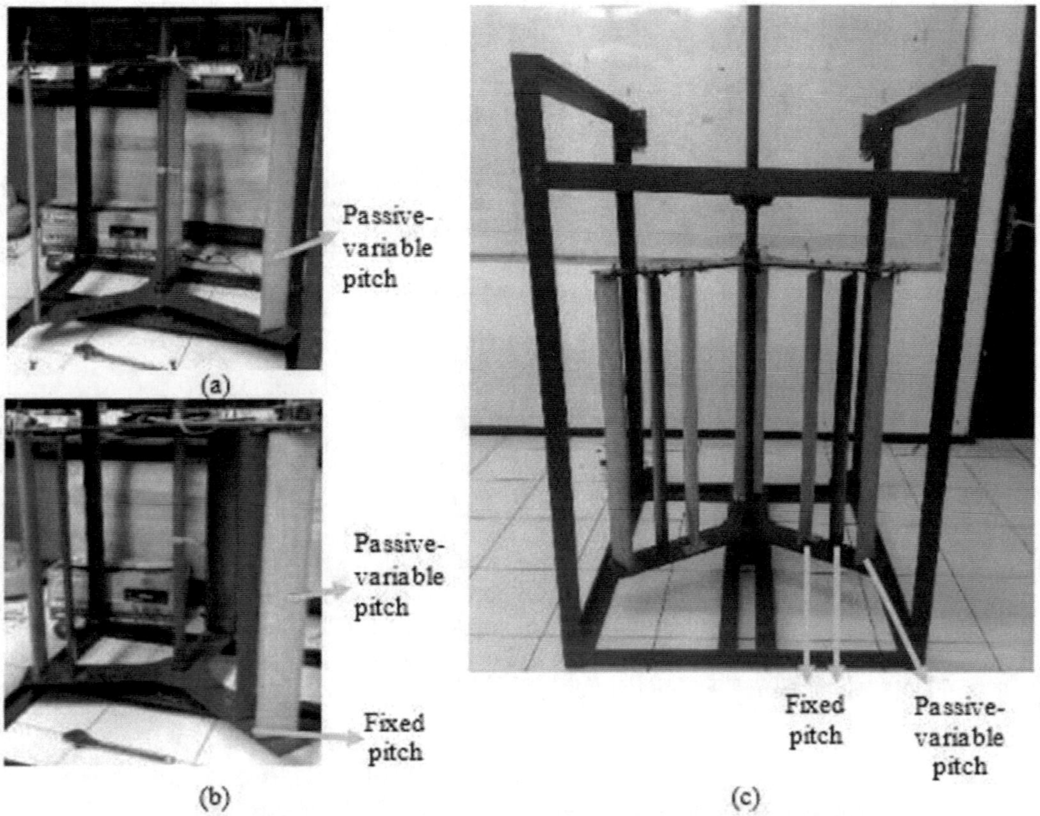

Figure 2. (a) VAHT 3 Blades, (b) VAHT-SBC 6 Blades, (c) VAHT-SBC 9 Blades

2.2. Test location

Experimental study conducted in an open channel flow. The current velocity was 1.1 m/s – 1.3 m/s. Width of the channel is 1 m and depth is 1.7 m, as shown in figure 3 (a). Width and depth of the open channel was used to determine the design of the turbine and the frame.

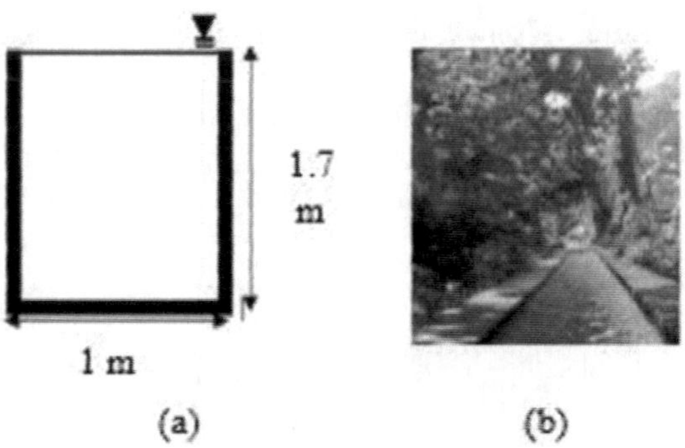

Figure 3. (a) Width and Depth of The Channel, (b) The Open Channel Flow

2.3. Experimental set-up

The current velocity data collection was carried out at 3 locations by using the current meter of Dentan CM-1BX series, as shown in figure 4. Propellers of the current meter ware placed in 0.5 m from the turbine frame, collected by 10 data. The data ware processed to obtain an average value. Average current velocity in 3 locations can be seen in table 2.

IOP Conf. Series: Journal of Physics: Conf. Series **1022** (2018) 012023 doi:10.1088/1742-6596/1022/1/012023

Figure 4. Current meter

Table 2. Current Velocity at 3 Different Location

Location	Average of current velocity (m/s)
Location 1	1.1 m/s
Location 2	1.2 m/s
Location 3	1.3 m/s

Turbine torque data on azimuth (Θ) 0°, 90°, 120°, 180°, and 240° were taken using the JTN 4603BN series torque wrench. Torque wrench was used by locking the tip of the torque wrench to the tip of the turbine shaft, as shown in figure 5. Each azimuth performed 10 torque data retrieval, thus for one model turbine and one variation of current velocity obtained a total of 50 torque data. The data ware processed to obtain average torque for 3 model turbines and for 3 variations of current velocity.

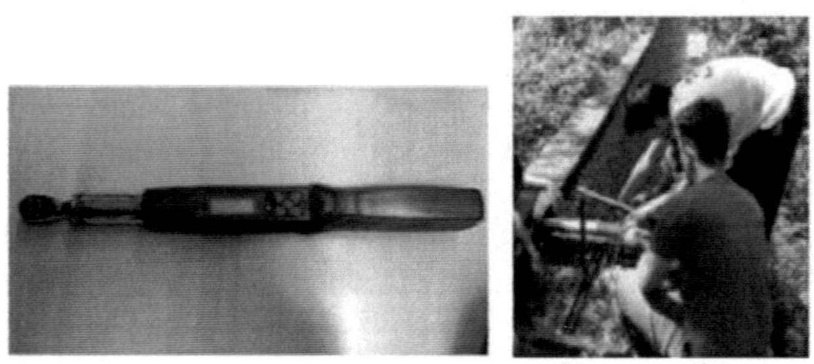

Figure 5. Torque Wrench and The Torque Data Collection

The RPM data (rotation per minute) of the turbine ware taken using the ONOSOKKI HT-3200 series tachometer. The tachometer was used by locking the end of the tachometer to the tip of the turbine shaft, as shown in figure 6. For one variation of current velocity and one model turbine, RPM data ware taken as many as 10 data. The data ware processed to obtain an average RPM for 3 model turbines and for 3 variations of current velocity.

Figure 6. Tachometer and The rpm Data Collection

3. Results and Discussion

Turbine perfomances can be presented in the form of Cp. Cp shows the amount of mechanical power that can be extracted from current power by the turbine. The larger the Cp, the greater the power that can be extracted by the turbine. Cp is a powerful variable to compare the performance of a turbine. Cp is expressed in equation (1) [10]. Where T is torque (N.m), ω is angular velocity (rad/s), ρ is water density (kg/m^3), A is swept area of the turbine (m^2), and V is current velocity (m/s).

$$C_P = \frac{T\omega}{0{,}5\rho A V^3} \qquad (1)$$

Tip Speed Ratio is the ratio between the tangential velocity of the blade and current velocity. The TSR value states how fast the turbine is spinning against a certain current velocity. TSR is expressed in equation (2) [10]. Where R is turbine radius (m), ω is angular velocity (rad/s), and V is current velocity (m/s).

$$TSR = \frac{R\omega}{V} \qquad (2)$$

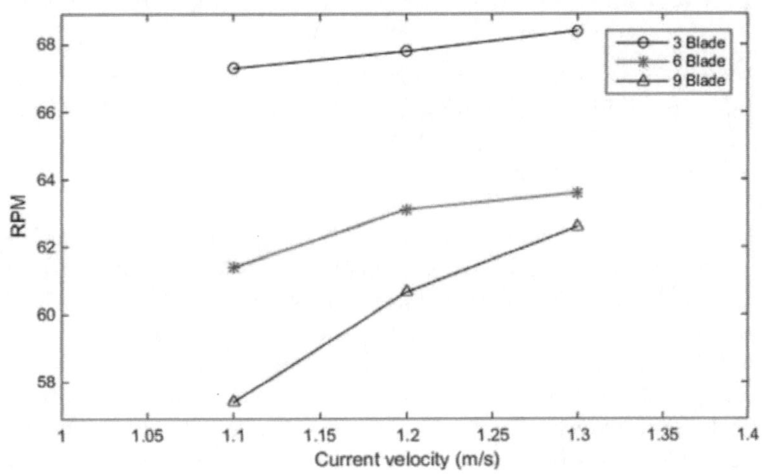

Figure 7. RPM-Current Velocity Curves for Three Model Turbines

Based on figure 7, the relationship between RPM and current velocity is proportional. The higher the current velocity, the higher the turbine RPM. VAHT 3 blades has the highest RPM on the three variations of current velocity, followed by VAHT-SBC 6 blades, then 9 blades. The highest RPM of 68.4 rpm is obtained at a current velocity of 1.3 m/s by VAHT 3 blades. The increase of current velocity has a positive effect on RPM, while the addition of blades has negative effect on RPM.

Moment inertia consists of mass (kg) and radius (m) variable. VAHT 3 blades, VAHT-SBC 6 blades, and VAHT-SBC 9 blades have the same radius. The variables that distinguish between the turbines is the mass. The turbine mass will increase as the number of blades increases. VAHT-SBC 9 blades has a higher mass and a higher moment inertia than VAHT-SBC 6 blades and VAHT 3 blades, respectively. Thus, VAHT-SBC 9 blades has the lowest RPM among the three other turbine variations, because it has the highest moment inertia.

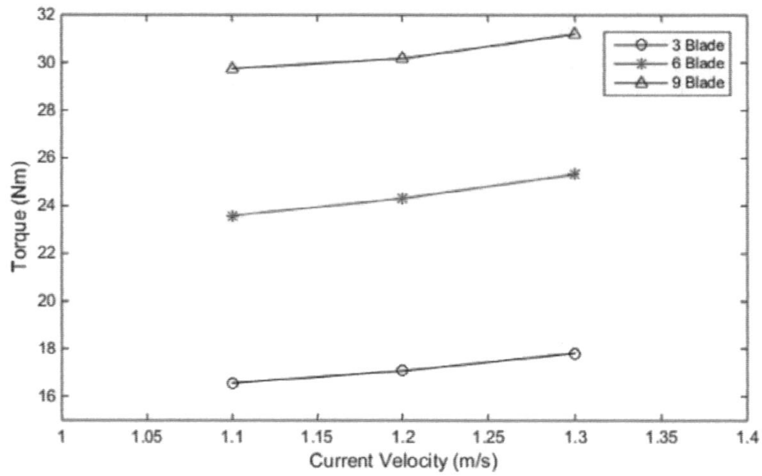

Figure 8. Torque-Current Velocity Curves for Three Model Turbines

Based on figure 8, the relation between current velocity and turbine torque is proportional. The higher the current velocity, the higher the turbine torque. VAHT-SBC 9 blades has the highest torque on all variations of current velocity followed by VAHT-SBC 6 blades and VAHT 3 blades. Increasing the number of blades and current velocity has a positive effect on the torque of VAHT-SBC 6 blades and VAHT-SBC 9 blades.

VAHT-SBC 9 blades has the highest torque between VAHT-SBC 6 blades and VAHT 3 blades for three variations of current velocity. VAHT-SBC 9 blades has a higher moment inertia than VAHT-SBC 6 blades and VAHT 3 blades. The turbine mass will increase as the number of blades increases. The increasing in mass will increase the turbine moment inertia and the torque.

The darrieus type VAHT in this study is a lift device, a turbine that utilizes a lift force from a blade to rotate. The addition of a blade causes an increase in the lift force acting on the turbine. VAHT-SBC 9 blades has more blade, so it gives the largest lift force and also the largest torque than VAHT-SBC 6 blades and VAHT 3 blades. It is because the higher the lift force, the higher the turbine torque. Through the innovation of SBC mechanism, VAHT torque increased without increasing the dimensions of the turbine.

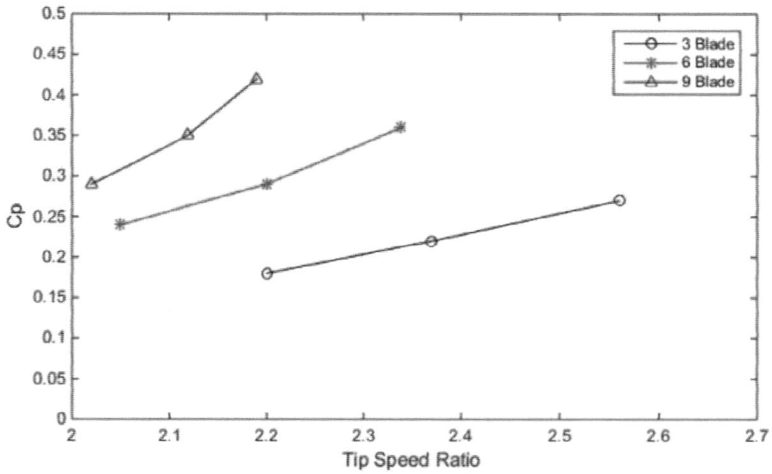

Figure 9. Cp - TSR Curves for Three Model Turbines

Based on the current velocity data, torque and RPM, the value of Cp and TSR can be obtained using the equations (1) and the equation (2). Figure 9 shows the graph of the relationship between TSR and Cp. The highest Cp was produced by VAHT-SBC 9 blades, followed by VAHT-SBC 6 blades and VAHT 3 blades. The highest Cp is 0.42 obtained on TSR 2.19 by VAHT-SBC 9 blades. The Cp value of 0.42 has reached 93% of the maximum theoretical Cp value. While the lowest Cp is 0.18 obtained on TSR 2.2 by VAHT 3 blades. The addition of the number and arrangement of straight blade cascaded has positive effect on Cp of the VAHT-SBC 6 blades and VAHT-SBC 9 blades.

4. Conclusions

In this study, an experimental study was conducted on an open channel to investigate an innovative turbine with the addition of the number and arrangement of straight blade cascaded (SBC). Average torque and average RPM of 3 model turbines (VAHT 3 blades, VAHT-SBC 6 blades, and VAHT-SBC 9 blades) on 3 variations of current velocity ware measured for coefficient of performance (Cp) and tip speed ratio (TSR) of 3 model turbines calculation. The VAHT 3 blades represents conventional turbines, while VAHT-SBC 6 blades and VAHT-SBC 9 blades represent the innovative turbines proposed in this study.

The higher the Cp, the greater the power that can be extracted by a turbine. The highest Cp produced by VAHT-SBC 9 blades (Cp of 0.42 at TSR 2.19), followed by VAHT-SBC 6 blades and VAHT 3 blades. The Cp value of 0.42 has reached 93% of the maximum theoretical Cp value.

The addition of SBC causes to an increase in lift force and moment inertia which has an impact on increasing torque and Cp. VAHT 3 blades, VAHT-SBC 6 blades, and VAHT-SBC 9 blades have the same diameter and height, so the swept area of the turbines is also same. The addition of SBC can increase energy extraction without increasing turbine dimensions. It was found that Cp and torque of VAHT-SBC 9 blades and VAHT-SBC 6 blades (using SBC mechanisms) higher than VAHT 3 blades (without SBC mechanism).

5. References

[1] Peter B and Martin W 2015 Performance Measurements of Cylindrical- and Spherical-Helical Cross-Flow Marine Hydrokinetic Turbines, with Estimates of Exergy Efficiency *J. Renewable Energy* **74** 318

[2] Samaraweera K K M N P, Pathirathna K A B, De Silva H E D and Sugathapala A G T 2010 Development of Darrieus-Type Vertical Axis Wind Turbine for Stand-Alone Applications *Proc. Int. Conf. on Sustainable Built Environment (Kandy)* p 421

[3] Ye Li and Sander M C 2010 Three-Dimensional Effects and Arm Effects on Modeling a Vertical Axis Tidal Current Turbine *J. Renewable Energy* **35** 2325

[4] Mohamed H M 2012 Performance Investigation of H-rotor Darrieus Turbine with New Airfoil Shapes *J. Energy* **47** 522

[5] Brian Kirke and Leo Lazauskas 2011 Limitations of fixed pitch Darrieus hydrokinetic turbines and the challenge of variable pitch *J. Ren. Ener.* **36** 893

[6] Ridho Hantoro, I Ketut Aria Pria U, Erwandi and Aries S 2011 An Experimental Investigation of Passive Variable-Pitch Vertical-Axis Ocean Current Turbine *ITB J.Eng.Sci.* **43** 27

[7] Sang-Hun Han, Jin-Soon Park, Kwang-Soo Lee, Woo-Sun Park and Jin-Hak Yi 2013 Evaluation of Vertical Axis Turbine Characteristics for Tidal Current Power Plant Based on In Situ Experiment *J. Ocean Engi.* **65** 83

[8] Fengmei Jing, Qihu Sheng and Liang Zhang 2014 Experimental Research on Tidal Current Vertical Axis Turbine with Variable-Pitch Blades *J. Ocean Engi.* **88** 228

[9] Ion Paraschivoiu, Francois Delclaux, Philippe Fraunie and Claude Beguier 1983 Aerodynamic Analysis of the Darrieus Rotor Including Secondary Effect *J. Ener.* **7** 416

[10] Sukanta Roy and Ujjwal K S 2015 Wind tunnel experiments of a newly developed two-bladed Savonius-style wind turbine *J. Appl. Ener.* **137** 117

IOP Conf. Series: Journal of Physics: Conf. Series **1022** (2018) 012024 doi:10.1088/1742-6596/1022/1/012024

Analysis of student's scientific attitude behaviour change effects blended learning supported by I-spring Suite 8 application

Rini Budiharti[1*] and Waras N S[2]

[1,2]Physics Education, Teacher Training and Education Faculty, Universitas Sebelas Maret

*Corresponding email: rini.budiharti28@yahoo.co.id

Abstract. This article aims to describe the student's scientific attitude behaviour change as treatment effect of Blended Learning supported by I-Spring Suite 8 application on the material balance and the rotational dynamics. Blended Learning models is learning strategy that integrate between face-to-face learning and online learning by combination of various media. Blended Learning model supported I-Spring Suite 8 media setting can direct learning becomes interactive. Students are guided to actively interact with the media as well as with other students to discuss getting the concept by the phenomena or facts presented. The scientific attitude is a natural attitude of students in the learning process. In interactive learning, scientific attitude is so needed. The research was conducted using a model Lesson Study which consists of the stages Plan-Do-Check-Act (PDCA) and applied to the subject of learning is students at class XI MIPA 2 of Senior High School 6 Surakarta. The validity of the data used triangulation techniques of observation, interviews and document review. Based on the discussion, it can be concluded that the use of Blended Learning supported media I-Spring Suite 8 is able to give the effect of changes in student behaviour on all dimensions of scientific attitude that is inquisitive, respect the data or fact, critical thinking, discovery and creativity, open minded and cooperation, and perseverance. Display e-learning media supported student worksheet makes the students enthusiastically started earlier, the core until the end of learning

1. Introduction

Implementation of learning in the curriculum 2013 is a learners-centred learning. The availability of good facilities and infrastructure including the internet facilities at most of schools need to be optimized for e-learning based learning. According to [9], e-learning is able to enhance the learning experience because students can learn anywhere and in any condition as long as they can access the internet without having a face to face learning. However, according to several research reveal that students would prefer a face to face learning to web based learning. Web based learning could be combined with face to face learning [2]. Therefore, there is needed to combine between e-learning and conventional learning in a class, called a blended learning model. At the high school level, this learning model is suitable to be applied because the student competent character still needs accompaniment from the learner. The blended learning is a learning strategy that integrates between face to face learning with a web based learning, which combines many varieties of media in a learning management system [13]. Blended learning integrates-or blends-learning programs in different formats to achieve a common goal [20]. Thus, blended learning model consists of online learning component and face to face learning component so that the two components implementation of the learning division is noteworthy on two components will be suitable with the learning competence target. Related to this, [4] presents the learning alternative options, one of them is pure class model. In this model all learning activities are done in the classroom, but there are given tasks for the learners to access the internet/web. The research results [7] which is done in junior high school of the Surakarta Residency region shows that science learning through blended learning type of pure class effective applied in learning compared to four other learning model such as Learning Cycle 7E. Joyfull Learning, ARCS and Cooperatif Learning Type of TAI.

There are three main learning theories that are used as the basis for web based learning, namely: behaviorism, kognitivism and constructivism [23]. Blended learning model is the solution for the limitation hour of lessons that a teacher has with puts involvement of learners in the learning agenda focus on learners or Student Centered Learning. The participants have positive perceptions of blended learning approach [18]. It supports the opinion of the Humazah that states that blended learning open learning paradigm shift from teaching to learning centered toward the new paradigm that is centralized to the learners. In addition Huzamah also states that the model based on the mindset of Technological, Pedagogical, and Content Knowledge (TPACK). TPACK in implementation of learning is the necessity of teacher to think supporting lesson technologies, the lesson strategies (including approaches, models and methods), as well as what materials will be taught in order to make students reach to competencies target.

In practice, [8] mentions five keys as the guidelines for implementing the learning using blended learning, namely: 1) Live Event: is a direct learning or face-to-face (instructor-led instruction) that is synchronal in time and in the same place (classroom) or in the same time but in the different places (such as the virtual classroom) or in the same time but different places (such as the virtual classroom). 2) Self-Paced Learning: is combining a independent learning (self-placed learning) that allows the learners to learn anytime, anywhere using a variety of contents (learning materials) that are specifically designed for independent study are either text-based or multimedia-based (video, animations, simulations, images, audio, or a combination of all of them). 3) Collaboration: is combining collaboration, either a collaboration of teachers, or a collaboration among learners who can both cross-school/college, such as chat rooms, discussion forums, e-mail, website/weblog, mobile phone. 4) Assessment: is that the designer must be able to arrange assessments either assessment test or a non-test assessment, or a test that is more authentic (authentic assessment/portofolio) in the form of projects, products and so on. Besides that, it also needs to be considered the arrangement between the forms of online assessment and offline assessment. 5) Performance Support Materials: make sure that the learning materials in digital forms are accessible to the participants to learn in both offline and online. Hence, in a simple definition blended learning model is the learning model which combines the online version learning and the class version or face-to-face learning.

Underlining what are presented by [15], media support in the implementation of blended learning is something important. The software supporting packaging learning media e-learning based is evolving so rapidly, one of them is i-spring suite 8. It is a software to create fast learning tools produced by i-sprng to develop e-learning professional learning on power point which is linked to quizzes, surveys, and interaction [17]. I-spring is one of the tools that change the presentation file into flash and SCORM/AICC forms, which are forms that commonly used in e-learning through Learning Management System (LMS). The program i-spring suite 8 has swf extension which can easily connect to the internet by embedded it, so that it directs the learning into interactive one. Through the phenomena or facts presented above, students are directed to actively interact with the media display animations or other forms, as well as interact with other students to discuss in order to get the concept of learning in each stage. It is asserted by [10] that the finding indicates that students taught by computer supported animations were better than those in traditional teaching group.

In fact, the physics is a clump of science includes the process, products and attitudes. Hence, the results of learning physics are not only seen from the results of domain cognitive but also include attitude. According to [6] attitudes are general evaluations people hold in regard to themselves, other people, objects, and issues. Scientific attitude in learning science is often associated with attitudes towards science [14]. According to [12] scientific attitude of learners are essentially no different from any other skills (cognitive, social, process and psychomotor). The scientific attitude is closely connected with the results of the learning learners. Bloom [6] refers to the importance of attitudes by stating the effect of affective domain behaviours in achieving cognitive domain behaviours in education. The scientific attitude is one of the factors that influence the results of the study [19]. Thus, it can be concluded that the scientific attitude is a form of intelligence or skills that are needed in the learning process and affect the results of his studies. The dimensions of attitude inquisitive, data/facts appreciative, critical thinking, discoveries and creativity, open minded and cooperation, diligence and environmental change sensitivity [14]. The description of the students attitudes in each dimensions is colored by learning techniques applied by the learners. Student Centered Learning (SCL) model will create a different description of students attitudes in Teacher Centered Learning (TCL). Student Centered Learning (SCL) seeks any dimensions will be developed optimally.

2. Methods

This research was conducted using a model Lesson Study. According to [16] lesson study is a systematic process used by teachers in Japan to test the teaching effectiveness in order to improve the learning outcomes. The systematic process is teachers work collaboratively to develop the learning plans and learning devices, doing observation, reflection and revision of learning plans in cycle and continuously. According to [16] lesson study is a model of the educators professions development through the learning teaching collaboratively and continuously based on the principle of kolegalitas and mutual learning to build a learning community in order to improve the teacher professionalism and the learning quality. Lesson study is a single instance of action research that consists of stages: 1) study: a Study curriculum and standard, consider long term goals and for student learning and development; 2) plan: select a research lesson, anticipate student thinking, data collection and lesson plan; 3)do research lesson: one team member, teaches other collect data;4) Reflect: share data, what was learned about student learning, what are implication for this unit and more broadly, what understandings and new questions do we want to carry forward in our work. According to Wikipedia, Lesson Study is done through four stages by using the concept of Plan-Do-Check-Act (PDCA) [1].

2.1. The Stages of Planning (plan)

This step begins with an analysis of needs over the problems found in the study, namely that indication that teacher centered learning as well as the internet utilization as an e-learning based medium is not yet optimal. Further, the research teams together with physics teachers in the schools collaborate in lesson study to arrange a lesson plan which reflects students-centered based and e-learning based, in this case the blended learning model is applied. Through the lesson plan learning process becomes a very ripe plan, which can anticipate all possibilities that will occur during the learning implementation, either in the early, the core up to the final stage of learning.

2.2. The Stages of Implementation (do)

In this stage, there are two main activities, namely: the implementation of learning activities conducted by the model teacher to practice the lesson plan that is arranged together, and the observation activities on the learning implementation that are done by another member or lesson study community that is research team. There are some things that need to be noticed in the stage of implementation, namely: 1) the teacher do the learning as instructed in the lesson plan. 2) the students are attempted to be able to experience the learning process in a proper and natural setting, not in a under pressure condition that is caused by the lesson study program. 3) during the learning activities, observer is not allowed to interfere the teacher and student' concentration. 4) the observers observes carefully towards the interaction of students, student-learning materials, students-teacher, students-other environments, using the observation instrument which has been previously prepared and compiled together. 5) observers should be able to learn from the learning and not to evaluate the teacher. 6) observers can record via a video camera or a digital photo for documentation and materials analysis purposes as long as the recording activities do not interfere with the learning process. 7) observers take a note about the students' attitude during the learning process takes place, for the example about the students' comments or discussions and it it possible to write down the students' name, there is a construction process of understanding of the related students. The note is made based on the guidelines and the sequences of students' learning experiences that are listed in the lesson plan.

2.3. Stages of Reflection (check)

The stages of reflection activities are done in the form of discussion between the model teacher and the observer (research team), as in a improvement effort for further learning process. The discussion starts with model teacher conveys their impression after finishing the lesson plan, for examples their general or special impression of the learning process such as the difficulties or problems when doing the arranged lesson plan, including the effectiveness of e-learning media. Furthermore, all observers wisely deliver the response or the suggestions towards the learning process that has been implemented (not towards the teacher). When delivering suggestions, the observer must be supported by evidence obtained from observations, not just based on the opinions. The various talks that are developed in the discussion can be used as feedbacks to all participants for the correction or improvement purposes of the learning process.

2.4. The Stages of Follow Up (act)

From the results of the reflection activities, those can be acquired a number of new knowledge or important decisions in order to improve and enhance the learning process which refers to the learning targets. Those are as materials for model teacher as well as observer (research team) to develop a better learning process. The subject of this research is the Eleventh grade of MIPA 2 at SMAN 6 Surakarta. The object of this research is the scientific attitudes about student behaviour on learning blended learning i-spring suite 8 based especially in the material of balance and rotation dynamic, which is done by two cycles of learning. The research data are obtained through observation techniques, then are validated through triangulation techniques namely interviews and document review. The data analysis technique that is used is descriptive analysis that refers to analysis models, [21] includes three components, namely 1) data reduction; 2) presentation of data; and 3) withdrawal and verification conclusion.

3. Finding and Discussion

Discussion of the student's scientific attitude is based on the dimension of the scientific attitude. These dimensions include dimension of attitude of inquisitive, data/facts appreciative, critical thinking, discoveries and creativity, open minded and cooperation, diligence and environmental change sensitivity.

3.1. The Dimensions of The Inquisitive Attitude

The change off scientific attitude of student's inquisitive attitude dimension on enthusiast indicator of looking for answer can be explained that in the pre-treatment learning process takes place through lectured method using Blackboard media, in the beginning of learning students get information about the material topic that will be taught, then teachers come into the core of learning, so that students tend to be passive. Different post treatment through blended learning based i-spring, in the beginning activities the lesson starts with showing e-learning media about phenomena related to the concept that will be taught, learners are enthusiastic about the problems, they sought an answer to the problem, and then the teacher directs the students to propose the hypotheses. The explained can be illustrated by figure 1,2.

Figure 1. Scientific attitude of learners pre treatment on the dimension of the inquisitive

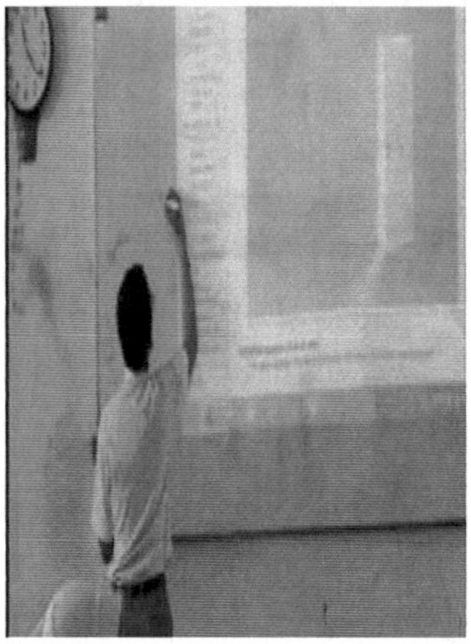

Figure 2. Scientific attitude of learners post treatment on the dimension of the inquisitive

3.2. The Dimensions of The Inquisitive Attitude

The change scientific attitude of student's data/facts appreciative attitude dimensions on indicator of explain the observation result objectively/honestly can be explained that the activity of communicating is an activity that the most endeared of the most learners as they can communicate the results of their discussion expressively. In this activity, learners are seen showing a pretty good behaviour namely learners honestly communicate the observations the same with results of the discussion.it also occurs in learning through blended learning-based i-spring, students honestly present or communicate the discussion results of the group, however the activities look more structured because each member of the group, however the activities look more structured because each member of the group is able to play in appropriate role and their responsibility. These explanations can be illustrated by figure 3,4.

Figure 3. Scientific attitude of learners pre-treatment on the dimension of data/facts appreciative attitude

Figure 4. Scientific attitude of learners post-treatment on the dimension of data/facts appreciative attitude

3.3. The Dimensions of The Critical Thinking Attitude

The change of scientific attitude of students' critical thinking attitude dimensions on the indicator of asking for any changes/new things can be explained that in the pre treatment although learning is informative but in each step of the learning that has been finish by the teacher always give the opportunity to the students to ask about learners understanding, however the quantity of students who ask still few and students tend to passively follow the learning process, they are less concerned with the learning situation. In the learning through blended learning, the learners have a discussions in a group with the help of e-learning media i-spring based and learners' activity sheet so that the learners in groups are practice to communicate their opinion with each other. Hence, the learners are being active to ask a friend or teacher related to the data analysis in learner activity sheet and new discovered things. These explanations can be illustrated by figure 5,6.

Figure 5. Scientific attitude of learners pre treatment on the dimension of the critical thinking attitude

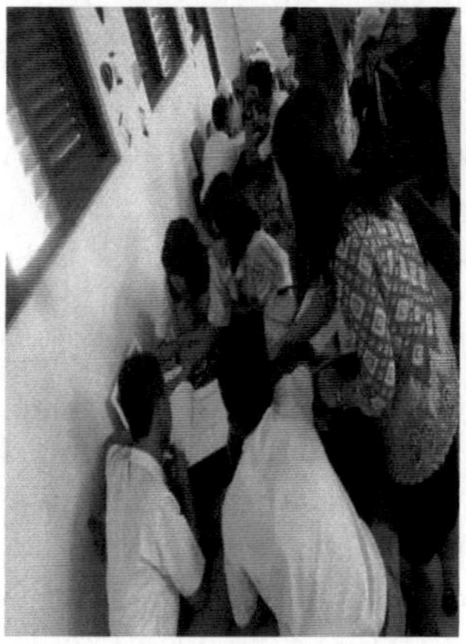

Figure 6. Scientific attitude of learners post treatment on the dimension of the critical thinking attitude

3.4. The Dimensions of The Discoveries And Creativity Attitude

The change of scientific attitude of students' discovery and creativity attitude dimensions on indicator of explaining the result of observations in with appropriate facts can be explained that in the pre treatment learning is informative, there is no chance to the students to have a discussion, students are less concerned with the learning situation, students' activities summarize and record all the information that are conveyed by the teacher and open the guidance book. In learning through blended learning, guided by e-learning media and I-Spring based directs the students to do the steps of learning in groups, furthermore the students can summarize the learning result with their own understanding based on acquired data/facts, guided by the learners' activity sheet that is explained by the teacher/downloaded in e-learning. These explanations can be illustrated by figure 7,8.

Figure7. Scientific attitude of learners pre treatment on the dimension of the discoveries and creativity attitude

Figure 8. Scientific attitude of learners post treatment on the dimension of the discoveries and creativity attitude

3.5. The Dimensions of The Open Minded And Cooperation Attitude

The change of scientific attitude of students' open minded and cooperation attitude dimensions on indicator of willingness to renew the opinion and the conclusion can be explained that the students' activity on the pre-treatment activities when class discussion, most of students tend to be passive when only accepting the class discussion results. It shows the learners' boredom in the learning activities. In contrast from the study through blended learning i-spring based, class discussion is done critically so that new things can be found. Students actively express their opinion based on logical analysis and accurate evidence. These explanations can be illustrated by figure 9,10.

Figure 9. Scientific attitude of learners pre treatment on the dimension of the open minded and cooperation attitude

Figure 10. Scientific attitude of learners post treatment on the dimension of the open minded and cooperation attitude

3.6. The Dimensions of The Diligence Attitude

The change of scientific attitude of students' diligence attitude dimensions on indicator of completing scientific activities can be explained that at the time the pre treatment learners are given a group task in the end of the learning which has function to clarify the taught concepts, although has not yet completely done but still need to be submitted. This occurs over the teacher's permission because the learning is oriented in weariness materials. In the study through blended learning i-spring based, the learning is oriented in the process where the learners are involved in the scientific learning steps which refers to the existing activity in learners' activity sheet. Thus the learners' task is not to end the learning process but it is integrated into the learning process based on the learning steps consecutively procedurally. These explanations can be illustrated by figure 11,12.

Figure 11. Scientific attitude of learners pre treatment on the dimension of the diligence attitude

Figure 12. Scientific attitude of learners post treatment on the dimension of the diligence attitude

3.7. The Dimensions of The Environmental Change Sensitivity Attitude

The change of scientific attitude of students' environmental change sensitivity attitude dimensions on indicator of actively participating in a group discussion can be explained that in the pre treatment learning process takes place using blackboard media as a learning medium and lecture method so that the learners tend to be less care in early learning. Different from the learning through blended learning i-spring based, supported by display e-learning media the learners become enthusiastic to learn by reading and asking the teacher about learners' activity sheet which their group accepts. That is because the learners interest in the learning activities. These explanations can be illustrated by figure 13,14.

Based on the analysis presented, it can be concluded that the blended learning model that is supported by i-spring media can effectively change the students attitude which covers all dimensions of scientific attitude. The conclusions above reinforcement from multiple research results as summed up by [5] concerning about the effectiveness of i-spring media in the learning process can be concluded that the media is overall effective and suitable as a learning medium. Student scientific attitude that are taught with the STAD type cooperative learning combine with e-learning more positive (higher) compared with the student scientific attitude that are taught with the STAD type cooperative learning [11]. In addition, based on [3] about the i-spring media concludes that the usage of the computer learning media i-spring has a positive effect or can improve the learning outcomes of the profession program students at STIKes Karsa Husada.

Figure 13. Scientific attitude of learners pre treatment on the dimension of the environmental change sensitivity attitude

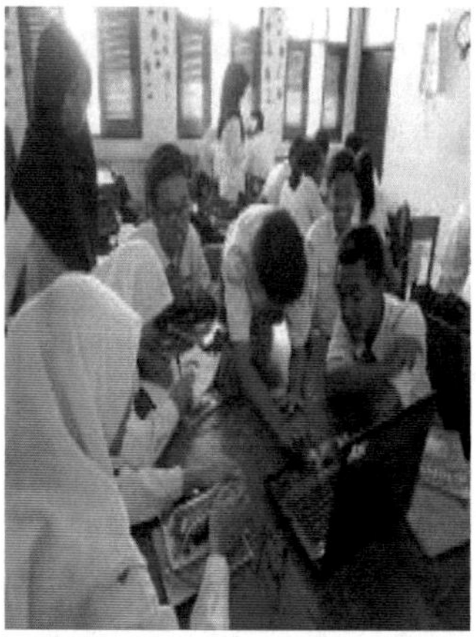

Figure 14. Scientific attitude of learners post treatment on the dimension of the environmental change sensitivity attitude

4. Conclusion and Recommendation

Based on the discussion, it can be concluded that the nature of Blended Learning model is a model of learning which combines online version learning and class or face to face leaning. Either online learning or face-to-face learning are carried out based on existing guidance or chosen learning alternative. There are three main learning theories that are used as the basic for web-based learning, namely: behaviourism, cognitivism and constructivism. In addition, the model is based on the mind set of Technological, Pedagogical, and Content Knowledge (TPACK). The implementation of blended learning need a media support, namely i-spring suite 8. Through the phenomena or facts presented in media impressions, students are instructed to actively interact with media animations slide or other forms, as well as interact with other students to discuss in order to get the concept of learning at each stage of learning. The learning using blended learning which supports media i-spring suite 8 is able to give behaviour change effects on all dimensions of scientific attitude of students namely attitude of inquisitive, data/facts appreciative, critical thinking, discovery and creativity, open minded and cooperation, diligence and environmental change sensitivity.

The results of this research can be used as the basic for further research and can be used as a joint effort among teachers, parents, students and schools in improving the quality of physics learning in a senior high school. The learning alternative or the learning guidance which is chosen in learning using blended learning must be adapted with the needs of the learning material, time allocation and personal computer facilities belonging to the learners.

5. References

[1] Administrator 2016 Pengertian dan tahapan-tahapan pembelajaran lesson study (Bandung: SMA BPI 2 Bandung) h 2

[2] Alessi S M and Trollip S R 2001 *Multimedia for Learning: Methods and Development* (Boston: Allyn and Bacon) p 10

[3] Alfiyansyah R 2016 Penggunaan media pembelajaran i-spring presenter untuk meningkatkan motivasi belajar dan hasil belajar pada mata kuliah keperawatan dasar nutrisi *Pedagogia Jurnal Ilmu Pendidikan* **14** 368-69

[4] Anitah S 2009 *Teknologi pembelajaran* (Surakarta: Inti Media Surakarta) p

[5] Aulia Y and Adeng S 2015 Perencanaan pembuatan media pembelajaran menggunakan metode computer based instruction pada pelajaran tik (kelas viii SMP n 39 Semarang) *Edu*

IOP Conf. Series: Journal of Physics: Conf. Series **1022** (2018) 012024 doi:10.1088/1742-6596/1022/1/012024

Komputika Journal **2**(48)

[6] Bekmezci M Celik I Sahin I Kiray A and Akturk A O 2015 Analysis of scientific attitude, computer anxiety, educational internet use, problematic internet use, and academic achievement of middle school student according to demographic variables *International Journal of Social, Behavioral, Educational, Economic, Business and Industrial Engineering* **9** 4002

[7] Budiharti R et al 2015 Kajian kualitas efektivitas *blended learning* ipa terpadu berbasis SETS di SMP wilayah eks karesidenan Surakarta *Jurnal Materi dan Pembelajaran Fisika (JPMF)* **5** (51)

[8] Carman J A 2005 Blended learning design Diperoleh 25 Januari 2017 dari http://www.gilantlearning.com/pdf/Blended- Learning-Design.pdf/

[9] Castle S and Trollip S R 2010 An analysis of student self-assesment of online, blended, and face-to-face learning environments: Implications for sustainable education delivery *International Education Studies* **3** 36-40

[10] Dasdemir I dkk 2008 The effect of animation technique on teaching of acids and bases topics *Journal of Turkish Science Education* **5 ..**

[11] Dina L 2016 Kumpulan Abstrak Hasil Penelitian Disertasi dan Tesis Program Pascasarjana UM 02

[12] Fakhruddin 2010 Sikap ilmiah siswa dalam pembelajaran fisika dengan penggunaan media komputer melalui model kooperatif tipe STAD pada siswa kelas X3 SMA Negeri 1 Bangkinang Barat *Jurnal Geliga Sains* **4**(2)

[13] Gunawan 2015 *Model Pembelajaran Sains Berbasis ICT* (Matararam NTB: FKIP UNRAM) h 23

[14] Herson A 2009 Penilaian sikap ilmiah dalam pembelajaran sains *Jurnal Pelangi Ilmu* **2**(3)

[15] Huzamah 2014 *Pembelajaran Bauran (Blended Learning)* Jakarta: Prestasi Pustakaraya

[16] Ibrohim 2010 *Panduan Pelaksanaan Lesson Study di KKG* (Malang: Universitas Negeri Malang) h 10

[17] Ispring 2012 Diperoleh 2 Maret 2017 dari http://www.ispringsolutions.com/blog/meet-ispring-suite-8-the-most-powerful-e-learning-authoring-tool-for-powerpoint/*Preprint* gr-qc/0125002

[18] Kosar G 2016 A study of EFL instructors perceptions of blended learning *International Conference on Teaching and Learning English as an Additional Language* 4

[19] Maretasari E 2012 Penerapan model pembelajaran inuiri terbimbing berbasis laboratorium untuk meningkatkan hasil belajar dan sikap ilmiah siswa *Unnes Physics Education Journal* **1** 2

[20] Rusman 2012 *Belajar dan Pembelajaran Berbasis Komputer* (Bandung: Alfabeta) h 303

[21] Sugiyono 2013 *Metode Penelitian Malang* (Bandung: Alfabeta) h 404

[22] Widodo H P 2015 *Penelitian Tindakan untuk Penulisan Tesis dan Disertasi* (East Java Indonesia: Peoyek Penelitian Tindakan dengan Cambridge University Press) h 25

[23] Wijaya M 2012 Pengembangan model pembelajaran e-learning berbasis web dengan prinsip e-pedagogy dalam meningkatkan hasil belajar *Jurnal Pendidikan Penabur* **11**(23)

Designing electronic module based on learning content development system in fostering students' multi representation skills

I Resita[1] and C Ertikanto[2]

[1]Student of Master of Physics Education Program, Faculty of Teacher Training and Education, University of Lampung, Indonesia,
[2]Physics Education Department, University of Lampung, Indonesia.

Corresponding email: [1]isni.resita.ir@gmail.com, [2]chandrafkip@gmail.com

Abstract. This study aims to develop electronic module design based on Learning Content Development System (LCDS) to foster students' multi representation skills in physics subject material. This study uses research and development method to the product design. This study involves 90 students and 6 physics teachers who were randomly chosen from 3 different Senior High Schools in Lampung Province. The data were collected by using questionnaires and analyzed by using quantitative descriptive method. Based on the data, 95% of the students only use one form of representation in solving physics problems. Representation which is tend to be used by students is symbolic representation. Students are considered to understand the concept of physics if they are able to change from one form to the other forms of representation. Product design of LCDS-based electronic module presents text, image, symbolic, video, and animation representation.

Key words: electronic module, learning content development system, multi representation

1. Introduction

Facing the rapid development of the 21st century era, educators and educational institutions are required to deliver students who have 21st century competence. Based on [1], there are ten types of 21st century competence, some of them are critical thinking, problem solving skill, decision making skill, effective communication skill, multi representation skills, information literacy, and ICT (Information and Communication Technology) literacy. In order to develop students' information literacy and ICT literacy competence, ICT integration in the learning process is needed. ICT integration, including internet, as a learning tool can support the increasing of students' information literacy and ICT literacy competence effectively [2]. Currently, ICT has been widely applied in the learning process. Many learning media are made by utilizing information and communication technologies such as computers. Even laboratory practice activities can now be done by utilizing ICT. Information and communication technology certainly has a large effect to overcome the limitations in the learning media.

Students' difficulties in understanding the lessons are influenced by some factors: (1) students only rely on one supporting book and it is the student worksheet. It shows that in terms of learning material, it is still very limited, (2) there are still only few supporting media for the learning process such as internet, computers, and other media, and (3) because of the limited media, students are rarely invited to practice directly. These three factors have caused the students' low interest and motivation in the learning process. Research findings in the recent years provide a strong evidence on the importance of learners' control in the learning process that they undergo [3]. It is because the quality of education has relevance with the quality of the graduates, while the quality of the graduates is determined by learning process. While for educational institutions, students' high achievement indicates the success of the institutions in the learning process. As a professional teacher, the making of learning media becomes relatively easier when the development of this technology is optimally developed and implemented in the classroom learning process. Teachers can integrate educational technologies which enable the efficiency of the academic learning [4].

Based on an observation on physics learning process in high school, it is discovered that many factors influence the learning process, one of them is the students themselves. Students have different learning styles that influence the absorption of information. Based on data, it shows that 80% of the students claimed that it is difficult to understand physics material. There are 50% of the students who claimed that they quickly get bored when they study physics because the media that are used in the learning process are monotonous and uninteresting. This is in contrast to the results of interviews with high school physics teachers who claimed that their students are enthusiastic when they study physics, only a few students who look unenthusiastic. Teachers also claimed that students are able to follow the learning process and able to answer questions well in the question and answer session. Whereas in fact, many students still have difficulty in understanding physics material, this may be because of the deficiency in the variety of the media that are used in learning process. 95% of students claimed that in physics subject, especially the images and the formulas in the optical geometry chapter are still difficult to be translated into verbal definitions. Furthermore, students only use one type of representation to solve physics problems.

Skilled students tend to use non-mathematical representation, while less skilled students tend to directly use mathematical representation [5]. Students' success in solving physical problems is influenced by the representation format of the problems. Students also claimed that there is a significant effect on the learning approach that is used on their representational skill [6]. During this time, physics teaching process tends to use mathematical approach. Teachers spend too much time on math problems. Mathematical approach is conducted by giving examples and practice questions. Samples of questions are given in the classroom, the samples are either gotten from the textbooks or done on the board directly. For the further practice, homework is given to the students. Questions for the homework are usually in form of essays of three to five questions and they are taken from the textbooks [7].

Associated to multi representations, using multiple representation to support complementary processes rests on the now extremely well known observation that even representations that are informationally equivalent still differ in their computational properties [8]. Representation is something that has, describes, or symbolizes an object or a process. Multi representation means re-representing same concepts with different formats including verbal, picture, graphic and mathematic [9]. Multiple representational formats in a particular concept learning process give an opportunity to communicate it well, as well as how they work with systems and processes of a particular physical concept [10]. Multi representation in geometry proving learning process proposes three forms of representation; problem representation, visual representation, and evidence representation [11].

Multiple representations can be used so that one representation constrains interpretations of another one. Often learners can find a new form of representation complex and can misinterpret it. In this case one might use a second, more familiar or easy to interpret, representation to support learners' understanding of new complicated representation [12]. Each student has different specific ability from one another, some show that their verbal ability is more prominent, some show that their spatial and quantitative ability are stronger, and the rest show that their visual and graphic ability are more solid. Performance of various representations when a concept is explained will certainly give an opportunity to the students to be able to understand the concept using their own specific abilities. This is in line with opinion [13], which states that multiple representations can support construction of a deeper conceptual understanding. Along with the development of computing technology field, representations of various physical quantity interaction in a phenomenon can be presented by using dynamic formats in form of animations and simulations [14].

One of the media that can present material in multi representation is module. Module is a set of learning media that is arranged systematically. It is a complete teaching unit designed to be used by a teacher or a small group of learners without the presence of instructor [15]. The purpose of this module is to facilitate the learning process without regular supervision, all elements of a course which are given by teachers normally should be formed into a set of printed, audiovisual, or computer-based materials (or any combination of them).

Although module can be used independently by students as learning media, it should be still combined with face to face learning activities with teacher. Learning method that combines between e-learning system with conventional or face to face method is called blended learning. Blended learning is a combination of various learning models aimed to optimize learning process either it is distance, traditional, media using, or even computer based learning process [16]. Students who are taught by using blended learning method have higher achievement than those who use traditional one [17]. Students who use blended learning method also have higher average score than those who use traditional learning methods [18].

One of the programs that can be used to construct interactive modules that can be used in blended learning process and to allow the development of multi representation skills is LCDS. Based-on [19], Microsoft Learning Content Development System (LCDS) is a free tool that enables Microsoft training and certification community to create high-quality, interactive, online courses and Microsoft Silverlight Learning Snacks. LCDS allows anyone in Microsoft training and certification community to publish e-learning courses and Learning Snacks by completing easy-to-use LCDS forms that seamlessly generate highly customized content, interactive activities, quizzes, games, assessments, animations, demos, and other multimedia. LCDS allows anyone in a specific community or organization to publish e-learning using LCDS easily with customizable content like interactive activities, quizzes, games, exams, animations, demos, and other multimedia [20]. LCDS helps to create content with text and pictures, interactive activities, quizzes, games, assessments, animations, demos, and other multimedia. It also provides the decision to embed pictures, movies, audio files, flash and silverlight content into the course. These features of LCDS and its support for teaching and learning gives a clear indication that would be appropriate to design and develop a suitable learning set for teaching and learning of electrical and electronic technology concepts [21].

Based on the description above, the researchers made an alternative way by designing an interactive learning media in form of electronic modules that can present a more attractive physics material by utilizing the development of information technology. The researchers conducted a development research aimed to produce a design of an interactive module based on Learning Content Development System (LCDS) to cultivate students' multi representation skills in Geometry Optical materials.

2. Method

This research can be classified as education development research, which is a development of LCDS-based electronic physics module especially in optical geometry material to foster multi representation skills. The product of this development research is an electronic module based on LCDS especially in optical geometry material to foster multi representation skills.

This research uses research and development method based on media development research design by Sugiyono [22]. Sugiyono's research design step consists of 10 steps: 1) potentials and problems, 2) information gathering, 3) product design, 4) design validation, 5) design improvements, 6) product trials, 7) product revisions, 8) usage trials, 9) design revisions, and 10) mass production. Sugiyono's research design step that is used in this study is limited only from the first to the third steps. Design validation, product testing, until mass production are not implemented because of the researchers' limited time and fund.

Data collection techniques which are used by the researchers in this research is obtained through questionnaire, observation, and interview. Questionnaires, observations, and interviews are used to analyze students' initial needs at the needs analysis stage. Questionnaires were given to 90 students and 6 physics teachers in 3 different Senior High Schools in Lampung Province. Questionnaire data analysis technique was done by: 1) coding or classifying data; 2) performing data tabulation based on the classification made; 3) scoring the respondents' answers; 4) processing total score of the respondents' answers; 5) calculating questionnaire response percentage on each item; 6) calculating questionnaires' mean percentage; 7) visualizing the data; and 8) interpreting the score thoroughly.

3. Result and Discussion

The result of the development research that has been done is the LCDS-based electronic module to cultivate students' multi representation skills especially in geometric optical materials. This research is conducted through some steps based on development procedures that have been modified. The details of the results from each step are explained as follows:

3.1 Potentials and Problems

Based on the research which was conducted in 3 different Senior High Schools in Lampung Province, it is founded that methods which are used by teachers in the learning process is not attractive and monotonous, then it often makes students feel bored. 80% of the students claimed that it is difficult to understand physics materials, and 50% of the students claimed that they quickly get bored during the physics learning process because of the monotonous and uninteresting learning media. This is in contrast to the results of interviews with high school physics teachers who claimed that their students are enthusiastic during the physics learning process, only a few students who are not enthusiastic. Many students still have difficulty in understanding the physics material. It is because of the deficiency in the variety in the learning media . 95% of the students claimed that in physics subject, especially images and formulas in the optical geometry chapter are still difficult to be translated into verbal definitions. Furthermore, students only use one type of representation to solve physics problems, which is symbolic or mathematical representation.

3.2 Gathering Information

Based on the potentials and problems, information is collected until solution is gotten by developing an LCDS-based electronic module design to cultivate multi representation skills. The developed module presents material exposure with various representations, which are image, symbolic, and verbal.

3.3 Product Design

Initial product or module design is made by identifying module's material and format that are going to be produced . Product is developed by using LCDS program that is supported with other various softwares such as Macromedia Flash, iSpring quizMaker, and Wondershare Quiz Creator. The developed electronic module contains information that is presented with various representations such as light properties, light reflection, and light refraction.

The developed product is an interactive module with LCDS program. The said module is an interactive module to cultivate multi representation skills in Geometry Optical material. Before conveying the learning content, there is a cover page which was created by using introduction template with the module title, image associated with the material, author label, and the agency of the interactive module publisher label. After the cover, there is main page that is divided into three points, such as "User Guide", "Profile", and "Introduction". User manual submenu contains some buttons and its functions that are included in the interactive module. Content was made by using animation template. Then moving to profile submenu, it contains a short profile about the module author. This section was created by using animation template. Next button is an introductory submenu that explains the purpose of creating the interactive module and also explains more about the material or contents of the developed module. This section was created by using introduction template.

After the main page menu, students then find a preliminary menu, which includes the core competence, basic competence, and indicators of learning competence achievement. All three submenus were created by using introduction template. After seeing the competence achievement indicators submenu, students directly enter the first learning material which is properties of light. The first material includes a brief explanation about properties of light. Properties of light menu is intended to remind students about the material they have learned at the intermediate school level. The five properties of light are presented by using introduction template as they consist of simple images and text.

After completing the first material, the light reflecting material is presented. The presentation of this material consists of prerequisite ability test, apperception, delivery of learning objectives, reflection law, and reflection on various types of mirror (flat mirror, concave mirror, and convex mirror). The prerequisite ability test submenu was created by using multiple choice template because it consists of five multiple choice questions about the material that has been studied, such as properties of light. Apperception and learning objectives for this material were created by using introduction template by displaying a simple image of rearview mirror. The next material is about regular reflection and reflective reflection that were created by using animation template because it consists of animations that describe the differences between regular reflection and reflective reflection.

Meanwhile, the reflection law material presentation begins with the presentation of a simple simulation by using the demonstration template. So, students can conclude the law of reflection . After understanding the law, students learn to apply the law on different types of mirrors. The first mirror is a flat mirror. This sub material presents the process of shadow forming on a flat mirror using a simple animation presented by using an animation template. In addition, on the next page, simple pictures and videos about the relationship between angle that is formed between two flat mirrors with the number of shadows that are formed provided in demonstration template. A formula is then given to calculate the number of shadows that are formed if two flat mirrors form a certain angle, along with the sample questions which are presented by using introduction template. The next submenu is presented by using animation template, in this experiment result submenu, a link to open a video about an experiment of two flat mirrors that form an angle is given. The video is used as a guidance for the module users to do the experiment, then input the observation data on animation which contains a set of questions that have been made by using QuisMaker iSpring software.

After that, the material is continued to the convex mirror material, which begins with convex mirror characteristic material which is consisted by a convex mirror and a concave mirror by using an animation in a demonstration template. After that, three special rays on the convex mirror are shown by using animation template. It is continued to a page that contains space numbering on convex mirror which is presented by using animation template. Still about convex mirror, the formation and characteristics of shadow on the convex mirror based on location of the object are displayed.

The presentation of this material is presented by using the animation template as well. Learning continues on the applied calculations on the convex mirrors material that presents the formulas and samples of problems in the convex mirrors shown by using animation template. The next submenu is Sample Problem submenu showing a sample problem about reflections on the convex mirror thoroughly. This submenu was created by using demonstration template. The last submenu on the light reflection material is comprehension test of the material presented in demonstration template as well. After completing the light reflection material, learning continues on refraction of light material. Learning begins with a prerequisite test about light reflection which consists of five items. This prerequisite ability test is presented by using Multiple Choice template. The second submenu is apperception of refraction.

Apperception is conveyed by using a picture of a spoon phenomenon that appears to be broken in a water containing glass with introduction template. Then, the delivery of learning objectives is also presented by using introduction template. After that, the first material of refraction definition is conveyed through simple pictures by using introduction template, then students learn about refraction law which is delivered by using simulation template. This submenu consists of Willebrord Snellius images, texts about refraction, and an animation about refraction law. After that, the next material which is about refractive index formula and examples of the case are presented in simulation template. In the next submenu, students are exposed to the various refractive index materials through Table template. The material is continued on lens types which are displayed by using introduction template. The next page shows an animation about lens properties consisting of convex lenses and concave lenses through animation template. After that, three special rays on the lens are presented by using Click Table Animation template.

Before entering the formation of shadows on the lens material, students learn about spaces numbering in the lenses which is presented by using animation template. The learning is continued to shadow properties and formation on the lens which is presented by using Simulation template. Furthermore, an animation about focus of shadows which are formed on the screen is presented by using animation template. The study of refraction is completed with calculation on the lens material which presents applied formulas and sample problems by using animation template. The presentation about sample questions in refraction is continued on the demonstration template which presents some sample questions with discussion. The study of light refraction is ended by comprehension test presented by using demonstration template.

After you've finished developing your course, the LCDS gives you a few options for creating a distribution package. If you plan to host your course on a Learning Management System, you can create it as a SCORM package, which is a standard for e-learning content. Otherwise, you can copy the course files onto a CD or Web site. Based on the statement, we can know that there are four stages in writing the learning content on LCDS, they are create, review, refine, and delight. After finishing the development of a module with LCDS, LCDS will give some options to publish the product. If publishing the product on the Learning Management System is preferred, the files can be created in form of SCROM which is a standard on Learning Management System. The developed product can also be stored on CD.

The product of this development has some advantages, one of them is that this interactive module can be used as an alternative solution for students and teachers in facing the limitations of instructional media as well as providing a various and attractive learning resources in form of interactive module with LCDS program in Optical Geometry material for students. Students can learn independently to master Geometry Optical material, and this module is also completed with comprehension test in each sub material and the results can be automatically showed to the students, so that students know their abilities in mastering the material. This module is presented with multi representation by using text, image, symbolic, animation, and video which is presented attractively, so it is expected to be able to optimize the students' concept knowledge. The weakness of this interactive module is that it has not been developed to be accessed via online and every computer that is used to display this interactive module must install microsoft silver light. Another weakness of this module is the assessment for the student's learning achievements can not be done automatically. The assessment still has to be done manually, this is because the module has not been able to make an output in score list tablature form, and only limited to individual score.

Theoretically, the model of this electronic module can provide real problems that are packaged in multi representations in form of video, animation, or multimedia that give opportunity in providing the accessible and systematic material. Non-linear teaching materials provide students' opportunities in constructing meaning, providing lab models, and providing problem solving presented in hypertext, image, video, and animation, which can facilitate students in acquiring an understanding of concepts and learning achievements [23]. Deep understanding about the concept is the first step in the acquisition of the learning achievements.

4. Conclusion
It has successfully developed a product of electronic module design based on learning content development system in optical geometry material to cultivate students' multi representation skills. The developed product is designed by presenting the material with multi representation such as text, images, symbolic, video, and animation which are presented attractively.

5. References
[1] Binkley M, Erstad O, Herman J, Raizen S, Ripley M, Miller-Ricci M and Rumble M 2012 Defining twenty-first century skills In Assessment and Teaching of 21st Century Skills *Springer Netherlands* 17-66
[2] Yilmaz M B and Orhan F 2010 High School Students Educational Usage of Internet and Their Learning Approaches *Wor. J. on Edu. Tech.* **2** 100-112
[3] McLoughlin C and Lee M J W 2010 Personalised and Self Regulated Learning in the Web 2.0 Era: international exemplars of innovative pedagogy using social software *Aust. J. of Edu. Tech.* **26** 28-43
[4] Birch D and Burnett B 2009 Bringing academic on board: Encouraging institution wide diffusion of learning environments *Aust. J. of Edu. Tech.* **1** 117-34
[5] Harper K A 2006 Student Problem Solving Behavior *The Physics Teacher* **44** 250-51
[6] Kohl P B and Noah D F 2006 Physical Review Special Topics *Physics Education Research* **2** 010102
[7] Lindenfeld P 2002 Format and Content in Introductory Physics *Amer. J. of Phys.* **70** 12
[8] Lasry N and Aulls M W 2007 The effect of multiple internal representation on context-rich instruction *Amer. J. of Phys.* **75** 1030-37
[9] Waldrip B Prain V dan Carolan J 2006 Learning Junior Secondary Science through Multi-Modal Representations *Electronic Journal of Science Education* **11** 87-107
[10] Meltzer D E 2005 Relation Between Students Problem Solving Performance and Representational Format *Amer. J. of Phys.* **73** 463
[11] Wong W K Yin S K Yang H H and Cheng Y H 2011 Using Computer-Assisted Multiple Representations in Learning Geometry Proofs *Educational Technology & Society* **14** 43–54
[12] Ainsworth S 2006 DeFT: A conceptual framework for considering learning with multiple representations, Learning and instruction *J. Lear. and Inst.* **16** 183-98
[13] Mayer R E 2003 The promise of multimedia learning: Using the same instructional design methods across different media, learning and instruction *J. Lear. and Inst.* **13** 125-39
[14] Zacharia Z and Anderson O R 2003 The effect of an interactive computer-based simulation prior to performing a laboratory inquiry based experiment on students' conceptual understanding of physics *Amer. J. of Phys.* **71** 618-29
[15] Smaldino S E Deborah L L and James D R 2011 *Instr. Tech. and Med. for Lear.* Indonesia Kencana
[16] Brinkmann D T 2005 Corporate developments and strategic alliances in e-learning *J. of Edu. and Train.* **43** 256-67
[17] Darmawan D 2014 *Pengembangan e-Learning Teori dan Desain* Bandung PT Remaja Rosdakarya
[18] Yapici U I and Akbayin H 2012 The Effect of Blended Learning Model On High School Students' Biology Achievement And On Their Attitudes Towards The Internet *The Turk. Onl. J. of Edu. Techn.* **11** 228-37
[19] Kazu I Y and Mehmet D 2014 Effect Of Blended Learning Environment Model On High School Student's Academic Achievement *The Turk. Onl. J. of Edu. Tech.* **13** 78-87
[20] Taufani D R and Iqbal M 2011 *Membuat Konten E-learning dengan Microsoft Learning Content Development System (LCDS)* Bandung Mugi Unikom

[21] Aremu A and Efuwape B M 2013 A Microsoft Learning Content Development System (LCDS) Based Learning Package for Electrical and Electronics Technology-Issues on Acceptability and Usability in Nigeria *Amer. J. of Edu. Res.* **1** 41-48
[22] Sugiyono 2014 *Metode Penelitian Pendidikan (Pendekatan Kuantitatif, Kualitatif, dan R&D)* Bandung Alfabeta
[23] Candiasa M 2005 Implementasi Jaringan Semantik dengan Hypermedia *J. Pend. Tek. dan Kej.* **2** 64-7

IOP Conf. Series: Journal of Physics: Conf. Series **1022** (2018) 012026 doi:10.1088/1742-6596/1022/1/012026

Integration of living values into physics learning based on local potentials

S Sarah[1*], Z K Prasetyo[2] and I Wilujeng[3]

[1]Physics Education, Tarbiyah Faculty, Sains Al Quran University indonesia.
[2,3]Science Education, Pascasarjana Program, Yogyakarta State University indonesia.

*Corresponding email: siti.sarah@unsiq.ac.id

Abstract. Living values are the principles and beliefs that influence the way of life and behavior of people in society. These values are defined to determine the individuals' characteristics in the physical, intellectual, social-emotional, and spiritual dimensions. Such values could be acquired through physics learning. Therefore, the study concerned here was aimed at determining the difference in the living values acquired between students of the grade officially termed Grade X at a state senior high school referred to as SMAN 1 Selomerto, Central Java, Indonesia, who learned physics by using content based on local potentials and those who learned physics without using that content. A quasi-experiment with the control group pre-test post-test design was conducted to collect the data. The data were analyzed by using tests of normality, homogeneity, and different. The results indicate no difference in the living values acquired between students learning physics by using local-potential content and those learning physics without using that content.

1. Introduction

BNN (*Badan Narkotika Nasional* or the National Narcotics Agency in Indonesia) shows that the number of drug users in the said country until November 2015 had increased to 5.9 million [1]. In addition, cases of corruption are currently in occurrence in almost all lines in Indonesia. The annual report of the Corruption Perceptions Index shows that Indonesia was ranked 90th among the 176 countries reported in it in 2016. Moreover, a research study states that the level of honesty among students is very low [2]. If traced, such a situation could have been found to be caused by learning processes that have not been in accordance with the purposes of education. It shows that living values are not yet adequately embedded in students. It is in spite of the fact that living values are recognized as basic values in life.

Well-implanted living values are the embryos of character formation. The application of living values contributes to 26% of students' character formation [3]. Character education should become an integral part of school life [4]. Therefore, efforts to cause living values to grow in students are necessary to save the nation's morals.

IOP Conf. Series: Journal of Physics: Conf. Series **1022** (2018) 012026 doi:10.1088/1742-6596/1022/1/012026

Living values, as a phrase, is also used as the name of a value education program. The program presents various experience activities and practical methodologies to teachers and facilitators to help children and adolescents explore and develop key personal and social values [5]. By means of the LVE (Living Values Education) model, students become accustomed to activities at school that exert a sufficiently significant impact of developing character up to 42.1% [6]. Living values also constitute the main factors forming the character of the nation because they could reduce conflicts in Indonesian society, which possesses multicultural characteristics [7].

The habit of nurturing living values should be done continuously in all environments whether they are at home, at school, or in society. The implanting of students' living values at school could be done through various activities like some innovative learning activities and extracurricular activities [8].

Living values in oneself are considerably influenced by the characteristics of the environment or society where one finds oneself. There are 13 living values that could be developed in Indonesia, namely, being religious, honest, tolerant, well-behaved, disciplined, hard-working, creative, independent, democratic, filled with nationalism (love for one's country), appreciative of achievement, cooperative, and responsible [3].

There are influences of students' attitudes and behaviors upon themselves and others as impacts of values and programs of emotional literacy [9]. Based on some of the above, the following important points could be made. Living values are the values that should grow, develop, and live in a person. To form these values takes considerable time. It also requires the participation of the people in a system to create them together.

The school is one of the effective systems to foster the values of life in students. This should be done by all stakeholders in the school. The teacher is the spearhead of the embedding of living values in students. Therefore, teachers are needed not only to teach the subject matter but also to instantiate and instill living values by teaching.

A lesson at school would more possibly provide a clear and relevant picture if the theme of the learning is unearthed from local potentials. Considerable research indicates that local potentials could be integrated into learning at school [10, 11, 12, 13].

A local potential is a specific resource potential possessed by a region. Local potentials cover natural resources, human resources, technology, and culture [14]. Culture is anything concerning how the local society views life in relation with its belief, productivity, work, staple food, creativity, values, and norms [15].

The SETS (science, environment, technology, and society) approach is in line with the utilization of local potentials. The SETS approach is a difficult one but it is an approach that could reach HOTS (high-order thinking skills) [16]. HOTS itself refers to the thinking abilities of students who are able to handle the multi-dimensional problems of social, economic, technological, and environmental systems. Through the integration of local potentials into learning physics would be a solution to the matter of growing the living values among students. The objective of the research concerned here was to determine the difference in living values between students learning physics by using content based on local potentials and those learning physics without using that content.

2. Method

The research instruments consisted of some learning equipment, a questionnaire, and an observation sheet. The learning equipment consisted of a syllabus, lesson plan, lesson textbook, students' worksheet, and evaluation sheet. The instruments were validated by experts' judgement. The questionnaire and observation sheet were used to obtain data concerning living values. Interviews with physics teachers were done to dig up information more deeply. Triangulation was done on the data obtained from the observation and questionnaire and from the results of the interviews.

The subject matters of local potentials were, among others, *opak* (a traditional food), the plastic mulch cover of farmland and plant nurseries, and natural resources in the form of mountains, tea from Tambi, and well water. The physics materials were about temperature and heat.

The research subjects were students of Grade X (the official term for the 1st grade of SMA, or *sekolah menengah atas* 'senior high school', i.e., Level 2 of the two-level high school system in Indonesia) at an SMAN (SMA *Negeri* or State SMA), namely, SMAN 1 Selomerto, Central Java, Indonesia, with the said students being divided into two classes. One class was to serve as control class and the other was to serve as experimental class.

The research was quasi-experimental in nature and it used the control-group pre-test post-test design. In the experiment, the experimental class learned physics with local-potential content by means of the guided inquiry method. Meanwhile, the control class learned physics by means of demonstration and lecture methods without local-potential content. The students in the experimental class and the control class were divided into several groups. The data analysis used (1) a test of equality in initial competence, (2) a test of normality, (3) a test of homogeneity, and (4) hypothesis testing.

3. Findings

Prior to the experiment, the two classes were tested for equality in mean scores for living values. By means of a t-test, it was found that the data for the living values under investigation had a value of significance (p) > 0.05. It means that the two classes were equal in the mean scores for honesty, cooperation, and responsibility.

After treatment, the data obtained were tested for the normality concerned.

Table 1. Results of the Test of Normality

		Kolmogorov-Smirnov[a]			Shapiro-Wilk		
	Class	Statistic	Df	Sig.	Statistic	df	Sig.
Honesty	Experimental	.165	24	.092	.934	24	.121
	Control	.157	27	.085	.959	27	.356
Cooperation	Experimental	.137	24	.200*	.956	24	.358
	Control	.189	27	.014	.935	27	.093
Responsibility	Experimental	.188	24	.028	.929	24	.093
	Control	.191	27	.012	.934	27	.086

a. Lilliefors Significance Correction

*. This is a lower bound of the true significance.

Table 1 shows that data for cooperation and responsibility as a pair had different respective levels of significance, namely, $p < 0.05$ (not normally distributed) and $p > 0.05$ (normally distributed). Therefore, the test of homogeneity was not conducted on the data for cooperation and responsibility. In consequence, the data for cooperation and responsibility were analysed by using nonparametric statistics.

Conversely, the data for honesty of both classes had a value of significance (p) > 0.05. It means that the data were normally distributed so that the test of homogeneity was then conducted on the data.

Table 2. Results of the Homogeneity Test on the Data for Honesty

		Levene Statistic	df1	df2	Sig.
Honesty	Based on Mean	1.209	1	49	.277
	Based on Median	.725	1	49	.399
	Based on Median and with adjusted df	.725	1	47.837	.399
	Based on trimmed mean	1.131	1	49	.293

Table 2 shows that the resulting value of significance (p) was 0.277, which was > 0.05. It means that the sample was taken from a homogenous population. Therefore, the data for honesty would be analyzed by using parametric statistics.

The hypothesis testing was done in different ways in accordance with the fulfillment of parametric or nonparametric requirements.

a. Honesty

In the case of the honesty aspect, the hypothesis testing conducted used the t-test because the parametric requirements were fulfilled. The t-test showed that the level of significance (p) was 0.277, which was > 0.05. It means that there was no significant difference in the level of honesty between the students who learned physics by using local-potential content and those who learned physics without using that content.

b. Cooperation and Responsibility

The hypothesis testing related to cooperation and responsibility used the Wilcoxon test.

Table 3. Results of the Test of Difference on Cooperation and Responsibility

	Control Class. Cooperation – Experimental Class. Cooperation	Control Class Responsibility -- Experimental Class Responsibility
Z	-.677[a]	-.525[a]
Asymp. Sig. (2-tailed)	.498	.600

a. Based on positive ranks.

b. Wilcoxon Signed Ranks Test

Table 3 shows that there was no significant difference in cooperation and responsibility between the students who learned physics with local-potential content and those who learned physics without that content. It could be seen from the values of significance (p) of 0.498 (for cooperation) and 0.600 (for responsibility), both of which were > 0.05.

4. Discussion

The content based on local potentials in learning physics initially motivated the students to learn. It could be seen from the students' enthusiasm at the beginning of the learning session. In the first meeting, there were almost no sleepy students nor students amusing themselves. The students were a little surprised but happy with the unusual learning method and, moreover, with the use of the content based on local potentials. So the students' interest in the learning was aroused. However, with the passing of time, the students' enthusiasm lessened. They had difficulties in several steps of the guided inquiry method. The

IOP Conf. Series: Journal of Physics: Conf. Series **1022** (2018) 012026 doi:10.1088/1742-6596/1022/1/012026

difficulties occurred at the time of identifying a problem, collecting data, and analyzing data. The students became very reluctant to comply when asked to read the material in the lesson book. It caused the teacher to spend much time explaining the purpose of the experiment. Such a situation is in opposition with the guided inquiry method, which requires students' independence in learning. At the time of collecting data, many students still did not understand yet what independent variables, control variables, and dependent variables were. There were also some students who still found it difficult to do correct measuring activities so that they needed to be trained. The following presents the results of the students' experiments.

Table 4. The Results of Experiment 2 in the Experimental Class

Group	Initial term (°C)	Final term (°C)	Time
1	25	30	00:03:01
	25	30	00:02:42
2	24.5	29,5	00:02:23:13
	25	30	00:02:30:48
3	22	27	00:02:28:94
	21	26	00:02:28:68
4	25	30	00:03:35:86
	25	30	00:03:00:69
5	24	29	00:02:26:06
	24	29	00:02:08:33
6	24	29	00:02:01:88
	24	29	00:02:05:50

Table 4 shows that Group 2 and Group 6 got data that were not sufficiently accurate. The difficulties experienced by the students since they began the problem identification and the data collection caused them to have difficulties in analyzing the data.

According to the original research design, there were supposed to be seven face-to-face sessions, each of which requiring two lessons' time (2x45 minutes). During the seven sessions, seven experiments ought to have been accomplished. However, only five experiments came into being. In view of the limited time given to the researcher, the research was put to an end after five experiments were done.

A series of the experiments were purposely designed to implant several living values, namely, honesty, responsibility, and cooperation. Possibly the lack of effectiveness of the learning became a cause of the nonexistence of difference in living values between the students learning physics by using local-potential content and those learning physics without using that content.

If all schools in the Regency of Wonosobo, Central Java, Indonesia, are divided into three groups, namely, those of high, middle, and low categories, SMAN 1 Selomerto would fall into the low category. The categorization is based on students' initial competence on entry, school accreditation, and school age.

In general, the students who study in SMAN 1 Selomerto are those not accepted at other schools. Even if there are students with high levels of competence, they are very few in number. SMAN 1 Selomerto had been running for only three years at the time the research was made. Automatically, it was still in the process of getting organized in matters of facilities and resources, the number of teachers, the number of students to be recruited, administration, and others. Therefore, it was quite possible that the aspect of implanting living values was not yet one of the matters given priority by the school.

The implantation of living values requires making one get used to new matters. And it is to be done not only at school but also in the family and society. The results of the interviews with the physics teacher and principal of SMAN 1 Selomerto indicate that in general, their students come from families that are not sufficiently well off. A part of the students come from families where one or both parents are away working in another part of the country or even in another country. So such students live together with their grandparents or other relatives. It indicates that the implantation or growing of living values in the family is very much at a minimum. A survey by Ainis Mufarika and Suhanadji in five poor families in the District of Mojokerto, Indonesia, shows that the poor parents use a combination of democratic and laissez-faire parenting to guide their children. The democratic parenting style is characterized by the encouragement of parents to children, the parents' attention, their deliberation, and the good communication between parents and children. The laissez-faire parenting style is characterized by parents giving freedom to their children to hang out or play. In addition, however, parents are less knowledgeable about what their boy has done [17].

How to overcome this problem? The way is to intensify the growth of good values in the family, the school, and the community. Research by Sukiyani and Zamroni and also by Anisah indicates that the outlook of the families to character education is influenced by the expectations of parents on their children. Parents teach character through good parenting, model behaviour, and habituation, giving an explanation for the action, the application of high standards and being realistic for the child, and involving children in decision making. Results of character education in the family show that children who grow up in a complete family's affection feel more fulfilled than children from single-parent families [18, 19].

5. Conclusion

The results show that there is no difference in living values between students who learn physics by using local-potential content and those learning physics without using that content..

6. References

[1] Kompas com 2016 Buwas: Pengguna Narkoba di Indonesia Meningkat hingga 5,9 juta Orang. http://regional.kompas.com/read/2016/01/11/14313191/Buwas.Pengguna.Narkoba.di.Indonesia.Meningkat.hingga.5.9 Juta Orang

[2] Utomo E, Nurlaeli L and Simangunsong H 2011 Indonesian Students' Opinion, Knowledge of, and Attitude toward Topics and Values in the Civic Education Subject *Journal of Education Research and Policy* **3** 154-76

[3] Komalasari K 2012 The Living Values-Based Contextual Learning to Develop the Students' Character *Journal of Social Sciences* **8**(2) 246-51

[4] Romanowski M H 2005 Through the Eyes of Teachers: High School Teachers' Experiences With Character Education *ProQuest Education Journal* 34 6-23.

[5] Tillman D 2012 *Living Green Values Activities for Children and Young Adults: A Special Rio+20 Edition* Switzerland: Association for Living Values Education International (ALIVE)

[6] Saripudin D and Komalasari K 2015 Living Values Education in the School Habituation Program and Its Effect on Students' Character Development *The New Educational Review* **39**(1): 51-62

[7] Wening S 2012 Pembentukan Karakter Bangsa Melalui Pendidikan Nilai *Jurnal Pendidikan Karakter* Tahun II I 55-65

[8] Komalasari K Saripudin D and Masyitoh I S 2014 Living Values Education Model in Learning and Extracurricular Activities to Construct Students' Character *Journal of Education and Practice* **5**(7)

[9] El Hassan Karma and Rula K 2005 The Effect of Living Values: An Educational Program on Behaviors and Attitudes of Elementary Students in a Private School in Lebanon *Early Childhood Education Journal* **33**(2)

[10] Ardan A S *et a.* 2015 Needs Assessment of the Development of the Biology Textbook for Grade X at High School Based on the Local Wisdom of Timor *International Education Studies* **8**(4)

[11] Kesiman M W A and Ketut A 2012 The Implementation of Hypertext-Based Learning Media for a Local Culture-Based Learning *Journal of Information Technology Education: Innovations in Practice* **11** 377-85

[12] Chusorn P Ariratana W and Chusorn P 2014 Strategy Challenges in the Local Wisdom Application Sustainability in Schools *Procedia - Social and Behavioral Sciences* **112** 626 – 634

[13] Sahlberg P 2007 Educational Policies for Raising Students' Learning: The Finnish Approach *Journal of Educational Policy* **22**(2) 147–171

[14] Hariyadi P 2010 Penguatan Industri Penghasil Nilai Tambah Berbasis Potensi Lokal Peranan Teknologi Pangan Untuk Kemandirian Pangan *Jurnal pangan* **19**(4)

[15] Arowolo D 2010 The Effects of Western Civilization and Culture on Africa *Afro-Asian Journal of Social Sciences* **1**(1) Quarter IV

[16] Zoller U 2013 Science, Technology, Environment, Society (STES) Literacy for Sustainability: What Should It Take In Chem/Science Education? *Educ. quím.* **24**(2) 207-2 1Ů

[17] Mufarika A and Suhanadji 2013 Pola Pengasuhan Anak Pada Keluarga Miskin (Studi Kasus 5 Keluarga Miskin di Desa Kebontunggul, Kecamatan Gondang, Kabupaten Mojokerto) *Jurnal Penelitian*

[18] Sukiyani F and Zamroni 2014 Pendidikan Karakter dalam Lingkungan Keluarga *SOCIA* **11**(1) 57-70

[19] Anisah A S 2011 Pola Asuh Orang Tua Dan Implikasinya Terhadap Pembentukan Karakter Anak *J.Pendidikan Universitas Garut* **05**(1) 70-84

The profile of students' problem-solving skill in physics across interest program in the secondary school

S K Jua[1,2], Sarwanto[1,3] and Sukarmin[1,4]

[1] Faculty of Teacher Training and Education, Sebelas Maret University
Jl. Ir. Sutami, Kentingan, No. 36, Surakarta, Central Java, Indonesia

Corresponding email: [2]kostaosry@gmail.com, [3]sarwanto@fkip.uns.ac.id, [4]karmin.abdulkarim@gmail.com

Abstract. Problem-solving skills are important skills in physics. However, according to some researchers, the problem-solving skill of Indonesian students' problem in physics learning is categorized still low. The purpose of this study was to identify the profile of problem-solving skills of students who follow the across the interests program of physics. The subjects of the study were high school students of Social Sciences, grade X. The type of this research was descriptive research. The data which used to analyze the problem-solving skills were obtained through student questionnaires and the test results with impulse materials and collision. From the descriptive analysis results, the percentage of students' problem-solving skill based on the test was 52.93% and indicators respectively. These results indicated that students' problem-solving skill is categorized low.

1. Introduction

Is it a problem? The Big Indonesian Dictionary defines the word problem as something to be solved. For some people especially college students, physics is a problem. As evidence, many students are more interested in choosing IPS or social majors than IPA or science in order to avoid physics. In addition, many students are not interested in continuing lectures in physics and do not want to work in physics-related jobs or physics fields [1]. In fact, physics plays an important role in the development of the world, especially in the field of technology, and innovation in medicine [2]. Physics is closely related to human life. Physics is rich in its concepts, theories, and equations. Physics speak the language of mathematics in qualitative and quantitative calculations. In addition, dominant in physics is the nature of problem-solving, meaning almost all aspects of physics are concerned with problem-solving [3]. Physics needs to be solved through problem-solving because the problem solving of physics is a tradition that has been done by physicists, such as Newton who discovered his laws [4]. Thus, physics education should create students who are adept at problem-solving.

All science develops knowledge that directs students to be proficient at problem-solving. Problem-solving is the goal of modern education so that college students as individuals who can solve the problems encountered in everyday life [5]. Slavin (2011: 29) defines problem-solving as the application of knowledge and skills to achieve certain goals [6]. Gok and Silay (2008: 254) stated that problem-solving is the ability of students to use existing information or information obtained to determine what to do in a particular situation [7]. In this case, students must know what to do (know how to solve problems) in situations, students do not know how to solve the problem using existing information. Thus, the educational process should enable students to use their knowledge in solving problem, referring to the effort required to achieve its goals or solutions.

Students' problem-solving ability in physics learning is still low due to internal and external factors. Based on research conducted by some researchers [8], the internal factors which cause low problem-solving skill were: 1) when students are given the opportunity to solve the physics problem, students often directly use the equation without doing problem analysis; 2) the students guess the formula to solve the problem; 3) the students memorize the settlement of the problem that has been resolved to solve a new problem; 4) lack of students' ability to remember appropriate equations; 5) students do not understand the basics of physics related to definitions, principles, laws in physics concepts; 6) they are weak in term of math skill to analyze physics problems; 7) students are confused to understand, remember, and convert units; 8) the students do not understand the problem given by the teacher; 9) students quickly give up when having trouble or difficulties in solving problems, there are also students who immediately give up when facing difficult physics problems.

The external factors of students' low skills are: 1) learning in the classroom that is still teacher centered; 2) the material of physics is taught by lecture, discussion, and assignment method; 3) lack of problem-solving practices during the classroom learning process; 4) lack of motivation and experience of physics teachers; 5) lack of exercises related to physics concepts; 6) lack of books and guidance on problem-solving materials in physics learning; and 7) lack of laboratory practice related to the concepts of physics studied. 6) they are weak in term of math skill to analyze physics problems; 7) students are confused to understand, remember, and convert units; 8) the students do not understand the problem given by the teacher; 9) students quickly give up when having trouble or difficulties in solving problems, there are also students who immediately give up when facing difficult physics problems. The external factors of students' low skills are: 1) learning in the classroom that is still teacher centered; 2) the material of physics is taught by lecture, discussion, and assignment method; 3) lack of problem-solving practices during the classroom learning process; 4) lack of motivation and experience of physics teachers; 5) lack of exercises related to physics concepts; 6) lack of books and guidance on problem-solving materials in physics learning; and 7) lack of laboratory practice related to the concepts of physics studied.

Based on the facts found above it is necessary to do a learning process that orientation on problem-solving so that students are accustomed to solving physics problems and problems-in life faced. The problem-solving process is a linear hierarchy process because each step of problem-solving is a continuation of the previous problem-solving step [5]. Troubleshooting requires defining, investigating, reviewing, and processing information about the problem you want to solve.

The most popular troubleshooting steps are the four Polya (1945: xvii) the step which is: recall (problem understanding), planning (strategy planning for problem-solving), implementation (implementing strategy), and evaluation (checking again) [9]. In addition, Heller, Keith, and Anderson (1991: 628-629) stated the problem-solving steps consisting of visualizing the program, describing its physics description, planning a solution, execute the plan, checking and evaluating [10]. Based on the steps proposed above, then the problem-solving steps in the lesson shown in Table 1.

Before the students followed the steps of problem-solving in the learning process, the research had done to see the profile of problem-solving skills of high school students who followed the across interests program of physics. The results of the study were further used as guidelines for developing the appropriate learning stages to improve problem-solving skill. In addition, the results of this study also used to select the indicator as a step of problem-solving that need to be trained more intensely by students during the problem-solving learning.

Table 1. The Steps of Problem-Solving and Its Indicators Adapted from The Problem-Solving Steps of Polya, Heller, Keith, and Anderson

Steps	Indicators	Code
Understanding the problem	Identifying the known variables[a] and converting the support units of problem-solving;	A1
	Identifying the variables which are being asked;	A2
	Visualizing the problem and/or identifying the supporting variables of problem-solving.	A3
Planning the strategy	Determining the theory and equation to solve the problem.	B1
Implementing the strategy	Using teory or equation to solve the problem asked.	C1
Rechecking	Rechecking the truth of the answer;	D1
	Rechecking the appropriateness of theory/quantity/unit used.	D2

[a]Statements of physics, concept, theory, pictures, units, equation, and etc.

2. Method of Research

The type of this research was descriptive research. The subjects of this research were the students of class X, Social program 4, Senior High School or SMAN 1 Sambungmacan Sragen, semester 1, academic year 2016/2017. The research stage was done by observation of the learning process which done before starting the learning process. The students were asked to fill in questionnaire related to problem-solving skill. The instruments used in this research was a questionnaire of problem-solving skill and the test results with impulse and momentum materials and essay-based collision of 4 items.

The data were analysed using descriptive analysis by mean, the number of students' scores were categorized into the criteria shown in Table 2.

Table 2. The category of problem solving skill of students

Interval Category	Category
$94,25 < X$[b]	Very high
$79,75 < X \leq 94,25$	High
$65,25 < X \leq 79,75$	Enough
$50,75 < X \leq 65,25$	Law
$X \leq 50,75$	Very Low

[b]Total of students' scores

Furthermore, students' scores were counted using this equation:

$$X = \frac{Total\ student\ score\ per\ stage/indicator}{Maximum\ score\ per\ stage/indicator} \times 100 \quad (1)$$

3. Results and Discussion

In addition, to collect data through questionnaires and analysis of students' test results, researchers also observed the learning process of students for two hours of lessons. Some of the results of the observation in the learning process are: 1) students were not ready or not prepared to follow physics lessons; 2) the students forgot the physics concept which learned earlier; 3) the students' mathematical ability is weak; 4) students did not dare to ask the teacher about the problem or concept of physics that they understood; 5) students were still difficult to distinguish between the size and unit; and 6) students were still confused when converting units.

Based on the students' answers in the questionnaire, it was found that: 1) some students couldn't understand and solve problems because students pay attention and listen to teacher's explanations about the concepts of physics being studied; 2) some students assumed that gathering information is critical in solving problems so that students could find ideas to solve problems; 3) students usually plan and implement strategies to solve problems when students work together or discuss in groups. This is in line with the statement of Ikwanuddin, Jaedun, and Purwantoro (2010: 216) which stated that problem-solving skills were excellent for being trained in groups because in groups the students can give their contribution [11].

Table 3. Summary of students' test analysis of problem-solving skills

Step of problem-solving skills	Indicators	Category	%
Understanding the problem	A1	Enough	64.00
	A2	Low	51.29
	A3	Very low	0.00
Planning strategy	B1	High	73.49
Implementing strategy	C1	Enough	60.34
Recheking	D1	Very low	26.51
	D2	Low	52.37
		Low	52.93

Based on the analysis of students' test results on the collision materials momentum and impulse find some things such as 1) some students know and can write the amount or concept provided in the problem or problem; 2) most students can determine or know the exact equation to solve the problem; and 3) students can apply the equation to solve the problem.However, the average problem-solving skill of the students was analyzed based on the test results were categorized as "low" (52.93%). The highest percentage achieved by 73.49% students on the indicator determines the theory and/or equation to solve the problem. Meanwhile, the lowest percentage of problem-solving skill is 0.00% in the indicator to visualizing the problem and/or identifying the supporting variables of problem-solving.

3.1. The Profile of Students with "Very Low" Problem-Solving Skills

Student problem-solving skills that are categorized as very low are obtained on the result of student test analysis that is in understanding the problem skill, especially on indicators visualizing the problem and/or identifying the supporting variables of problem-solving. In addition, re-examination skills, especially on indicators re-examining the truth of answers are also categorized very low. This can be due to: *first*, the student's focus that always leads to known variables and the exact equations to solve the problem. As a result, many students do not find the right answer in problem-solving [12]. Visualizingand identifying the problem is very important in the troubleshooting process. The reason, when students know the right content in the problem, then students can plan appropriate steps to solve the problem [3]. *Second*, students assume the results already obtained are the correct answer. After getting the results from the calculation, students move directly to work on other questions, without paying attention to his work step from the beginning. As a result, many student calculations are wrong, although the equations used are correct. This is in line with Pardimin and Widodo's (2016: 394) study which states that students are only able to carry out problem-solving steps until implementing a plan to solve the problem. For students, checking back slows down troubleshooting time. This is why students' problem-solving skills have not been satisfactory [13]. According to Kurniawan, Handayanto, and Parno (2016: 108-109) students do not collect information and do not check the results of problem-solving, because students do not understand the underlying concept of problem-solving [14]. If students master the concept well, then the process of gathering information and re-examining problem-solving can be done well.

3.2. The Profile of Students with "Low" Problem-Solving Skills

Based on the results of the analysis of the data above, the problem-solving skill of low categorized students is at the stage of understanding the problem and checking again. Based on the answers from the student questionnaire, the low comprehension skills caused by the students did not pay attention to the initial explanation of the teacher, so that the students did not understand the problem posed by the teacher. Based on the test results, the problem-solving skills of the students are low at the stage of

IOP Conf. Series: Journal of Physics: Conf. Series **1022** (2018) 012027 doi:10.1088/1742-6596/1022/1/012027

understanding the problem especially on the indicators identifying the variables asked. While the skills to check back is also categorized low, especially on the skills to check the suitability of quantities and units.

The low student problem-solving skills can be caused by several things such as: *first*, the student does not prepare before doing the physics lesson, for example, re-reading the physics concept learned earlier. However, this is not done by the students, so the students have difficulty solving the problems posed by teachers. Habits of students who are not prepared, including not re-learning the concepts learned before, cause students difficulty in solving problems [15]. According to Selçuk, Çalışkan, and Erol (2008: 156-157), the troubleshooting process requires students to use previously learned knowledge to solve the problem. Thus, self-preparation in the form of reading and recalling physics concepts that have been studied, it is necessary to be done to help students in solving problems [5].

Secondly, the students do not understand the basics of physics related to definitions, principles, laws in the concept of physics. For example, students are still difficult to distinguish between the magnitude and the units.Other than that, students do not remember the exact equations to solve the problem. For example, students use the momentum equation to solve the perfect collision problem.If students are weak in the mastery of the concept of physics, resulting in low problem-solving skills [16]. Doctor et al (2015: 11) who examines'the role of conceptual knowledge in problem-solving and its impact on students' finds that conceptual understanding is a "starting point" for problem-solving, helps students to differentiate some applicable principles to solve problems, and helps students set their mindset when dealing with problems [17].

Third, students do not understand the problem asked about the problem (Figure 1) and/or writing a false mark on a known variable, so students are confused when planning the problem-solving strategy. According Ikwanuddin, Jaedun, and Purwantoro (2010: 222) understand the problem of not just writing down what is known and asked with the correct notation. If the student is superficial in understanding the problem, resulting in failure to plan problem-solving strategies that impact on the failure of problem-solving [18].

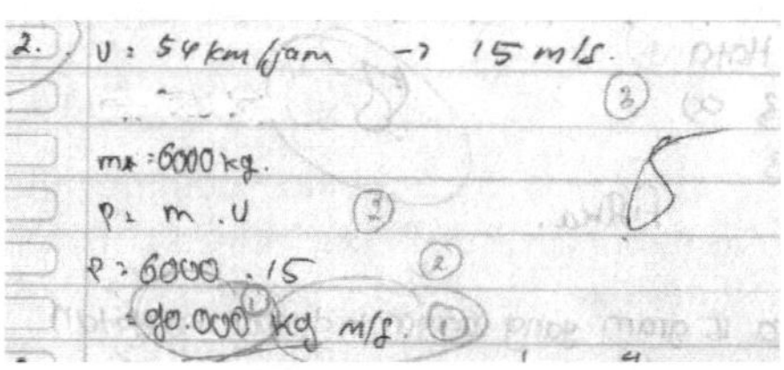

Figure 1. Students did not identify or write the variables which are being asked.

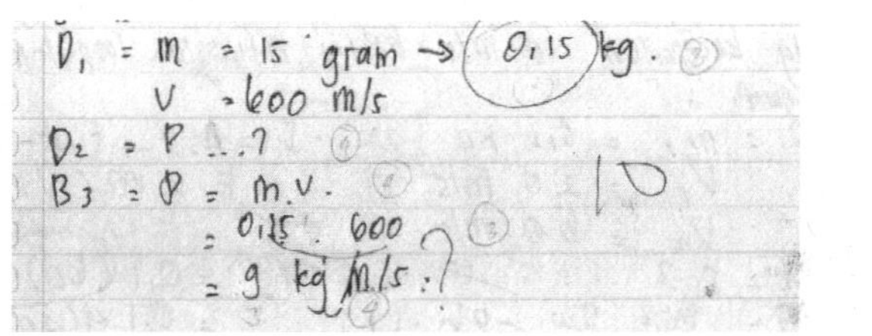

Figure 2. Students still incorrectly convert units. For example, the conversion of units from gram (g) to kilogram (kg) should be divided by 1000, but students multiply by 1000 or divide by 100.

Fourth, students are still wrong and/or do not convert units when identifying problems (Figure 2), consequently, the result of problem-solving and unit writing is still wrong. According to Ogunleye (2009: 90), the unit plays an important role in the calculation of physical problems. If the unit and unit conversion are wrong, then many errors are experienced in the calculation of physical problems. Thus, students need to be taught about the use of the right units and how to convert units [3].

Figure 3. Some students directly use the equation without analyzing and identifying the problem-solving support components.

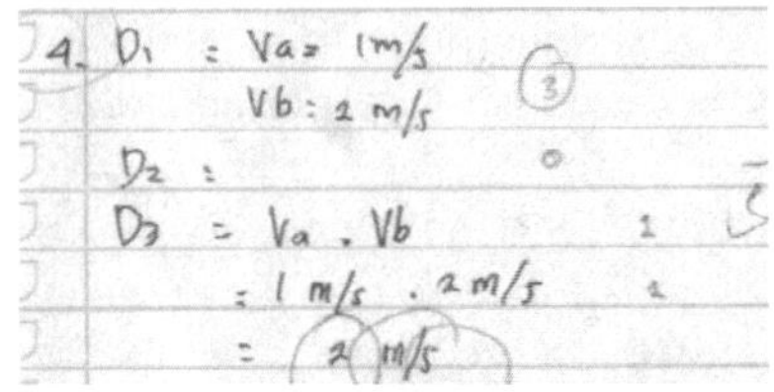

Figure 4. Almost all students do not solve the problem that is considered difficult, although the equation used is correct.

Fifth, students do not analyze the variables being questioned and directly use the equations without doing problem analysis. Students do not read or interpret the problem thoroughly so that students cannot identify the sign (which indicates direction) appropriately. For example, on the concept of impulse and collision other than numbers, the direction also has value or meaning in solving the problem.This is in line with the research of Pardimin and Widodo (2016: 394) which states that students do not write what is being asked what problems they are facing, so teachers are difficult to guess whether students understand the problem posed [11]. When students directly use the equation without analyzing or identifying the problem first, it can produce causal relationships that are not in accordance with the purpose of solving the problem to be solved [14]. In addition, problem-solving without prior analysis may lead to low problem-solving abilities [8].

3.3. The Profile of Students with "Enough" Problem-Solving Skills

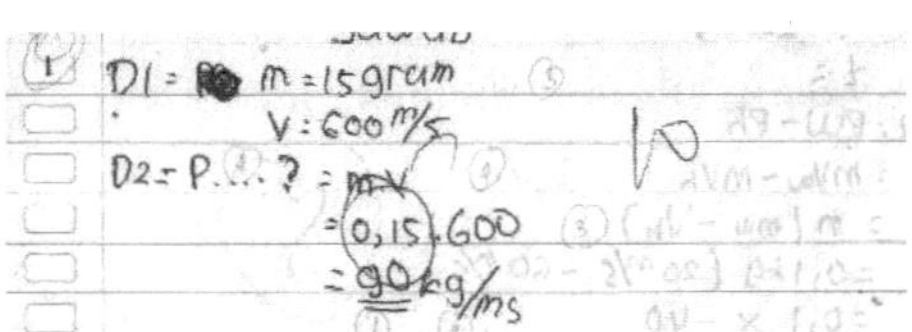

Figure 5. Students do not convert units, but when using equations to solve student problems using incorrect unit conversion results and students do not identify the variables that support troubleshooting.

Based on test results, students' skills in understanding the problem especially on indicators identifying the known variables and converting the support units of problem-solving and students' skills in implementing the strategy are also categorized as "sufficient". This can be due to a low student's mathematical ability. When analysed further, many students did not solve the problem that was categorized as difficult. The cause is a complex mathematical solution and the result is that students quickly give up to solve the problem. According to Brad (2011: 27-28), students quickly give up because students do not have many ways to solve the problem [18]. But according to Walsh, Howard & Bowe (2007: 9), students quickly give up because students are accustomed to using a strategic approach or equation [12]. Thus, when faced with complex problems, the approach does not apply to solve the problem.

3.4. The profile of students with "high" problem-solving skills.

Based on the test results, the students' skills in identifying the known variables and determining the theory and/or equations for solving the problem are categorized as "high". This shows that most students know the variables expressed in the problem. In addition, most students also know the exact equations in solving the proposed problem. According to Ogunleye (2009: 90), it is important for students to remember and know the exact equation because physics is never separated from the equation to solve the problem [3].

External factors causing low problem-solving skills are the time of physics lessons in the afternoon. This is in line with the results of Lestari research (2015: 116-123) which states that study time has a significant influence with students' mathematics learning outcomes and student learning outcomes have increased when lessons are held in the morning [19]. Mathematics and physics are hard and difficult lessons. Study time in the morning can help students to concentrate more and focus on lessons. If the learning is done in the afternoon or afternoon, the concentration and focus of the students have decreased because the students have activities and thinking from the morning. So, students may be tired of thinking to solve the physics problem.

Based on data results and data analysis above, then the teacher needs to implement the learning by emphasizing the whole stages of problem-solving. Meanwhile, when the teacher trains students problem-solving skills in the form of quantitative questions, teachers need to train students to understand the variables asked, identify the variables of problem-solving support, convert the units precisely and thoroughly, and students are trained to re-examine the correctness of the problem either the final value and the units. In addition, teachers must also train students to understand the meaning of every word and sentence expressed in the matter, since the identification of the problem is not just limited to the values or quantities stated in the question.

4. Conclusions

Based on the results and discussion above, the problem-solving skill of Social Grade X student 4 SMAN 1 Sambungmacan is categorized still low. This is evidenced by the average percentage of problem-solving skills analysis of student test results (52.93%). When faced with the problem of quantitative problems, students are very skilled in implementing the strategy (73.49%). However, students are still less skilled at visualizing the problem and/or identify the variables of problem-solving support (0.00%) and re-examine the correctness of the problem (26.51%). The low student problem-solving skills are mostly due to internal factors of the students. Thus, students should continue to be trained to solve problems with the stages of problem-solving skills, especially on problem-solving skills understanding the problem and re-examination.

5. References

[1] Gulo Y S and Sulandari S A 2016*Prosiding Pertemuan Ilmiah XXX HFI Jateng & DIY*vol 30(Salatiga: Fakultas Sains dan Matematika Universitas Kristen Setya Wacana) 93
[2] Reddy M and Panacharoensawad B 2017*J. of Education and Practice* **8** 59
[3] Ogunleye A O 2009 *J. of College Teaching and Learning* **6** 85
[4] Heller K and Heller P 2010 *Cooperative Problem Solving in Physics A User's Manual*USA: University of Minnesota
[5] Selçuk G S, Çalışkan S and Erol M 2008*Latin American J.of Phys. Education* **2** 151
[6] Slavin R 2011 Jakarta: Indeks (The original book was published in 2009)
[7] Gök T and Silay İ 2008 *J. of Theory and Practice in Education***4** 253
[8] Rahmat M, Muhardjito and Zulaikah S 2014*Jurnal Fisika Indonesia* **54** 108
[9] Polya G1945 *How To Solve It: A New Aspect of Mathematical Method* New York: Doubleday Anchor Book
[10] Heller P, Keith R and Anderson S 1991 *American J. Phys.***60** 627
[11] Ikwanuddin, Jaedun A and Purwantoro D 2010 *Jurnal Kependidikan***40** 215
[12] Walsh L N, Howard R G and Bowe B 2007*American Physical Society: Physical Review Physics Education Research* **3** 1
[13] Pardimin and Widodo S A 2016 *J. of Education andLearning* **10** 390
[14] Kurniawan B, Handayanto A and Parno 2016 *Prosiding Seminar Nasional IPA Pascasarjana UM* **1** (UM) 105
[15] Azizah R, Yuliati L and Latifah E 2015 *Jurnal Penelitian Fisika dan Aplikasinya* **5** 44
[16] Kurniawan B, Handayanto A and Parno 2016 *Prosiding Seminar Nasional IPA Pascasarjana UM* **1** (UM) 105
[17] Doctor J L, Strand N E, Mestre J P and Ross B H 2015 *American Physical Society: Physical Review Phys. Education Research* **11** 1
[18] Brad A 2011*ERIC***4** 21
[19] Lestari I 2015*Jurnal Formatif* **3**115

IOP Conf. Series: Journal of Physics: Conf. Series **1022** (2018) 012027 doi:10.1088/1742-6596/1022/1/012027

Appendix

The problem used to test the students' problem-solving skills:

1. Right: A 15 gram bullet moved at a speed of 600 m/s. Calculate its collision or momentum!
   Left: A 40 gram bullet moved at 80 m/s. Calculate the momentum or its collision!
2. A car moved at a speed of 54 km/h. The mass of car was 2000 kg. Calculate the speed of a truck that has the same momentum/collision as the car, but its mass was 6000 kg.
3. The 0.1 kg of massive ball was thrown horizontally to the right at a speed of 20 m/s. Then, the ball was hit and moved at 60 m/s to the left. Calculate the value of the impulse.
4. Two balls (A and B) with the same mass, they moved closer to each other with velocity $V_A = 1$ m/s and $V_B = 2$ m/s. Ball A to the right, while the ball B to the left. As a result, both balls experienced a perfect tumble collision. What were the direction and the speed of ball A soon after the collision?

IOP Conf. Series: Journal of Physics: Conf. Series **1022** (2018) 012028 doi:10.1088/1742-6596/1022/1/012028

Synthesis and characterization of *Mono*-Aqua-*Pentakis* (Isoni-Cotinic Acid) Nickel (II) Sulfate Trihydrate

H Syaima[1], S B Rahardjo*[2] and N Amanati[2]

[1]Chemistry Graduate Program, Graduate School, Universitas Sebelas Maret, Surakarta, 57126, Indonesia
[2]Department of Chemistry, Faculty of Mathematics and Natural Sciences, Universitas Sebelas Maret, Surakarta, 57126, Indonesia

*Corresponding email: husna_syaima@yahoo.com, *sentotbr@yahoo.com

Abstract. A complex of nickel (II) with isonicotinic acid (asint) was successfully obtained. The complex was synthesized in 1:2 mole ratio of metal to the ligand in methanol. The percentage of nickel was 6.91% determined by Atomic Absorption Spectroscopy (AAS). Therefore, the predicted formula was $Ni(asint)_5SO_4(H_2O)_4$. The molar conductivity of the complex was measured by conductivity meter corresponding to 1:1 electrolyte. The thermal analysis of the formed complex was determined by Differential Thermal Analysis (DTA) indicating that the complex contains four water molecules as ligand and hydrates. The magnetic susceptibility measurement showed that the complex was paramagnetic with μ_{eff}= 3.30 B.M. Electronic spectra of the formed complex appeared at two transition peaks on λ= 394 nm and 659 nm. The infrared spectra of the complex showed a shift of tertiary N-group absorption in 1234 and 1338 cm^{-1} compared to isonicotinic acid at 1149 and 1331 cm^{-1}. In addition, the shift also appeared in the -OH group absorption which was to the lower wavenumber at 3371 cm^{-1} from 3425 cm^{-1} (isonicotinic acid). This fact indicated that the functional groups were coordinated to the central metal ion. The possibility formula of the complex was $[Ni(asint)_5(H_2O)]SO_4 \cdot 3H_2O$ with octahedral structure.

1. Introduction

Metal complex is a compound composed of a metal as a central atom and ligand as an electron donor. The study of complex formation becomes an interesting thing to learn because the complex formed may provide many benefits in scientific applications such as clinical, analytical, photocatalytic, and medical biology. One example of a complex compound that is considered is the complexity of nickel metal ions. Nickel has 28 atomic numbers and $3d^8$ electron configuration. Nickel is a class VIIIB transition element and most of its existence is as Ni(II) [1]. Complexes of Ni(II) is synthesized to form multiple geometries with coordination number 4, 5 and 6 include a square-planar [2-3] or tetrahedral [4-5], square-pyramidal [6], trigonal-bipyramidal, [7] and octahedral configurations [8-9]. Tetrahedral and trigonal-bipyramid geometries are very rare. The most common nickel (II) complex geometry is octahedral and square-planar [1]. Ni(II) is interesting to learn due to its cheap price, tends to be stable and abundant in nature.

Isonicotinic acid (asint), shown in Figure 1, a multifunctional chelating and/or bridging ligand, has proved to be very powerful for the construction of multi-dimensional metal–organic coordination

IOP Conf. Series: Journal of Physics: Conf. Series **1022** (2018) 012028 doi:10.1088/1742-6596/1022/1/012028

networks [10]. The complexes with asint as ligands have raised interest in fluorescence probing with numerous potential applications for studies of microsecond diffusion and dynamics of membranes [11]. Asint has -N=, -OH and C=O which are possible to donate electrons to Ni(II). Sahin et al. [12] reported a complex synthesis of nicotinamide with Zn^{2+} ions forming $[ZnBr_2\ (C_6H_6N_2O)_2]$ with an octahedral structure as shown in Figure 2. Asint has a structure that is similar to isonicotinamide, therefore, it is possible to form a complex with Ni(II).

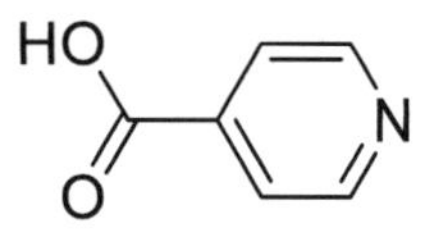

Figure 1. The Structure of Isonicotinic Acid

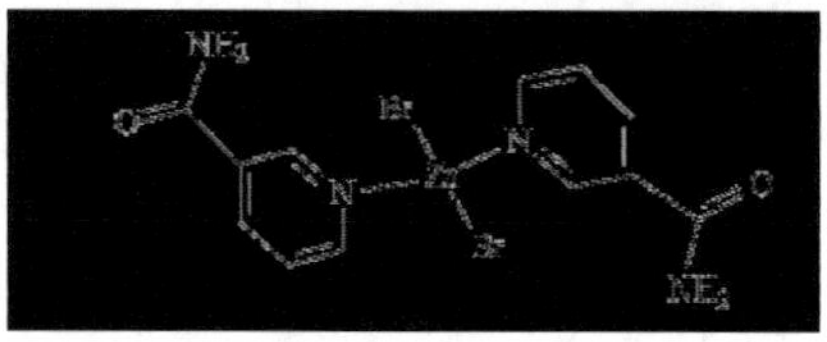

Figure 2. Complex Structure $[ZnBr_2(C_6H_6N_2O)_2]$

In this study, we synthesized a novel complex of Ni(II) with isonicotinic acid and characterized it with the analytical and spectroscopic methods.

2. Experimental

2.1. Materials

The chemicals and solvents were of reagent grade and used without further purification. All chemicals such as $CuSO_4{\cdot}5H_2O$, isonicotinic acid, $NiSO_4{\cdot}6H_2O$, $FeSO_4{\cdot}7H_2O$, $CuCl_2{\cdot}2H_2O$, $NiCl_2{\cdot}6H_2O$, $CuSO_4{\cdot}6H_2O$, $AlCl_3{\cdot}6H_2O$, and nitric acid were purchased from E. Merck.

2.2. Physical Measurements

Spectra UV-VIS of metal complex and $NiSO_4{\cdot}6H_2O$ were recorded in methanol solvent using UV-Vis Double Beam Shimadzu PC 1601 spectrophotometer. The nickel content was determined by Atomic Absorption Spectrometer (AAS) Shimadzu AA-6650. The presence or absence of H_2O molecules in the complex was estimated from the results of thermal analysis using Differential Thermal Analyzer Shimadzu 50. Infrared spectra were recorded on Prestige-21 Shimadzu spectrophotometers as KBr pellets in the frequency range of 4000–450 cm^{-1}. Molar conductivity (Λ*m) of 1 mM complex and standard solution in water were measured on Jenway CE 4071 conductivity meter at 25 °C. The magnetic moment was measured using Auto Sherwood Scientific 10169 Magnetic Susceptibility Balance.

2.3. Synthesis of Ni(II) Complex

$NiSO_4{\cdot}6H_2O$ (0.263 g; 1 mmol) was dissolved in methanol (5 mL) then added dropwise on isonicotinic acid (0.269 g; 2.2 mmol) in hot methanol (20 mL). The mixture was refluxed for ± 2 hours. The solution was concentrated to ± 20 mL then allowed to stand for 2 days. The obtained greenish solids were filtered off, washed with methanol and then dried in a desiccator.

3. Result and discussions

3.1. Formation of The Complex

The electronic spectra of the formed complex in methanol are shown in Figure 3.

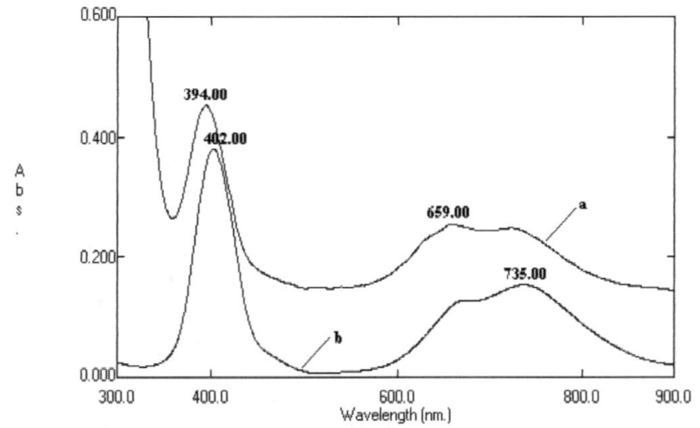

Figure 3. Electronic Spectra The Complex (a) and $NiSO_4·6H_2O$ (b) in Methanol

Figure 3 shows the shift of the maximum wavelength from 735 and 402 nm ($NiSO_4·6H_2O$) to 659 (15175) and 394 (253807) nm (Ni-asint). It indicated that asint as the ligand change molecules of water and bind to nickel ion. Thus, the shifting of wavelength toward smaller showed the formation of the complex.

3.2. Analysis of The Amount of Nickel with Atomic Absorption Spectroscopy (AAS)

The result of nickel content analyzed by AAS in the Ni(II)-asint complex was 6.92%. If the result of the measurement is compared to the theoretical analysis shown in Table 1, it can be predicted the formula of Ni(II)-asint complex. The different value of nickel content based on theoretic and experiment is generally happen in the research [13]. This case happened in the percentage of nickel in the complex of Ni(II) with 6-amino benzothiazole derivative. Theoretically, it was 7.41 % but based on experiment 7.49 % [15]. It needs more methods to confirm the number of presence of water molecules as well as the state whether as hydrates or ligands, such as thermogravimetry and differential thermal analysis. Therefore, the formula of the formed complex is close to $Ni(asint)_5SO_4(H_2O)_4$.

Table 1. Percentage of Nickel in The Complex Theoretically

Empirical Formula	Mr	% Ni
$Ni(asint)_5SO_4(H_2O)_3$	824.37	7.12
$Ni(asint)_5SO_4(H_2O)_4$	**842.38**	**6.97**
$Ni(asint)_5SO_4(H_2O)_5$	860.40	6.82

3.3. Analysis of Electrical Conductivity

The results of molar conductivity measurement of the standard solution and complex in water are shown in Table 2. If both results compared, the conductivity of the Ni(II)-asint complex is close to the electrical conductivity of $FeSO_4·7H_2O$, $CuSO_4·5H_2O$, and $NiSO_4·6H_2O$. The value of the measurement is consistent with the 1:1 electrolite, indicating that the SO_4^{2-} ions are not coordinated to Ni^{2+}.

Table 2. Molar Conductivity of Metal Salts and Ni(II)-asint in Water

Solution	Λ_m (S cm^2 mol^{-1})	Cation: Anion Charge
Water	0	-
$FeSO_4{\cdot}7H_2O$	103.00 ± 1.00	1:1
$CuSO_4{\cdot}5H_2O$	173.67 ± 1.82	1:1
$NiSO_4{\cdot}6H_2O$	147.33 ± 1.82	1:1
$CuCl_2{\cdot}2H_2O$	216.00 ± 0.47	2:1
$NiCl_2{\cdot}6H_2O$	252.00 ± 0.82	2:1
$AlCl_3{\cdot}6H_2O$	315.00 ± 0.82	3:1
Ni-asint	**146.83 ± 4.18**	**1:1**

3.4. Thermal Analysis

Differential Thermal Analysis (DTA) showed the peak of endotherm at temperature 74.37°C (*Heat = -0.55 J*) and 148.01°C (*Heat = -0.69 J*), shown in Figure 4. It indicated the evaporation of water molecules as hydrates and the ligand. It is reported that decomposition step at 60-100 °C is assignable to the removal of co-crystallized water molecules [15-16]. Complexes of Ni(II) are known in each of the common geometries: octahedral, tetrahedral and square planar. Therefore, nickel may be coordinated to five molecules of asint and one water forming six coordinated complex. The possible structure is $[Ni(asint)_5(H_2O)]SO_4{\cdot}3H_2O$.

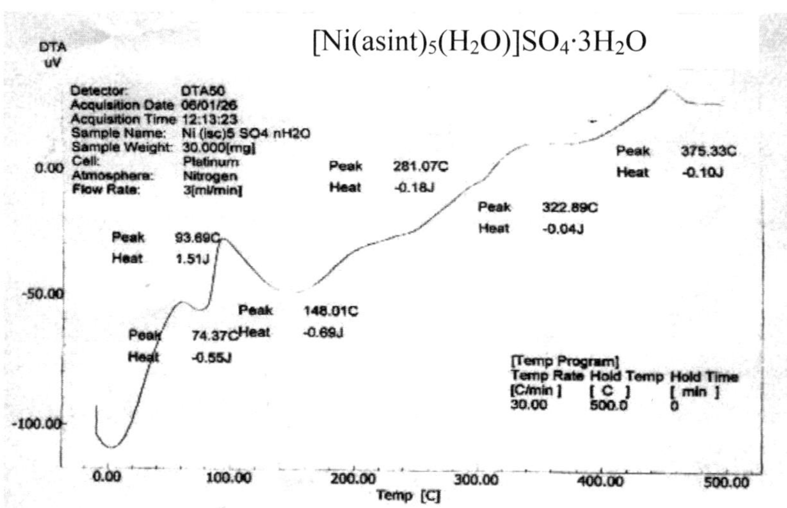

Figure 4. Thermogravimetric/Differential Thermal Analysis Spectra

3.5. Infrared Analysis

Table 3 shows the infrared analysis of isonicotinic acid and Ni-asint complex. It shows the absorption of C = N groups of cyclic pyridine chains in free ligands (1149 cm^{-1}; 1330 cm^{-1}) having a shift towards larger wavenumbers (1234 cm^{-1}; 1338 cm^{-1}) after forming the Ni-isonicotinic acid complex. The same result occurs in the complex of metal with Salicilydene-Cefotaxime. It has a pyridine infrared absorption at 1600 cm^{-1} then shift to 1640 cm^{-1} after complexing [17]. From the result, we can confirm the coordination of C=N to nickel ion. This indicates the N-pyridine group is coordinated to Ni(II) central atom, as in the Zn(II)-nicotinamide complex which originally had a free ligand absorption in the region of 1592 cm^{-1} and 1573 cm^{-1} shifted to 1603 cm^{-1} and 1570 cm^{-1} [10]. Meanwhile, the infrared absorption of -OH in the complex shifted toward lower wave number compared to the ligand. It indicated that -OH also binds to the nickel ion as the ligand. Thus, the spectra of other functional groups do not experience a shift after the formation of the complex.

Table 3. Infrared Spectra of Functional Groups Isonicotinic Acid and The Complex

Compounds	υ OH	υ NH bending	υ COH	υ C=O cyclic	υ C=N aromatic
Isonicotinic acid	3425	1616	1411	1716	1149 ;1331
Ni(II)-asint	3371	1616	1411	1716	1234;1338

3.6. Magnetic Properties (μ_{eff})

The effective magnetic moment value (μ_{eff}) of the Ni(II)-asint complex is 3.30 B.M. The effective magnetic moment value (μ_{eff}) shows the paramagnetic complex with two unpaired electrons. The result of measuring the magnetic moment value in the complex $[Ni(L)_2]\cdot 2H_2O$ with L = Schiff Base ligand from 1-amino-5-benzoyl-4-phenyl-1H pyrimidine-2-one with 2-hydroxy-naphthaldehyde in the range 2.8-3.5 B.M. octahedral structure [18]. It is reported by Mukherjee et al. [19] that nickel complexes with a value of magnetic moment 3.0, 3.2, and 3.3 B.M. forming octahedral geometry shown in Table 4. Therefore, the Ni (II)-asint complex is also assumed to be octahedral.

Table 4. Magnetic Moment Value of Nickel Complexes

Complexes	Magnetic Moment Value (B.M.)	Geometry
$[Ni(HSalpn)(NCS)(H2O)]\cdot H_2O$	3.3	Octahedral
[Ni(HSalpen)(NCS)]	3.0	Octahedral
$[\{Ni(Salpen)\}_2Ni(NCS)_2]$	3.3	Octahedral
$\{Ni(Salen)\}_2Ni(NCS)_2]$	3.2	Octahedral

3.7. Electronic Spectra

The solution UV–Vis spectrum of the formed complex showed two peaks at 659 and 394 nm shown in figure 3. The octahedral complex Ni(II) has three electronic transitions $^3A_{2g}\rightarrow {}^3T_{2g}$ (F), $^3A_{2g}\rightarrow {}^3T_{1g}$ (F) dan $^3A_{2g}\rightarrow {}^3T_{1g}$ (P) which appear as three peaks in the UV-Vis spectrum [1]. The three peaks of the UV-Vis spectra of the octahedral complex Ni(II) are in the vicinity of 10,000 nm, 600nm and 400nm [20]. However, the Ni(II)-isonicotinic acid only shows two transition peaks. The transitions are not visible because they are outside the range of the UV-Vis spectrophotometer used. Tetrakis (2-amino-3-methylpyridine)di(aqua)nickel(II) also showed two transitions at 620 nm (16116 cm-1) and 375 nm (26630 cm-1) indicating octahedral geometry [21]. Another octahedral complex of nickel shows two transition peaks of UV-Vis as well [22].

3.8. Proposed Structure

From all of the characterization analysis above, the proposed structure of the complex shown in Figure 5.

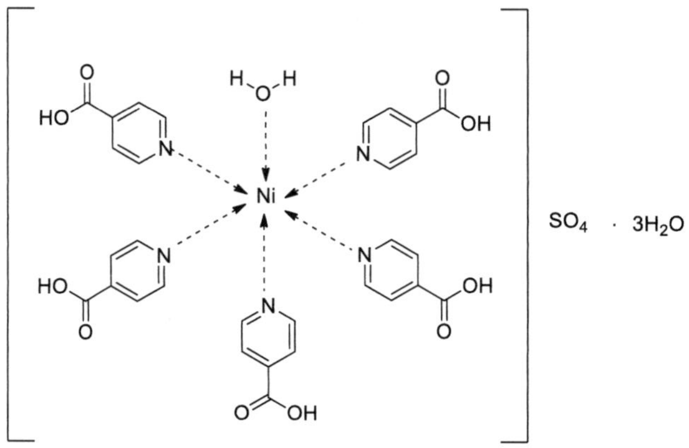

Figure 5. Suggested Structure $[Ni(asint)_5(H_2O)]SO_4{\cdot}3H_2O$.

4. Conclusion

The Ni(II) complex can be synthesized from $NiSO_4{\cdot}6H_2O$ and isonicotinic acid in refluxing methanol with a mole ratio of metal to ligand = 1:2. The complex formula is $[Ni(asint)_5(H_2O)]SO_4{\cdot}3H_2O$, namely mono-aqua-pentakis (isonicotinic acid) nickel(II) sulfate trihydrate. The ligands are coordinated to central metal ion through N-pyridine in the isonicotinic acid and -OH group of water. The complex is paramagnetic and has two peaks at maximum UV-Vis absorption of 659 and 394 nm. It is indicated that the complex forming octahedral geometry.

5. References

[1] Lee J D 1994 *Concise Inorganic Chemistry* Fourth Edition (London: Chapman and Hall)
[2] Routaray A, Mantri S, Nath N, Sutar A K and Maharana T 2016 *Polyhedron* **119** 335-41
[3] Takjoo R, Centore R, Akbari A and Ahmadi M 2014 *Polyhedron* **80** 243-49
[4] Routaray A, Mantri S, Nath N, Sutar A K and Maharana T 2015 *Spectrochim Acta A Mol Biomol Spectrosc.* **139** 96-404
[5] Masoud M S, Ali A E, Elasala G S and Kolkaila S A *2017 Journal of Chemical and Pharmaceutical Research* **9**(4) 171-79
[6] Allali M, Ichou Y, Allali N, Zarrouk A and Habbadi N 2016 *J. Mater. Environ. Sci.* **7**(6) 2042-50
[7] Chastain B B, Rick E A, Pruett R L and Gray H B 1968 *J. Am. Chem. Soc.* **90**(15) 3994–4000
[8] Pahonţu E, Julea F, Chumakov Y, Petrenco P, Roşu T and Gulea A *2017 Journal of Organometallic Chemistry* **836–837** 44–55
[9] Nikiforov A A, Eremin E A, Gurzhii V V, Misharev A D, Medvedskii N L, Ponyaev A I and Belyaev A N 2017 *Russian Journal of Coordination Chemistry* **43** 269–77
[10] Liu B, Xu L and Guo G 2006 *Journal of Solid State Chemistry* **179** 883–90
[11] Li L, Castellano F N, Gryczynski I and Lakowicz J R 1999 *Chem.Phys. Lipids* **99** 1
[12] Sahin E, Ide S, Atac S and Yurdakul S 2002 *Journal of Molecular Structure* **616** 253-58
[13] Ilies D, Pahontu E, Shova S, Gulea A and Rosu T 2013 *Polyhedron* **51** 307–15
[14] Daravath S D, Kumar M P, Rambabu A, Vamsikrishna N, Ganji N and Shivaraj 2017 *Journal of Molecular Structure* **1144** 147-158
[15] Abdolmaleki S and Ghadermazi M 2017 *Inorganica Chimica Acta* **461** 221–32
[16] Kavitha N and Lakshmi P V A 2015 *Journal of Saudi Chemical Society* **21** S457-S466
[17] Anacona J R, Johan C and Ovidio A A 2013 *International Journal of Inorganic Chemistry* http://dx.doi.org/10.1155/2013/108740
[18] Sonmez M 2003 *Inorganic Chemistry* **77** 397-402
[19] Mukherjee P, Biswas C, Drew M G B and Ghosh A 2007 *Polyhedron* **26** 3121–28

[20] Sharpe A G 1992 *Inorganic Chemistry 3rd Edition* (New York: John Willey and Sons Inc.)
[21] Rahardjo S B and Aditya H P 2017 *IOP Conf. Series: Materials Science and Engineering* **17** 012048 doi:10.1088/1757-899X/172/1/012048
[22] Vashi R T, Shelat C D and Patel H 2010 *International Journal of Current Pharmaceutical Research* **2(3)** 57-60

Acknowledgement
The authors gratefully express acknowledgments to Ministry of Research, Technology and Higher Education (RISTEKDIKTI) of the Republic of Indonesia for financial support.

Effect of water volume based on water absorption and mixing time on physical properties of tapioca starch – wheat composite bread

I K Prameswari[1*], G J Manuhara[1], B S Amanto[1] and W Atmaka[1]

[1]Department of Food Science and Technology, Faculty of Agriculture, Sebelas Maret University

*Corresponding email: ixorakiranap@gmail.com

Abstract. Tapioca starch application in bread processing change water absorption level by the dough, while sufficient mixing time makes the optimal water absorption. This research aims to determine the effect of variations in water volume and mixing time on physical properties of tapioca starch – wheat composite bread and the best method for the composite bread processing. This research used Complete Randomized Factorial Design (CRFD) with two factors: variations of water volume (111,8 ml, 117,4 ml, 123 ml) and mixing time (16 minutes, 17 minutes 36 seconds, 19 minutes 12 seconds). The result showed that water volume significantly affected on dough volume, bread volume and specific volume, baking expansion, and crust thickness. Mixing time significantly affected on dough volume and specific volume, bread volume and specific volume, baking expansion, bread height, and crust thickness. While the combination of water volume and mixing time significantly affected for all physical properties parameters except crust thickness.

Key words: Substituted bread, water, mixing time, physical properties, texture

1. Introduction

White bread is one of bakery product with high carbohydrate content and nowadays white bread becomes alternative source of carbohydrate aside from rice in south east asia countries. Normally, white bread is made from wheat flour, however the needs of wheat flour in South East Asia countries only can be filled through import. In order to reduce the dependency on wheat flour, the use of wheat flour should be diverted with the local commodities such as tapioca. The utilization of tapioca is because the price is cheaper than wheat flour and with a neutral flavor. Eriksson et al., (2014), states that the higher application of cassava flour in bread making to subtitute wheat flour, the harder bread will be produced, while the elasticity and the bread volume are lower [1].

However, the disadvantage of using tapioca is because it has no gluten content [2] which needed to expand the dough in bread making process. Using flour without gluten content increases the hardness level of the bread because of imperfect expansion and it also reduces the bread's volume which causes bread's textures becomes hard and dense [3]. This aspect can reduce the level of consumer's acceptance. Thus, to improve the characteristic of the bread, modification in bread processing are needed, such as the control of water amount, and sufficient mixing time in dough making, especially when the dough consists of different kinds of flour.

According to Jacob (1951), the amount of water that added in bread making process depends on the flour's capability to absorb water and the expected characteristic of the bread [4]. The normal amount of water added in white bread making is 64-66% of the flour's portion. Absorption level of the flour shows the percentage of water that can be absorbed by the flour after the dough becomes dull. Wheat flour and tapioca have different level of water absorption [5]. The change of flour's water absorption caused by flour substitution results to difference volume of water which added differently in different composition of flours of composite flour in order to make dough with optimal characteristic [6] (Adekele and Odedeji, 2010 ; Rauf, 2015).

The ingridients mixing is finished when the dough is formed well (Nur'aini, 2011). In process of mixing, gluten function is to maintain the air inside the dough and gas produced by the yeast while fermenting, which makes the dough expanding [7]. However to Utami (2010) the excessive time in mixing cause the structure of gluten broken and the capability of the dough to hold the gas loses and so the bread does not expand perfectly [8]. On contrary, if the mixing time is too short, the dough will not expand and its texture become stiff. Thus, time management in mixing process must be investigated to produce optimal dough expanding and also the desired texture of the bread. This research is conducted to identify the effects of water amount and mixing time on tapioca starch – wheat composite bread physical characteristic.

2. Research Methods

2.1. Instruments

The instruments are divided into two, first are the tools that used to make white bread such as agitator, oven, basin, analytic scale, pan, spoon, spatula, cloth, plastic, rolling pin, electrical stove and the second one are tools that used to do the analysis such as scale, container with measurement 23 cm x 15 cm x 12 cm, scanner Canon Printer MP287.

2.2. Ingredients

Ingredients that used to make white bread are hard wheat flour, tapioca starch from local producer, bread improver, margarine, yeast, skimmed milk, refined sugar, salt and shortening. Meanwhile, there is white millet to analyze the physical characteristic of the bread.

2.3. Research Stages

2.3.1. Determining Water Volume and Mixing Time. Determining the water volume depends on water absorption of the flour. Mixed wheat flour and tapioca starch, both are tested along side with the formula to analyze the water absorption level. The result of the analysis used to determine the water volume in the dough. Then, there was consideration to determine the exact time of mixing using mixer until the dough is formed perfectly.

2.3.2. Determining Initial Formula of Products. The ratio between wheat flour and tapioca starch were taken from the preliminary research resulted in 70 : 30 ratio (wheat : tapioca starch). The ingredients to make white bread were 175 gram wheat flour, 75 gram tapioca starch, 3.75 gram salt, 10 gram shortening, margarine, 1 gram bread improver, 5.5 gram yeast and water (111.8 ml; 117.4 ml; or 123 ml).

2.3.3. Making White Bread with Tapioca Starch Substitution. The stages in made white bread as follows; first weigh all ingredients based on formula. Before the other ingredients are added, mixed the wheat flour and tapioca in the mixer until evenly distributed. Then, added 15 gram of refined sugar, 3.75 gram of salt, 5 gram of skimmed milk, 10 gram of shortening, 1 gram of bread improver and 5.5 gram of yeast into the mixer alongside with the mixed wheat flour and tapioca. Added water gradually according to the treatment (111.8 ml; 117.4 ml; 123 ml) and mixed the dough until it formed perfectly according to the proper mixing time (16 minutes, 17 minutes and 36 seconds, 19 minutes and 12 seconds). After the dough formed perfectly, formed it into round shape and wraped it in the plastic for 15 minutes in room temperature. After that, formed the dough into flat shape by using rolling pin then roll it and put it into pan which already been coated with margarine. Steamed the dough for 45 minutes with temperature between 45-50 °C and RH 75-80%. The last stage baked it inside the oven for 40 minutes at temperature 150°C.

2.4. Analysis

The analysis of physical characteristic consists of dough volume and spesific volume, bread volume and specific volume, baking expansion, height of dough and bread, weight loss and thickness of crust.

2.5. Analysis of Data

The data is analyzed by using *Two Way Analysis of Variances* (ANOVA). If the effect of treatments are found, then the analysis continued with *Duncan Multiple Range Test* (DMRT) in significant degree 5% ($p \leq 0.05$).

3. Result and Discussion

Table 1. Physical Characteristics of Dough and Bread with 30%-Tapioca-Substitution.

	Parameters								
	VD (cm^3)	SVD (cm^3/g)	VB (cm^3)	SVB (cm^3/g)	BA (%)	DH (cm)	BH (cm)	WL (%)	CT (mm)
A	395,71	1,00	952,37	2,67	240,70	5,03	6,43	9,15	6,20
B	366,41	0,91	958,87	2,55	261,69	5,33	6,60	5,82	6,27
C	379,37	0,94	948,40	2,50	249,94	5,27	6,65	6,62	5,67
D	384,79	0,94	1077,37	2,80	279,78	5,27	7,20	5,99	5,73
E	407,54	1,00	997,50	2,65	244,80	5,10	6,77	7,62	5,67
F	368,55	0,91	1014,14	2,66	270,31	5,23	6,67	7,79	6,13
G	411,98	1,00	997,62	2,60	242,15	5,20	6,43	6,63	5,93
H	390,89	0,94	977,81	2,56	250,50	5,30	6,23	7,11	5,33
I	398,15	0,97	1266,06	3,32	318,00	5,07	8,43	6,73	5,53

[a]Notes : A : DSA-Coherence; B : DSA-Coherence+10%; C : DSA-Coherence+20%; D : DSA+5%-Coherence; E : DSA+5%-Coherence+10%; F : DSA+5%-Coherence+20%; G : DSA+10%- Coherence; H : DSA+10%- Coherence +10%; I : DSA+10%- Coherence +20%; VD : Volume of Dough; SVD : Specific Volume of Dough; VB : Volume of Bread; SVB : Specific Volume of Bread; BA : Baking Expansion; DH : Dough Height; BH : Bread Height; WL : Weight Loss; CT : Crust Thickness

3.1. Volume and Specific Volume of Dough

The result of *two way* ANOVA analysis showed that the varied water volume and mixing time affected the volume of white bread dough with p value 0.0002 ($p<0.05$) and 0.003 ($p<0.05$). **Table 1.** showed that the increased of dough volume was in line with water volume and in contrary with mixing time. Hera *et al.* (2013), said that the larger amount of water that added in the dough, made more flour particles hydrated, and it affected the characteristic range of dough in process of mixing [9]. Water hydrates the ingredients to make dough and expand the starch granule. Thus, the water hydrates starch granule sufficiently and it made the granule starch bigger and expands [10]

Excessive time in mixing process can reduced the volume of the bread which the consistency of the dough also getting lower and sticky. According to Gomez et al. (2011), excessive time of mixing makes the dough softer but it has low consistency and becomes sticky [11]. It was marked by spatial power and the deceased of dough endurance level. The result showed that the mixing of the bread

IOP Conf. Series: Journal of Physics: Conf. Series **1022** (2018) 012029 doi:10.1088/1742-6596/1022/1/012029

dough in more than 24 minutes lead to broken structure of the gluten. The both combinations affected on dough volume with p value 0.00006 ($p<0.05$).

The result of *two way* ANOVA analysis showed that the varied water volume did not have significant effect with p value 0.065 ($p>0.05$) and the mixing time had the real impact on p value 0.005 ($p<0.05$) towards specific volume of white bread dough. **Table 1**. showed that adding larger amount of water increased the specific volume of dough. The specific volume of the dough was a comparison between volume of the dough and dough's weight [12]. Therefore, specific volume of the dough was in line with dough's volume. The combinations of water volume and mixing time affected on dough specific volume with p value amount 0.00004 ($p<0.05$).

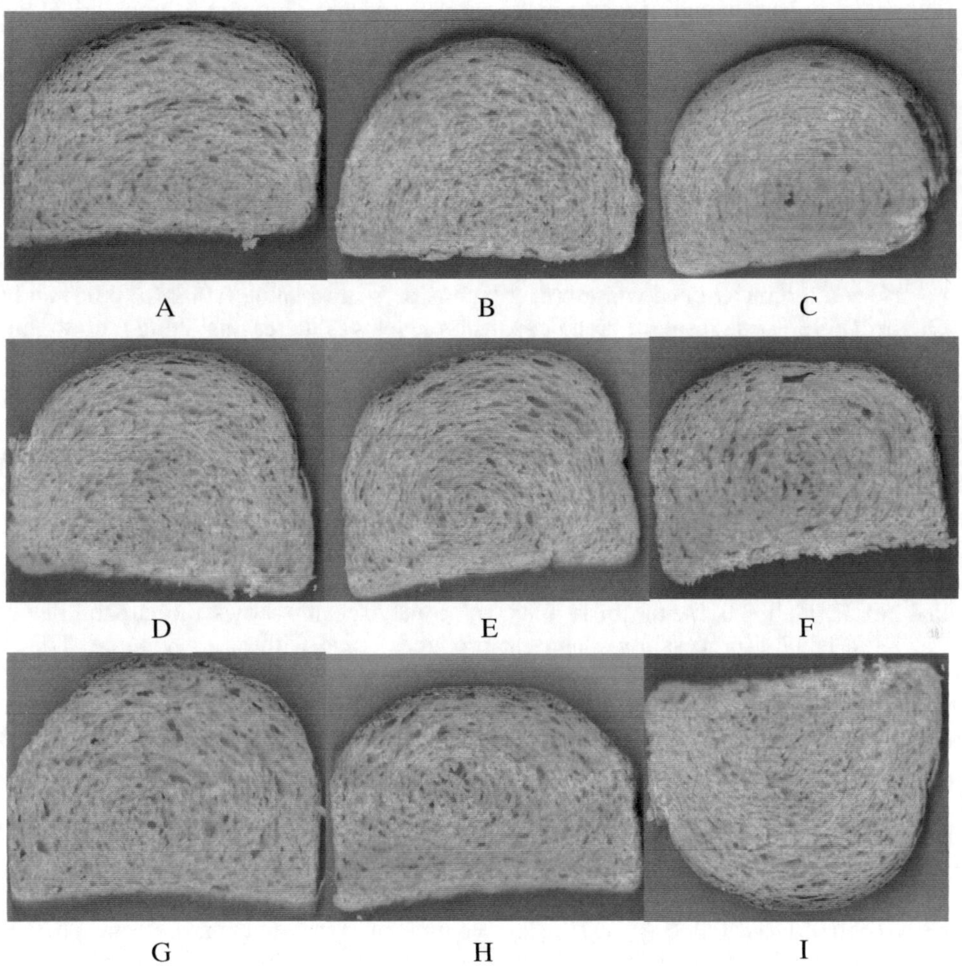

Figure 1. Bread with 30%-Tapioca-Substitution

3.2. Volume and Specific Volume of Bread

The *two way* ANOVA analysis showed that the varied water volume in process of mixing affected on white bread's volume with p value 0.00002 ($p<0.05$) and 0.0004 ($p<0.05$). **Table 1**. told that the more water in mixing increased bread volume. The application of tapioca starch reduced the number of gluten. The existed capacities of flour have to absorb large amount of water in order to bind the starch with water. Therefore, the considerable amount of starch can raise its viscosity followed by the swollen starch granule [13].

The increasing of swelling power increased the volume of bread. The water and gas holding are achieved which results to higher bread volume. Moreover, the more numbers of mixing quantity in longer period of mixing caused gluten matrix were formed because the flour absorbs enough water and it allows more expansion [14]. Meanwhile, the combination of mixing time and water volume affected on bread volume with p value 0.000007 ($p<0.05$).

The statistic result of two way ANOVA showed that the varied water volume and mixing time affected the specific volume of white bread with p value 0.0001 ($p<0.05$) and 0.002 ($p<0.05$). **Table 1.** told that the more water volume increased the specific volume of bread. Bread specific volume was in line with the bread volume in contrary with the weight of bread [12]. Generally, there was an assumption that the higher amount of water makes higher specific volume of the bread. However the crumb will be larger if the amount of water is overly added in longer time of proofing. Besides, overly hydrated bread produces weak final structure and difficulties in slicing [15]. However, the combinations of mixing time and water volume affected on bread spesific volume with p value 0.000007 ($p<0.05$).

3.3. Baking Expansion

The analysis result of *two way* ANOVA showed that the varied water volume and time mixing affected on the level of white bread expanding with p value 0.001 ($p<0.05$) dan 0.00001 ($p<0.05$) and mixing time affected on the level of white bread expansion with p value 0.001 ($p<0.05$) and 0.00001 ($p<0.05$). On the **Table 1.**, the result of bread expansion level was increasing along with the increasing of water volume.

The capability to hold gas inside tapioca starch – wheat coamposite bread was lower compared to wheat flour because of less gluten to develop bread volume. By bonding more amount of water, the additional water that was needed to expand the bread because tapioca starch has large amount of amylopectin which make the structure of starch easier to open then water is easy to be absorbed. The expansion of starch granule can be seen through the large number of apex value in flour viscosity. The apex value of flour viscosity is 5387.94 mPas higher compared to the apex value of wheat flour 2375.25 mPas (Imanningsih, 2012).

According to Faridah (2015), the high apex value in viscosity shows the capabilities of starch granule to swell in heating process [16]. Thus, it produces bread with larger volume. The sufficient mixing time allowed yeast to adapt with its medium and the gas distribution into the dough [11]. The more amount of mxing time makes gluten can expand and maintain gas better. Meanwhile, the combinations of mixing time and water volume affected on baking expansion with p value 0.0000004 ($p<0.05$).

3.4. Dough and Bread Height

The result of the analysis using *two-way* ANOVA showed that experiments involving different volume of water and mixing time did not have big impact in changing height of white bread dough, as it p value was.902 ($p>0.05$) and 0.360 ($p>0.05$). The volume of water and mixing time unsignificantly affect the increasing height of bread dough. Adding amount of water and increasing mixing time likely influenced the dough height to decrease. It was because high water content in the dough make dough mass increase and the dough tend to widen, filling the space in pan because its consistency was low. According to Gomez et al. (2011), excessive time of mixing can make dough softer but it lowers dough consistency and makes it stickier [11]. Besides, combination of mixing time and water volume significantly affected the dough height with p value of 0.0009 (<0.05).

The analysis result of *two-way* ANOVA indicated that the varied water volume did not give any significant effect meanwhile mixing time affected height of white bread significantly with p value of 0.057 ($p>0.05$) and 0.003 ($p<0.05$). Height of bread increased as the volume of water was added. This was in line with treatment of mixing of which the longer dough is mixed, the dove height also increased. Moreover, the high bread height is related to the rising of dough. Height of bread is also related to volume of bread. In the process of baking, there was a quick increasing volume which caused by the evaporation of water from the starch molecules so that the volume after baking increases [17]. The increasing of bread volume resulted in the increasing height of bread. However, the combination of mixing time and water volume significantly affected the height of bread with p value of 0.00003 ($p<0.05$).

3.5. Weight Loss

The analysis result using *two-way* ANOVA showed that volume of water and mixing time did not contribute significant impact in terms of white bread weight loss with p value of 0.615 ($p>0.05$) and 0,598 ($p>0{,}05$). Bread weight loss decreased as volume of water and mixing time increased although did not represent significant result. According to Sundari et al. (2015), high level of heat affected the decreasing water content of dough in large quantities [18]. Besides, fat and protein would be damage if more heat is given.

Moreover, Roos et al. (2011) states that bread weight loss in baking was associated with water-binding capacity of bread matrix. Tapioca application in the white bread contributed in reducing amount of gluten as starch binder so that water shall be increased to help gluten bind more starches. Tapioca mainly consists of amylopectin. According to Muflihati et al. (2015), amylopectin has a structure that makes it absorb water easily that the water will be bound after absorbing [17]. Water content prevent dehydration of bread so it also prevent the bread from weight loss [19]. High rising bread likely has a chance to get a slight weight loss [20]. However, combination of mixing timeand water volume significantly affected with p value of bread weight loss 0.001 ($p<0.05$).

3.6. Bread Crust Thickness

The analysis result of *two-way* ANOVA showed that varied volume of water and mixing time did not significantly influence on crust thickness of white bread with p value of 0.066 ($p>0.05$) and 0.478 ($p>0.05$). Crust thickness of bread decreased as volume of water was increased. It was in line with treatment of mixing time of which long duration also made crust thickness decreased. According to Gallagher et al. (2003), adding for about 10% or 20% water to dough reduce crumb firmness and thickness of crust which is caused by decreased starch retrogradation [21].

According to Charley (1982), the high quality bread performed when its convex shaped, not too thick with golden color crust [22]. Good crust layer showed the appearance of white bread cohesive. However, combination of mixing time and water volume did not affect the thickness of bread crust with p value of 0.084 ($p>0.05$).

4. Conclusions

Varied treatments of water volume showed significant effect to dough volume, volume and specific volume of bread, level of rising, and crust thickness. Varied treatments in mixing time indicated significant impact to volume and specific volume of dough, volume and specific volume of bread, baking expansion, bread height, and crust thickness. Moreover, combination of water volume and mixing time affected all parameters of physical characteristics except crust thickness. The more water and the longer mixing time, volume and specific volume of bread, baking expansion, and bread height also significantly increased.

References

[1] Eriksson E, K Koch, C Tortoe, P T Akonor and C Oduro Yeboah 2014 Evaluation of the physical and sensory characteristics of bread produced from three varieties of cassava and wheat composite flours *J. Food and Pub. Heal.* **4**(5) p 214-22

[2] Arlene A, Judy R and Maria F 2009 *Pembuatan Roti Tawar dari Tepung Singkong dan Tepung Kedelai National Symp. RAPI VIII*

[3] Wijayanti Y R 2007 *Substitusi Tepung Gandum (Triticum aestivum) dengan Tepung Garut*

(Maranta arundinaceae L) pada Pembuatan Roti Tawar (Yogyakarta: Gadjah Mada University)

[4] Jacobs M B 1951 *The Chemistry and Technology of Food and Food Product vol. 2* (New York: Interscience Publisher)

[5] Widaningrum S W and Soewarno T S 2005 Pengayaan Tepung Kedelai pada pembuatan Mie Basah dengan Bahan Baku Tepung Terigu yang Disubstitusi Tepung Garut *J. Pasc.* **2**(1) p 41-48

[6] Rauf R and Dwi S 2015 Daya serap air sebagai acuan untuk menentukan volume air dalam pembuatan adonan roti dari campuran tepung terigu dan tepung singkong *J. Agrit.* **35**(3)

[7] Koswara S 2009 *Teknologi Pengolahan Roti Seri Teknologi Pangan Populer (Teori dan Praktek)* (Jakarta: eBookPangan.com)

[8] Utami I S 2010 *Memahami Proses Pengolahan Roti* (Jakarta: Food Review Indonesia **5**(5))

[9] Hera E D L, Cristina M R and Manuel G 2013 *Effect of Water Content and Flour Particle Size on Gluten-free Bread Quality and Digestibility* Food Tech. Area (University of Valladolid)

[10] Dewi M N H and Cakrawati 2012 *Bahan Pangan Gizi dan Kesehatan* (Bandung: Alfabeta)

[11] Gomez A, C Ferrero, A Calyelo, M C Anon and M C Puppo 2011 Effect of mixing time on structural and rheological properties of wheat flour dough of bread making *Int. J. Food Properties* **14**(3) p 583-98

[12] Nwosu U L, Elochukwu C U and Onwurah C O 2014 Physical characteristics and sensory quality of bread produced from wheat and african oil bean flour blends *Afr. J. of Food Sci.* **8**(6) p 351-55

[13] Kurniawan S 2010 *Pengaruh Lama Fermentasi dan Konsentrasi $Ca(OH)_2$ untuk Perendaman Terhadap Karakteristik Tepung Mocaf (Modified Cassava Flour) Varietas Singkong Pahit* (Surakarta: Sebelas Maret University)

[14] Priyati A, Sirajuddin H A and Guyup Mahardhian D P 2016 Pengaruh kecepatan putar pengadukan adonan terhadap sifat fisik roti *Jurnal Ilmiah Rekayasa Pertanian dan Biosistem* **14**(1)

[15] Gomez M and Lorena S S 2015 *Gluten Free Bakery Products and Pasta* (Barcelona: Omnia Science) p 565-604

[16] Faridah H M 2015 *Pengaruh Jumlah Air dan Jenis Hidrokoloid Terhadap Formula Roti Tawar Mini Bebas Gluten Berbasis Tepung Beras, Pati Jagung dan Pati Singkong* (Bogor: Bogor Agricultural Institute)

[17] Muflihati I L, Birgitta N, Afriyanti and Reny M 2015 *Efek Substitusi Tepung Terigu dengan Pati Ketan Terhadap Sifat Fisik Cookies National Conf.* (Yogyakarta: PGRI University)

[18] Sundari D, Almasyhuri and Astuti L 2015 Pengaruh proses pemasakan terhadap komposisi zat gizi bahan pangan sumber protein *Litbangkes Med.* **25**(4)

[19] Midle L B, Laura A R and Maria C P 2012 Gluten-free bread based on tapioca starch: testure and sensory studies *J. Food Bioprocess Tech.* **5** p 888-896

[20] Raikos V, Madalina N, Wendy R and Garry D 2014 Comparative study of the functional properties of lupin green pea, fava bean, hemp and buckwheat flours as affected by pH *J. Food Sci. and Nutr.* **2**(1) p 6

[21] Gallagher E, T R Gormley and E K Arendt 2003 Crust and crumb characteristics of gluten free breads *J. Food Eng.* **56** p 153-161

[22] Charley H 1982 *Food Science 2nd Edition* (New York: John and Sons)

[23] Moorthy S N 2004 *Tropical Sources of Starch* (Florida: CRC Press)

Characteristic of total suspended particulate (TSP) containing Pb and Zn at solid waste landfill

M A Budihardjo[1*], K Noveandra[1] and B P Samadikun[1]

[1]Department of Environmental Engineering, Universitas Diponegoro, Jl. Prof. H. Sudharto SH, Tembalang, Semarang, Indonesia, 50275

*Corresponding email: m.budihardjo@ft.undip.ac.id

Abstract. Activities conducted at municipal solid waste landfills (MSWLs) potentially cause air pollution. Heavy vehicles in MSWLs release various pollutants that can have negative impacts for humans. One noticeable pollutant at MSWLs is airborne total suspended particulate (TSP) which may contain heavy metals such as Pb and Zn and can cause disease when inhaled by humans. In this study, TSP from a landfill in Semarang, Indonesia was collected and characterized to quantify the concentration of Pb and Zn. Meteorological factors (i.e. temperature, humidity and wind velocity) and landfill activities were considered as factors affecting pollutant concentrations. TSP was sampled using dust samplers while the concentrations of heavy metals in TSP were analyzed using an Atomic Absorption Spectrophotometer (AAS). Pb concentration ranged from 0.84 to 1.78 $\mu g/m^3$ while Zn concentration was from 7.87 to 8.76 $\mu g/m^3$. The levels of Pb were below the threshold specified by the Indonesian Government. Meanwhile, the threshold for Zn has not yet been determined.

1. Introduction

Semarang, a large city in Indonesia, generates approximately 3,500 $m^3$of solid domestic waste per day which is collected and transferred to landfill for processing. The only landfill in Semarang, Jatibarang, started operating in 1992 and handles up to700 tons solid waste per day. The site has an area of more than 40 ha and consists of two zones, each equipped with liners to contain leachate which is treated to prevent it harming the environment [1].

The activities in the landfill involve the operation of heavy vehicles and equipment including waste disposal by garbage trucks, waste excavation and stockpiling using excavators, and waste distribution and solidification using loaders. All these activities produce air pollution such as dust, which is also generated from waste degradation, the use of internal combustion engines to generate power for mechanical recycling, composting activities and waste sorting [2, 3].

Dust or total suspended particulate (TSP) are particles with a diameter less than 100 μm. Some characteristics possessed by the particulates are the form of density, size, size distribution, adhesiveness, corrosive nature, reactivity, and toxicity. Of these characteristics, particle size was considered an important parameter that needed to be taken into account [4]. Tests on the relative toxicity of five groups of pollutants showed that the most threatening pollutant for health was particulate, followed consecutively by NO_2, SO_2, hydrocarbon, and Carbon Monoxide [5]. TSP also contain heavy metals, such as lead (Pb) and Zinc (Zn) [6].

Inside the human body, Pb is able to obstruct enzyme activities involved in the formation of haemoglobin (Hb). Pb is also retained in the kidneys, liver, nails, fat tissue, and hair. At low concentrations, Pb can influence psychological and nerve function. Exposure at high levels can damage whole organ systems such as the central nerve system, kidney and blood, and could be fatal [7]. Meanwhile, exposure to Zn can lead to deficiencies in other minerals. Excessive intake of Zn can also cause nausea, vomiting, fever and even liver and kidney disorders [8].

Up to now, there has been no available information related to Pb and Zn concentration in TSP at Jatibarang landfill; therefore, there is a need to explore the amount of heavy metals (Pb and Zn) at Jatibarang landfill and provide information on whether these are at harmful levels for human health.

2. Methods

This study was conducted at Jatibarang landfill, Semarang. Three sampling locations were determined based on consideration of dominant wind direction (Figure 1). The TSP sample was taken using a dust sampler and tested in the laboratory of Environmental Engineering, Diponegoro University, Semarang, using a gravimetric method. The heavy metal (Pb and Zn) contents were analyzed using a wet destruction method involving atomic absorption spectroscopy (AAS).

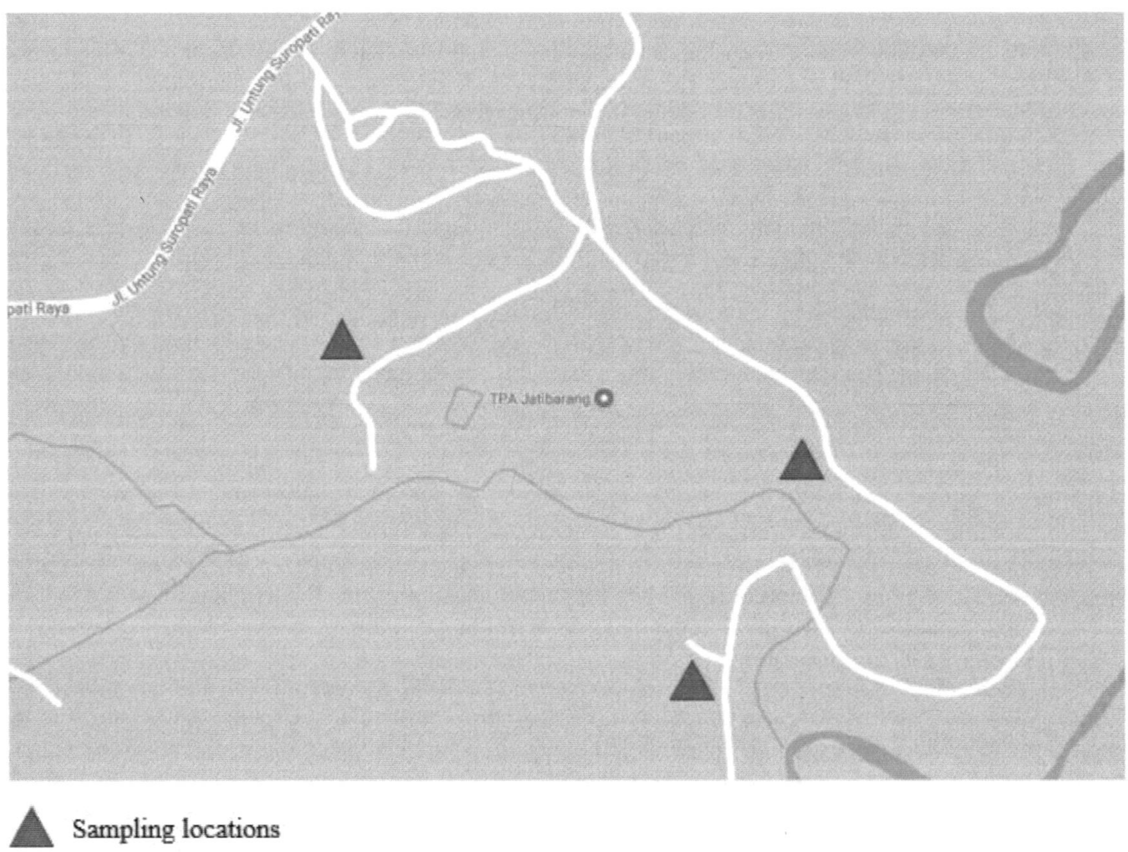

Figure 1. Sampling Locations

The main equipment used in this study were: (1) dust sampler DS 600-MVS to collect TSP; (2) an analytical balance to measure the filter weight before and after sample taking; (3) Fiber Glass Filter, a media for collecting the dust particles (TSP); and (4) atomic absorption spectroscopy (AAS), to analyze the concentration of Pb and Zn in the TSP. Meteorological factors such as temperature, air pressure, and wind velocity were also measured during sampling.

3. Results and Discussions

TSP samples were taken in Jatibarang landfill after considering the specific landfill activities, dominant wind direction and land contours. Samples were taken for six days to obtain results that represented the daily TSP concentration in the landfill. The meteorological conditions, such as temperature, humidity and wind velocity, were also monitored. During the sampling process, temperatures were between 32.6 to 35.7°C with humidity of 46.17-54.82% and the air pressure ranged from 754.1-758.61 mmHg.

Figure 2 shows the concentration of Pb in TSP over the six sampling days. The minimum Pb concentration was 0.84 µg/m^3 and the maximum was 1.78 µg/m^3. Pb concentration in TSP was below the air quality threshold standard regulated by the Indonesian Government.

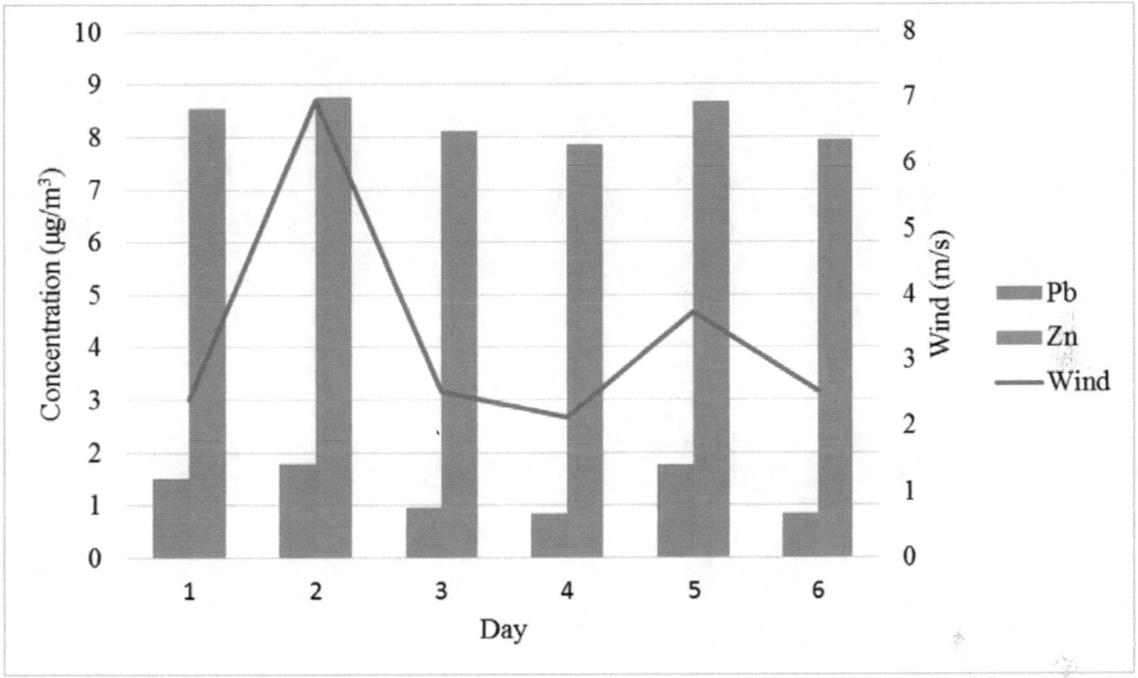

Figure 2. Concentration of CO

The sampling result for the Zn content in TSP was between 7.87 µg/m^3 to 8.76 µg/m^3 as shown in Figure 2. However, Zn content in TSP has not yet been regulated by the Indonesian Government so there is no standard to compare these figures with. Figure 2 also indicates that Pb and Zn concentration tended to increase as wind velocity increased.

4. Conclusions

This study on Pb and Zn concentration in TSP was conducted at Jatibarang landfill, Semarang to clarify the contribution of landfill activities to air quality, especially TSP containing heavy metals. The maximum concentration of Pb in TSP was 1.78 µg/m^3, while the concentration of Zn reached 8.76 µg/m^3. These results indicate that the concentration of Pb in TSP meets the air quality standard stated by the Indonesian Government, while the air quality standard of Zn in TSP is not yet regulated.

5. References

[1] Budihardjo M A 2016 *Adv. Mater. Sci. Eng.* 2016
[2] Koshy L, Jones T and BéruBé K 2009 *Biomarkers* **14**(supl) pp 49-53
[3] Paoli L, Corsini A, Bigagli V, Vannini J, Bruscoli C and Loppi S 2012 *Environ. Pollut* **161** pp

70-75
[4] Yang Y, Fang W, Xue M, Xu Z and Huang C 2016 *J. Mater. Cycles Waste* **18**(2) pp 296-302
[5] Chan C K and Yao X 2008 *Atmos. Environ* **42**(1) pp 1-42
[6] Hu X, Zhang Y, Ding Z, Wang T, Lian H, Sun Y and Wu J 2012 *Atmospheric Environment* **57** pp 146-152
[7] Kampa M and Castanas E 2008 *Environ. Pollut* **151**(2) pp 362-367
[8] Zheng N, Liu J, Wang Q and Liang Z 2010 *Sci. Total Environ* **408**(4) pp 726-733

IOP Conf. Series: Journal of Physics: Conf. Series **1022** (2018) 012031 doi:10.1088/1742-6596/1022/1/012031

Utilization of fish bone as adsorbent of Fe^{3+} ion by controllable removal of its carbonaceous component

M Nurhadi[1], R Kusumawardani[1], I I Widiyowati[1], Wirhanuddin[1] and H Nur[2]

[1]Department of Chemical Education, Mulawarman University, Kampus Gunung Kelua, Samarinda, 75119, East Kalimantan, Indonesia
[2]Center for Sustainable Nanomaterials, Ibnu Sina Institute for Scientific and Industrial Research, Universiti Teknologi Malaysia, 81310 UTM Skudai, Johor, Malaysia

Abstract. The performance of fish bone to adsorb Fe^{3+} ion in solution was studied. Powdered fish bone and carbonized fish bone were used as adsorbent. All absorbents were characterized by X-ray diffraction (XRD), IR spectroscopy, nitrogen adsorption, scanning electron microscopy (SEM) and TG analysis. Powdered fish bone and carbonized fish bone were effective as adsorbent for removing Fe^{3+} ion in solution. The metal adsorptions of Fe^{3+} ion were 94 and 98% for powdered fish bone and fish bone which carbonized at 400 and 500 °C.

1. Introduction

Metal elements which have a density greater than 5 g cm^{-3} and high molecular weight are classified as heavy metals. One kind of essential heavy metals is Fe element which highly soluble in water at pH less than 5 and to be toxic to human when excess. The high content of Fe metal will impact on human health such as can cause poisoning (vomiting), intestinal damage, premature aging to sudden death, Arthritis, birth defects, bleeding gums, cancer, kidney cirrhosis, constipation, diabetes, diarrhea, dizziness, tiredness, hepatitis, hypertension, insomnia.

Many methods have been used by researcher to remove heavy metal from water such as ion exchange, solvent extraction, complexation, adsorption, coagulation, phytoextraction, chemical precipitation, chemical oxidation or reduction, evaporation and reverse osmosis. The most cost-effective method which often was used by many researchers is adsorption due to easy to operate, high adsorption efficiency and low maintenance [1, 2]. Otherwise, the other methods have some shortcomings such as need a lot of reagent and energy, incomplete metal removal, low selectivity, high operational cost and unfriendly to environment [1].

Many researches have created to adsorb and remove heavy metals from solution by using absorbent. Zinc (II) was removed by using dried animal bones and powdered fish bones absorbent [1, 3]. Cu (II) ion was absorbed with absorbent from pretreated fish bones, rice husk and modified rice husk [4, 5]. Cd (II) was removed by using sweet potato (*Ipomoea batatas* L.) peel absorbent [6]. Cr (III) was absorbed by using activated carbon from alga *Ulva Lactuca* [7]. Pb (II) and Ni (II) were absorbed with natural bentonite absorbent [8].

In this research, adsorbent from Belida (*Chitalalopis*) fish bone which modified with by controllable removal of it carbonaceous component was used to adsorption and removal Fe^{3+} ion from

IOP Conf. Series: Journal of Physics: Conf. Series **1022** (2018) 012031 doi:10.1088/1742-6596/1022/1/012031

solution. Adsorbent from the Belida fish bone were investigated from characterization properties and adsorption efficiency for Fe^{3+} ion removing.

2. Materials and Methods

2.1. Raw Materials

The fish bone materials were collected from waste of Belida (*Chitalalopis*) fish from East Kalimantan, Indonesia.

2.2. Carbon Preparation

The fish bone from Belida fish waste was washed with boiling water to separate bone and impurities. The fish bone was dried in an oven at 110 °C for 24 hours. Furthermore, it was crushed and sieved to get a powder with size of 100 mesh and it was labeled as FB. The fish bone powder was carbonized in a furnace at variations in temperature of 300, 400, 500, 600, 700 and 900 °C for 1 hour. The results are labeled in Table 1.

Table 1. Codes and Treatments Done to The Samples

Samples	Temperature of carbonization (°C)	Time of carbonization (h)
FB	-	-
CFB-300	300	2
CFB -400	400	2
CFB -500	500	2
CFB -600	600	2
CFB -700	700	2
CFB -900	900	2

2.3. Preparation of Stock of Metal

$Fe(NO_3)_3 .9H_2O$ (Merck) was used to prepare stock solution of iron. That compound was dissolved in distilled water until concentration of Fe^{3+} ion 1000 mgL^{-1}. Another concentration of Fe^{3+} ion which used in experiment was prepared by diluting of the stock solution.

2.4. Samples Characterization

The samples were characterized by XRD, FTIR, SEM, TGA and N_2 adsorption. The crystallinity of the samples were investigated base on X-ray diffraction (XRD) pattern which recorded with a Bruker AXS Advance D8 difractometer using Cu K_α radiation (λ = 1.5405 Å, 40 kV and 40 mA). The functionalize groups in the samples were determine from IR spectra which recorded on a Perkin Elmer Fourier transform infrared (FTIR) spectrometer, with a spectral resolution of 2 cm^{-1}, scans 10 s, at temperature 20 °C. The surface textures in the samples were determined by using scanning electron microscopy (SEM), JEOL JSM-6701F instrument with an accelerating voltage of 15 kV. The amount of carbon in the samples was calculated from thermal gravimetric Analysis (TGA) data which collected using STA Linseis PT1600 instrument with heat rate 10 °C min^{-1} and target temperature 900 °C Surface area of the samples was calculated by N_2 adsorption at 77 K using a Quantachrome instruments version 11.0 and the pore size distribution was calculated with the Barrett-Joyner-Halenda (BJH) model.

2.5. Adsorption Test

The sorption performance of the samples was tested by contacting 0.5 g sample with 50 mL Fe^{3+} (100 mgL^{-1}) in 250 mL Erlenmeyer flasks at room temperature under stirring for 1 h. The suspension was centrifuged to separate solid and filtrate. The filtrate was analyzed for metal ion concentration by using Atomic Absorption Spectrophotometric analysis (Analysis 200 AA, Perkin Elmer, USA). The effect of pH for sample performance to metal sorption was investigated in the range of 1.0 – 11.0. The mixing 0.1 M HCl and 0.1 M NaOH was used to arrange of the pH of metal solution. The amount of metal adsorption in the adsorbent was determined by the equation:

$$\text{Metal adsorption } (\%) = \frac{(C_i - C_f)}{C_i} \times 100\%$$

where C_i and C_f are initial and final concentration of metal ion (mgL^{-1}) in the solution [8].

3. Results and Discussion

3.1. Physical Properties

The XRD pattern of (a) FB, (b) CFB-300, (c) CFB-400, (d) CFB-500, (e) CFB-600, (f) CFB-700 and (g) CFB-900 are shown in Figure 1. All of the solid samples showed the angles corresponding to the peaks at $2\theta = 25.9$, 31.8, 39.8, 46.8, 49.6, 53.4 and 64.4 which are represented to the crystalline phase of hydroxyapatite [9]. The intensity of the peaks of hydroxyapatite increase when the carbonization temperature is increased.

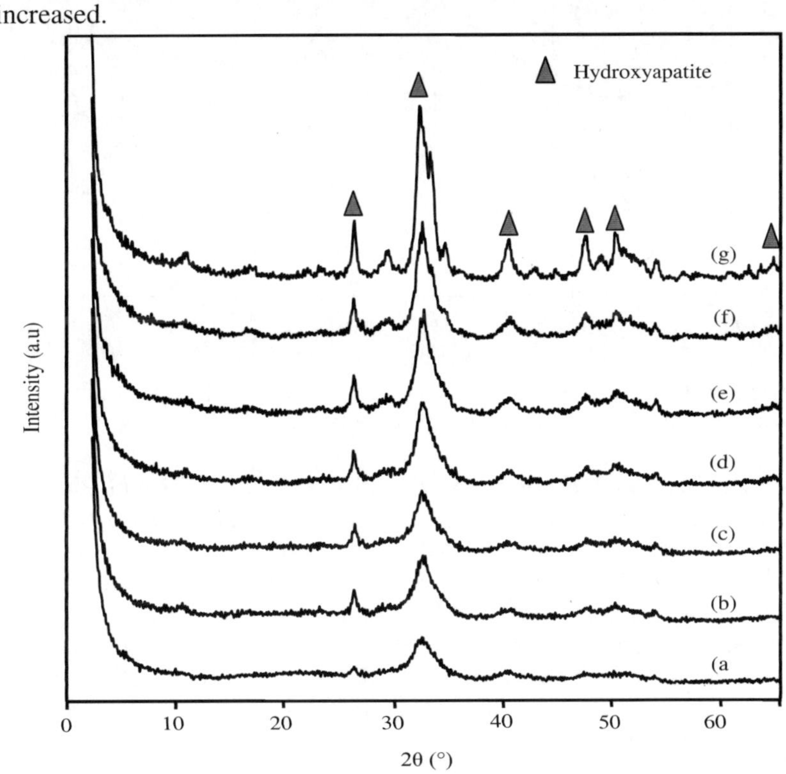

Figure 1. XRD pattern of (a) FB, (b) CFB-300, (c) CFB-400, (d) CFB-500, (e) CFB-600, (f) CFB-700 and (g) CFB-900

The FTIR spectra for the surface functional groups of (a) FB, (b) CFB-300, (c) CFB-400, (d) CFB-500, (e) CFB-600, (f) CFB-700 and (g) CFB-900 are shown in Figure 2. The IR absorption a broad peak at around 3424 cm^{-1} and 1653 cm^{-1} were attributed to hydroxyl stretching mode which indicated adsorbed water in the samples [10]. The intensity of hydroxyl groups decreased when the carbonization temperature increased. Carbonization process caused the appearance of peaks at around 2925 and 2851 cm^{-1} were assigned to the absorption of organic material as C–H and N–H group symmetric and asymmetric stretching mode. The absorption peak at around 2014, 1435 and 876 cm^{-1} were indicated of carbonate ion substitution [8]. Carbonate ion substitution clearly appear when the carbonization temperature of samples were increased more than 600 °C. All samples showed the peak at around 1043, 600 and 564 cm^{-1} were attributed intensive phosphate group.

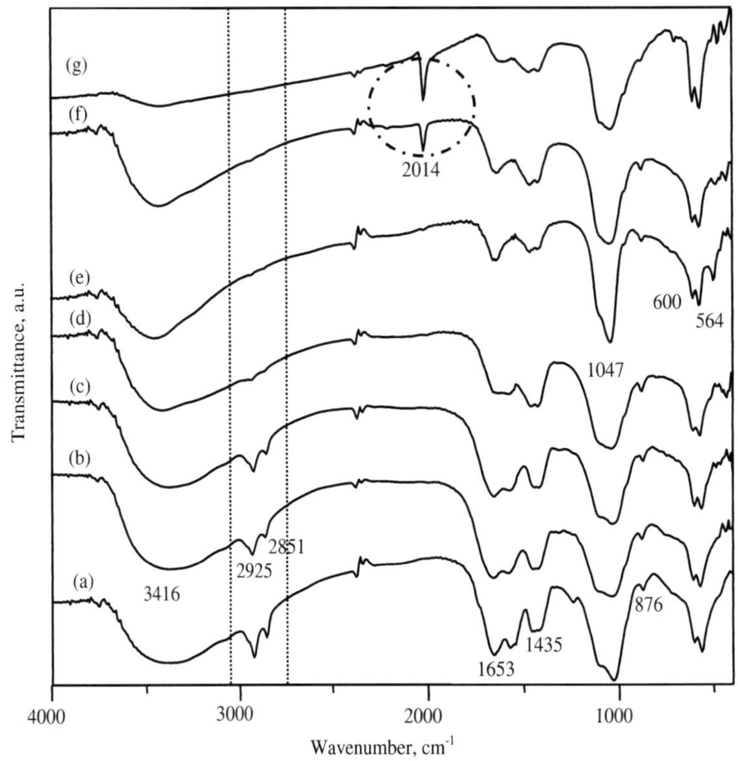

Figure 2. FTIR spectra of(a) FB, (b) CFB-300, (c) CFB-400, (d) CFB-500, (e) CFB-600, (f) CFB-700 and (g) CFB-900.

N_2 adsorption-desorption isotherms and pore size distributions of FB and CFB-500 are shown in Figure 3. The Isotherm of FB sample (Figure 3(a)) is III type in the IUPAC classification which indicated non porous or macro porous and no identifiable monolayer. The pore size distribution of FB sample indicated the presence of uniform pores with pore radius ~18 Å. The isotherm of CFB-500 sample (Figure 3(b)) is IV type in the IUPAC classifications of mesopores materials with clear hysteresis loops in the relative pressure range ~0.350 – 0.992 which is associated with the capillary condensation of nitrogen in the pores. According the IUPAC classification, the CFB-500 sample also possesses H1 which often associated with porous materials exhibiting a narrow distribution of relatively uniform (cylindrical-like) pores.

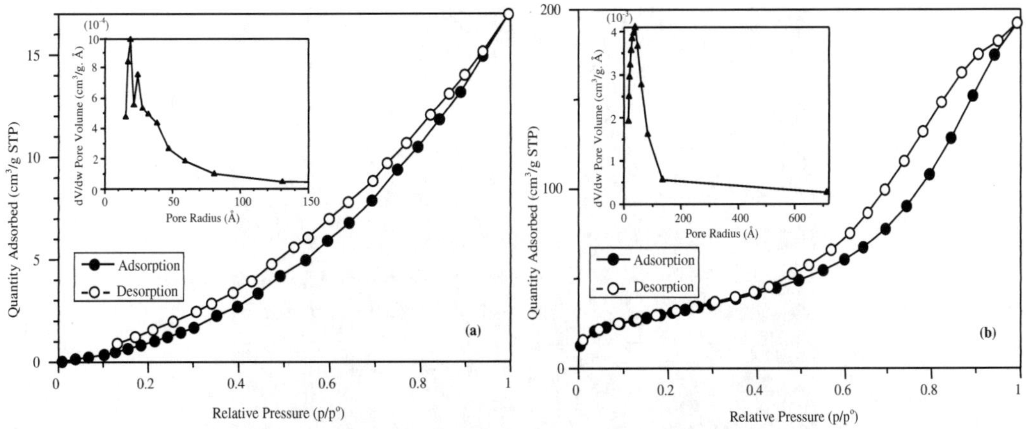

Figure 3. N_2 Adsorption-Desorption Isotherms and Pore Size Distribution of (a) FB and (b) FB-500

The pore size distribution indicates the presence of uniform mesopores with pore radius ~29 Å. The surface area of FB and CFB-500 samples were 16.0 m^2/g and 162.9 m^2/g, respectively. Base on pore radius and surface area, CFB-500 possesses pore radius and surface area larger than FB. This indicating carbonization process can cause opening and collapsing of pores. Physical properties of FB and CFB-500 samples are listed in Table 2.

Table 2. Physical Properties of FB and CFB-500

Samples	Pore Radius (Å)	Pore Volume (cc/g)	Surface Area (m^2/g)
FB	18.1	0.029	16.0
CFB -500	28.7	0.315	162.9

The morphology of FB, CFB-300, CFB-500 and CFB-900 were collected by scanning electron microscopy (SEM). A representative SEM image of samples is shown in Figure 4. From the SEM image, it can be seen that the FB before carbonization process have clean surfaces and sharp-edged porous structures. The morphology of fish bone after carbonization process in varying temperature such as CFB-300, CFB-500 and CFB-900 have roughness surfaces.

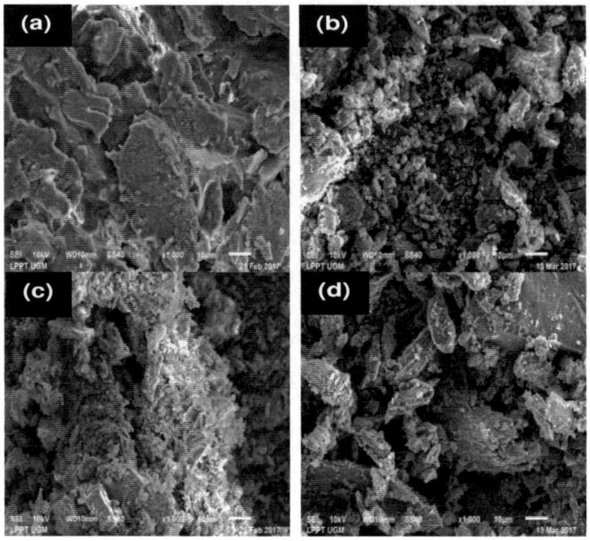

Figure 4. SEM images of(a) FB, (b) CFB-300, (c) CFB-500 and (d) CFB-900

Figure 5 shows the TGanalysis plot of (a) FB (b) CFB-500 and (c) CFB-900. The weight losses of the samples were shown in two temperature ranges. In the first temperature range 32 – 196 °C, the TGA plot of all samples displayed a slight weight loss for FB (~ 9.5%), CFB-500 (9.7%) and CFB-900 (7.8%). This temperature range can be correlated with the evaporation of the water adsorbed on samples. In the second temperature range 196 – 522 °C, the TGA plot saw weight lose of FB, CFB-500 and CFB-900 are 31.8, 11.9 and 7.9%, respectively. The weight loss in the second temperature range can be associated with the decomposition of organic compounds. Decreasing of weight lose percentage in the second temperature rang also can be correlated with carbon contain in the sample which it decrease when the temperature carbonization increased.

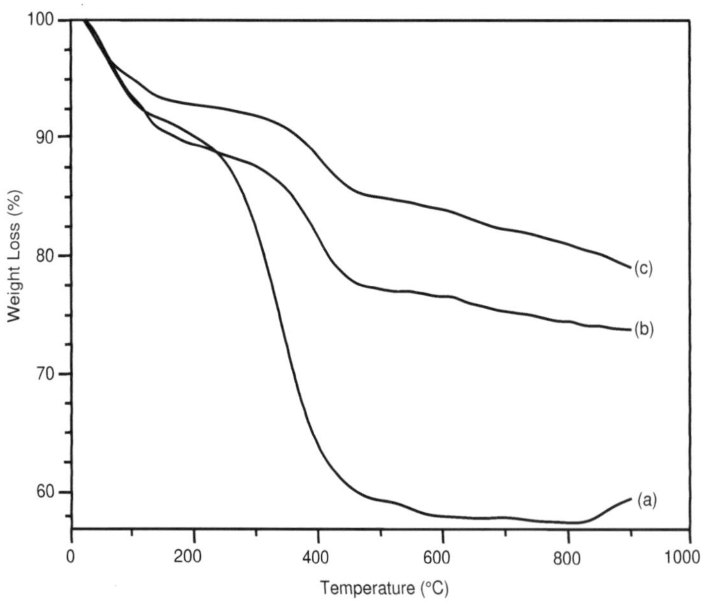

Figure 5. TG Analysis of (a) FB, (b) CFB-500 and (c) CFB-900

3.2. Effect of Carbonization Temperature on Fe^{3+}Adsorption

Figure 6 shows the histogram of the effect of carbonization temperature to Fe^{3+} ion adsorption. We could see that the metal adsorption 87% when fish bone powdered (FB) was used as adsorbent. The metal adsorption increased when the fish bone was carbonized in varying temperature. The metal adsorption 94% when used adsorbent of the fish bone that carbonized at 300 °C (CFB-300). The highest metal adsorption reached 98% when used absorbent of the fish bone was carbonized at 400 °C (CFB-400) and the same metal adsorption was obtained when the fish bone was carbonized at 500 °C (CFB-500). Carbonization of fish bone at 400 and 500 °C caused all organic compounds in fish bone changed to carbon element that it useful as adsorbent. Carbonization of the fish bone at 400 and 500 °C also cause to increasing the surface area of the adsorbent from 16.0 m^2/g (FB) to 162.9 m^2/g (CFB-500). The metal adsorption decreased when carbonization of the fish bone was created more than 500 °C. The metal adsorption 95% when carbonization of fish bone was performed at 600 °C (CFB-600) and the metal adsorption almost no change when carbonization was created at 700 and 900 °C. It was caused when the carbonization temperature increased the carbon release from samples become increase during the carbonization process and the amount of carbon inside sample becomes decreased.

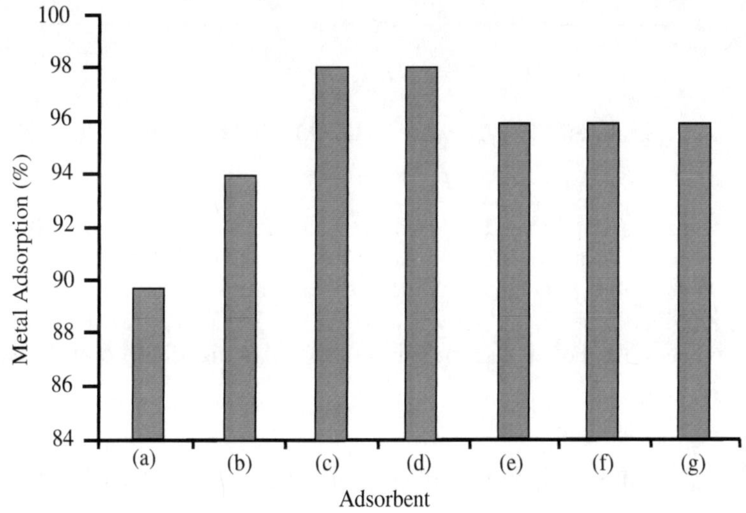

Figure 6. Metal adsorption of (a) FB, (b) CFB-300, (c) CFB-400, (d) CFB-500, (e) CFB-600, (f) CFB-700 and (g) CFB-900.

3.3. Effect of Solution pH on Fe^{3+}Adsorption

Figure 7 shows the graph of the effect solution pH on Fe^{3+} adsorption. The experiments were carried out in Fe^{3+} solution (100 mgL^{-1}; 50 mL) with 0.5 g absorbent (CFB-500) at room temperature with duration of contact time 1 h at different pH solution (pH 1 – 11). The data show that the solution pH almost did not influence to Fe^{3+} ion adsorption. Absorbent can adsorb Fe^{3+} ion with metal adsorption more than 96%. Even, metal adsorption reached 98% when the experiments were performed at pH 7 and 8.

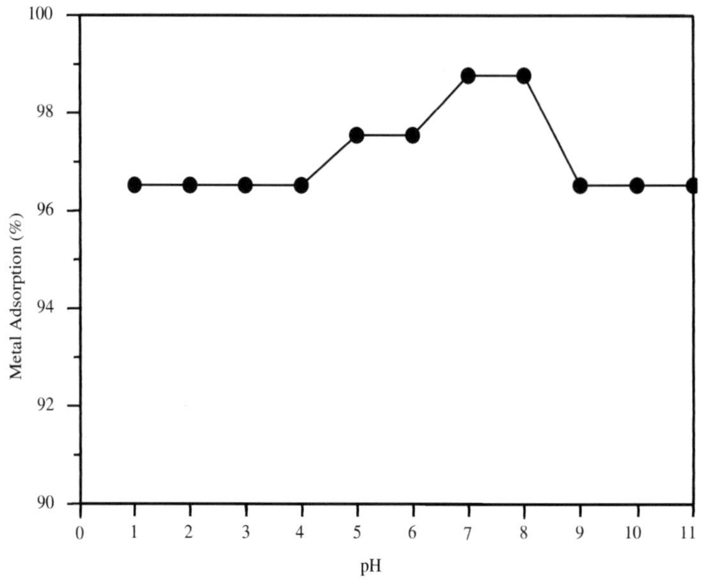

Figure 7. The Effect of pH to Metal Adsorption

4. Conclusions

The results of this research asserted that the powdered fish bone and carbonized fish bone at varying temperature can be promoted as adsorbent for adsorption and removing Fe^{3+} ion from solution. Metal adsorption reached 94% when powdered fish bone was applied as adsorbent. The best metal adsorption 98% was obtained when used adsorbent from fish bone which carbonized at 400 and 500 °C. The adsorption and removal Fe^{3+} ion from solution by carbonized fish bone as adsorbent was not influenced by pH condition.

5. References

[1] Lima H K, Tenga T T, Ibrahima M H, Ahmad A and Chee H T 2012 *APCBEE Procedia* **1** 96 – 102
[2] Ehab M Z, Hesham H S, Hassan M A and Ahmad M F 2013 *Ecol. Eng.* **61** 390–93
[3] Fawzi B, Sameer A, and Fadhel M 2000 *Sep. Purif. Technol*. **21** 155–64
[4] Bayram K, Adem T and Yusuf D 2010 *Desalination* **264** 37–47
[5] Ayhan D 2008 *J. Hazard. Mater.* **157** 220–29
[6] Edidiong D A and Alastair D M 2016 *J. Environ. Chem. Eng.* **4** 4207–28
[7] Amany E, Ahmed E N, Azza K and Ola A 2007 *J. Hazard. Mater* **148** 216–28
[8] Donat R, Akdogan A, Erdem E and Cetisli H 2005 *Colloid Interface Sci.* **286** 43–52
[9] Chakraborty R and Chowdhury DR 2013 *Chem. Eng. J.* 215–216 and 491–99
[10] Nurhadi M 2017 *Bull. Chem. Reac.Eng. Catal.* **12**(1) 55-61

Acknowledgements
The author is grateful to the funding from government by Ristekdikti Grant, Indonesia.

Natural wrapping paper from banana (*Musa paradisiaca* Linn*)* peel waste with additive essential oils

Widiastuti Agustina E.S. [1*] **and Elfi Susanti V.H**[2]

Department of Chemistry Education, Faculty of Teacher Training and Education, Universitas Sebelas Maret, Jl. Ir. Sutami 36A Kentingan Surakarta 57126 Indonesia

[*]Corresponding email : widiastuti_aes@staff.uns.ac.id

Abstract. The research aimed to produce natural wrapping paper from banana (*Musa Paradisiaca* Linn.) peel waste with additive essentials oils. The method used in this research was alkalization. The delignification process is done with the use of NaOH 4% at the temperature of 100°C for 1.5 hours. Additive materials in the form of essential oils are added as a preservative and aroma agent, namely cinnamon oil, lemon oil, clove oil and lime oil respectively 2% and 3%. Chemical and physical properties of the produced papers are tested included water content (dry-oven method SNI ISO 287:2010), pH (SNI ISO 6588-1.2010), grammage (SNI ISO 536:2010) and brightness (SNI ISO 2470:2010). Testing results of each paper were compared with commercial wrapping paper. The result shows that the natural paper from banana peel waste with additive essential oil meets the standard of ISO 6519:2016 about Basic Paper for Laminated Plastic Wrapping Paper within the parameter of pH and water content. The paper produced also meet the standard of ISO 8218:2015 about Food Paper and Cardboard within the grammage parameter (high-grade grammage), except the paper with 2% lemon oil. The paper which is closest to the characteristic of commercial wrapping paper is the paper with the additive of 2% cinnamon oil, with pH of 6.95, the water content of 7.14%, grammage of 347.6 gram/m^2 and the brightness level of 24.68%.

1. Introduction

The development of science and education increase the demand for paper which has been growing every year. It is proved by the demand for paper in 2005 that reaches almost 5.60 tons, and it increased to 6.45 tons in 2009. The paper is used for various needs, for instances, to write, to read, or to wrap the food. One of the paper types which is produced in Indonesia is the wrapping paper. Its production number increases from 88.940 to 90.930 tons during 2009 to 2011 [2]. The paper which is commonly used is made of the wood. It then causes the deforestation to be widely spread. Moreover, the pulping process remains using chemical material that is hard to be degraded naturally. The government currently finds it hard to overcome the scarcity of wood as the raw material of paper production in the significant number. As a result, the paper company produces the paper through illegal logging that destroys the nature [3].

The high number of paper demand causes the need for alternative raw material that is other than wood for wrapping paper industry. The production starts with the process of pulp making. The raw material to produce pulp is the cellulose in the form of fiber in which almost all of the plants contain the cellulose that can be used as the raw material of pulp production. One of the plants containing high cellulose is the banana.

Banana (*Musa paradisiaca* Linn.) is one of the tropical fruit thriving throughout Indonesia. It is one of the useful commodities that lives in a short time, is easy to be cultivated and can be harvested along the year. The most widely used part of banana is the flesh of the fruit, meanwhile, the outer skin is only used for animal feed and organic fertilizer. It indicates that the banana peel waste has not been optimized yet.

The data from Institute for Research and Development of Cellulose Industries shows that the banana peel waste contains more fine fibers than those that are made of wood with high cellulose (60-65%), hemicellulose (6-8%) and lignin (5-10%). Meanwhile, the most commonly used softwood to make the conventional raw material of the paper only contains cellulose (41%), hemicellulose (24%) and lignin (27.8%) [4]. It indicates that the cellulose in the banana peel is far higher than that in softwood. Moreover, banana peel only contains lignin (5-10%) that makes the cellulose separation process easier to do. Thus, the banana peel is potentially used to be the raw material of papermaking to reduce the dependence of wood as the raw material of wrapping paper. The use of banana peel waste also increases the optimization of the waste itself by increasing the economic value.

To produce the pulp, lignin that contains in banana peel should be hydrolyzed and separated through a delignification process. Alkalization method is suitable to make the pulp by using the material which is fine and has the short fiber as banana peel [5]. The cooking solution of NaOH is widely used due to its economical factor [6]. Sodium hydroxide is the strong alkali compound. In the process of paper production, the addition of NaOH solution is functioned to dissolve lignin in the pulping process. It also speeds up the process of fiber separation and extraction. Based on the previous research, banana peel cooked in NaOH of 4% for 1.5 hours in 100^0C results the natural paper in which the chemical characteristics (pH and water content) resemble the commercial paper very closely [8].

The qualified papers contain not only the fiber but also the additional material [9]. This research uses the additional additive material which is tapioca and essential oil consisting of cinnamon oil, lemon oil, clove oil, and lime oil. Tapioca acts as the sizing agent and adhesive [10]. Meanwhile, the essential oil is functioned to preservative and aroma agent due to the unpleasant smell produced by the alkalization process. The essential oil is proved to be the antibacterial, antifungal, insecticidal properties and antiviral antioxidants. Some of them have also been being used as cancer treatment, food preservative, aromatherapy and perfume [11]

2. Methods

2.1 Tools and Materials

The research used a blender, oven, filter paper, baking pan, rotary evaporator, electric cooker, pipettes, pot, balance scales, sample test, covered cup, furnace, analytical balance, cup clamp, glassware and pH Meters. The materials of this research were Kepok banana peel, NaOH, tapioca, essential oil consisting cinnamon oil, lemon oil, clove oil as well as lime oil, potassium chloride as well as distilled water.

3. Research Procedure

3.1 Pulp Making

The pulp making was carried out by alkalization with NaOH solution of 4% in a 6: 1 ratio to banana peel mass. The cooking is done at 100^0C for 1.5 hours. After being cooked, it then should be filtered and cleaned with distilled water. The bleaching process using H_2O_2 of 10%.

3.2 Addition of Additive Substance

The addition of additive includes tapioca of 10% and essential oil, consisting cinnamon oil, lemon oil, clove oil and lime oil with the concentration of 2% and 3% of the pulp mass.

3.3 Printing

The process of pulp printing was done in glass plate of 20 x 20 cm. Afterward, the drying process was done in the oven of 40-50°C for 24 hours.

3.4 Chemical Characterizations
The water content test was conducted by the dry-oven method based on SNI ISO 287: 2010. The pH test was done based on SNI ISO 6588-1: 2010 on how to test pH in the water extract-part 1: Cold extract.

3.5 Physical Characterization
Physical characterization was conducted at The Institute of Pulp and Paper. Grammage test was performed based on SNI ISO 536: 2010; meanwhile, the brightness test was done based on SNI ISO 2470: 2010.

4. Result and Discussions
4.1 Paper from Banana Peel Waste
Banana peel can be used as raw material in the making of paper because of its high concentration of cellulose. Mashing process of banana peel is essential as banana peel should be brushed, lacerated, and pushed to expand the fibers. While expanding, the outer layer of the fiber will be decomposed, and the fibrils come out. The process which causes the fiber surface to get larger is necessary to tighten the fiber of the paper. In the process of cooking, the temperature should not exceed 100°C since the high temperature in cooking process will decrease the pulp quality. It then causes the pulp not to last for the long period. Moreover, cooking process will shade the color of banana peel from yellow to dark brown indicating the process of lignin extraction from the cellulose.

The bleaching process was done to remove the lignin contained both in pulp and fiber. It then results in the high and stable brightness. In this process, the H_2O_2 solution was used due to the characters of hydrogen peroxide which has strong oxidation properties with the bleaching concept of Totally Chlorine Free [12]. Also, hydrogen peroxide also has a weak acid, is easily decomposed when heated, and releases the oxygen. Due to its ability to release the oxygen, hydrogen peroxide is effective as the bleaching material. The O_2 will act as the oxidizer to bleach the material.

In bleaching process, the pulp shaded from dark brown to yellow; this color shows that lignin has been removed. Moreover, the heating process generated bubbles from the decomposition reaction of the hydrogen peroxide into oxygen and water.

4.2 The Process of Adding Additives
The resulted pulp is added by 10% of tapioca and essential oil consisting of cinnamon oil, lemon oil, clove oil, and lime oil with the concentration variation of 2% and 3%. Tapioca is functioned as the sizing agent and the adhesive; meanwhile, the essential oil is functioned as preservative and aroma agent. It is due to the unpleasant smell of the paper which is caused by alkalization process. After being added by additive substance, the pulp became lighter, and the resulted aroma was based on the added essential oil.

4.3 Paper Printing
This is the finishing process in which the formed pulp was printed to produce the paper. The form of the paper is shown in Figure 1. and Figure 2. Physically, all of the papers were yellowish white in which they have the different level of color and slightly rough texture.

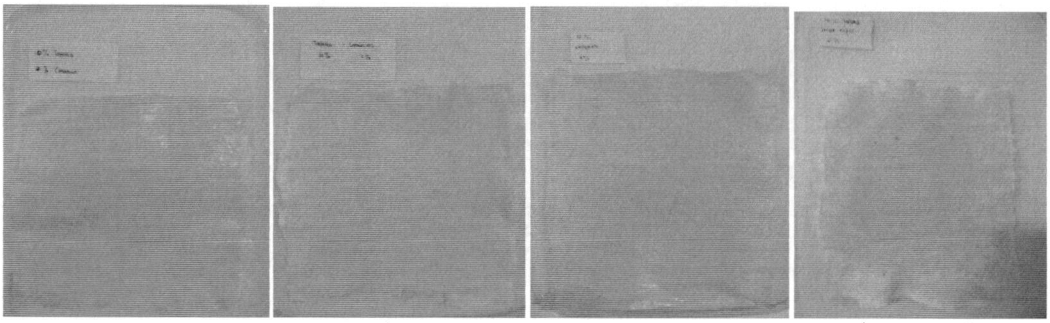
a b c d

Figure 1. The Paper with The Addition of 2% Essential Oil
a. Cinnamon, b. Lemon, c. Clove, d. Lime

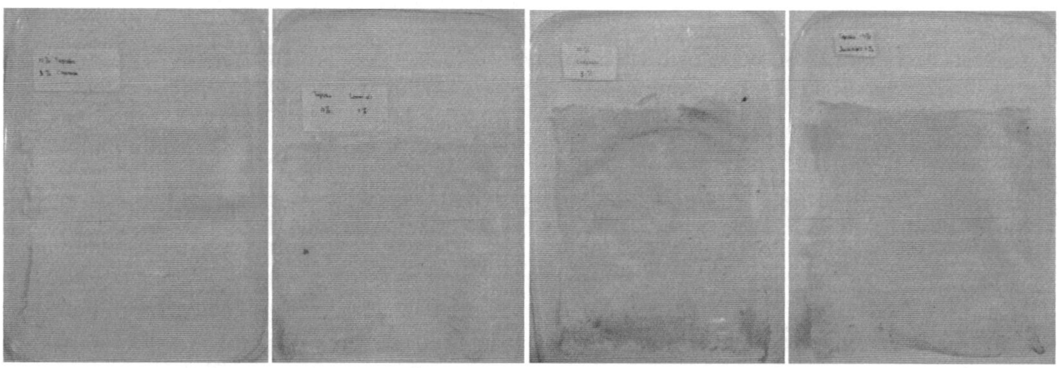
a b c d

Figure 2. The Paper with The Addition of 3% Essential Oil
a. Cinnamon, b. Lemon, c. Clove, d. Lime

4.4 Chemical Characterization

The chemical characteristics of paper, include water content and pH, are shown in Table 1.

Table 1. Chemical Properties of Natural Wrapping Paper

No	Essential Oil Variation	Water Content (%)	pH
1.	No essential oil	5.470	7.95
2.	Cinnamon oil 2%	7.140	6.95
3.	Cinnamon oil 3%	7.295	6.90
4.	Lemon oil 2%	7.351	6.85
5.	Lemon oil 3%	7.642	6.75
6.	Clove oil 2%	7.600	6.90
7.	Clove oil 3%	7.725	6.85
8.	Lime oil 2%	7.402	6.75
9.	Lime oil 3%	7.647	6.70
10.	Commercial wrapping paper	6.093	7.55
11.	SNI	≤ 8	6 - 9

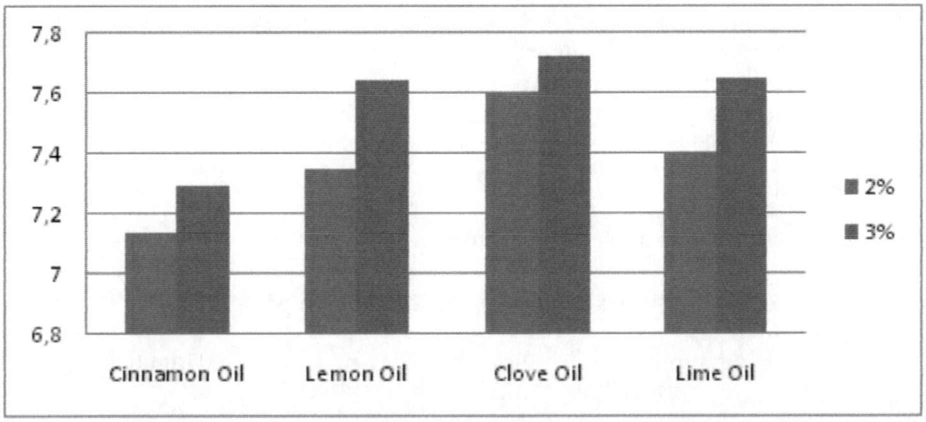

Figure 3. Water Content of Paper with Additive Essential Oils

SNI ISO 6519:2016 states that the water content of the basic paper for Laminated Plastic Wrapping Paper should not exceed 8%. All of the various papers of banana peel with essential oil meet the standard of water content stated by SNI. The result of water content test indicates that the concentration of essential oils affects the water content of the paper produced. The higher the concentration of essential oils added the higher water content of the paper produced (Figure 3.). Essential oil as additive agent increases the amount of bond between fiber, so it reduces the amount of pore water evaporation as a path. Physically, it can be seen from the surface of the paper with additives that are more subtle than the paper without additives due to filling the empty spaces on the bond between fiber. Thus the water content of the paper produced is also increasing. Moreover, there may be another factor affecting the water content, namely drying process, room humidity, air flow, air pressure, the numbers of sample and the thickness of the sample.

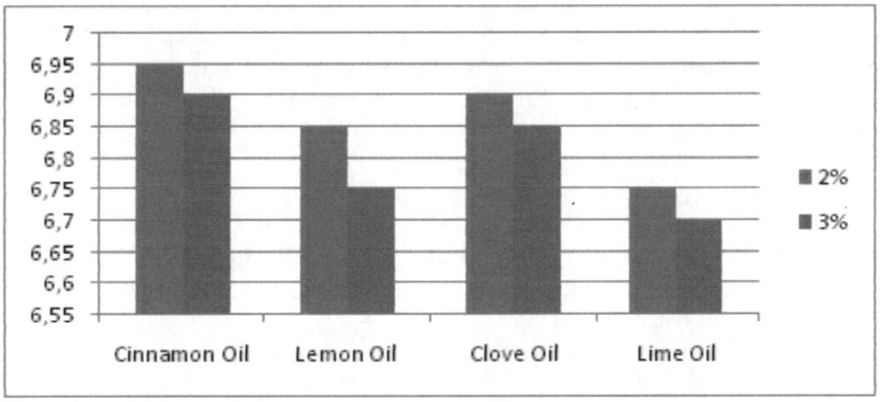

Figure 4. pH Value of Paper with Additive Essential Oils

The pH value is essential to be noticed since it affects the skin response when contacting the paper. SNI ISO 6519:2016 states that the Basic Paper for Laminated Wrapping Food Plastic must have pH value in the range of 6-9. All of the variation of banana peel paper with additive essential oil meet the standard quality. The result showed that the concentration of essential oils affects the pH value of the paper produced. Essential oils decrease the pH value of paper produced. The higher concentrations of essential oils added the higher acidity of the produced paper, which was marked by a decrease in pH values (Figure 4.). This relates to the chemical content of essential oils used. Cinnamon oil and clove oil has a high concentration of eugenol while lemon oil and lime oil has a high concentration of limonene. Both eugenol and limonene are acids so that the addition of the essential oils into the pulp could increase the acidity of the paper produced.

The result of the research on chemical characteristics parameters shows that banana peel paper with additive essential oil meet the standard of ISO 6519:2016 about Basic Paper for Laminated Plastic Wrapping Paper on the parameter of pH and water content.

4.5 Physical Characterization

The physical characteristics of paper, include grammage and brightness, are shown in Table 2.

Table 2. Physical Properties of Natural Wrapping Paper

No	Essential Oil Variation	Grammage (g/m^2)	Brightness (%)
1.	No essential oil	487.4	28.80
2.	Cinnamon oil 2%	347.6	24.68
3.	Cinnamon oil 3%	356.6	54.43
4.	Lemon oil 2%	651.6	47.22
5.	Lemon oil 3%	319.2	34.34
6.	Clove oil 2%	474.8	26.99
7.	Clove oil 3%	275.2	21.96
8.	Lime oil 2%	375.6	27.53
9.	Lime oil 3%	238.2	38.01
10.	Commercial wrapping paper	83.0	22.82
11.	SNI	255 - 500	-

Grammage is the mass of paper of a particular unit area. According to SNI, paper and cardboard grammage for food packaging range from 255-500 g / m^2. Grammage is essential to be measured due to the function of parameter value on the tensile index and tear measurement. All variations of banana peel paper with the additive essential oil are meet the standard of SNI 8218:2015 about Paper and Cardboard for Food Packaging on the parameter of grammage (Food packaging with high grammage) except the paper with 2% lemon oil (651.6 g/m^2). Grammage related to the strength of the tear. Paper with high grammage shows the high number of the fibers contained. The amount of fiber will play a role in the process of tearing because each fiber will draw up a sheet of pulp on the particular grammage that contribute to the overall energy needed to tear.

Table 2. shows that the brightness level of banana peel with additive essential oil tends to decrease as the concentration of essential oil increases. However, it will not occur in the paper of additive cinnamon oil. It indicates that the concentration of essential oil affects the brightness level of the produced papers. The more the concentration of essential oil is used, the less the brightness level of the paper is produced.

In addition to being adapted to ISO standards, the produced paper is also compared to the characteristics of commercial wrapping paper. All of the natural wrapping paper produced have higher water content and lower pH value than commercial wrapping paper. The results show that natural wrapping paper from banana peels with 2% cinnamon essential oil additives has the closest characteristics of commercial wrapping paper. Natural wrapping paper with additive 2% cinnamon oil has pH of 6.95, the water content of 7.14%, grammage of 347.6 gram/m^2 and the brightness level of 24.68%.

4. Conclusions

The natural paper from banana peel waste with additive essential oil meet the standard of ISO 6519:2016 about Basic Paper for Laminated Plastic Wrapping Paper within the parameter of pH and water content and ISO 8218:2015 about Food Paper and Cardboard within the grammage parameter (high grade grammage), except the paper with 2% lemon oil.The result which is closest to commercial wrapping paper is the paper made of banana peel waste within the additive cinnamon of 2%, containing pH of 6.95, the water content of 7.14%, grammage of 347.6 gram/m^2 and the brightness level of 24.68%.

5. References

1] Pusat Grafika Indonesia 2012 HTI Industri Kertas dan Industri Grafik
[2] Indrawan D A, Efiyanti L, Tampubolon R M and Roliadi H. 2015 *J. Pen. Has. Hut.* **33** No.04:283-302
[3] Manurung E G T and H H Sukaria 2000 Industri Pulp dan Kertas: Ancaman Baru Terhadap Hutan Alam Indonesia
[4] Tjahyono Yudi 1998 Proses Pembuatan Pulp. Bandung: Balai Besar Penelitian dan Pengembangan Industri Selulosa
[5] Prabawati S Y and Wijaya A G 2008 *Jurnal Aplikasi Ilmu-Ilmu Agama* **9**(1) 44-56
[6] Saleh A, Pakpahan M D M and Angelina N 2009 *Jurnal Teknik Kimia* **3**(16)
[7] Sucipto, Wijana S and Wahyuningtyas E 2009 *Jurnal Teknologi Pertanian* **10**(1) 46-53
[8] Novianti P and Setyowati W A E 2016 *Proceeding of Science Education National Conference*
[9] Syafii W 2000 *Jurnal Teknologi Industri Pertanian* **10**(2) 54-55.
[10] Syamsu K, Roliadi H, Candra K P and Arsyadi A J 2014 *Jurnal Teknologi Pertanian* **9**(1) 16-25
[11] Prabuseenivasan S, Jayakumar M and Ignacimuthu S 2006 *BMC Complementary and Alternative Medicine*
[12] Greschih T, Basta J, Zhang C, Zhan H and Jiang S 2008 *2nd International Paper Making and Environment Conference* (Tianjing, China)

Acknowledgement
This research was supported by Grant of Fundamental Research, PNBP, Universitas Sebelas Maret 2017 (Contract Number: 623/UN27.21/PP/2017)

IOP Conf. Series: Journal of Physics: Conf. Series **1022** (2018) 012033 doi:10.1088/1742-6596/1022/1/012033

Senior high school students' need analysis of Three-Tier Multiple Choice (3TMC) diagnostic test about acid-base and solubility equilibrium

Ardiansah[1*], M Masykuri[2] and S B Rahardjo[3]

[1] Science Education Student in Sebelas Maret University of Surakarta, Indonesia
[2] Science Education Department in Sebelas Maret University of Surakarta, Indonesia
[3] Chemistry Department in Sebelas Maret University of Surakarta, Indonesia

*Corresponding email: ardiansah@student.uns.ac.id

Abstract. Students' conceptual understanding is the most important comprehension to obtain related comprehension. However, they held their own conception. With this need analysis, we will elicit student need of 3TMC diagnostic test to measure students' conception about acid-base and solubility equilibrium. The research done by a mixed method using questionnaire analysis based on descriptive of quantitative and qualitative. The research subject was 96 students from 4 senior high schools and 4 chemistry teachers chosen by random sampling technique. Data gathering used a questionnaire with 10 questions for student and 28 questions for teachers. The results showed that 97% of students stated that the development this instrument is needed. In addition, there were several problems obtained in this questionnaire include learning activity, teacher's test and guessing. In conclusion, this is necessary to develop the 3TMC instrument that can diagnose and measure the student's conception in acid-base and solubility equilibrium.

1. Introduction

Student conceptual understanding is the major prerequisite knowledge that must be obtained by the student after learning chemistry. However, a student sometimes doesn't achieve this goal completely [1–3]. A student can hold his/her misunderstandings either in high school or campus. They might difficult to understand and use scientific model. This difficulty of conceptual comprehension can cause misconception in students.

The misconceptionis a concept that different from science conception accepted by science society [4] while the concept is a set of specific objects, symbols, or events which are grouped together on the basis of shared characteristics and which can be referenced by a particular name or symbol. The misconception definition included the conception which is not complete. It means that conception with missing pieces would direct to misunderstanding in full scientific concept. It also caused by incomplete understanding or interpretation by students in the complex subject [5–7].

Acid-base was one of the subjects that have higher complexity level. It contained a sub-microscopic, macroscopic, and symbolic level of chemistry[8,9]. Based on Johnstone theory, sub-microscopic level in this subject include reaction mechanism in acid and base, the collision of acid and base molecules, the interaction of hydronium ion with indicator ion, etc. The macroscopic level of Johnstone's theory observed by our eye. For instance, the changing colour of an indicator when titrating process between hydrochloric acid and sodium hydroxide. The last level, symbolic level includes the symbol of a molecule, the equation counting of hydronium ion concentration, etc.

Chemistry researchers generally agree that acid-base and solubility are the complex subjects in chemistry. Several types of research about acid-base and solubility equilibrium (ABSE) concept already done. For instance, about acid-base [10–20] that include of acid and base concept [21,22], titration concept [23], salt hydrolysis concept [24], buffer concept[25]. Solubility equilibrium researcherswere conducted in stirring and temperature effect [26,27], solution of chemistry [28,29] and solubility concept [30]

Measurement of acid-base complexity can be developed using various instruments [31,32]. There were used interviews (53%), open-ended tests (34%), multiple-choice tests (32%) and multiple tier tests (13%) to diagnose misconception [33]. Multiple tier test includes of two-, three-, and four-tier[34–36]. The three-tier multiple choice constructed from original multiple-choice. The first tier was original multiple-choice; the second tier was a reason for the first answer, and the third tier was a scale of certainty. With this instrument, we can determine the conceptual comprehension. It includes science knowledge, false positives, false negatives, lack of knowledge, lucky guess, and misconception. The consideration of this result easy to calculate although use common processing data like Microsoft Excel and open office Calc. With this instrument, the teachers' also can determine student certainty with a confidence level [37]. Further, only two research related to chemistry done by 2014 that used three tier multiple choice test.

The instrument, 3TMC, was chosen because of a number of strengths[33]. These instrument strengths were: (1) time efficient in administering, (2) immediately scored, (3) objectively scored, (4) validity evidence is strong, (5) applied to a large number of subjects, (6) gives an opportunity to decide the proportions of false positives and false negative, and (7) determines the answers given to the first two tiers are due to misconception or a mistake due to lack of knowledge. However, there was two weakness of this instrument that underestimates the proportions of lack of knowledge since cannot decide whether the responder is sure for his/her answer in the first tier, in the second tier or in both tiers, and overestimates students' scores [34,35,38].

Three-tier multiple choice diagnostic test is the potential test to disseminate in Indonesia. Review article conducted by Gurel, Eryilmaz, McDermott [33] stated that no similar research done by Indonesia's researchers'. It gives the chance for our further research about this scope. Although there was an acid-base based three-tier multiple choice conducted before [39], the misconception Indonesia is needed by students' and teachers' need analysis. This research will enhance teacher reference to obtain their students' concept and concept comprehension of ABSE subject. Author's research elicits student's concept mapping in ABSE subject. Using 3TMC, the conceptual comprehension from student might find during the Treagustten steps of research [40,41].

2. Method

This is a mixed method of research which include of descriptive quantitative. This research aims to explore students' need about three-tier multiple choice diagnostic test. The instrument developed to determine their misconception and concept comprehension about acid-base and solubility equilibrium. The subject in this instrument includes acid and base concept, titration concept, salt hydrolysis concept, buffer concept, and solubility equilibrium.

2.1. Participants

Participants used in this research were students who learned acid-base and equilibrium solubility chapter in chemistry subject. They come from four schools located in Bengkayang Residence, West Kalimantan, Indonesia. The subjects of research were chosen using purposive sampling. There were 96 students involved in this study that already learned about ABSE subject in their classes. Furthermore, we collected four chemistry teachers taught in representative school to make sure the data valid.

2.2. Data Collection

Data were collected from the questionnaire given to the students and the teachers. Validity used in the Aiken's formula which value .92 – 1.00 and average for .97that can interpret as valid for six validators from four Likert relevance scales [42]. The students' questionnaire consisted of 10 questions which three questions were open ended question and other provided 2 to 6 options each. The sixth question has a large number of choices because we need to cover as many as possibilities for students' response about the type of test used by the teacher in class. Also, we gave possibilities the students to choose their choice more than one. Question number 1 to 4 about students' problem in learning chemistry generally that is learning manner in chemistry (1st question), reason used that manner (2nd question), the problem in learning chemistry and settle it. Question number 5 asked students' opinion about concept content in acid-base and solubility equilibrium. In line with this need analysis, we ask the student about the test in teachers' daily evaluation which is found in 6th to 8th question. Furthermore, to obtain information about student guessing when answering the question, we made the 9th question. At the end of the questionnaire, 10th question, we asking about the need for a diagnostic test to obtain their conception.

The other questionnaire gave to teachers. This questionnaire consists of 28 questions. The indicator used are teaching and learning process, subject book and substance, evaluation instrument, and student need of a diagnostic instrument. There were 18 questions in open ended question and the others were close ended question.

2.3. Data Analysis

Descriptive analysis on investigating students' need analysis was carried out using descriptive manner. Each option described student need and their manner without a correct and incorrect decision. Further, students' analysis of need was determined when we analyse all of the questions in the questionnaire. The complementary data, international journal, collected using the relevant scope of this research. The journal analyses related to this research. The data collected using some journal search engine such as: Scopus, Science direct, ERIC, EBSCO, and google scholar. Keywords used in this searching were student's misconception, ABSE, 3TMC, need analysis, and misconception in Indonesia

3. Result and Discussion

3.1. Student Questionnaire Result

Based on this questionnaire result, several data were obtained. Every question was interpreted to show students' need analysis about this instrument. There were seven aspects that we got with this research. The aspects were:

1) learning method when studying at home,
2) acid-base and solubility equilibrium was the conceptual subject,
3) test question format used by the teacher,
4) type of question that student prefer to,
5) most of the test type used by the teacher to evaluate student concept comprehension,
6) student action when cannot answering the question, and
7) students' need of the misconception instrument.

Each aspect contained one question so that there were seven questions made. Other three aspects, students' problem in learning chemistry, located in #2 until #4 question.

Students were asked about (question #1) learning method when studying at home. The result showed that almost a half of student chosen to understand as his/her learning method when studying chemistry at home. It indicated that chemistry still becomes memorising subject for the student. In addition, this result showed that teacher meaningful learning application tends to the memorising system. However, this problem becomes a general problem in Indonesia's student [43]. Further, he stated that teachers tend to have the students memorise contents of the textbooks and teach them techniques about how to answer multiple choice questions by giving them drills. These students do not learn and understand mathematics and science, but merely memorise mathematical and scientific formulas for the examinations.

IOP Conf. Series: Journal of Physics: Conf. Series **1022** (2018) 012033 doi:10.1088/1742-6596/1022/1/012033

Question number two until four especially used for student problem in learning chemistry generally. These questions were open ended question. So that, the student can write his/her problem, his/her settle for providing his/her problem. Student's problems obtained in this question are:
1) the teacher only explains about formula,
2) student memorise the formula given by his/her teacher without knowing the usage of formula,
3) student memorise the conceptual subject but did not understand the concept,

In line with this question, we give a student to asking their conceptual comprehension to decide the acid-base subject whether it conceptual or algorithmic subject. This question was written in question #5. With this question, the student gives his/her opinion about this subject. Although most of them give the conceptual as their reason, there was about 36% student do not realise that this subject was conceptual. In this question, can be concluded that student assumes that algorithmic comprehension is emphasised in students' mind.

A sixth question asking about test question format used by his/her teacher. This question is an open question that student can choose more than one question as for their choice. In this question, there was five options given that is generally used by a teacher that is: yes-no question, short answering, multiple choice, essay, and multiple-tier. The essaywas the most test format used by a teacher with 63%. In addition, the student chose about 23% in multiple choice student. The teacheris not preferredto the yes-no question, short answering and multiple-tier format of the test.

The seventh and eighth question requires thestudent to give an opinion about the popular type of teacher test whether it algorithmic or conceptual. This question almost same as the previous question. However, the surprising result was obtained that related each other. In #7 question, most of the student (about 60%) chosen conceptual test as his/her type of question prefer to. It seems described the next question that teachers were used the algorithmic test to evaluate student conceptual comprehension. There were 93% students chosen algorithmic test.

The ninth question required student opinion about his/her action when cannot answer the question in multiple choice question. Two third of students were chosento guessas for their choice. It proved that student uses the weakness of guessing possibility as their answer. In high school level, there were five choices of multiple choice that means students' correct possibility to answer about 20%. With this possibility, the teacher must provide the effective distractors to increase students' guessing. Only 3% student chose left blank as his/her choice while 29% of student chose other activities such as cheating and counting another choice. With 3TMC instrument, the guessing probability decrease from 13% (using 8 option) until 4% (using 25%). This instrument can cover student guessing cultures when answering the question.

The last question, we obtained that most of the student (97%) need this instrument about the three-tier multiple-choice diagnostic test in acid-base and solubility equilibrium subject to measure and describe their misconception and concept comprehension. This means that student needs this instrument. There only 3% of student need not this misconception instrument.

This table show student needs questionnaire result.

Table 1. Questionnaire Result of 3TMC Instrument Need

Aspect	Question	Percentage
Learning method when studying at home	Memorising	23
	Understanding	43
	Both of them	34
Acid-base and solubility equilibrium were the conceptual subjects	Yes	64
	No	36
Test question format used by teacher	Yes-no question	4
	Short answering	7
	Multiple choice	23
	Essay	63
	Multiple-tier	4

Aspect	Question	Percentage
Type of question that student prefer to	Conceptual	40
	Algorithmic	60
Most of the test type used by teacher to evaluate student concept comprehension	Algorithmic	93
	Conceptual	7
Student action when cannot answer the question	Guessing	68
	Left blank	3
	Other, ex: cheating	29
Students' need for the misconception instrument	Yes	97
	No	3

3.2. Teachers' questionnaire result

The result of teachers' questionnaire was obtained several results. Every aspect interpreted to show students' need analysis about this instrument from teachers. The indicator used are teaching and learning process, subject book and substance, evaluation instrument, and student need of a diagnostic instrument.

The first aspect, we asked the teacher about teaching and learning process. This aspect located in question number one until eight. Question number one until four asking about the approach used by the teacher in teaching and learning process while the number five until eight about teachers' strategy. We obtained that direct approach was the most approach used by the teacher. Another approachto science, technology, and society approach, constructivist approach, contextual approach, and problem-based approach seldom used by teacher caused by the complexity of this approach. However, they used the scientific approach as their standard that must be applied in 2013 curriculum of Indonesia's education system.

Teachers' strategy in teaching student varied. Most of the teacher used direct instruction strategy when teaching their students. They assume that their subject can be transferred fast through this strategy. Another strategy that is a discussion, cooperative and project are not preferred by teachers. Those strategies consume much time due to lack of limited estimation time. Also, students are not familiar with those strategies.

The secondaspect obtained teachers' book used in teaching and learning process and references. This aspect included in question number #9 until #13. The result showed that teacher chosen their book or reference developed by his/herself and from authorised publisher. Teachers generally used two or three books as their references when teaching chemistry. Furthermore, they used until four books when teaching in class. It enhanced teacher and student knowledge about the subject in chemistry class.

The most important aspect in this questionnaire is the third aspect. It described teachers' evaluation instrument. It showed in question #14 until #25. This aspect can strengthen the student aspect of evaluation done by his/her teacher. Also, this aspect can describe reality in teachers' evaluation. This aspect includes four sub-aspects that are emphasising of concept in teaching, student difficulties in chemistry, teacher settle this problem and part of evaluation measured by the teacher.

In first sub-aspect, teacher emphasising conceptual aspect. However, the algorithmic aspect gets a little part. They clarified that student can count by his/herself using the formulae given. This statement in line with student opinion that acid-base and solubility equilibrium concept were the conceptual subjects. Teacher measure student conceptual comprehension by asking some of them randomly due to lack of time. Thus, this is an efficient decision for limited available time for teaching and learning process

The second sub-aspect, teacher stated that 30%-50% of students were having difficulty in conceptual comprehension. This statement also in line with student learning method at home. This difficulty was caused by the complexity of chemistry terminology, student learning method, and lack of understanding the basic concept or prerequisite concept. In student open ended question, they usually memorise the molecule's name, formula, and the name of the element. It becomes a difficulty if they forgot what they're memorised.

The third sub-aspect require teacher attention to settle the problem. The students' problem also become teacher problem. The teachers settle the problem using two ways. First, they repeat and remedial the subject that did not pass the standard. It can be done using improve the method and teaching strategy. Further, the teacher gave the coaching question for under criteria indicator. The second, they relate the subject with other previous or prerequisite subject. It can increase student comprehension with relating other subjects.

The last sub-aspect, evaluation measurement system used by the teacher. Teachers measure cognitive domain more than another aspect those are affective and psychomotor. In cognitive domain, a teacher using multiple choice and open-ended question to assess student comprehension. A teacher using 21% of remembering aspect, 45% of understanding aspect, 16% applicating and 18% of higher order thinking skill as their test aspect. This result can be concluded that teacher using in lower order thinking the skill of Bloom taxonomy Anderson revision.

In other two domains, a teacher using standard measurement. In affective domain measurement, the teacher used questionnaire and observation sheet. This measurement is measured only in specified time. In the other word, the affective domain used only several times in a semester. This term also did by the teacher to measure psychomotor domain. They only measure this domain at lab activity or lab practice.

The last aspect is student need of a diagnostic instrument. We obtained two results about this aspect. First, cognitive measurement tools used by the teacher do not tend to a diagnostic test. This problem caused teacher need a diagnostic instrument to describe and measure students' conceptual comprehension and misconception. Second, all of teacher certain this instrument can measure and map the student misconception.

4. Conclusion

This instrument is needed by the student. With some strength of this instrument, we assume that students want to map their conceptual comprehension include: know concept, false positive, false negative, lucky-guess, misconception, and don't know concept. This instrument also can be used by the teacher to assess their students' weakness in concept comprehension. The 3TMC instrument can become a diagnostic tool.

This instrument used only focused on student's and teacher's need about 3TMC diagnostic test in ABSE chapter. The result of misconception and concept comprehension are not described. The instrument result might be varied depends on the subject. However, the tendency of misconception sometimes can be predicted. Furthermore, these two questionnaires cannot represent other instrument needs such as formative and summative.

5. References

[1] Akram M, Surif J Bin and Ali M 2014 Conceptual difficulties of secondary school students in electrochemistry *Asian Soc. Sci.* **10** 276–81
[2] Tümay H 2016 Reconsidering learning difficulties and misconceptions in chemistry: emergence in chemistry and its implications for chemical education *Chem. Educ. Res. Pr.* **17** 229–45
[3] Park E and Choi K 2013 Analysis of student understanding of science concepts including mathematical representations: pH values and the relative differences of pH values *Int. J. Sci. Math. Educ.* **11** 683–707
[4] Nakhleh M B 1992 Why Some Students Don't Learn Chemistry Chemical Misconceptions *J. Chem. Educ.* **69** 191–6
[5] Garnett P J and Treagust D F 1992 Conceptual difficulties experienced by senior high school students of electrochemistry: Electrochemical (galvanic) and electrolytic cells *J. Res. Sci. Teach.* **29** 1079–99
[6] Peterson R F, Treagust D F and Garnett P 1989 Development and application of a diagnostic

instrument to evaluate grade-11 and -12 students' concepts of covalent bonding and structure following a course of instruction *J. Res. Sci. Teach.* **26** 301–14

[7] Schaffer D L 2013 *The development and validation of a three-tier diagnostic test measuring pre-service elementary education and secondary science teachers' understanding of the water cycle* (the University of Missouri)

[8] Taber K S 2013 Revisiting the chemistry triplet: drawing upon the nature of chemical knowledge and the psychology of learning to inform chemistry education *Chem. Educ. Res. Pract.* **14** 156

[9] Johnstone A H 2000 Teaching of Chemistry - Logical or Psychological? *Chem. Educ. Res. Pract.* **1** 9–15

[10] Özmen H, Demırcıoğlu G, Coll R K K, Demircioğlu G, Coll R K K, Demırcıoğlu G and Coll R K K 2009 A comparative study of the effects of a concept mapping enhanced laboratory experience on Turkish high school students' understanding of acid-base chemistry *Int. J. Sci. Math. Educ.* **7** 1–24

[11] Özmen H, Demircioğlu G, Burhan Y, Naseriazar A and Demircioğlu H 2012 Using laboratory activities enhanced with concept cartoons to support progression in students' understanding of acid-base concepts *Asia-Pacific Forum Sci. Learn. Teach.* **13** 1–29

[12] Şendur G, Özbayrak Ö and Uyulgan M A 2011 A study of determination of pre-service chemistry teachers' understanding about acids and bases *Procedia Comput. Sci.* **3** 52–6

[13] Kelly R M and Akaygun S 2016 Insights into How Students Learn the Difference between a Weak Acid and a Strong Acid from Cartoon Tutorials Employing Visualizations *J. Chem. Educ.* **93** 1010–9

[14] Chiu M 2007 A National Survey of Students' Conceptions of Chemistry in Taiwan *Int. J. Sci. Educ.* **29** 421–52

[15] Ültay N and Çalik M 2016 A comparison of different teaching designs of "acids and bases" subject *Eurasia J. Math. Sci. Technol. Educ.* **12** 57–86

[16] Adadan E and Savasci F 2012 An analysis of 16-17-year-old students' understanding of solution chemistry concepts using a two-tier diagnostic instrument *Int. J. Sci. Educ.* **34** 513–44

[17] Hoe K Y and Subramaniam R 2016 On the prevalence of alternative conceptions on acid-base chemistry among secondary students: insights from cognitive and confidence measures *Chem. Educ. Res. Pract.* **17** 263–82

[18] Hand B M and Treagust D F 1988 Application of a conceptual conflict strategy to enhance student learning of acids and bases *Res. Sci. Educ.* **18** 53–63

[19] Ross B H 1989 *High-school students' concepts of acids and bases* (Queen's University)

[20] Schmidt H-J 1991 A label as a hidden persuader: chemists' neutralisazion concept *Int. J. Sci. Educ.* **13** 459–71

[21] Bretz S L and Mcclary L 2014 Students ' Understandings of Acid Strength: How Meaningful Is Reliability When Measuring Alternative Conceptions? *J. Chem. Educ.* **92** 212–9

[22] Drechsler M and Driel J Van 2008 Experienced Teachers ' Pedagogical Content Knowledge of Teaching Acid – base Chemistry 611–31

[23] Sheppard K 2006 High school students ' understanding of titrations and related acid- base phenomena **7** 32–45

[24] Seçken N 2010 Identifying student's misconceptions about SALT *Procedia - Soc. Behav. Sci.* **2** 234–45

[25] Orgill M and Sutherland A 2008 Undergraduate chemistry students ' perceptions of and misconceptions about buffers and buffer problems *Chem. Educ. Res. Pract.* **9** 131–43

[26] Blanco A and Prieto T 1997 Pupils' views on how stirring and temperature affect the dissolution of a solid in a liquid: a cross-age study (12 to 18) *Int. J. Sci. Educ.* **19** 303–15

[27] Prieto T, Blanco A and Rodriguez A 1989 The ideas of 11 to 14-year-old students about the nature of solutions *Int. J. Sci. Educ.* **11** 451–63

[28] Çalýk M, Ayas A and Ebenezer J V 2005 A Review of Solution Chemistry Studies : Insights

into Students ' Conceptions *J. Sci. Educ. Technol.* **14**
[29] Ebenezer J V. and Erickson G L 1996 Chemistry students' conceptions of solubility: A phenomenography *Sci. Educ.* **80** 181–201
[30] Önder İ and Geban O 2006 The Effect of Conceptual Change Texts Oriented Instruction on Students' Understanding of the Solubility Equilibrium Concept *Hacettepe Univ. J. Educ.* **30** 166–73
[31] Ardiansah 2016 Identifikasi Konsep Alternatif Pada Guru Kimia : Sebuah Kajian Literatur *Pros. Semin. Nas. Pendidik. Sains "Peningkatan Kualitas Pembelajaran Sains dan Kompetensi Guru melalui Penelit. Pengemb. dalam Mengahadapi Tantangan Abad-21"* 49–54
[32] F02110032 A, Lestari I and Enawaty E 2014 Miskonsepsi guru SMA negeri pada materi ikatan kimia menggunakan certainty of response index (CRI) *J. Pendidik. dan Pembelajaran* **3** 1–18
[33] Gurel D K, Eryılmaz A and Mcdermott L C 2015 A Review and Comparison of Diagnostic Instruments to Identify Students ' Misconceptions in Science *Eurasia J. Math. Sci. Technol. Educ.* **11** 989–1008
[34] Caleon I and Subramaniam R 2010 Development and application of a Three-Tier Diagnostic Test to assess secondary students' understanding of waves *Int. J. Sci. Educ.* **32** 939–61
[35] Caleon I S and Subramaniam R 2010 Do students know what they know and what they don't know? Using a four-tier diagnostic test to assess the nature of students' alternative conceptions *Res. Sci. Educ.* **40** 313–37
[36] Ardiansah A 2017 Penggunaan tes diagnostik bertingkat dalam mengidentifikasi miskonsepsi siswa: Sebuah kajian literatur *Prosiding Seminar Nasional Pendidikan Sains PPs Unesa 2017* (Surabaya: Program Studi Pendidikan Sains Unesa) p C 114-C 119
[37] Sia D T, Treagust D F and Chandrasegaran A L 2012 High school students' proficiency and confidence levels in displaying their understanding of basic electrolysis concepts *Int. J. Sci. Math. Educ.* **10** 1325–45
[38] Caleon I and Subramaniam R 2010 Development and application of a Three-Tier Diagnostic Test to assess secondary students ' understanding of waves 146–55
[39] Muchtar Z and Harizal 2012 Analyzing of students' misconceptions on acid-base chemistry at senior high schools in Medan *J. Educ. Pract.* **3** 65–74
[40] Damanhuri M I M, Treagust D F, Won M and Chandrasegaran A L 2016 High school students' understanding of acid-base concepts: An ongoing challenge for teachers *Int. J. Environ. Sci. Educ.* **11** 9–27
[41] Peterson Treagust, D. F. R F 1989 Grade-12 students' misconceptions of covalent bonding and structure *J. Chem. Educ.* **66** 459–60
[42] Aiken L R 1985 Three coefficient for analyzing the reliability and validity of ratings *Educ. Psychol. Meas.* **45** 131–42
[43] Hendayana, S, Supriatna, A and Imansyah H, Hendayana S, Asep S and Imansyah H 2010 Indonesia's issues and challenges on quality improvement of mathematics and science education *J. Int. Coop. Educ.* 41–51

Acknowledgement
This work was supported by Indonesia Endowment Fund for Education from the Indonesia Ministry of Finance. Any opinions, findings, and conclusions or recommendations expressed in this material are those of the authors and do not necessarily reflect the views of the funding donator.

IOP Conf. Series: Journal of Physics: Conf. Series **1022** (2018) 012034 doi:10.1088/1742-6596/1022/1/012034

Comprehensive understanding of mole concept subject matter according to the tetrahedral chemistry education (empirical study on the first-year chemistry students of Technische Universität Dresden)

D W Prabowo[1], S Mulyani[1], K-H van Pée[2] and N Y Indriyanti[1,3*]

[1]Department of Chemistry Education, Sebelas Maret University, Indonesia
[2]Department of Chemistry and Food Chemistry, Technische Universität Dresden, Germany

*Corresponding email: nurma.indriyanti@staff.uns.ac.id

Abstract. This research aims to apprehend: (1) the shape of tetrahedral chemistry education which is called the future of chemistry education, (2) comprehensive understanding of chemistry first-year students of Technische Universität Dresden according to the chemistry education's tetrahedral shape on mole concept subject matter. This research used quantitative and qualitative; paper and pencil test and interview. The former was conducted in the form of test containing objective test instrument. The results of this study are (1) learning based on tetrahedral shape of chemistry education put the chemical substance (macroscopic), symbolic representation (symbol), and its process (molecular) in the context of human beings (human element) by integrating content and context, without emphasis on one thing and weaken another, (2) first-year chemistry students of Technische Universität Dresden have comprehensively understood the mole concept associated with the context of everyday life, whereby students are able to find out macroscopic information from statements that are contextual to human life and then by using symbols and formulas are able to comprehend the molecular components as well as to interpret and analyse problems effectively.

1. Introduction

The development of human life increases rapidly in line with the complexity of human life problems. Issues of energy resources, food crisis, global warming, environmental damage, and much more. With the increasing of the problems of human life, education, science and technology have also to be developed in order to be able for solving these problems.

Education is defined as the process of human's reciprocal adjustment to nature, to his fellows, and to the ultimate nature of the cosmos [1]. Science and technology is an intellectual system of human life resulted from the education system in order for adapting to the challenges of nature. Education is an attempt to help children to perform the tasks of their life in order to be independent and responsible [2]. Therefore, education has also to be developed to contribute in the development of human life.

Chemistry is one of the subjects of science that has a big role and influence for human life. Applications of chemistry have contributed significantly to the progress of human civilization [3]. Chemistry education has a very important role in the development of chemistry from time to time. Chemical education gives more attention to the provision of knowledge regarding the nature, components, changes and interactions between matter [4].

Today, a popular shape of chemistry education is a planar triangle. This shape has been used effectively to help us understand the three levels of learning, which are symbolic, macroscopic and sub-microscopic or molecular, required by students to understand chemistry. The three levels of learning form the three corners in the triangle. No one form is considered more dominant than the others, but each one complements the other. These forms of the subject are (a) the macro and tangible: what can be seen, touched and smelt; (b) the sub-micro: atoms, molecules, ions and structures; and (c) the representational: symbols, formulas, equations, mathematical manipulation and graphs [5].

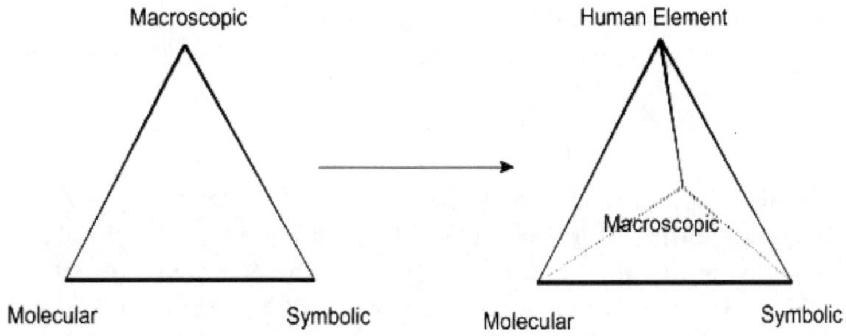

Figure 1. Tetrahedral Chemistry Education. Transformation of Chemistry Education

However, various forces have shaped the teaching and learning of chemistry at the beginning of the 21st century. These include fundamental changes in chemistry as defined by new research areas; changes in our understanding on how students learn; application of computer technology and information in studying complex scientific phenomena; and external influences, such as global concerns about energy and water resources and environmental issues. In response to these forces, the shape of a triangle chemical education have to be emphasized into a tetrahedral shape, in which the fourth peak represents the human context in studying chemistry, that is the human element [6].

As a new shape of chemistry education, only has a little research been done on this topic. In contrast to the planar triangle chemistry education which has proven to be a prefix in the introduction to the first-year chemistry reader [7]. This triangle has also become a benchmark for national science education standards in the United States [8].

In chemistry education, there are several topics in chemical education that are important and fundamental discussions, ones is the concept of mole. The concept of mole is at the heart of all quantitative chemical calculations and is used by almost all laboratory chemistry experimental materials including limiting reagents, ideal gas law, stoichiometry, reaction equations, titration, electrolysis, chemical kinetics, thermodynamics, and others [9]. There may be no concept given in the first-year of learning chemistry is more important to be understood by students in addition to the concept because the concept of the mole is very essential for the study of the stoichiometry [10]. Although the concept of mole is an important concept in chemistry education, many studies reveal that the concept of mole is also a subject that is difficult to be understood. In many studies of the concept of mole, there are such a consensus that students are weak in understanding the concept of mole [11].

In connection with this, a survey with a large sample size including high school students (aged 16 years) to first-year students (aged 19 years) reported that there was an increase in the proportion of incorrect answers related to the concept of moles, where the answer did not match the definition of the IUPAC, so it was concluded that there was a superficial learning about the concept of mole [12]. Another study in the United States found the fact that second graders at high school and first-year students at the university did not have a comprehensive understanding of 'independent units' in the definition of the concept of mole [13].

IOP Conf. Series: Journal of Physics: Conf. Series **1022** (2018) 012034 doi:10.1088/1742-6596/1022/1/012034

By combining two issues about the tetrahedral shape of chemistry education as well as the mole concept, researchers desire to examine the tetrahedral shape of chemical education on the concept of mole subject matter.

2. Research Methods

This research used an qualitative research design that is using data or empirical evidence to answer the problem. The research was conducted at Technische Universität Dresden, Germany. Research subjects were first-year chemistry students. We used a convenience sample in one lecture class.

The technique used to collect data is in the form of survey, interview and documentation. The survey was conducted in the form of questionnaire containing objective test instrument. Collecting data using a focus group interview gives richer and more research data. The curriculum documents represent a type of document analysis used in order to extract themes in regard to the topic being evaluated.

3. Discussion

The text of your paper should be formatted as follows:

Based on the results of cognitive test, first-year chemistry students at the Technische Universität Dresden (TUD) reaches a high enough score, which is the average correct answer correctly, is 75.96%; wrong answer 13.33%; and unanswered ones is 10.71%. The most difficult experience was in item number 6which only are 44.44% of students able to answer correctly. While for item number 1 and 2, students are able to answer correctly with the highest percentage of 88.89%. Recapitulation of cognitive test results is presented in table 1.

In the cognitive test instrument, students read the material about Liquefied Petroleum Gas (LPG). Thus, students are expected to be more interested in learning chemistry because the cognitive test instrument has a correlation with everyday human life.

Table 1. Cognitive Test Results

Item Number	Percentage (%)		
	Correct	Wrong	Unanswered
1	88,89	11,11	-
2	88,89	11,11	-
3	86,67	13,33	-
4	64,44	15,56	20,00
5	86,67	13,33	-
6	44,44	20,00	35,56
7	82,22	17,78	-
8	68,89	11,11	20,00
9	73,33	8,89	17,78
10	80,00	6,67	13,33
11	71,11	17,78	11,11
Average	75.96	13,33	10,71

In the beginning. of the instrument (item number 1 and 2), students are simulating to read a brochure about a gas product, which is bought from a company; they therefore have to look for some information from the brochure. After finding the mass information of each gas and its mole mass, then students then look for the mole value of each gas.

IOP Conf. Series: Journal of Physics: Conf. Series **1022** (2018) 012034 doi:10.1088/1742-6596/1022/1/012034

A company produces LPG for commercial with the following specification :

Total Weight	:	37 kg
Tare Weight	:	18 kg
Composition	:	60% propane gas & 40% butane gas
Volume	:	44.5 litres
CG ref. (regulator)	:	CG2

Figure 2. Beginning Part of The Cognitive Test

Item number 1 and 2 involve 4 elements simultaneously in tetrahedral chemistry education, which are macroscopic, molecular, symbolic and human elements. Macroscopic is what can be seen, touched and kissed. In this case, it includes gases in LPG cylinder, because the mass can be calculated using the balance. Molecular or sub-micro is atoms, molecules, ions and structures. The propane molecules and the butane molecules inside the LPG cylinder including the amount of each molecule are molecular components that cannot be identified directly. The representational is symbols, formulas, and equations used by students so that students can find the amount of substances of each molecule inside the cylinder. While the LPG theme is the human element because LPG is very wide in everyday life.

In this part, we emphasized a new dimension on teaching chemistry to pay attention on the role of chemistry in everyday life. It is known that students who are able to answer correctly on item number 1 and 2 have understood comprehensively the concept of mole associated with the context of everyday life, whereas students are able to find macroscopic information from statements that are contextual to human life then by using symbols and formulas can understand the molecular components.

Item number 3 and 5 aim to determine the level of students' understanding on the definition of mole. These items were developed based on Menis & Novick's research [14]. Item number 4 is aims to calculate the amount of substances by the elementary entities of the atoms according to Avogadro's hypothesis.

Item number 6 is calculation of gases volume related to the amount of substance according to Avogadro's hypothesis and the ideal gas law. In this case, students have to calculate the volume of gases inside the LPG cylinder if the gases occupy a large blank space. Item number 7 aims to determine the level of students' understanding on the state of the molecules inside LPG cylinder. Item number 8 discusses the relationship among the mass of matter, the amount of substances and the molar mass. Interestingly, beside determining the amount of substances, students also got an interesting information related to the history and facts of the unpleasant odor from LPG cylinder when it leaks. Item number 9, 10 and 11 aim to understand the use of the mole concept on the quantitative calculation of stoichiometry. Students have to not only quantitatively calculate, but also understand the use of LPG and other fuels (kerosene and methane) in everyday life.

All items in this study involve the macroscopic, molecular, symbolic and human element. In the last part of the cognitive test, students are given a text about the issue of global warming. This is certainly relevant to the theme of LPG because the combustion reaction of fossil fuels, including LPG, kerosene, methane and much more, produce carbon dioxide. Carbon dioxide can be concentrated in the atmosphere. In fact, carbon dioxide is the main cause of the greenhouse effect, which is able to capture and re-radiate infrared radiation to the surface of the earth. Thus, the increased concentration of carbon dioxide in the atmosphere can increase the temperature of the earth's surface.

IOP Conf. Series: Journal of Physics: Conf. Series **1022** (2018) 012034 doi:10.1088/1742-6596/1022/1/012034

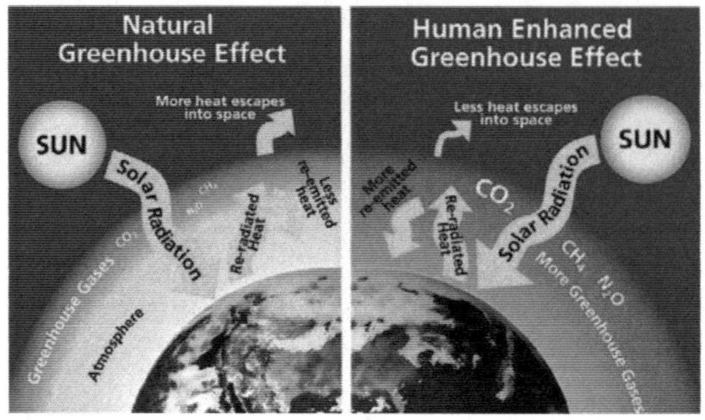

Figure 3. Explanation of Global Warming Issue in The Last Part of The Cognitive Test

Item number 12 is able to attain students' responses or comments after doing cognitive test for items number 1 through number 11. Students are required to illustrate an image that represent their comments or responses. It is expected that students be able to have an understanding related to the human elements.

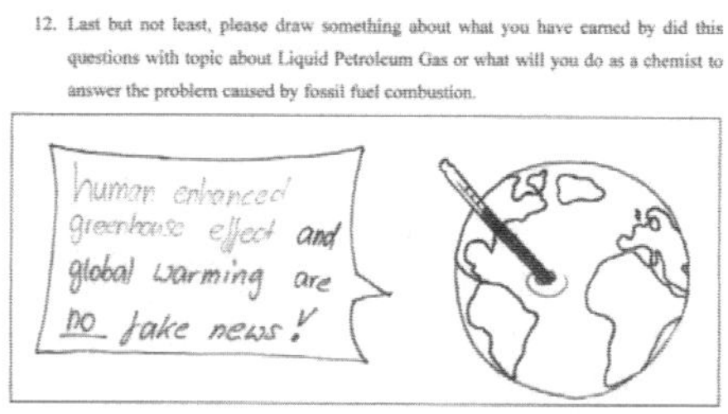

Figure 4. Image Illustrated by The First Student

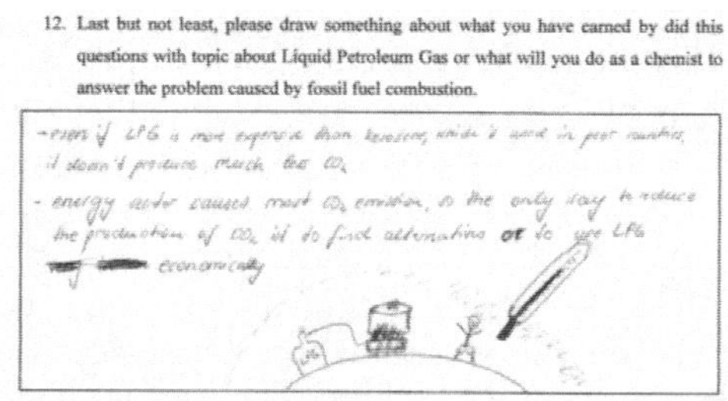

Figure 5. Image Illustrated by The Second Student

The first student illustrated an image in figure 4 where the surface temperature of the earth is measured using a giant thermometer showing a high temperature. This indicated that the first student is interested in the information at the last part of the cognitive test explaining the rising of earth surface's temperature. Similar to first student exposure, the second student tried to give the same explanation but with a different illustration in figure 5.

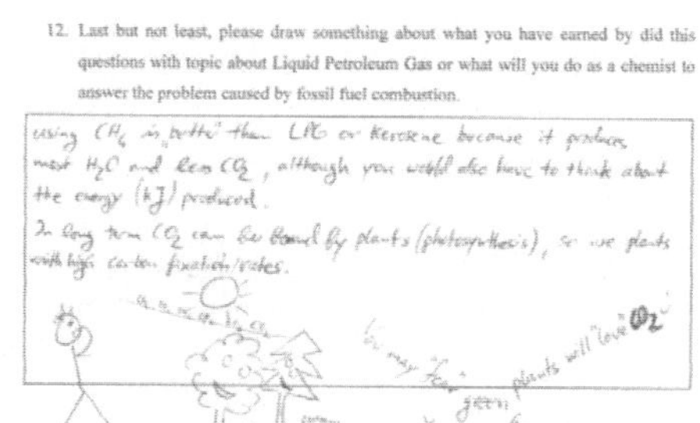

Figure 6. Image Illustrated by The Third Student

The third student interested in the issue of the amount of carbon dioxide (CO_2) produced from fossil fuels combustion in figure 6. Based on his observations related to examples which appear in the cognitive test (item number 9,10 and 11), he concluded that the use of methane (CH_4) is better than LPG (C_3H_8 and C_4H_{10}) and kerosene ($C_{12}H_{26}$) because of less carbon dioxide produced.

The image by the fourth student in figure 7indicates that he was interested in the history of the gas explosion at London School of New London in Texas, USA. The explosion was caused by leakage of natural gas pipeline explained in the item number 8. This explosion also brought to a new invention of using odor to detect gas leakage.

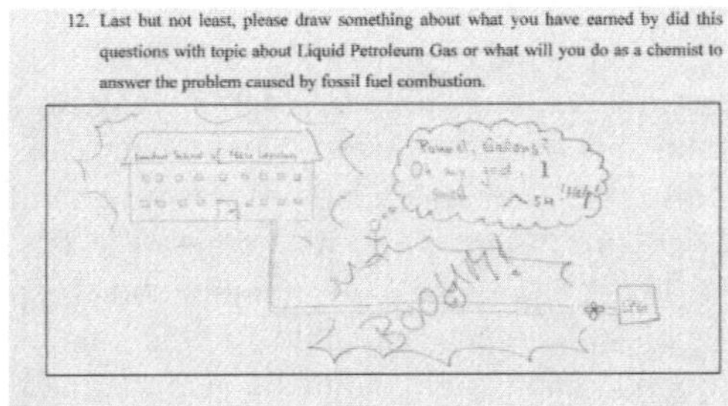

Figure 7. Image Illustrated by The Fourth Student

The fifth student illustrated all the things obtained during the completion of the cognitive test in figure 8. He accommodated all the information from LPG composition, gas properties in LPG, usefulness, as well as the impact caused by the using of LPG as fuel.

Figure 8. Image Created by The Fifth Student

In this research, interview was conducted using focus group interviews. Subjects in this focus group were the first-year chemistry student. Focus group interviews were conducted to acquire informations on: (1) the difficulties experienced by the first-year chemistry students during the cognitive test; and (2) understanding from cognitive test related to everyday life.

Based on interviews and discussions, it is known that the first-year chemistry student at the Technische Universität (TU) Dresden suffered two difficulties during the cognitive test. The first is item number 6 of the questionnaire turned out to be quite difficult for the students, since they had not learned about ideal gases, yet. The second issue was a language problem. The TU Dresden students had difficulties with understanding English that was quite surprising and embarrassing for us and tells us that this needs to be changed.

In addition, the results of interviews and discussions shows that students were very interested in learning many things related to the using of LPG in everyday life. The students also became aware that the use of fossil fuel energy such as gases and oils could affect the environment. This is due to the combustion of fossil fuels to produce carbon dioxide. This carbon dioxide is the main gas enhancing the greenhouse effect that brings a rising of global surface temperature or global warming. Thus, students understood not only the concept of mole cognitively, but also the use of LPG in many countries as well as became aware of the issue of global warming caused by human activities in utilizing fossil fuels as an energy source.

Based on these results, it can be concluded that the first chemistry student of Technische Universität Dresden has comprehensively understood the concept of mole associated with the context of everyday life whereas students are able to find macroscopic information from statements that are contextual to human life then using symbols and formulas are able to understand the molecular components. Thus, students have gained a coherent understanding of the definition of the concept of mole and the importance of interpreting chemical phenomena as well as the ability to analyse mole-related issues effectively.

4. Conclusions

Based on the results of the research, it can be concluded that (1) learning based on tetrahedral shape of chemistry education put the chemical substance (macroscopic), symbolic representation (symbol), and its process (molecular) in the context of human beings (human element) by integrating content and context, without emphasis on one thing and weaken another, (2) first-year chemistry students of Technische Universität Dresden have comprehensively understood the mole concept associated with the context of everyday life, whereby students are able to find out macroscopic information from statements that are contextual to human life and then by using symbols and formulas are able to comprehend the molecular components as well as to interpret and analyse problems effectively.

5. References

[1] Brubacher J S 1969 *Modern Philosophies of Education* (Michigan: McGraw–Hill Book)
[2] Langeveld M J 1979 *Beknopte Theoretische Pedagogiek* (Groningen: Wolters-Noordhoff)
[3] Sjöström J 2007 The Discourse of Chemistry (and Beyond) *Hyle – International Journal for Philosophy of Chemistry* **13** 83–97
[4] Nnamdi B 2014 The Role of Chemistry Education in National Development *The International Journal Of Engineering And Science* **3** 12–17
[5] Johnstone A H 2000 Teaching of Chemistry-Logical or Psychological *Chemistry Education : Research and Practice in Europe* **1** 9–15
[6] Mahaffy P 2004 The Future Shape of Chemistry Education *Chemistry Education : Research and Practice* **5** 229–45
[7] Kotz J & Treichel P Jr 2000 *Chemistry and chemical reactivity 5th ed.* (Philadelphia: Saunders College Publishing) pp 11-21
[8] National Research Council (NRC) 1996 *National science education standards* (Washington: National Academies Press) p 177-208.
[9] Ramette R W 1988 The mole concept is useful *Journal of Chemical Education* ***65*** 376
[10] Kolb D 1978 The mole *Journal of Chemical Education* **55** 728-32
[11] Gabel D L & Bunce DM 1994 Research on problem solving : Chemistry (New York: MacMillan Publishing)
[12] García J P, Pizarro A M, Pereira F, Martín M J & Bacas P 1990 Ideas de los alumno sacerca del mol : Estudio curricular*Enseñanza de lasCiencias* **8** 111-19
[13] Krishnan SR & Howe AC 1994 The mole concept developing an instrument to assess Conceptual understanding *Journal of Chemical Education* **71** 653-55
[14] Novick S & Menis J 1976 A Study of Student Perceptions of the Mole Concept*Journal of Chemical Education* **53** 720–22

Acknowledgments

We would like to express my thanks and deep gratitude to Prof. Dr.rer.nat. Karl-Heinz van Pée as Head of the Institutfür Biochemie, Technische Universität Dresden on occasion, permission and assistance during this research process and hospitality for 3 months in Dresden, Germany.

IOP Conf. Series: Journal of Physics: Conf. Series **1022** (2018) 012035 doi:10.1088/1742-6596/1022/1/012035

Students' misconceptions on solubility equilibrium

H Setiowati [1a], S B Utomo[1] and Ashadi[1]

[1] Postgraduate program of Science Education, Sebelas Maret University, Ir. Sutami Street Number 36 A Jebres, Surakarta, Central Java 57126 Indonesia

[a]Corresponding email: hanifahsetiowati@student.uns.ac.id

Abstract. This study investigated the students' misconceptions of the solubility equilibrium. The participants of the study consisted of 164 students who were in the science class of second year high school. Instrument used is two-tier diagnostic test consisting of 15 items. Responses were marked and coded into four categories: understanding, misconception, understand little without misconception, and not understanding. Semi-structured interviews were carried out with 45 students according to their written responses which reflected different perspectives, to obtain a more elaborated source of data. Data collected from multiple methods were analyzed qualitatively and quantitatively. Based on the data analysis showed that the students misconceptions in all areas in solubility equilibrium. They had more misconceptions such as in the relation of solubility and solubility product, common-ion effect and pH in solubility, and precipitation concept.

1. Introduction

Chemistry is one of the most important branches of science; it reviews the composition, structure and nature, transformation, dynamics, and energetics of matter. Chemical learning often combines abstract concepts, which become the next center of learning in chemistry and other sciences [1][2]. Chemistry is often considered as a difficult subject because it discusses complex material structures and requires intellectual understanding. The difficulty in studying chemistry can be seen from the characteristics of chemical subjects that include three levels of thinking, such as: macroscopic, sub-microscopic, and representational level [3]. The macroscopic level is a phenomenon that can be seen, touched, and smelled. The sub-micro level is what can't be observed visually. The representational level is a sub-micro representation into a symbol, formula, equation, molarity, mathematical manipulation, and graph. Each level of thought is complementary.

Chemistry learning in high school emphasizes mathematical calculations (representational level) rather than conceptual learning (macroscopic and sub-microscopic level). The presence of imbalance at the macroscopic, sub-microscopic, and representational level can cause students to have difficulty in understanding the concept of chemistry as a whole even in the most basic concepts. If the student does not understand certain basic concepts, then he will have difficulty in understanding more complex concepts and can lead to an understanding which is beyond the concept itself or misconception. Misconception is part of a larger knowledge system that involves many interrelated concepts that students use to make sense of their experiences [4]. Misconceptions can occur due to the limitations of students in constructing or building an understanding of a concept they receive during the learning process.Generally, there are five factors that can cause misconceptions, such as: students, teachers, textbooks, contexts, and teaching methods. Misconceptions that continuously occur in students will result in decreased their achievement. The results of previous research indicated that misconceptions of students in studying chemistry included basic concepts of chemistry [5], chemical bonding [1][6], chemical equilibrium [7][8], mole concepts [9], and nuclear chemistry [10].

One of the chemicals considered as difficult by students is solubility equilibrium. Solubility equilibrium materials have basic abstract, mathematical calculations, and complex concept that can cause misconceptions, and they are interconnected with one another. Students had more problems in understanding the areas such as dynamic nature of the solubility equilibrium, common-ion effect, selective precipitation, solving the problems of solubility product, understanding and interpreting graphs related to the solubility equilibrium concept [11]. Although chemistry students can solve various calculation problems on solubility equilibrium, for example solubility and solubility product calculation (K_{sp}), it does not guarantee that the student can understand the concepts in the material [12].

Misconceptions can be identified by diagnostic tests. Two-tier multiple choice diagnostic test is an effective diagnostic instrument for identifying student misconceptions [13]. The use of a two-tier diagnostic instrument at the beginning or end of teaching of a particular subject may help science teachers to get a better picture of students' understanding and the presence of misconceptions in a particular part of the topic being taught [14]. The two-tier multiple choice diagnostic test is one of the diagnostic tests which is a matter of two-tier problems. The first level consists of questions with five answer options, while the second level consists of five reason options that refer to the answer at the first level [14]. Students are considered master if they can answer both items correctly. Conceptual understanding can be divided into four categories: understand, misconception, understand little without misconception, and not understand [15][7]. This grouping is based on possible responses from students' answers using two-tiered questions. Based on the above description, the researcher wants to identify the misconception of high school students on the concept of solubility equilibrium by using Two-Tier diagnostic test.

2. Method

The subjects of the study were 164 eleventh grade of High School science program students in three school categories: high, medium, and low. The research was conducted in school year 2016/2017. The research was a descriptive study. Data collection techniques used tests, interviews, and document studies. The test instrument used two-tier diagnostic instrument consisted of 15 question items. Responses were marked and coded into four categories[7][15]:

Table 1. Possible Patterns of Students' Answers and the Categories

No	Patterns of Students' Answers	Category of Understanding
1.	The main answer is correct – the reason is correct	Understand
2.	The main answer is correct – the reason is incorrect	Misconception
3.	The main answer is incorrect – the reason is correct	Misconception
4.	The main answer is incorrect – the reason is incorrect	Not understand
5.	The main answer is incorrect – the reason is not answered	Not understand
	The main answer is correct – the reason is not answered	Understand little without misconception
6.	The main answer and reason are not answered	Not understand

Based on these categories, learners misconceptions and how many learners (percentage) are experiencing misconceptions can be determined on the test items. The percentage of each item can be calculated in:

$$\% = \frac{the\ number\ of\ the\ patterns\ of\ students'answers\ in\ each\ category}{the\ number\ of\ students} \times 100\%$$

Semi-structured interviews were conducted to 45 students according to their written responses which reflected different perspectives, to obtain more elaborated source of data. The data obtained from the results of test and interview were analyzed. Data analysis was aimed to give meaning to the data that has been collected from the research. Descriptive analysis techniques were used to determine the types of misconception experienced by students on solubility equilibrium.

3. Result and Discussion

Misconceptions on Solubility Equilibrium

3.1. Difficulty of Soluble Salt Equilibrium

There were 37.80% of students in misconception group. Most students who had misconceptions ignored the law of charge neutrality in solution to the concept of soluble salt equilibrium. It was in accordance with a statement that the amount of ions in the over saturated solution was greater than the saturated solution without regarded to the law of neutrality [16]. The correct concept was that the amount of charge of ions from the ionized salt must be zero, so that the solution is neutral. In addition, some students who had misconceptions ignored the difficulty of soluble salt phases in equilibrium reactions. The correct concept was that in equilibrium reaction for solution with soluble salt had a solid phase (s) and the ions had aqueous phase (aq).

Another misconception was that when the concentration of produced equilibrium ion was equal to the salt concentration had by 27.44% of students. They assumed that equilibrium mean between the salt and ions in the same rate so that it had the same concentration. This misconception corresponded to the result which stated that the resulting ion concentration was equal to the salt concentration at equilibrium [16]. The correct concept was that the equilibrium, concentration of the ions and concentration of salt (salt solubility) could be determined by the mole ratio of the equilibrium system.

3.2. The Concept of Solubility

This study revealed that few students had misconceptions on the concept of solubility. There were 29.27% of students in misconception group. Students experiencing misconceptions assumed that the solubility value was calculated from mole of salt under saturation conditions. The correct concept of solubility was the number of salt moles dissolved in a liter of solvent and yielded a saturated solution. Thus, the solubility value could be calculated from the molarity of the salt under saturation [17].

Other misconceptions found was about the relationship of K_{sp} value with solubility. The researcher gave a question about the most soluble salt with different number of ions based on the given K_{sp} value. There were 85.37% in misconception group. The number of students experiencing misconception was quite large. Most students assumed that a large salt solubility had a large K_{sp} value regardless of the amount of ions in the salt. The correct concept was that in the salt which had the same number of ions, the greater the K_{sp} value the greater the solubility. However, if the salt had a different amount of ions, the solubility of each salt must be first calculated.

3.3. The Concept of Common Ion-Effect

Concerning this concept, there were 44.52% of students assumed that the solubility of soluble salt would increase if the equilibrium system shifted toward the reactant (the soluble salt). In this concept students understood about the concept of equilibrium was the addition of ion names on the equilibrium salt soluble would cause the ion concentration increased so that the equilibrium shifted toward the salt (reactants). However, students could not understand that if the reaction shifted toward the salt then the salt solubility would decrease. The correct concept was that if the same ion-containing compound was added to the soluble salt-soluble equilibrium, it would cause the concentration of soluble salt-soluble ions to increase and the system shifted to the left (salt formation). This caused less salt to dissolve in a solution containing the same ion.

3.4. The Concept of Precipitation

There were 71.95% students experiencing misconceptions assumed that saturated solution occurred when the product concentration was less than K_{sp}. The correct concept was that the saturated solution occurred when the product ion concentration was greater than K_{sp} while the unsaturated solution occurred when the product concentration product was smaller than K_{sp}.

3.5. The Relation of Solubility and pH Concept

There were 68.29% of students assumed that the hydroxide ions used to calculate the K_{sp} value of the base solution were from the negative antilog of pH. The correct concept was it used to calculate the K_{sp} value of an alkaline solution and to be noted in particular OH^- concentration.

4. Conclusions

Based on the description of results and discussion of the present research of misconception of high school students on solubility equilibrium it can be concluded that there are misconceptions in all areas in solubility equilibrium. They had more misconceptions such as in relation of solubility and solubility product, common-ion effect and pH in solubility, and precipitation concept. Teachers should pay attention to sub-materials that often lead to misconception when they are teaching solubility equilibrium in order improve their understanding.

5. References

[1] Ozmen H 2004 Some student misconceptions in Chemistry **13**

[2] Sirhan G 2007 Learning difficulties in Chemistry: An overview *J. of Turk. Sc. Edu.* **4** 2–20

[3] Johnstone A H 2000 Teaching of Chemistry - Logical or Psychological? *Chemistry Education Research and Practice* **1** 9

[4] Southerland S A, Abrams E, Cummins C L and Anzelmo J 2001 Understanding students' explanations of biological phenomena: Conceptual frameworks or p-prims? *Sci.Edu.* **85** 328–48

[5] Chiu M H 2007 A national survey of student's conceptions of chemistry in Taiwan *International Journal of Science Education* **29** 421–52

[6] Galley W C 2004 Exothermic bond breaking: A persistent misconception *J. of Chem. Edu.* **81** 523–5

[7] Salirawati D 2011 Chemical Equilibrium Misconception

[8] Bergquist W and Heikkinen H 1990 Student ideas regarding chemical equilibrium: What written test answers do not reveal *Journal of Chemical Education* **67** 1000

[9] Menis J and Fraser B J 1992 Chemistry achievement among grade 12 students in Australia and the United States *Research in Science & Technological Education* **10** 131–41

[10] Tekin B B and Nakiboglu C 2006 Identifying students' misconceptions about nuclear chemistry. A study of Turkish high school students *Journal of Chemical Education* **83** 1712

[11] Mohideen J M, Karunaratne S and Wimalasiri K M S 2013 Students' Understanding of the Solubility Equilibrium **17** 2013

[12] Alexander J 2001 Resources for student assessment assessing students' conceptual understanding **78** 629–31

[13] Chandrasegaran A L, Treagust D F and Mocerino M 2007 The development of a two-tier multiple-choice diagnostic instrument for evaluating secondary school students ' ability to describe and explain chemical reactions using multiple levels of representation **8** 293–307

[14] Treagust D 2012 Diagnostic assessment in science as a means to improving teaching, learning and retention *Australian Conference on Science* 1–9

[15] Abraham M R, Grzybowski E B, Renner J W and Marek E A 1992 Understandings and misunderstandings of eighth graders of five chemistry concepts found in textbooks *Journal of Research in Science Teaching* **29** 105–20

[16] Onder I and Geban O 2006 The effect of conceptual change texts oriented instruction on students' understanding of the solubility equilibrium concept *Hacettepe University Journal of Education* **30** 166–73
[17] Brady J E 2008 *Kimia Universitas, Asas dan Struktur, Jilid satu*. Yogyakarta: Binarupa Aksara.

Acknowledgement
Headmasters and chemistry teachers of SMA N 1 Simo, SMA N 1 Teras, and SMA N 1 Banyudono.

IOP Conf. Series: Journal of Physics: Conf. Series **1022** (2018) 012036 doi:10.1088/1742-6596/1022/1/012036

Concept cartoons for diagnosing student's misconceptions in the topic of buffers

I A Kusumaningrum [1a], Ashadi [1] and N Y Indriyanti[1]

[1]Postgraduate program of Science Education, Sebelas Maret University, Ir Sutami Street Number 36 A Jebres, Surakarta, Central Java 57126 Indonesia

[a]Corresponding email: idhaayukusumaningrum@student.uns.ac.id

Abstract. Student's misconceptions have been concerned over twenty years in the chemistry education research. It influences students to learn new knowledge and gain a correct concept. The buffer solution is found as a difficult topic due to student's misconception. However, the research related this subject are still rare. Concept cartoon has been used as one of the effective tools to diagnose misconceptions. This study aims to identify the effectiveness of concept cartoon to diagnose them. The concept cartoon consists of three concept questions. 98 students of grade 11 as respondents of this research and followed by interview for selected students. The data obtain of the study are analyzed by using a scoring key. The detected misconceptions are about what buffers do, what buffers are, and how buffers are able to do what they do. Concept cartoon is potential as a basic tool for remedial teaching.

1. Introduction

Over the past decades, students interest and achievement in chemistry have declined [1]. Many students assumed that chemistry is a challenging, complex, and an abstract subject that requires particular intellectual to be understood [2] [3] [4] [5]. There are many reasons chemistry difficult to learn. The first reason is teacher used a regularly common teaching method that makes student passive, not thinking, the information receiving role [4], and other reason student knowledge of chemistry is incomplete and incoherent. On the other hand, many students merely memorize chemistry concepts without profoundly learning them [6]. Students need to understand into sub-micro levels where the characteristic of substances is learned regarding the unseen (atoms, molecules, ions) and recorded in some representational level (symbols, notation) to learn chemistry [7]. Moreover, students do not come to the classroom with blank slates, but they have had a previous concept from their experiences which is described as pre-concept, and they used it to learn new ideas. When student's pre-concept are different from the views of scientist that called misconceptions [8]. Misconception not only from the students but also commonly caused by inappropriate teaching methods, materials, and solves for a full understanding that can be called as school-made misconceptions [9].

Deficiency ability of students in science makes occurrence of misconceptions. It happens to students at all levels of school [10]. In chemistry, many concepts that are frequently misapprehended especially in buffer solution. When understanding of buffers, students must relate the concept from macroscopic, microscopic, and symbolic perspectives [7]. They also must relate to another concept such as chemical equilibrium, acid/base chemistry, the particulate nature of matter, stoichiometry, chemical reactions, solution chemistry [2] [11]. However, in fact many research find that misconceptions occur in chemical equilibrium [12] [13] [14] [15] [16] [17] [18] and acid/base chemistry [2][19][20] [21] [22]

[23] [24]. While students' misconceptions in buffer solution find in some journals which are presented in Table 1.1.

Table 1. Student's Misconceptions in Buffer Solution

Sub Topic	Description
Definition of buffer solution	Buffer is a solution that contains an acid and its salt; a reactions can increase or decrease the acidity [25]
Component of buffer solution	-Buffer is formed from two chemicals that are any acid and base, in any proportion, not necessarily a weak acid/base and its conjugate in approximately equal amounts [11] -Buffer is formed by a weak acid and its salt, not its conjugate base [26] [27]
How do buffer solution work	Buffers maintaining neutral pH (pH = 7), if acid/base is added to buffer solution [26] [27]
pH of buffer solution	The strength of buffer solution (capacity) is determined by the strength of its components acid and base. A buffer solution has a strong base and a strong acid in a solution that can compensate. pH of a weak acid in solution is equal to its pKa [11]

The misconception is one of the problems in learning especially in chemistry. It must be removed because it can disturb the student for learning new knowledge. Many ways can be used to explore student's misconception, but in this research, we use concept cartoon. Concept cartoon is visual tool consist of three or more characters (one of the characters proposes right ideas, and other characters show wrong ideas) whereas it discusses a subject of concept in daily life [5] [28] [29]. Concept cartoons as a strategy to incorporate and integrate five-knowledges, they are pedagogy, content, context, research, and knowledge (PCCRK) [30]. Cartoon concept can make student inquires their knowledge [31]. Furthermore, concept cartoon can be used as a tool to identify students' misconceptions and effective for finding out student's idea without affecting of other ideas [5] [32] [30] [33] [29] [34] [35] [36]. Considering the fact of concept cartoons, it can find out student ideas and identify students' misconceptions. Thus, it is expected that student's misconceptions of the buffer solution can be detected with it. Although, there are many types of research in literature related to use of concept cartoon in physic, chemistry, and mathematics education. However, the researches of concept cartoons in detecting students' misconceptions of buffer solution are highly limited. This research aimed to identify the effectiveness of concept cartoon on detection student's misconception in the topic of buffer solution.

2. Method

This research is mixed-method study. There are three concept cartoons prepare related to buffer solution that developed based on literature and curriculum of the chemistry. Two lecturers who had some experience about concept cartoon to consult in the preparation of the concept cartoon and check the content validity, both regarding content and it's compliance with the rule of developing concept cartoons. The concept cartoons produced is applied to 98 participants attending to the 11th grade of High School in science program 2016 to 2017 second semester in three school categories high (labeled as A), medium (labeled as B), and low (labeled as C).

In this research, Teacher has made the instrument of true concept consist of the definition of buffer solution, the component of buffer solution, how do buffer solution works, function of buffer solution and calculating the pH of buffer solution to check student's concept. The buffer solution is taught in three meeting. In the first stage, the teacher gives the general information about definition and component of buffer solution. The second, how do buffer solutions works. The last, calculating the pH of buffer solution. For this purpose, concept cartoon is used to detect misconceptions related to the

IOP Conf. Series: Journal of Physics: Conf. Series **1022** (2018) 012036 doi:10.1088/1742-6596/1022/1/012036

concept. The concept cartoons are presented in the student's worksheet. In the first part, students are asked to choose a character that they think correct in the concept cartoon and explain their reasons for selecting the characters in the second part. Then, discussing about their ideas and presenting their ideas. At the same time, the teacher asks questions that encouraged the students to be active participants in learning and help student to find the correct concept. Thus, the student will change their conceptions.

The answer in the first and second parts are analyzed using an answering key which developed by the researchers to determine their understanding of the concepts [33]. Then classified according to the criteria of student's level knowledge can be seen in Table 1.

Table 2. Criteria of Student's Level Knowledge [37] [34]

Answering	Criteria
Correct Answer-Correct Explanation	Understanding
Correct Answer-Partially Correct Explanation	Partial understanding
Correct Answer-wrong Explanation	Misconception
wrong Answer-Correct Explanation	Misconception
wrong Answer-wrong Explanation	Not understanding

The student who identified misconceptions will be interviewed. After that, calculating the percent of student's level of knowledge according to the equation (1),

$$\text{\% student's level of knowledge} = \frac{number\ of\ student\ for\ each\ criteria}{total\ of\ student} \times 100 \quad \text{............} \quad (1)$$

3. Result

The first concept cartoon discusses the definition of buffer solution. Concept cartoon question "How is pH in mouth if food that we eat contains excessive acid?". The examples of first concept cartoon can be seen in figure 1 below.

Example of Reasons:

a. pH of a solution is constant because buffer solution is a solution that resists changes in pH when acid or base is added.

b. pH of solution <7, because pH in the mouth is acid, if excessive acid is added, pH in the mouth will constant, buffer solution can resist change in pH

c. pH of solution =7, solution will be neutral, because buffer solution can make the solution neutral

d. pH of solution >7, solution will be base because the properties of buffer solution must change to resists changes in pH when excessive acid is added

Figure 1. Concept Cartoon about Definition of Buffer Solution

IOP Conf. Series: Journal of Physics: Conf. Series **1022** (2018) 012036 doi:10.1088/1742-6596/1022/1/012036

The result of the interview (T as teacher and S as a student with misconceptions) is presented below.

T: Which character do you think is correct to the question "How is pH in the mouth if the food that we eat contains excessive acid?" Why do you choose that?

S2: None. I think pH in mouth is acid and buffer solution can resist changes in pH when acid is added in any amount, so pH value <7

T: Are you sure?

S2: Yes, of course, because buffer solution contains a component that can resist change in pH

T: How about you? What is your answer?

S3: I think None. Because substance which added is acid, so the solution becomes acid and pH value < 7

Second, concept cartoon about how do buffer solutions works. The concept cartoon question "how does buffer work maintaining a constant blood pH when the small amount of base added?". The example of second concept cartoon can be seen in figure 2.

Example of Reasons:

a. Increasing the amount of CO_3^{2-} ions because base will react with CO_3^{2-} ions

b. Reacting base with water, because water can dissociate to H^+ ion and OH^- ion and base will react with H^+ ion, so pH of solution will constant

c. Excreting base from the body so base will be removed from body

d. Reacting base with acid to produce salt and water. Because base always reacts with an acid, so pH=7, the properties of solution become neutral. adding base will not cause shifting the equilibrium

Figure 2. Concept Cartoon About How Do Buffer Solution Work

The result of the interview (T as teacher and S as a student with misconceptions) is presented below.

T: Which character do you choose S2 to the question "how does buffer work maintaining a constant blood pH when the small amount of base added?"

S2: None. I think in the blood, base react with acid producing salt and water. It makes pH of solution = 7 and solution will be neutral.

T: Are you sure?

S2: Yes, I am sure. Adding base also will not cause shift the equilibrium, so the pH is constant

T: How about you S3? What is your answer?

S3: I think None. In the blood, the small amount of base is added will react with acid so the base will remove and pH value in the blood is constant.

Third, concept cartoon about calculates the pH of buffer solution. The concept cartoon questions "Whether the volume of the solution affects the capacity of a buffer solution?" The example of third concept cartoon can be seen in figure 3.

Example of Reasons:

a. It does not affect because the concentration of weak acid/base and its conjugation that used in the Henderson-Hasselbalch equation. So, capacity of buffer solution is only affected by concentration.

b. It depends on the substances because not all substances can be reacted with an acid or based is added.

c. It affects, because if the volume of buffer solution is large. The buffer capacity also will large.

d. It affects because volume of buffer solution can be affected by acid/base is added

Figure 3. Concept Cartoon About Calculating The pH of Buffer Solution

The result of the interview (T as teacher and S as a student with misconceptions) is presented below.

T: Which characters do you choose to the question "whether the volume of the solution affects the capacity of the buffer solution?"

S2: None. I think volume of solution and the strength of its components acid and base affected the capacity of buffer solution

T: Do you think it is correct?

S2: Yes, I do because strong acid and base can resist the change pH

T: How about u S3?

S3: I think None, but I don't understand about it

The result of % student's level knowledge of concept cartoon 1, 2, and 3 can be seen in Table 2.

Table 3. The Result of Student's Level Knowledge of Concept cartoon 1, 2 and 3

School	Criteria											
	Understanding (%)			Partially understanding (%)			Misconception (%)			Not understanding (%)		
	1	2	3	1	2	3	1	2	3	1	2	3
A	31,3	12,5	31,3	15,6	15,6	21,9	46,9	40,6	34,3	6,2	31,3	12,5
B	20,6	10,9	30,4	11,8	13,8	15,7	44,1	36,3	32,4	23,5	39	21,5
C	21,9	6,3	21,9	9,4	9,3	21,9	40,6	34,4	34,3	28,1	50	21,9

4. Discussion

The research conduct, the concept cartoon which is used as detection student's misconception tools relates to buffer solution. Concept cartoon is used to stimulate discussion before, during, and after the investigation to arrange conclusion. There are various ways of using concept cartoon in sciences courses. Researcher uses concept cartoon as instructional material in science courses in the form of concept cartoon worksheet. Concept cartoons need to prepare and design following the characteristics: minimal amounts of text, the scientific idea applied in everyday situations, the alternative ideas put forward based on research that identifies common areas of misunderstanding. The scientifically acceptable viewpoints will be included among the alternatives. The alternatives put forward all appearing to be of equal status. Thus learner cannot work out which option is correct from the context [29].

Observation of research shows that students prefer enthusiastic in the learning. Thus, the lesson becomes more interactive and student-centered as students are actively involved in their learning. At the beginning of new lesson, the concept cartoon appears to be an effective stimulus for student to create their idea and focus on learning. They must choose one of the characters that true statements and discuss what they think about the ideas given. Then explain why they have those ideas. After that, they present their argument, and other students provide their argument, so it can motivate the student in discussing and increasing student's understanding. During the discussion, they look comfortable to express their idea. According to interview result, using concept cartoon in learning makes student enjoy their conversation, enhance their motivation, and probe student's misconception. These findings are parallel to the ideas of Gafoor [5] who state that concept cartoon makes student increasing their understand, attention and interest, improve motivation toward learning.

The first, concept cartoon shows the definition of buffer solutions. The buffer solution is a solution that resists changes in pH when small amount of an acid or alkali is added. The question of concept cartoons "How is pH in mouth if food that we eat contains excessive acid? The correct answers are pH of solution <7; solution will be more acid because of the amount of $[H^+]$ increase. The average pH range for saliva is considered to 5.6 to 7.9 according to the International Journal of Drug Testing. If excessive acid is added, then the amount of $[H^+]$ will increase. In this research, there are 46,9% students of A, 44,1% students of B, and 40,6% students of C detected misconception. Some students assume that pH of solution <7, solution will be acid because the normal pH in the mouth is acid and buffer solution can resist changes in pH when acid is added in any amount. This misconception parallel to the research of Muttlu & Sesen [27]. Other students assume that pH of solution <7, solution will be acid because substances which added is acid then the solution become acid and pH value < 7.

The second, concept cartoon shows how do buffer solutions work. A pH buffer contains molecules that can bind to other molecules in an acid or a base to neutralize them. The question of concept cartoon is "how does buffer work maintaining a constant blood pH when small amount of base added?". The correct answer: The components of buffer in blood are carbonic acid (H_2CO_3) and hydrogen carbonate ion (HCO_3^-). When small amount of bases added, base will react with carbonic acid (H_2CO_3) produce hydrogen carbonate ions (HCO_3^-) and water. Carbonic acid binds to the loose hydroxide molecules. In this manner, the hydroxide ion (OH^-) is removed from blood, preventing the pH of blood become base. Furthermore, adding base will shift the equilibrium to the left. Thus, the amount of H^+ is constant, and pH is constant. In this research, there are 40,6% students of A, 36,3% students of B, and 34,4% students of C detected misconception. Many students assume that base will react with acid to produce salt and water. pH of solution = 7, solution will be neutral. Therefore, buffer can maintain a constant blood pH. Adding base will not cause shift the equilibrium. These findings are parallel to the research of Orgill & Sutherland[11]; Mutlu & Sesen, [27].

The third, concept cartoon shows calculates the pH of buffer solution. The concept cartoon question "Whether the volume of solution affects the capacity of buffer solution?". The correct answer is the volume of solution affects the capacity of buffer solution because the capacity of buffer solution depends on mole of weak acid/base and its conjugate. The mole can be determined from concentration x volume. Thus, if amount of volume is large, the capacity of buffer solution is also large. In this research, there are 34,3% students of A, 32,4% students of B, and 34,3% students of C detected misconception. Some students assume that volume of solution and the strength of its components acid and base affected the capacity of buffer solution. These findings are parallel to the research of Orgill & Sutherland [11].

5. Conclusions

The findings of this research show that concept cartoons are practical tools to detect student's misconception in buffer solution. Many student's misconceptions have detected by concept cartoon. It has been remedied after investigation, discussion and interview about concept. Misconceptions find in this research are similar to the misconception in the literature. Another finding research shows that concept cartoons can be used in science learning especially in chemistry because it makes student active in learning, increases motivate to discuss and inquire their knowledge. It is suggested that concept cartoon should be used before learning to identify student's misconception and after learning to remove misconception. In another hand, each should give their ideas about the question of concept cartoon then brings ideas for discussion with friend.

6. References

[1] Osborne J and Collins S 2000 *Pupils ' and parents ' views of the school science curriculum* (London: King's College London)
[2] Demircioglu G, Ayas A and Demircioglu H 2005 Conceptual change achieved through a new teaching program on acids and bases *J. Chem. Educ. Res. Pract.* **6** 36–51
[3] Sirhan G 2007 Learning Difficulties in Chemistry: An Overview *J. Turkish Sci. Educ.* **4** 2–20
[4] Cardellini L 2012 Chemistry : Why the subject is difficult ? *Educ. quím* 1–6
[5] Gafoor K A and Shilna V 2013 Role of concept cartoons in chemistry learning *National Seminar On Learning Science by Doing - Sciencing* (Kannur. Kerala, India: PKM College of Education, Madampam) pp 1–9
[6] Gafoor K A and Shilna V 2013 Perceived difficulty of chemistry units in Std IX for students in Kerala stream Calls for further innovations Innovations in pedagogy and curriculum : from theory to practice Perceived difficulty of chemistry units in Std IX for students in Kerala stream *national seminar on Innovations in pedagogy and curriculum: from theory to practice* (Kerala) pp 1–7
[7] Johnstone 2000 Teaching of chemistry-logical or psychological? *Chem. Educ. Res. Pract. Eur.* **1** 9–15
[8] Mondal & Chakraborty A 2013 *Misconceptions in chemistry its identification and remedial measures* (deutschland/Germany: LAMBERT Academic Publishing)
[9] Barke H D, Hazari A and Yitbarek S 2009 *Misconceptions in chemistry: Addressing perceptions in chemical education* (Berlin, Heidelberg: Springer - Verlag)
[10] Basman T, Arifin A and Muris M 2016 The development of discovery-inquiry learning model to erduce the science misconceptions of junior high school students *Int. J. Environ. Sci. Educ.* **11** 5676–86
[11] Orgill M and Sutherland A 2008 Undergraduate chemistry students' perceptions of and misconceptions about buffers and buffer problems *Chem. Educ. Res. Pract.* **9** 131
[12] Kousathana M and Tsaparlis G 2002 Students errors in solving numerical chemical-equilibrium problems *Chem. Educ. Res. Pract. Eur.* **3** 5–17
[13] Pekmez E S 2010 Using analogies to prevent misconceptions about chemical equilibrium *Asia Pacific Forum Sci. Learn. Teaching,* **11** 1–35
[14] Şendur G, Toprak M and Pekmez E S 2010 Analyzing of students ' misconceptions about chemical equilibrium *International Conference on New Trends in Education and Their Implications* (Antalya-Turkey) pp 1–6
[15] Sendur G, Toprak M and Pekmez E S 2011 How can secondary school students perceive chemical equilibrium? *e-Journal New World Sci. Acad.* **6** 1512–31
[16] Al-balushi S M, Ambusaidi A K, Al-shuaili A H and Taylor N 2012 Omani twelfth grade students ' most common misconceptions in chemistry *Sci. Educ. Int.* **23** 221–40
[17] Demircioglu G, Demircioglu H and Yadigaroglu M 2013 An investigation of chemistry student teachers ' understanding *Int. J. New Trends Educ. Their Implic.* **4** 185–92
[18] Karpudewan M, Treagust D F, Mocerino M, Won M and Chandrasegaran A L 2015 Investigating high school students ' understanding of chemical equilibrium concepts *Int. J. Environ. Sci. Educ.* **10** 845–63
[19] Halstead S E 2009 *A critical analysis of research done to identify conceptual difficulties in acid-base chemistry* (University of KwaZulu-Natal, Pietermaritzburg)

[20] Muchtar Z and Harizal 2012 Analyzing of students misconceptions on acid-base chemistry at senior high school in Medan *J. Educ. Pract.* **3** 65–74
[21] Bayrak B K 2013 Using two-tier test to identify primary students ' conceptual understanding and alternative conceptions in acid base *Mevlana Int. J. Educ.* **3** 19–26
[22] Wan Y 2014 *Assesing college students' understanding of acid base* (Clemson University)
[23] Damanhuri, Muhd I M, Treagust D F, Won M and Chandrasegaran A L 2016 High school students ' understanding of acid-base concepts : an ongoing challenge for teachers *Int. J. Environ. Sci. Educ.* **11** 9–27
[24] Labobar H, Setyosari P, Degeng I N S and Dasna I W 2017 The effect of cognitive conflict strategy to chemical conceptual change *Int. J. Sci. Res.* **6** 2350–2
[25] Cyril N 2015 *An investigation of malaysian secondary school students ' mental models of acid-base chemistry* (University of Waikato)
[26] Sesen B A and Tarhan L 2011 Active-learning versus teacher- centered instruction for learning acids and bases *Res. Sci. Technol. Educ.* **29** 205–26
[27] Mutlu A and Şeşen B A 2016 Evaluating of preservice science teachers' understanding oF general chemistry concepts by using two tier diagnostic test *J. Balt. Sci. Educ.* **15** 79–96
[28] Webb, P., Williams, Y., Meiring L 2008 Concept cartoons and writing frames: developing argumentation in South African science classrooms? *African J. Res. Math. Sci. Technol. Educ.* **12** 5–17
[29] Ekici F, Ekici E and Aydin F 2007 Utility of concept cartoons in diagnosing and overcoming misconceptions related to photosynthesis *Int. J. Environ. Sci. Educ.* **2** 111–24
[30] Yitbarek S 2011 Chemical reaction: diagnosis and toward remedy of misconceptions *African J. Chem. Educ.* **1** 10–28
[31] Oskay 2016 The effect of school suspension on academic achievement *J. Educ. Instr. Stud. World* **6** 49–54
[32] Kabapinar F 2005 Effectiveness of teaching via concept cartoons from the point of view of constructivist approach *Educ. Sci.* **5** 135–46
[33] Yong C L and Kee Z 2017 *Overcoming students' misconceptions in science* (Penang, Malaysia: Springer)
[34] Salirawati D 2011 Chemical equilibrium misconception *Penelit. dan Eval. Pendidik.* **15** 232–49
[35] Ingec S K 2008 Use of concept cartoons as an assessment tool in physics education *Educ. Rev.* **5** 47–54
[36] Taslidere E 2013 Effect of conceptual change oriented instruction on students ' conceptual understanding and decreasing their misconceptions in DC electric circuits *Sci. Res.* **4** 273–82
[37] Abraham M R, Grzybowski E B, Renner J W and Marek E A 1992 Understandings and misunderstandings of eighth graders of five chemistry concepts found in textbooks *Res. Sci. Teach.* **29** 105–20

Acknowledgments
Thank you to Mr. Dr. Mohammad Masykuri, M.Si as chairman of the Department of Science Education has given the author permission to conduct this research and all of the people that support me in doing this research.

IOP Conf. Series: Journal of Physics: Conf. Series **1022** (2018) 012037 doi:10.1088/1742-6596/1022/1/012037

The use of an android–based-game in the team assisted individualization to improve students' creativity and cognitive achievement in chemistry

K H Sugiyarto[1] ,J Ikhsan[1*] and I R Lukman[1]

[1]Departement of Chemistry Education, Graduate School of Yogyakarta State University, Karangmalang, Yogyakarta, 55281, Indonesia

[*]Coresponding email : jikhsan@uny.ac.id

Abstract. The use of information and communication technology (ICT) in learning process resulted in positive impact to students' output. This research investigated the difference of improvement of students' creativity and cognitive achievement due to the use of android-based games on Chemistry Nomenclature in learning method of team-assisted individualization (TAI) into the improvement of students' creativity and cognitive achievement. This was an quasi experiment research with non-equivalent pretest-posttest control group design involving 2 groups of students of X grade of a senior high school in Yogyakarta, Indonesia, SMAN 1 Seyegan, Sleman. The groups were experiment and control which were chosen randomly, involving 32 students in each group. The difference of learning model in the two groups were the use of android-based games within learning model of TAI in the experiment group, but it was only the use of TAI model in control group. The android-based games were developed and validated previously in this investigation, and were excellent in quality for the use in Chemistry learning process, and were reported separately. The data of both students' creativity and cognitive achievement were measured before and after learning process. Data of students' creativity were collected with the instruments of questionnaire and observation sheets, and the data of cognitive achievement were collected with a set of test. Statistical analysis of MANOVA was used to analyze data to measure the difference of the improvement of students' creativity and cognitive achievement between experiment and control groups. The results showed that the improvement of creativity and cognitive achievement of students in the experiment group was higher significantly than that in control group.

1. Introduction

Lifelong learning can occur at anywhere and anytime. Learning process can be conducted and obtained from the formal environment as usually done, but it can be from informal situation. Infrastructure and facilities to support lifelong learning for students should be provided by combining formal and informal learning environment to result in optimum outcome [1]. Experience in the formal learning environment has been gained by all students in the school in Indonesia through face-to-face process of teaching and learning. While in the informal learning, learners can gain learning experience from various facilities, including mobile devices to encourage and to support learning process and activities. Mobile devices offer the possibility of spontaneous learning with unstructured situations [1]. In addition, mobile technologies should be viewed as media that can be utilized in an informal learning activity and can be elaborated in learning scenarios [2]. Mobile devices offer new opportunities to create a flexible, diverse and interactive learning environment, and enable lifelong learning possible by anyone from anywhere. The advantages of the use of mobile devices as learning tools should encourage educators to develop mobile based instructional media. It is due to large number of mobile phone users in Indonesia, including senior high school students. The media can be used as instructional media for independent learning activities to improve cognitive learning achievement and creativity of students. Sawyer (2006) argued that creativity is an important factor in both formal and informal learning [3]. Peters [4] stated:

"Much of the literature concerning education and the creativity and the prototype of economic development and the need for building forms of cultural, social, and public entrepreneurship. The problem is that beyond the formulation of concepts such as' creative cities', 'creative cities', and' creative classes', have been made of creativity in schools apart from frosting instrumental version of creativity or simply regarding 'education, training, And skills' as one aspect of the creative economy".

From the observation in some schools in Yogyakarta, student' creativity need to be improved. Improvement of creativity is important to provide creative learning for students [4], and to enhance better learning outcome. Cheung (2012) stated that there were two issues that must be considered to achieve creativity in education, (1) teachers' creative meaning and (2) practices of teachers' creativity in the classroom [5]. Simonton (2012) stated that to convey creative learning in the classroom, a teacher should be able to make students "surprised", such as wearing T-shirts related to learning materials to be delivered [6]. An alternative solution to improve students' creativity is learning a topic with assistance of an android-based media. Android instructional media can stimulate students to do exercises enjoyably. With the android-based media, students are challenged to play and to learn the game in several levels as students learning speed.

In this global era, android games are widely available in market and can be downloaded freely. The android-based instructional media used in this research is also available in Play Store, with keyword of search "Chemondro". The media were developed based on Indonesian latest curriculum, "2013 Curriculum". The quality of media was considered to be excellent for the use in Chemistry learning, based on the reviews of three experts; expert of learning media, of learning subject materials, and chemistry teachers, as well as based on the responses of students as a step of the implementation in the research and development model. The screen capture of the example of some menus in the android-based-game implemented in this research was given in Figure 1.

Figure 1. Screen Capture of Android-Based-Games Developed and Used

The media have characteristics as follows: (1) the instructional media is an application with the file extension of apk. The media can be played by Android-OS devices, (2) the content of media is Nomenclatures of Chemical Compounds for grade X of senior high schools (SMA), (3) the media can be used both in a face-to-face classroom learning and in a non-face-to-face chemistry learning activities, (4) the media are able to visualize learning materials. Thus, the use of android instructional media that was developed can make students gain better comprehension of chemical compound nomenclature with fun and easy. This research investigated the effect of the use of android instructional media to the improvement of high school students' creativity and learning achievement.

2. Method

This research is a quasi-experimental design, with non-equivalent control-experiment group design. The population was Grade X senior high school students (SMA) in Yogyakarta, with the sample consisted of 64 students from 2 classes of Grade X, SMA N 1 Seyegan, control class and experiment classes. Those 2 classes as the samples were determined by random sampling. Learning was conducted following the lesson plan on Compound Nomenclature that was written based on the 2013 Curriculum, and was integrated into learning model of the TAI. Learning activities in both experiment and control groups were carried out with the same number of meetings, materials, learning methods and models, as well as evaluation. The different treatment in the 2 classes and research design are shown in Table 1.

Table 1. Non Equivalent Control-Experiment Group Design

Variable	Students Group	Pre-test	Treatment		Post-test
			Team Assisted Individualization	Android-based instructional media	
Creativity	Experiment	Yes	Yes	Yes	Yes
	Control	Yes	Yes	No	Yes
Cognitive	Experiment	Yes	Yes	Yes	Yes
	Control	Yes	Yes	No	Yes

2.1. Data Collection Instrument

The instruments used in this study were (1) questionnaires of creativity for students answered observation sheet to assess students' creativity during the face-to-face learning in the classroom, (2) a learning outcome test with a type of reasonable multiple-choice questions to measure students' cognitive achievement, and (3) a teacher's observation sheet to confirm the process of learning as on the lesson plan.

Learning outcomes and creativity of students from both experiment and control groups were accessed twice; e.g. through pre-test and post-test. Improvement of learning creativity and cognitive achievement was analysed using normality test of gain values. The normality test of the gain of each variable, students' cognitive achievement or creativity can be calculated using the following equation [7].

$$g = \frac{S_f - S_i}{100 - S_i}$$

Note: Sf = *final test*
Si = *initial test*
g = *gain*

The interpretation of the criteria of the gain of students' learning creativity and cognitive achievement was based on the gain value with the criteria was presented in Table 2 [7].

Table 2. Gain Value Criteria

Gain Value	Criteria
G ≥ 0.70	High
0.30 ≤ g < 0.70	Medium
G < 0.30	Low

The difference of learning creativity and cognitive achievement of students from control and experiment groups; because of the use of android-based games on Chemistry Nomenclature which was integrated into the learning method of TAI was statistically analysed using Multivariate Analysis of the Variance (MANOVA) using SPSS V.23. The test of normally and homogeneity toward the data of creativity and learning achievement was completed before the analysis of MANOVA was applied. The normality test was performed using Mahalonobis value using SPSS V.23. The Mahalonobis value obtained was then compared with *chi (q)* value. Data was considered normal at significance level of 0.05 when 40-60% of the data in each group has Mahalonobis values less than *chi (q)* values.

The homogeneity test conducted was a homogeneity test of dependent variable variance-covariance simultaneously. Data for homogeneity test was the data of creativity increase (gain) and cognitive learning outcomes of the students. The variance-covariance homogeneity test for the data was performed by Box's M test by use of the computer program, SPSS V.23. The matrix of variance-covariance was considered homogeneous if the *sig.* value > α. MANOVA analysis was performed to answer the research statistical hypotheses as following:

H_o: There is not any significant difference in the improvement of creativity (Y1) and learning outcomes (Y2) between students who learnt with Android-based instructional media using TAI learning model and those with LKS (Student Worksheet) using TAI learning model.

H_a: There is significant difference in the improvement of creativity (Y1) and learning outcomes (Y2) students who learn with Android-based instructional media using TAI learning model and those with LKS using TAI learning model.

3. Result and Discussion

The android-based game that was good in quality based on the reviews of media expert, material expert, chemistry teachers and students especially on the aspects of learning, materials, language, audio visual and software engineering was used in Chemistry learning. The Learning process was performed by utilizing android-based game in the integrated TAI, and the effect of learning was compared to that of learning without the use of android-based game. The data of learning achievement from both pre-test and post-test are given in Table 3.

Table 3. Students' Learning Achievement (Pre-test and Post-test)

Class	Total Student	Pretest		Post-test		gain	category
		Average Value	completeness (%)	Average value	completeness (%)		
Experimental	32	6.125	0	87.06	100	0.862	High
Control	32	5.156	0	83.25	96.8	0.82	High

Table 3 shows an increase of the average value of pre-test and post-test of students from both experimental and control classes; from 6.125 to 87.06 for the experimental class, and from 5.156 to 83.25 for the control class. It means that the gain of students' achievement in the experimental class which utilizes android-based game was higher than that in the control class. Students' learning by utilizing the android-based game integrated to TAI model gained significant results. This is because the instructional media can stimulate students' motivation and assist students to understand learning materials. This was stated by Sakat *et.al.* (2012) that the use of technology-based media can improve cognitive learning outcomes [8]. Media can also facilitate teachers to better deliver learning materials. The same thing was also reported by Jabbour (2014) that the use of mobile learning can result in the increase of students' outcomes [9].

The field test as an important step in the research and development model was also performed to check the improvement of students' learning creativity and cognitive achievement in both experimental and control classes before and after learning process. The data of creativity improvement were collected both by observation during learning process and by a set of questionnaire that was validated constructely and logically by expert. Results of the observation were qualitative data in the form of average increase from the whole students. Data of the questionnaires on students' learning creativity were presented in Table 4.

Table 4. Data of Students' Creativity

Class	Total Student	Pretest		Post-test		gain	category
		Average Value	completeness (%)	Average value	completeness (%)		
Experimental	32	122.34	74.90	133.1	88.72	0.55	Medium
Control	32	103.41	68.94	124.03	82.69	0.44	Medium

Table 4 shows an increase of creativity of students from both classes. In the experimental class there is a gain of 0.55 which is considered as medium gain category, while in the control class there was a gain of 0.44 which was medium gain category too. Those data show that the experimental class has higher students' creativity improvement than the control class does. The use of android-based game using the learning model of TAI in the learning process resulted in the increase of cognitive learning outcomes and a better creativity than without using the android-based game. This is in line with the finding by Ulfa *et.al.* (2017) that there was an increase in creativity and cognitive learning outcomes between students who using the android-based chemistry instructional media integrated into Learning Together and those who use Learning together without the android-based instructional media [10].

3.1. Analysis of Prerequisite Test

Analysis of prerequisite tests included data normality and population homogeneity. Normality test was performed on N-gain from data of the improvement of students' learning creativity and cognitive achievement of students from both the experimental and control classes. The Mahalonobis value resulted from the SPSS program in each class can be seen in Table 5

Table 5. Analysis of Multivariate Normality

Class	Percentage $d_i^2 \leq \chi^2_{p,0.5}$	Conclusion
Experimental	46.875.%	Normally Distributed
Control	53.125%	Normally Distributed

Based on the above listed in the Table 5, it can be noticed that Mahalonobis value was ≤ chi squared value or $d_i^2 \leq \chi^2_{p,0.5}$ (1,38) was in the range of 40 - 60%, which can be concluded that data were multivariate normally distributed.

Homogeneity analysis was done using SPSS program. Results of the homogeneity test can be seen in Table 6.

Table 6. Analysis of Multivariate Normality

Effect	Significance	Conclusion
Box's M	0.298	Homogeneous Data

The Box's M test obtained a significant value greater than 0.05, so it can be concluded that the variance-covariance matrix of the population is the same or homogeneous.

3.2. Hypothesis Testing

After the normality and homogeneity prerequisite tests were done, the hypothesis. This hypothesis analysis was performed, to study the difference in creativity and learning outcomes between students who learnt with Android-based instructional media using TAI learning model with those with LKS (print out students' sheet) using TAI learning model. The conclusion whether the hypothesis was significant or not was obtained by interpreting the value of Manova significance from the calculation obtained from SPSS program. If the value of significance is <0.05 ($p < 0.05$), then Ho is rejected and Ha is accepted. Hypothesis testing was done using N-gain data obtained from data of the increase of students' creativity and cognitive learning outcomes with SPSS. The analysis results of Manova test are presented in Table 7.

Table 7. Manova Analysis Results

Effect	Significance	Explanation	Conclusion
Hotelling's Trace	0.039	Ho is rejected	There is a difference in the increase of creativity and cognitive learning outcomes

TAI learning model was able to support and to enrich face-to-face learning in the classroom. It was able to transform learning from teacher-centered to student-centered. Thus, it brings about the improvement of creativity and cognitive achievement of students. The media included learning materials of introduction game or basic learning materials which should be learnt and mastered by students first. The game required students to learn the introduction games and basic materials before continuing to higher level of learning materials and game. On the last game, students were facilitated to play games with competition among the other students. It made the students more enthusiastic and challenged. Based on this study, learning process was expected to be more varieties and enjoyable, by combining TAI model with android-based-game. Therefore, students were expected can understand Chemistry materials better.

4. Conclusion

The use of android based media on the Nomenclature of Compound in TAI learning model can bring about higher students' gain of creativity and learning achievement compared to that who used worksheet (LKS) in TAI learning model.

5. References

[1] Jong T D, Specht M and Kopper R 2008 Contextualised media for learning *Educational Technology & Society* pp 41-53
[2] Hoppe U 2007 *How can we integrate mobile devices with broader educational scenarios?* (Nottingham: University of Nottingham)
[3] Sawyer R K 2006 Explaining creativity *The Science of Human Innovation* (Oxford: Oxford University Press)
[4] Tanggaard L 2014 A situated model of creative learning *Eur. Edu. Res. J.* **13** 107-116
[5] Cheung R H 2012 Teaching for creativity: Examining the beliefs of early childhood teachers and their influence on teaching practices *Aust. J. of Earl. Chil.* **37** 43-52

IOP Conf. Series: Journal of Physics: Conf. Series **1022** (2018) 012037 doi:10.1088/1742-6596/1022/1/012037

[6] Simonton D K 2012 Teaching creativity: Current findings, trends, and controversies in the psychology of creativity *Teaching of Psychology* **39(**3) 217-22

[7] Hake R R 1998 Interactive-engagement versus traditional methods: A six-thousand-student survey of mechanics test data for introductory physics courses *Amer. J. of Phy. Res.* **66(**1) 64-74

[8] Sakat A A, Zin M Z, Muhamad R, Ahmad A, Ahmad N A and Kasmo M A 2012 Educational technology media method in teaching and learning progress *Amer. J. of Appl. Sci.* **9(**6) 874-78

[9] Jabbour K K 2014 *An* analysis of the effect of mobile learning on Lebanese higher education *Infor. in Edu.* **13** 1-15

[10] Ulfa A M, Sugiyarto K H and Ikhsan J 2017 *The effect of the use of android-based aplication in learning together to improve students' academic performance* (AIP Confrence Proceedings 1847:1)

[11] Borg W R and M D Gall 2007 *Educational Research : An Introduction Eighth Edition* (New York: Longman) p 590

[12] Thiagarajan, Sammel D S and Semmel M I 1974 *Instructional development for training teacher of exceptional children* (Bloomington: Indiana University)

Chemistry teachers' understanding of science process skills in relation of science process skills assessment in chemistry learning

N Hikmah[1*], S Yamtinah[1], Ashadi[1] and N Y Indriyanti[1]

[1]Science Education Department, Sebelas Maret University, Indonesia

[*]Corresponding email: nailihikmah@gmail.com

Abstract. A Science process skill (SPS) is a fundamental scientific method to achieve good knowledge. SPS can be categorized into two levels: basic and integrated. Learning SPS helps children to grow as individuals who can access knowledge and know how to acquire it. The primary outcomes of the scientific process in learning are the application of scientific processes, scientific reasoning, accurate knowledge, problem-solving, and understanding of the relationship between science, technology, society, and everyday life's events. Teachers' understanding of SPS is central to the application of SPS in a learning process. Following this point, this study aims to investigate the high school chemistry teachers' understanding of SPS pertains to their assessment of SPS in chemistry learning. The understanding of SPS is measured from the conceptual and operational aspects of SPS. This research uses qualitative analysis method, and the sample consists of eight chemistry teachers selected by random sampling. A semi-structured interview procedure is used to collect the data. The result of the analysis shows that teachers' conceptual and operational understanding of SPS is weak. It affects the accuracy and appropriateness of the teacher's selection of SPS assessment in chemistry learning.

Key words: Science process skills, science process skills assessment, chemistry teacher

1. Introduction

The purpose of science education is to help students to understand scientific knowledge and build students' abilities using scientific approach [1]. SPS is an important aspect of teaching to achieve good knowledge. SPS skill in the learning process is not only served as the basis of scientific methods but also learning about the characteristics of knowledge. SPS can be categorized into two levels, which are basic and integrated levels. Basic science process skills consist of observing, classifying, measuring, using numbers, using space and time relationship, inferring, predicting, and communicating. Integrated science process skills consist of identifying variables, formulating hypotheses, defining variables operationally, experimenting, and interpreting data and drawing conclusions[2]. Changing hypotheses, building and interpreting data and images will help children to grow as individuals who are capable not only to access knowledge but also determine it [3]. Individuals with higher mental development can describe, portray and solve both individual and social problems. Also, SPS has a great influence in

IOP Conf. Series: Journal of Physics: Conf. Series **1022** (2018) 012038 doi:10.1088/1742-6596/1022/1/012038

developing ones higher mental processes including critical and creative thinking and they can transfer this ability to other disciplines [4].Teachers are expected to have a good understanding of SPS so they can transfer it to their students [5] and they are expected to be able to measure SPS from students so that students' achievement in mastering SPS can be appropriately evaluated. From the results of this evaluation, teachers are expected to develop SPS learning method to their students.

2. Methodology

The methodology of this research is qualitative analysis. The sample in this study consists of 8 chemistry teachers, with the detail of four chemistry teachers with 1-5 years of teaching experience and four chemistry teachers with teaching experience of more than five years. Data is collected by semi-structured interviews with six questions present. From the interview, data is classified based on the suitability of the answers. Data is presented in a table form for each question. Data analysis is conducted by connecting the interview results.

3. Findings

The Results from semi-structured interviews are shown in the following tables 1-6.

Question 1: What do you know about SPS? Explain your answer.

Table 1. Chemistry Teachers' Opinion about SPS

Category	Frequencies	Chemistry teachers' ideas
Scientific Method	4	SPS is a skill in using scientific methods
Theoretical understanding	2	SPS is student's ability to understand science
learning approaches	2	SPS is a learning approach which is designed for students to discover facts and concepts

Question 2: How do you teach SPS to students in chemistry learning?

Table 2. Chemistry Teachers' Ideas About Teaching SPS to Students in Chemistry Learning.

Frequencies	Chemistry Teachers' Ideas
7	Practical work/ Laboratory performance
1	Science project
2	Problem-based learning
1	Discussion
1	Visiting industry
1	Lecture

Question 3: What SPS indicators that are trained to students?

Table 3. Chemistry Teachers' Ideas About SPS Indicators That are Trained to Students

Category	Frequencies	Chemistry Teachers' Ideas
Scientific Method	5	Skills in practical work in laboratory such as observing, formulating problems, hypothesizing, designing experiments, analyzing data, drawing conclusions, and communicating
Skills using equipment	1	Skill in utilizing laboratory equipment
Concept	1	Understanding of scientific concepts in chemistry
Learning activities	1	Students' active performance in learning process

Question 4: In chemistry, how do you assess student's SPS?

Table 4. Chemistry Teachers' Ideas About How to Assess Student's SPS

Category	Frequencies	Chemistry Teachers' Ideas
Observation of practicum	4	Observations from laboratory performance
Paper and Pencil Test	2	Written test about conceptual understanding
Observation class	2	Observations during learning in the classroom

Question 5: What SPS indicators are assessed in chemistry?

Table 5. Chemistry Teachers' Ideas About Indicators are Assessed in Chemistry

Category	Frequencies	Chemistry Teachers' Ideas
Scientific Method	5	Skills in laboratory performance such as observing, formulating problems, hypothesizing, designing experiments, analyzing data, drawing conclusions, and communicating
Skills using laboratory equipment	1	Skill in utilizing laboratory equipment
Concept	2	Understanding of scientific concepts in chemistry

Question 6: What is the difficulty in assessing SPS?

Table 6. Chemistry Teachers' Ideas About The Difficulty in Assessing SPS

Category	Frequencies	Chemistry Teachers' Ideas
Instrument of assessment	5	There is no special assessment instrument to assess all SPS indicators, so the assessment is subjective
Students ability	2	Students are not accustomed to doing problem-solving, so SPS is difficult to perform
Assessment method	1	Teachers can't asses students' SPS one by one accurately because of the weakness of the observation method

4. Discussion and Conclusion

In table 1. It shows that four chemistry teachers have an understanding that SPS is a skill in using scientific methods. Meanwhile, two chemistry teachers understand SPS as a concept, and the rest two chemistry teachers understand SPS as a learning approach. This fact shows that chemistry teachers still have a weak understanding of SPS.

In table 2. It shows that the majority of chemistry teachers believe that SPS can be trained to students through practice in laboratory, science projects and problem-based learning. However, chemistry teachers who argue that discussions, industry visits, and lecture methods can be used as SPS learning methods. This argument doesn't follow the definition of SPS. SPS cannot be trained through discussions, industry visits, and lecture method.

In table 3. It shows that most of the chemistry teachers train SPS to students as a scientific skill that included in laboratory practice. However, some teachers who instruct SPS less appropriately. They train SPS by enhancing student's ability in using equipment in the laboratory, while in fact, SPS is not only about the skills of using equipment, they also teach SPS as mere conceptual understanding and consider that student's actual performance during learning process as a part of SPS.

In Table 4. It shows that four chemistry teachers assess SPS through observations during practical work in the laboratory, two chemistry teachers using written test, and two teachers assess SPS through observations in the classroom. Those form of assessments is in line with their understanding of SPS. Observation during practical work in the laboratory is appropriate to assess SPS, but the written test of concept comprehension is not. The written test cannot be used to assess SPS if the question indicators do not refer to SPS. Meanwhile, classroom observation is not suitable to assess SPS.

In table 5. It shows that chemistry teachers assess SPS based on the learning activities that are employed in training SPS. Majority of chemistry teachers assess SPS indicators as integrated skills in the laboratory. One chemistry teacher assesses SPS only through skills in using laboratory equipment, and two chemistry teachers are inappropriately assessing SPS by assessing students' conceptual understanding and attitudes during the learning process.

In table 6; It shows that chemistry teachers assess SPS based on subjective observation because there is no particular instrument is available to be used. The observation instrument can't be used to assess SPS accurately, so it is not the best and suitable one to assess SPS. Meanwhile, there is teacher's misconception in understanding SPS, so their assessment of the students' SPS is not legit.

From the results of the chemistry teacher's answers to the question of SPS and its assessment, the majority of the teachers do not fully understand SPS. They can adequately define SPS but cannot mention skill indicators in the SPS. Some chemistry teachers already understand the concept of SPS, but they lack in SPS operational practices. Most of the chemistry teachers train SPS to students through practical work in the laboratory. SPS is considered taking place in a science laboratory and the practical work only about using apparatus and following instructions [6], and many of the skills associated with experimental investigations are rarely taught explicitly [7]. This happens because teachers assume that students can obtain SPS only through experience doing practical work in the laboratory. The assumption facilitates the acquisition of the operational aspects of the SPS but does not promote a conceptual understanding of the accuracy of the scientific inquiry process involved in the investigation. Students may be able to observe procedurally, but they do not have a clear understanding of the purpose of observation in scientific investigation. Thus, creativity and originality, which are the characteristics of scientific inquiry, will also be difficult to develop with the limited acquisition of conceptual understanding required in SPS [8]. Chemistry teachers' should have a strong understanding of SPS and demonstrate competency in SPS both operational and conceptual so SPS can be effectively and meaningfully taught to students[9].

Chemistry teachers understand that observation can be used as an SPS method of assessment, but in practice, teachers observe without using appropriate assessment instruments, so the science process skills are not measured accurately and adequately. Besides, the teachers who assess SPS through the written test but only measure the concepts and theories instead of SPS. There are also teachers who measure SPS only from students' accuracy in using tools or instruments. Teachers' difficulty in assessing SPS by observation method lies in large classes with 30 students. Teachers will find it difficult to accurately assess the SPS of each student. These conditions prompted the efficient and objective SPS measurement. One of the possible solutions is using written test [10]. The test subject may take written test even though a tool is often required to complete the test subject. The process of science is inseparable in practice from the conceptual understanding involved in learning and the application of science [11].

One of the role the chemistry teacher as a science teacher is to facilitate and help students to understand the content of science knowledge. One of the contents of science knowledge is the skill of science process. Chemical teachers are required to have an in-depth understanding of the science process's skills as well as how to teach the science process. Content and pedagogical knowledge should be integrated to create new knowledge [12]. It can be defined as pedagogical content knowledge (PCK). PCK is a special knowledge that teachers have about how to teach certain content to learners through a strategy that leads to understanding. PCK is the ability of teachers to understand and apply how to help a group of students understand the specific subject by using some instructional, representational, and assessment strategies by working within the contextual, cultural, and social constraints in environmental learning [13].

PCK is described as the knowledge that used to transform the content of subjectinto a form thateasier understood by students [14][15][16][17][18]. In this case, PCK development involves a shift in the teachers' understanding of the ability to understand the subject matter for themselves, and can explain the subject matter in new ways, rearrange every part of the material in activity and emotion, in metaphors and exercises, and in the example and demonstrations, so that it can be understood by students [17].

Seven categories of knowledge according to PCK that teachers should have to manage to learn effectively. The seven categories are [19]:

Table 7. Seven Components of PCK.

Components	Elements
Knowledge of science	Science content, scientific practice, the nature of science, scientific process
Knowledge of goals	Scientific literacy, real-life application, integrated understanding
Knowledge of students	Different levels, needs, interests, prior knowledge, ability, learning difficulties, misconceptions
Knowledge of curriculum organization	State and local standards, state and local standardized tests, making connections between lessons and units, organizing lessons in specific order, making decisions about what to teach, flexible design
Knowledge of teaching	Various teaching methods, use of motivating activities, ability to select effective activities.
Knowledge of assessment	Formal and informal ways of assessment, skills for students' discussion and questioning, immediate feedback
Knowledge of resources	Materials, activities, multimedia, local facilities, laboratory technology, science magazines.

Based on PCK exposure, the following arethe PCKchemistry teacher descriptionbased on seven components according to pedagogic competence and professional competence.

4.1. Knowledge of Science

All respondents have a chemical education background. However, not all chemistry teachers have an in-depth knowledge of SPS. 4 chemistry teachers are not right in understanding the SPS. They understand the SPS as the ability of students to use reason and creativity to understand the science and approaches to learning that students can find facts and concepts.

4.2. Knowledge of Goals
The teacher's knowledge of learning objectives is closely related to how the teacher does the learning and evaluates them. Achievement of learning objectives can be known from the evaluation of learning. One of the goals of SPS learning is students can solve problems in daily life. From the eight chemistry teachers, most ofthem less understandthe SPS. This has an impact on SPS learning. Some respondents put incorrect answers regarding SPS learning.

4.3. Knowledge of Students
Knowledge of students includes their interest in chemistry learning. Most of them think chemistry is hard to learn. In this case, the teacher must plan the lesson so that the material can be easily understood by the students. four chemistry teachers are less precise in planning SPS lessons. this can be seen from the form of learning they are doing: Problem-based learning, Discussion, Visiting industry, and Lesson with the lecture.

4.4. Knowledge of cCurriculum Organization
The curriculum includes teacher insight into the high school curriculum in Indonesia. Indonesia uses a scientific approach to science learning. 8 respondents know the curriculum in Indonesia is using the scientific approach. On the other hand, two teachers who argue that the scientific approach is the same thing as the science process skills. This suggests that teachers should better understand the scientific approach and science process skills.

4.5. Knowledge of Teaching
Teachers' understanding of SPS influences how teachers teach SPS to students. The SPS teaching form affects the classroom management conducted by the teacher. Teachers have their form of teaching based on their understanding of the SPS. When teachers think SPS is a skill in using scientific methods, they use the practical work/laboratory performance and science project to teach SPS to students. Teachers who argue that SPS is the ability of students to use reason and creativity to understand science, use Problem-based learning models and discussion methods to teach the KPS.

4.6. Knowledge of Assessment
Assessment has an important role to evaluate the achievement of learning objectives.four teachers are not appropriate in assessing SPS. They use Paper and Pencil Test and Observation class to rate SPS. and three chemistry teachers evaluated SPS with inappropriate indicators, i.e. Using laboratory equipment and Understanding concepts in chemistry. With improper assessment methods, the achievement of learning objectives can not be evaluated properly.

4.7. Knowledge of Resources
Some chemistry teachers used laboratories and classes to teach SPS. this will support SPS learning. but there are some teachers who only use the class. The using facility is related to how the chemistry teacher understands teaching SPS to the students.

Given the importance of PCK reflection, understanding the relationship between teachers' reflective capacity, PCK, and student learning will give a clearer picture of how student learning relates to the knowledge andthinking brought by the teacher [19]. The teacher's understanding of SPS will affect the learning process that has been planned and its evaluation.The process of science can not be separated in practice from the conceptual understanding involved in learning and the application of science so that appropriate instruments are needed to measure it. Knowledge of assessment is an important component of PCK. This component includes knowledge of specific instruments, approaches, or activities [20]. Also, teachers should know what students know about topics and areas

IOP Conf. Series: Journal of Physics: Conf. Series **1022** (2018) 012038 doi:10.1088/1742-6596/1022/1/012038

that may be experiencing difficulties. These components include knowledge of students' conceptions of a particular topic, learning difficulties, motivation, and diversity of abilities, learning styles, interests, developmental levels, and needs [15]. Also, the teachers'knowledge of SPS will affect students' understanding of SPS. Based on PCK analysis, chemistry teachers as respondents in this study still have a weak PCK.Therefore it is expected that teachers can improve their understanding of the SPS by learning independently or attending training.

5. References

[1] Shahali E H M and Halim L 2010 Development and validation of a test of integrated science process skills *Procedia - Soc. Behav. Sci.* **9** 142–6

[2] Kruea-in N and Thongperm O 2014 Teaching of science process skills in thai contexts : status, supports and obstacles *Procedia - Soc. Behav. Sci.* **141** 1324–9

[3] Lati W, Supasorn S and Promarak V 2012 Enhancement of learning achievement and integrated science process skills using science inquiry learning activities of chemical reaction rates *Procedia - Soc. Behav. Sci.* **46** 4471–5

[4] Maedor K S 2003 Thinking creatively about science suggestions for primary teachers *Gift. Child Today* **26** 25–9

[5] Kruea-In C, Kruea-In N and Fakcharoenphol W 2015 Science direct a study of thai in-service and pre-service science teachers' understanding of science process skills *Procedia - Soc. Behav. Sci.* **197** 993–7

[6] Ong S, Ismail Z and Fong S 2006 Development and validation of test for integrated science processes *Proc. Int. Conf. Meas. Eval. Educ.* 149–56

[7] Scanlon E, Morris E and Paolo T Studies in science education contemporary approaches to learning science : practical work *Stud. Sci. Educ.* 37–41

[8] Emereole H U 2009 Learners ' and teachers ' conceptual knowledge of science processes: the case of botswana

[9] Settlage J and Southerland S A 2007 *Teaching science to every child: using culture as a starting point* (New York: Taylor & Francis)

[10] Dillashaw F G and Okey J 1980 Test of the integrated science process skills for secondary science students *Sci. Educ.* **64** 601–8

[11] Harlen W 1999 Purposes and Procedures for Assessing Science Process Skills *Assesment Educ.* **6** 37–41

[12] Loughran J. Berry A. & Mulhall P. 2006 *Understanding and developing science teachers' pedagogical content knowledge* (Rotterdam, The Netherlands: Sense Publishers)

[13] Lee E and Luft J A 2008 Pedagogical Content Knowledge Experienced Secondary Science Teachers ' Representation of *Int. J. Sci. Educ.* **30** 1343–1363

[14] Geddis A N, Onslow B and Beynon C 1993 Transforming content knowledge : Learning to teach about isotopes *Sci. Educ.* **77** 575–91

[15] Grossman P.L. 1990 *he making of a teacher: Teacher knowledge and teacher education* (New York: Teachers College Press)

[16] Marks R and Marks R 1990 Theme : Pedagogical Content Knowledge Pedagogical Content Knowledge : *J. Teach. Educ.* **41**

[17] Shulman L. 1987 *Knowledge and teaching: Foundations of the new reform* (Harvard Educational Review)

[18] Shulman L. 1986 *Those who understand: Knowledge growth in teaching, Educational Researcher*

[19] Park S and Oliver J S 2008 Revisiting the conceptualisation of pedagogical content knowledge (PCK): PCK as a conceptual tool to understand teachers as professionals *Res. Sci. Educ.* **38** 261–84

[20] Tamir P 1988 Subject matter and related pedagogical knowledge in teacher education *Teach. Teach. Educ.* **4** 99–110

IOP Conf. Series: Journal of Physics: Conf. Series **1022** (2018) 012039 doi:10.1088/1742-6596/1022/1/012039

Using *Convective Stratiform Technique* (CST) method to estimate rainfall (case study in Bali, December 14th 2016)

Ayu Vista Wulandari[1*], Khafid Rizki Pratama[2] and Prayoga Ismail[3]

[123]Indonesian State College of Meteorology Climatology and Geophysics, Jalan Perhubungan I No 5, Pondok Betung, Pondok Aren, Tangerang Selatan, Banten 15221

[*]Corresponding email: ayu.vista10@gmail.com

Abstract. Accurate and realtime data in wide spatial space at this time is still a problem because of the unavailability of observation of rainfall in each region. Weather satellites have a very wide range of observations and can be used to determine rainfall variability with better resolution compared with a limited direct observation. Utilization of Himawari-8 satellite data in estimating rainfall using Convective Stratiform Technique (CST) method. The CST method is performed by separating convective and stratiform cloud components using infrared channel satellite data. Cloud components are classified by slope because the physical and dynamic growth processes are very different. This research was conducted in Bali area on December 14, 2016 by verifying the result of CST process with rainfall data from Ngurah Rai Meteorology Station Bali. It is found that CST method result had simililar value with data observation in Ngurah Rai meteorological station, so it assumed that CST method can be used for rainfall estimation in Bali region.

Key words: rainfall, CST, satellite, Himawari-8

1. Introduction

Indonesia is the largest archipelagic country in the world extending from the geographical latitude of 7° 20' N - 14° LS and 92° - 141° E with about 70% water area and 30% land area. Tropical rainfall in Indonesia is very important to be studied, other than as a source of life of high rainfall can be a source of flood natural disasters. Indonesia's rainfall plays an important role in the process of determining global weather and climate conditions because two-thirds of the total global rainfall occurs in the tropics (Tjasyono, 2013). One of the causes of high rainfall in the tropics is the year-round solar warming that results in the evaporation rate in Indonesia and the presence of convergence areas that have high cloud growth potential.

Geographically, Bali is located at 9° 0' - 7° 50' S and 114° 0' - 116° 0' E. Bali is included in tropical climates influenced by the Asian Monsoon and the Australian monsoon that led to Bali experienced two seasons of the dry and rainy season. The Asian monsoon brings wet air masses to Indonesia including Bali, causing the rainy season in Bali to occur around the period of November to March with the highest average rainfall in December and January. Ngurah Rai Meteorological Station Bali recorded the amount of rainfall on December 14, 2016 that occurred in was 78.8 mm/day. The rain events can be categorized as high intensity rainfall.

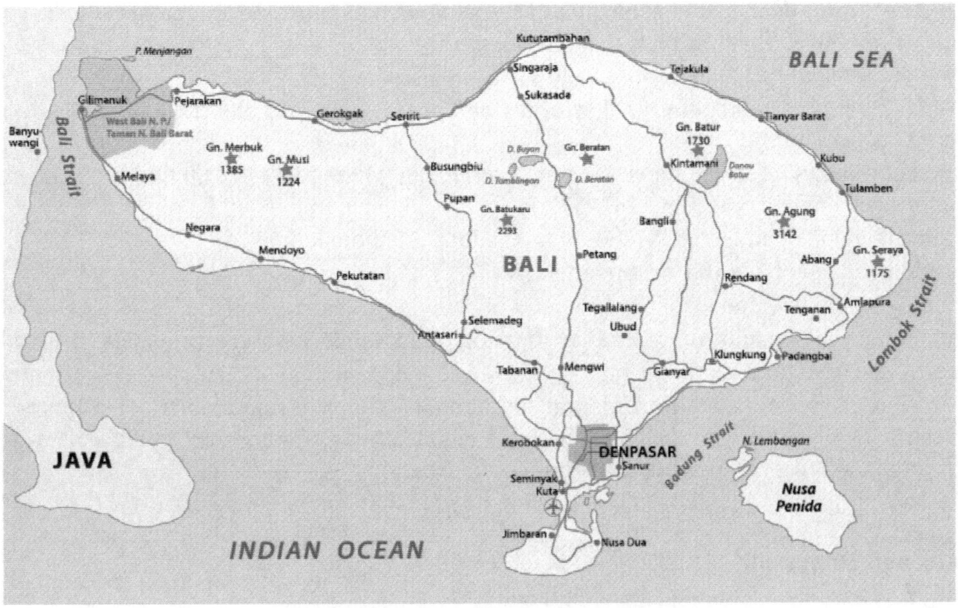

Figure 1. Area of Study (Bali)
(Source: https://www.bali.com/map.html)

Rainfall data report in weather information are carried out at every Meteorological Station in Indonesia. However, the availability of rainfall data is still limited because rainfall observations are not conducted in every region and rainwater damage can also be hampered in weather data reports. Rainfall estimation is one way in handling the problem so that used of satellite weather data that can reach all areas even areas that do not have observation stations and rain gauges. This weather satellite data can be obtained up to date and is not charged, but its utilization is still very limited among the public (Suwarsono et al, 2009; Endarwin, 2014). Utilization of Himawari-8 satellite data in estimating rainfall is done using Convective Stratiform Technique (CST) method. The CST method is applied by separating convective and stratiform cloud components using infrared channel satellite data. Cloud components are classified based on slope because the physical and dynamic growth of air movement and precipitation between the two components is very different (Goldenberg et al, 1990). This research was conducted to find out the comparison of rainfall estimation result with CST method in Bali area on December 14, 2016 by utilizing infrared channel data of Himawari-8 satellite image and verify result from CST method with rainfall data from Ngurah Rai Meteorological Station, Bali.

2. Data and Method

The location of the research was conducted at Ngurah Rai Meteorological Station with coordinates station 08 ° 44'46 "S and 115 ° 09'57" E representing Bali area on December 14, 2016. The method used is Convective Stratiform Technique (CST) method by using data as follows:

2.1. Himawari-8 Satellite Data

Satellite data in the form of PGM (Portable Gray Map) obtained from http://weather.is.kochi-u.ac.jp/sat/GAME and calibration data (CAL) obtained from http: //weather.is.kochi -u.ac.jp/sat/CAL is used for estimation of rainfall. In addition, satellite data taken from the Meteorological Climatological and Geophysical Agency in the form of infrared channels are also used to verify the results from satellite data PGM and CAL.

IOP Conf. Series: Journal of Physics: Conf. Series **1022** (2018) 012039 doi:10.1088/1742-6596/1022/1/012039

2.2. Observation Data

Weather observation data from Ngurah Rai Meteorological Station on December 14, 2016 to verify the results of rainfall estimation from the CST method.

The data processing in this study are as follows:

- Converting the Himawari-8 infrared satellite data in the .pgm and .dat file extensions to obtain the cloud top temperature.
- Set the coordinates of the research area at Ngurah Rai Meteorological Station with the coordinates of stations 08 ° 44'46 "S and 115 ° 09'57" E.
- Calculate the slope parameter (S). For data resolution 5.5 km x 5.5 km.
- Separation of convective and stratiform nuclei.
- Estimating hourly rainfall.
- Evaluation of the results of rainfall estimation with rainfall observation data from NgurahRai Meteorological Station.
- Application of spatial rainfall estimation for samples of heavy rain events per 3 hours and daily. The estimation results in the plot using Python programming and compared with rainfall observation data from NgurahRai Meteorological Station.

3. Results and Discussion

3.1. CST Method

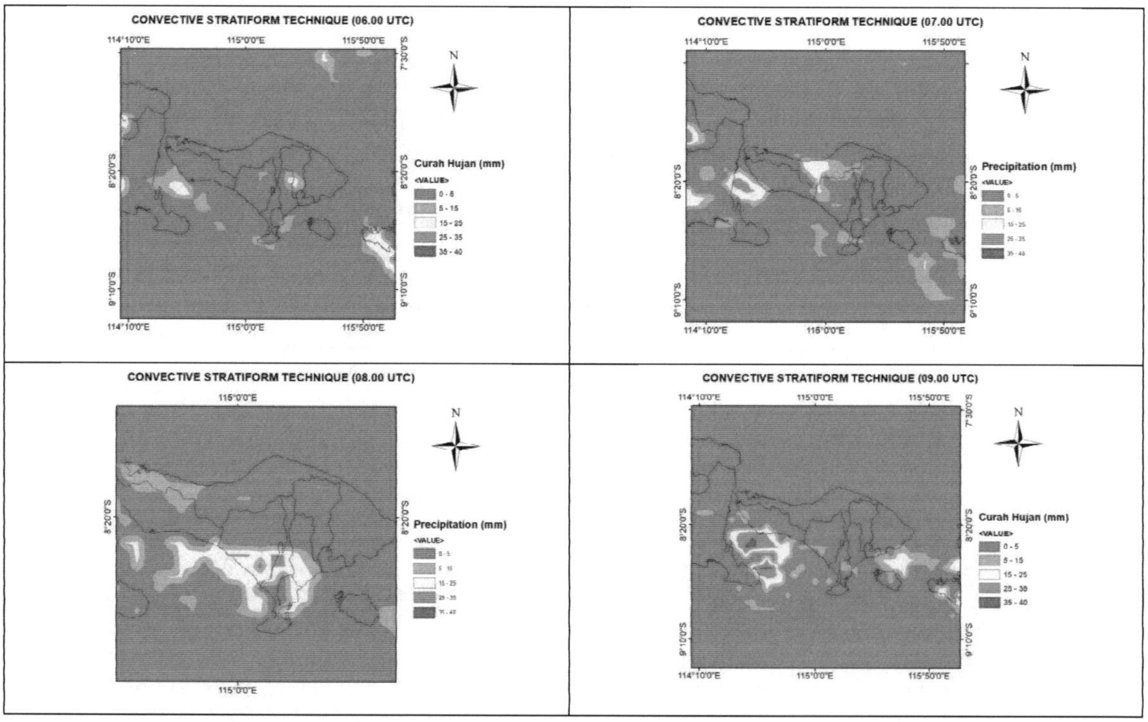

Figure 2. Rainfall Estimation in Bali using CST Method (06.00-09.00 UTC)

Through the utilization of CST method to estimate rainfall can be done. In this research, rainfall estimation is done for Bali area when there is moderate rain phenomenon on 14 December 2016. In accordance with the availability of weather satellite data, the estimation can be determined for at least one hour at a time. Furthermore, the accumulation of the hourly estimation results can be determined also can be applied for longer period rainfall, as well as daily to monthly rainfall estimation. In the

Convective Stratiform Technique (CST) method there is a separation to determine whether the rain to be estimated is from a stratiform or convective cloud type. In a grid of estimation, it is assumed that the convective cloud type is a region of high cloud cover or a towering cloud, whereas the stratiform cloud type is a stratiform cloud cover area or a cloud that grows horizontally. In this case, the stratiform type of rainfall values tends to be smaller than the convective.

The estimation results on the CST method shows the rainfall forecast on Bali area on December 14, 2016 at 06.00 UTC, almost evenly distributed throughout Bali Island. At 06.00 UTC, shows a contour with a value of rainfall estimation that is not too significant for heavy rain category. This is due to the separation that occurs is an estimate of stratiform clouds. While at 06.00 UTC and 07.00 UTC, there is a contour of rainfall estimation with a range of values ranging from 25 - 35 mm, these indicate in some areas of Bali there is a convective cloud that caused precipitation (rainfall). This estimation is based on the separation of convective clouds that show that clouds are dominant in the area of Bali. This result is assumed that in the morning in the Bali region is not too dominant by the growth of convective clouds but stratiform clouds that have a uniform cloud contour with small slopes. At 08.00 UTC, the convective contours are visible in the middle and most of the island, and this also indicates high precipitation with values ranging from 25 to 40 mm. And at 09.00 UTC, the visible cloud is a stratiform cloud, where most of the contours are evenly distributed, with estimation values ranging from 5 -15 mm.

3.2. Satellite imagery

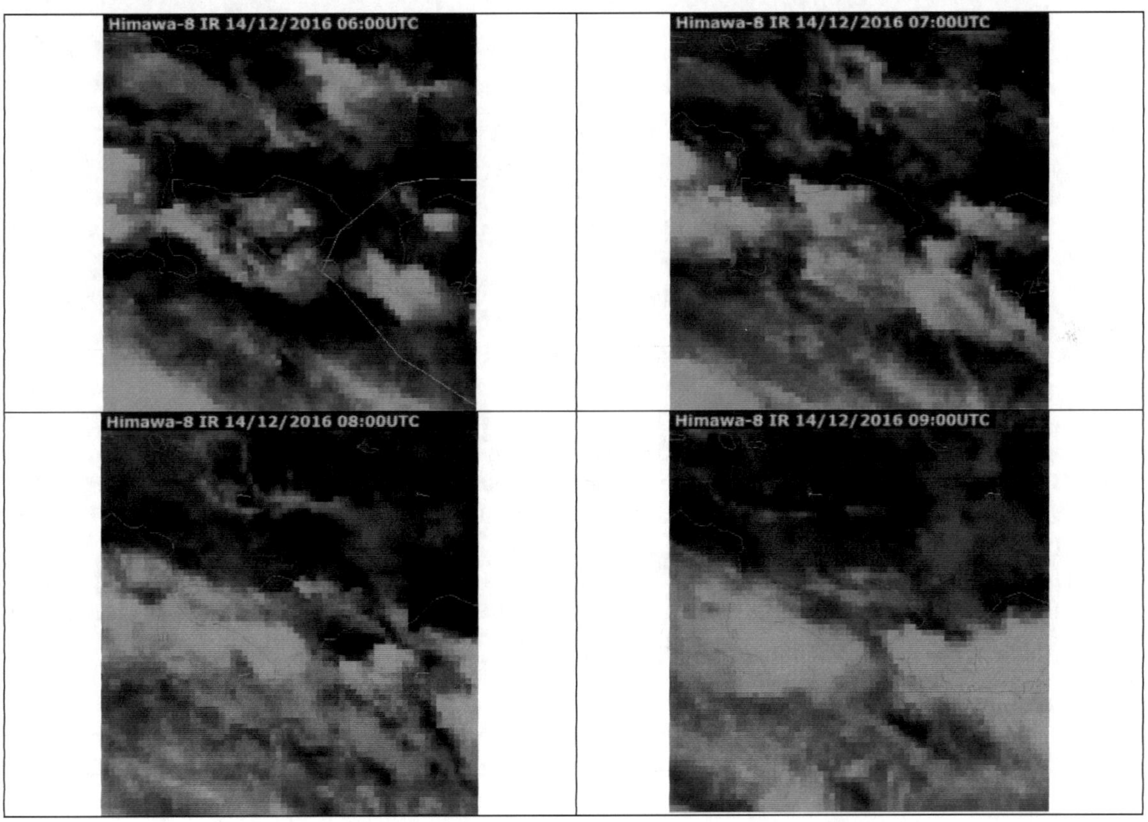

Figure 3. Himawari-8 Satellite Imagery in Bali (06.00-09.00 UTC)

The verification of rainfall estimation by using Himawari-8 satellite imagery data in Bali on 06.00 UTC - 09.00 UTC. In the results of the image of Himawari-8 satellite at 06.00 UTC shows the

IOP Conf. Series: Journal of Physics: Conf. Series **1022** (2018) 012039 doi:10.1088/1742-6596/1022/1/012039

presence of thick cloud contours in the southern region of Bali Island which indicates that in the region occurs precipitation (rainfall) is quite high. While at 07.00 UTC, the thick cloud contours that occur in the southern region of Bali island growing and expanding to some parts of the central island of Bali. At 08.00 UTC and 09.00 UTC, the thick clouds of the contour move westward so that the rainfall is reduced.

3.3. Meteorological Station Observation

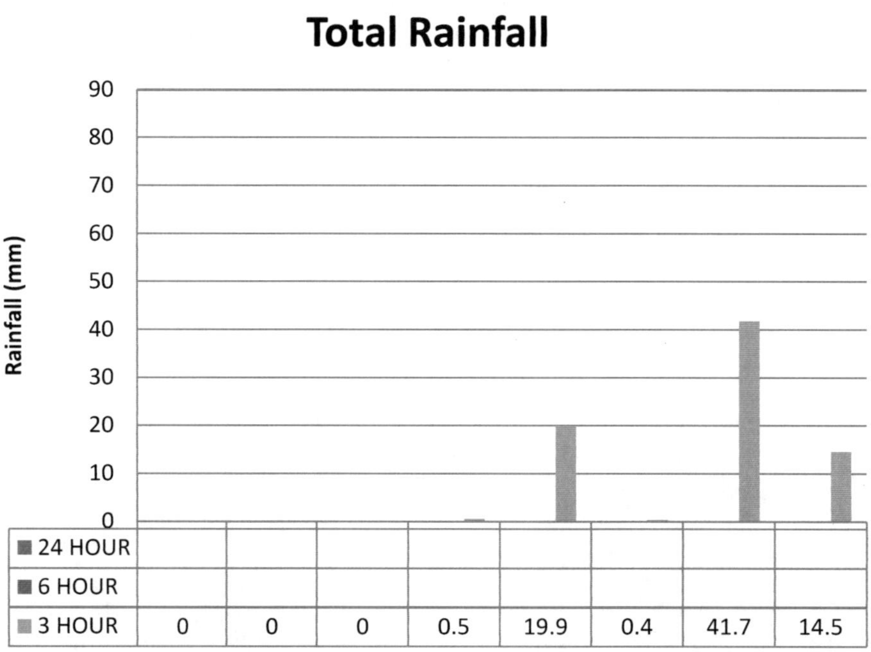

Figure 4. Total Rainfall in Bali on December 14, 2016

Ngurah Rai Meteorological Station observation data is used to verify the CST method rainfall estimation. The rainfall at 12.00 UTC measured about 20.4 mm, this indicates that in the preceding hour precipitation (rainfall) is quite high. This result corresponds to the accumulated estimation of precipitation on Convective Stratiform Technique (CST) method which shows rainfall values ranging from 15 to 20 mm per hour. While at other important hours such as at 18.00 UTC and 21.00 UTC shows rainfall that occurred in Bali region ranged from 30-40 mm.

Based on the data above, it can be seen that the results shown from the CST model almost approaching the value of the observation data. It proves that the CST model can be applied in estimating rainfall in areas that have not yet observed rainfall such as Bali region. The CST model can be used to improve gathering meteorological information of observers.

4. Conclusions

Based on the analysis of the application of rainfall estimation by CST method on December 14, 2016 at Ngurah Rai Meteorological Station that representing Bali region, can be concluded that CST method can be applied to make rainfall estimation in Bali area. It corresponds to rainfall observation data from the Ngurah Rai Meteorological Station. The CST method is quite consistent in estimating rainfall per hour. The good quality of these estimation shows that the method is suitable for year-round use in the monsoon-dominated Bali area, which can be used to monitor near-real time rainfall.

5. References

[1] Adler R F and Negri A J 1988 *J. of Appl. Meteo.* **27** 30–51
[2] Andani, Ashriah J P and Putra Richard M 2016 *Prosiding Seminar Hari Meteorologi Dunia 2016*
[3] Endarwin Hadi, S, Tjasyono B H K and Gunawan D 2013 *J. Meteo. Dan Geof.* **14** 19-24
[4] Endarwin 2014 *Dissertation* Program Studi Sains Kebumian, Institut Teknologi Bandung, Bandung
[5] Goldenberg S B, R A Houze, Jr. and Churchill D D 1990 *J. of the Meteo. Soc. of Jap.* **68** 37-63
[6] Hong, Yang R F, Adler A and Negridan Huffman G J 2007 *J. of Nat. Haz.*
[7] Islam M N, Islam A K M S, Hayashi T, Terao T and Uyeda H 2002 *J. of Nat. Disas. Sci.* **24** 83-89
[8] Panjaitan A 2012 *pemanfaatan citra satelit untuk informasi meteorologi penerbangan* (Jakarta: Materi Diklat Penerbangan)
[9] Suwarsono P, Kusumaning A D S and Kartasamita M 2009 *J. Peng. Jauh* **6** 32-42
[10] Tjasyono H K, Bayong, Harijono and Sri Woro B 2013 *Atmosfer Ekuatorial* (Jakarta: Puslitbang Badan Meteorologi Klimatologi dan Geofisika)

Acknowledgement

This research was supported by *Indonesian State College of Meteorology Climatology and Geophysics (STMKG).*The authors would like to thank to the college for funding our research and provide open source data for analysis to be used in this paper.

IOP Conf. Series: Journal of Physics: Conf. Series **1022** (2018) 012040 doi:10.1088/1742-6596/1022/1/012040

Enhancing effectiveness of agriculture group in supporting government program to increase food security

Daru Retnowati[1] and A H Subarjo[2]

[1] Faculty of Agriculture UPN "Veteran" Yogyakarta
[2] Mechanical Engineering Adisutjipto School of Technology Yogyakarta

Corresponding email: [1]daru.retnowati.upnveteranyk@gmail.com, [2]ab.haris.79@gmail.com

Abstract. Food Security is closely related to agriculture, including fisheries. Food is a basic necessity and indispensable to humans. Nowadays, there are many agricultural lands and fisheries are turning to settlements and offices. To overcome these obstacles, the government took the policy of forming farmer groups. Farmer groups are channeling the government assistance, whether capital, seeds, training, or technology and knowledge assistance. This research is qualitative. The population in this study were members of the fish farming group in Purwomartani, Kalasan, Sleman. The population in this study were 4 Farmers Group in Purwomartani, Kalasan, Sleman. The sample in this research is 1 farmer group with the largest number of members that is 31 people. For the other three groups of fish farmers the number of members is 20 people. The results show that farmer groups are effective in supporting government programs. The role of farmer groups is needed to support the successful management of agricultural land, improvement of knowledge and skills of fish farmers, renewal of agricultural technology and equipment, and marketing of agricultural products.

Key words: Food Security, Farmer Group, Technology, Management

1. Introduction

Food is one of the basic necessities for every living creature, every living thing trying to meet this food needs. Food demand is also a major factor for the success of the government programs. Food availability is also an indicator of wellbeing. Food can be the trigger of a conflict between regions and countries. Food needs for people can be met from agriculture, fishery, and plantation. In accordance with the narrative about the fulfillment of food needs in the district of Sleman, especially in the village of Purwomartani, there are several kinds of potential that has been cultivated by the residents. These potentials include paddy fields such as rice, plantation agriculture such as salak pondoh, land fisheries such as freshwater fish ponds. The existence of aquaculture business has a considerable economic function, it is seen with the existence of ponds that are located around the river. Even in some areas not only cultivated with minimal and small capital, but also attract large investors to invest in the form of fishing ponds and restaurants. Sleman Regency is geographically a land that is far from the coast, but this limitation does not make the creativity of its citizens stunted. This can be seen from the existence of fishery cultivation in Purwomartani village that can accommodate the economic potential of Sleman Regency in general, and Purwomartani Village in particular, most of the people are engaged in the informal trading sector.

Fish cultivation is an activity of breeding or fish enlargement aimed to increase production, productivity, and quality [7]. In supporting the production process of fishery the availability of quality and sustainable fishery resources is important, because the quality and quantity of resources will affect the pattern and direction of production and marketing. Therefore, management and utilization of water resources should be done wisely by taking into account aspects of production factors and benefits. [5].

Fisheries is a sub-sector of agriculture in the broad sense, thus development in the agricultural sector including development in the fishery sub-sector by increasing the production of fishery business also means increasing the income of fishermen and fish farmers. Therefore, all fishery resources can be utilized to improve the living standards and welfare of fishermen and fish farmers in particular, and the people of Indonesia in general [5].

In an effort to support the existence of creative business citizens in the field of fish cultivation through empowerment programs of small and medium enterprises, government of the District of Sleman actually has acted quickly by making various efforts to give support. Some of them are training the representatives of each group of farmers, coaching through extension workers (in Indonesia Petugas Penyuluh Pertanian or PPL), distributing equipment to each member of the group of cultivators, and soft loan distribution with interest enough to each group. Formerly the distribution of soft loans through the National Community Empowerment Program (PNPM) Mandiri, but currently the distribution of assistance through cooperatives. Therefore, in the village of Purwomartani also formed a cooperative "Mina Manunggal" which functioned as a medium distribution of government assistance programs. Distribution of government credit assistance through cooperatives, because the cooperative is a legal entity. Cooperative "Mina Manunggal" was formed in 2011. Members of the cooperative "Mina Manunggal" are the managers of farmer groups in the village of Purwomartani. At the time of the monetary crisis that began in 1997, nearly 80% of large businesses went bankrupt and made mass layoffs to their employees. However, the cultivation of fisheries conducted by farmer groups, most still can survive in crisis with all its limitations. In addition, as a business sector run in the lower and middle level, fishery cultivation conducted by communities in this farmer group has a major role in reducing unemployment. International donor agencies show an interest in supporting the development of small and medium enterprises.

Farmers are people who work on crops, animals, and fish as a daily livelihood. In Indonesian Big Dictionary, farmers are people whose livelihood is planting. In the Agricultural Extension System Legislation, the definition of the farmer is an individual Indonesian citizen and his family or cooperative who manages business in agriculture, agroforestry, minatani, agropasture, breeding of animals and plants, in and around the forest, including upstream business, farming, agro-industry, marketing, and supporting services [3].

Modern agriculture was developed in line with the history of agricultural extension in Indonesia, agriculturalization of agriculture with new technologies, introduced to farmers through the media of farmer groups. The development of farmer groups started with the establishment of rural extension centers. In 1948, Soewardjo sparked the establishment of a Village Community Education Center (BPMD) in each sub-district as a village community education center, from 1950 to 1956 a total of 375 BPMDs could be built in Indonesia. In 1959 there was established the Operation Command of the Prosperous Movement (KOGM) aimed at increasing the production of food, especially rice, and in the rural areas a farmer organization was formed to implement the five farming enterprises which later developed into farmer groups. During the first five-year development period of 1969-1974, farmers' groups grew and developed, Agricultural Counseling Center (BPP) was developed and during the second five-year development phase up to the fifth stage (1974-1994) agricultural extension institution was continuously improved and reproduced [3].

The definition of food according to Law No. 7 of 1996 is: *Everything derived from biological resources and water, whether processed or not processed, which is intended as food or drink for human consumption, including food additives, food raw materials, and Other materials used in the process of preparing, processing, and or preparing food and beverages* [1].

Food is one of the basic human needs that is part of human rights, as set forth in the Universal Declaration of Human Rights in 1948, and Law No. 7 of 1996 on Food. The definition of food is the ingredients that are eaten daily to meet the need for maintenance, growth, work, replacement of tissues and regulate the processes in the body. Staple food, which is food that is eaten regularly by a group of people in large enough quantities, to produce most of the energy source. Food is consumed by humans to get energy power to perform life activities (such as breathing, working, building, and replacing damaged tissue). Food is a fuel that serves as a source of energy [11].

Meanwhile, according to the Food and Drug Administration, food for consumption that is not only in the form of rice, but also vegetables, fruits, meat both poultry and cattle, fish, eggs, also water. Food security according to the Law no. 7 of 1996 on Food Article 1 paragraph 17 is the condition of the fulfillment of adequate food, both in quantity and quality, and safe, equitable, and affordable [11].

Food security according to the Rome Declaration and World Food Summit Plan of Action (1996) is "... when all people, at all time, have physical and economic access to sufficient, safe and nutritious food to meet their dietary needs and foods preferences for an active and Healthy life [11]. FAO (1992) defines food security as a situation where at all times it has adequate amounts of food that is safe and nutritious for a healthy and active life. In general, food security is a guarantee that the food and nutritional needs of every resident are the main condition in achieving the level of health and well-being sufficient [4].

2. Literature Review

As a literature review, the farmer group "Mina Perkasa has same type with farmer group in Cakuang district. The results of the empowerment of rice field agribusiness system (*Oryza sativa L.*) in Cangkuang are: 1) Increasing knowledge, attitude and skills of farmers about the optimal utilization of production facilities, and the establishment of cooperation between farmers and providers of production facilities; 2) Increasing knowledge, attitude and skill of farmer about improvement of rice cultivation technique; 3) Increasing knowledge and attitude of farmers about role and function of farmer group in supporting farming activity. Fostering farmer groups in an effort to empower the group is directed at all subsystems agroinput and agropoduct. The guidance of farmer groups is done through counseling about the role and function of farmer groups as learning rides, production units and cooperation rides between members in farmer groups, between farmer groups and other farmer groups, and between farmer groups and other agribusiness actors [2]. Farmer groups play various roles, as a forum for farming and organizational learning, cooperation vehicle, farming production unit, in addition farmer group also play a role in giving feedback about performance of a technology [8].

The same type also shown by Shinta Anggun Lowisada journal, the existence of farmer groups can be seen from the group members who join to get benefits such as fertilizer subsidy, the support of government and society, and the continuity of activities such as empowerment through regular meetings and mutual cooperation. Capital assistance from Gapoktan is also one of the things of existence, making it easier to get low-priced auction pads, giving members the ability to plan and implement in improving joint production facilities and ease of government aid information in the form of capital and production facilities. The collective action of collectivity of production which includes management of irrigation resources, the ease of provision of production facilities, the collectivity of capital, and accommodating the aspirations of members, gives more opportunity to the members of farmer groups to overcome its limitations and is proof of the existence of farmer groups [9].

Sa'diyah in his journal stated Joglo Tani greatly contribute in improving the role of youth, so that Joglo Tani youth can play an active role in the development of the agricultural sector in order to realize food security. The role of youth in the form of the role of production, the role of promotion and marketing, and the role of community empowerment. The role of Joglo Tani youth in the agricultural sector is not always easy, but there are some obstacles that are faced: (1) technical: lack of youth spirit, land degradation, uncertain weather conditions, minimal costs, lack of youth-based agricultural training, (2) lack of support Social community, (3) lack of government support. Strategic efforts made by youth Joglo Tani in realizing food security with reference to indicators of food security theory to meet food needs ranging from individual to state both the number and quality it can be concluded that youth Joglo Tani has contributed to realize food security. The role of youth Joglo Tani is a role implemented by farmer groups [6].

The market structure at the level of medium concentration oligopsoni traders who have a moderate strength in market performance. Market structure at the kecamatan trader level. Low oligopsony that shows traders have low levels of strength in market performance. Market structure at the whole seller level is a high concentration oligopsoni which indicates a high power in influencing the market, so that other institutions are needed as farmer groups to cut the dependence of farmers on traders [10].

3. Research Methods

3.1 Methods

The type of research used is survey research. In the survey research data obtained is the real data that exist in the field at the time of data retrieval research done, and not done a treatment on the object of research. This research is descriptive. Qualitative analysis approach to describe Farmer Group Effectiveness, especially on fish farmers in supporting government programs to improve food security.

3.2 Time of Research

Implementation of the study was conducted in January 2014 to May 2014 in Purwomartani Village, Kalasan District, Sleman Regency, of Yogyakarta Special Province. The location is expected to gain primary data generated from participants' observation techniques and interviews with farmers group managers of fish farmers.

3.3 Population Research

The population in this study were members of fish farming groups in Purwomartani, Kalasan, Sleman, Yogyakarta Special Province. In Purwomartani, Kalasan, Sleman there are four (4) fish farming groups, namely "Mina Suka Mandiri", "Mina Suka Makmur", "Mina Perkasa", and "Mina Rejo" fish farming group. These four fish farming groups have the same business, ie most of the group members in the field of fish cultivation, meaning that farmer group members buy fish seeds from other farmer groups or from seed traders, then they raise the fish for sale to consumers or restaurants. But there are also members of the group who concentrate on fish nurseries, although in terms of numbers there are still many more fish farmers trying in the liberation field.

3.4 Sample Research

The sample was taken by purposive sampling by taking the farmer group of fish farmer with the largest number of members among the four farmer groups in the population in order to obtain more information, giving better description of the role of farmer group of fish farmers in supporting the family economic resilience, and Play a role in maintaining food security. Therefore, fish farming group used in this research is farmer group of "Mina Perkasa" which has 31 members. For the other three groups of fish farmers the number of members is 20 people.

4. Discussion

Good food security in a region can maintain the resilience of the area well because food is a major requirement. We can imagine if there is food shortage in a region, it can have an impact on the vulnerability conditions in the region. Crime will increase, people will fight over food, strong and powerful will surely be the winner in this struggle. People become easily emotional because the affairs of the fulfillment of food needs are not guaranteed. The vulnerability of conditions in a region can cause a nation's national resilience condition to be threatened, people will tend to be selfish. With good food conditions, the nutritional fulfillment of citizens will be good too, and of course this can affect

IOP Conf. Series: Journal of Physics: Conf. Series **1022** (2018) 012040 doi:10.1088/1742-6596/1022/1/012040

the national resilience of a nation. If the nutrition of its citizens is fulfilled, then the physical and intellectual qualities of the citizens will be good as well.

The essence of National Resilience lies in the ability and resilience of a nation to maintain its existence towards a desired future. Applied to the nation of Indonesia, the essence of National Resilience is the ability and resilience of the Indonesian nation to maintain and develop the sustainability of a unitary state that covers the area from Sabang to Merauke with Pancasila as its ideological foundation. [2]

In GBHN 1978 there is a paragraph on National Resilience:
"National Resilience is a dynamic condition that is an integration of the conditions of each aspect of the life of the Nation and the State. In essence the National Resilience is the ability and resilience of a nation to be able to guarantee its survival towards the glory of the Nation and the State ".
"...... The success of the National Development will increase the National Resilience. Furthermore, resilient National Resilience will further encourage National Development. .... Insight in achieving National Development is the Insight of Archipelago ".[2]

Based on the description, good food security can have an impact on the resilience of an area and the most extensive impact on national resilience of a country. With guaranteed food, the existence of a country will be guaranteed. The country will have the power to face the things that threaten its existence.

Based on the interviews, it was found that the scope of the activities of the "Mina Perkasa" fish farmer group included 1) extension and training activities, and 2) field activities or pools (ie seed procurement, enlargement, maintenance, harvesting and marketing). In addition to these activities, there are also land processing activities, so that the land / pond used does not experience saturation and can be used again after harvesting fish. Before the land is used for the pond, also done the processing first, such as the installation of tarpaulins to prevent leakage, installation of pralon and drainage making for water in healthy ponds and oxygen in the pond awake. Therefore, it would be better to install a water mill to make the air in the pool healthier, and water hyacinth plants and other aquatic plants. This is as disclosed by Mr. Tri, member of the farmer group "Mina Perkasa".

> *He said: "Before the pond is used first processed mas, let me not smell the cement, wear tarpaulins, given water hyacinth, it is also for eyup-eyupan fish. Let the water be given by the sewer, every day checked, better to use the waterwheel to keep the water flowing, the healthy fish mas"*

The members of fish farming groups that concentrate in the field of enlargement more when compared to those who do nursery business. This is because of a greater risk factor if attempted in the nursery than if at enlargement. Seedlings are easily killed by disease and require special treatment, especially if done in a simple pond by the river as done by members of fish farming groups. This is because the age of the fish is younger so that the immune system is more susceptible to disease than the cultivators / enlargement. However, enlargement efforts should also be wary of attacks of fish diseases such as fungi, especially in transition season / seasonal changes from rainy season to dry season or vice versa. In that season, the air condition is easily changed so that the fish is a little difficult to adjust the condition. In this season, farmers usually sprinkle salt so that the condition of the pond water is better and so the mushrooms are not easy to grow and attack the fish.

> The secretary said: *"Usually when the season of transition weather is uncertain, so the fish easily got the disease, usually mushrooms Sir. Well to overcome it usually use salt in the pool, group members are more interested in the enlargement of fish than the nursery, because the risk of death if the seeds are larger, if many dead (we will be) loss later"*

To increase the fish production, Sleman district government also conducted various coaching and mentoring efforts to the fish farmers. This mentoring and guidance is usually done by related agencies, such as agriculture service, cooperative service, and trade. Coaching and training is conducted through counseling during the farmer group meeting. In relation to training and development activities of the Department of Agriculture, Fisheries and Forestry of Sleman Regency, Rewang Sukirman gives the following explanation.

> *He said: "The activities of fisheries agendas in particular in Sleman District, fisheries agendas are not only in" Mina Perkasa "but I am a subdistrict, but overall we are not just" Mina Perkasa "like apprentice farmers to successful farmers, Course cultivation, managerial every year. Every year there must be training-related programs, "Mina Perkasa" has been trained for standing eight years ago and must be asked by its representative".*

Agricultural extension is an extension of the government. To carry out their duties, agricultural extension officers coordinate with the farmer group managers. The government's guidance not only deals with technical matters on fish farming, but also deals with managerial and marketing, such as good marketing and good bookkeeping, to facilitate the management of farmer groups if at any time requested information by members and the government.

In large group meetings, generally one sub-district, in addition to conveying various information or policies in the field of fisheries, field extension officers also provide input on the problems that have been submitted by the farmer group's board before the meeting took place. It is as revealed by Rewang Sukirman as follows:

> *"Before the next 35-day group meeting there should be any material that will be submitted with regard to technical or related to feed, so what is really needed by the farmers, what our group wants. So, usually before the meeting is held, we (agricultural extension officers) prepare the material and study"*

> Field Extension Officers also conduct direct guidance on the group, as revealed by Rewang Sukirman: *"Which must have been a regular group meeting, the fish farmer group" Mina Perkasa" we focus on Wednesday night, that is usually the least activity we advise in relation to cultivation techniques, what we want, what the group wants, We follow, we prepare the material"*

Communication between the agency and field extension officers took place quite smoothly. Ismunawan said: *"It is usually if we fit the disease, sometimes there is a help from the service, sometimes the extension came (to inform)".*

In Kalasan District, there is a Communication Forum of Fish Cultivators such as Rewang Sukirman: The role of Field Extension Officers is considerable in terms of farmer consultation to resolve field constraints. If there are obstacles that are difficult to solve by the group members then the group management will contact the Field Extension Officers, especially in the case of cultivation so that at the regular meeting of the cultivation group in the future The Field Extension Officer can provide solutions based on his knowledge and experience in the form of guidance, advice and advice. Generally, farmers do not have formal knowledge in the field of fisheries.

> *"Coaching, in the field there is PPL so intensely we are doing the coaching so that farmers succeed in obtaining maximum production".*

Extension agents are instrumental in the success of the agricultural production process. This was disclosed by Mr. Parjo who is a member of the farmer group "Mina Perkasa".

> He said *"The agricultural extension worker is very helpful for the fish production of group members, the farmer group members are also very helpful to their members".*

The existence of coaching from the government through field extension officers can improve the success of fish farming, so that the harvest was pretty good in its quality and quantity. They feel

IOP Conf. Series: Journal of Physics: Conf. Series **1022** (2018) 012040 doi:10.1088/1742-6596/1022/1/012040

facilitate in cultivation and marketing with the existence of farmer groups. Their principle, if the problem is faced together, will definitely not feel difficult.

With the farmer groups they can save on production costs, some of which farmers do not need to buy their own water pump to drain the pond. The water pump is jointly owned, so it also saves on maintenance costs. To buy seeds are usually purchased collectively, even they also have group cash that can be borrowed by members for the purposes of fish production and personal purposes, of course this can improve the welfare of farmers. In terms of marketing, the role of farmers' groups is also quite large, where farmer group members provide services to their members by linking farmers and restaurants or fishing who will buy crops. This was revealed from an interview with Mr. Parjo who is a farmer member of a farmer group.

"For the purposes of seeds, feed and equipment and marketing, farmers group members are quite active in finding breakthroughs with our partners, we also work together"

Communication between farmer groups is also quite effective. Mr. Semi said:

"Because they both share the same river flow we can not mentang and need the most, we have to tolerate and maintain the river, let our ponds well also".

To maintain the cleanliness of the river sometimes we also do joint work and not only followed by one member of the farmer group, but also done by members of other farmer groups.

Mr. Semi said:"In order for the good river condition, they also prohibit the disposal of garbage in the river, because if someone throws waste in the river it is feared to contaminate the fish pond of the farmers, this can even lead to death in fish. But the use of rivers for fish ponds can also pose a risk to their business continuity, ie during the rainy season sometimes the river flow is quite heavy and the river water surface can overflow, which can affect the existing ponds. Because the capacity of the river is quite limited, so the river water will flood the pools around it. And this could have an impact on the existing fish becoming water-soluble, the farmers would lose money."

Cooperation between farmer groups is also seen from the handling of fish harvest, usually they help each other to harvest the fish, once the harvest can be up to 1-1.5 tons. At the time of harvest, usually the farmers group representatives will contact the fishing and restaurant. Each farmer harvest will be deducted Rp10.000, - per kuintal to support the farmers group cash, will be useful for the farmers themselves, because it can be used to buy the equipment used collectively and savings and loan between members. For harvesting is not done simultaneously all members/ all ponds, but gradually. This is to maintain the selling value and quality of fish to be consumed by consumers.

The "Mina Perkasa" fish farmers still use simple processing technology. This is due to limited capital factor, while to use modern technology would require a large capital.

Mr Semi said: "The technology (that we used) is still limited Mas, most pump. For cultivation of land we still rely on experience, "mouth to mouth" between farmers. In addition, our capital is also limited"

5. Conclusion and Suggestion

5.1 Conclusion

a. Fish farming by fish farmers can support food security

b. The role of farmer groups is very supportive in fish production.

5.2 Suggestion

a. Further research is needed on the role of agricultural extension workers in other farmer groups.

b. Need infrastructure such as adequate equipment for farmer group members so that fish production to be better quality and quantity and can support food security.

6. References

[1] ______ 1996 UU NO 7 Tahun 1996 Tentang *Pangan* (Jakarta)
[2] Amal I dan Armawi A 1995 *Sumbangan ilmu sosial terhadap konsepsi ketahanan nasional.* (Yogyakarta: Gadjah Mada University Press) p 8 90
[3] Hariadi,S S 2016 *Petani memahami kearifan lokal petani tradisional "Samin" dan petani modern*
[4] Sitanggang E M dan Burhan J E M 2007 *Perpektif pangan masa depan* (Jakarta: Sinar Harapan)
[5] Subarjo A H 2012 *Tesis* Peran usaha tani ikan dalam menunjang ketahanan ekonomi keluarga (Anggota Kelompok) (Studi Pada Usaha Tani Ikan "Mina Perkasa" Di Randusari, Purwomartani, Kalasan, Sleman) (Yogyakarta: Universitas Gadjah Mada) p 23-24
[6] Sa'diyah L 2015 Kontribusi Joglo Tani di Mandungan Margoluwih Seyegan Sleman Yogyakarta dalam peningkatan peran pemuda pada pembangunan sektor pertanian guna mewujudkan ketahanan pangan *Jurnal Ketahanan Nasional* VOLUME **21** No 3 28 Desember 2015 p 156-163
[7] Bardach JE JH Ryther and WO McLarney 1972 *Aquaculture: the farming and husbandry of freshwater and marine organisms* (New York:Wiley-Interscience) p 868
[8] Nuryanti S dan Swastika D KS 2011 Peran kelompok tani dalam penerapan teknologi pertanian *forum penelitian agro ekonomi* volume **29** no 2 desember 2011 p 115-128
[9] Lowisada S A 2014 Pemberdayaan kelompok tani dalam usaha meningkatkan pendapatan Usahatani Bawang Merah (Studi Kasus di Kelurahan Sukomoro Kecamatan Sukomoro Kabupaten Nganjuk) Fakultas Ekonomi dan Bisnis Universitas Brawijaya (Malang: Universitas Brawijaya)
[10]Setyaningsih C 2012 Kajian pengembangan agribisnis jagung di kabupaten Grobogan *jurnal perdesaan dan pembangunan* volume **III** No 1 Februari 2012
[11] Retnowati D dan Subarjo A H 2015 Peran penyuluh pertanian dalam menunjang ketahanan pangan *Proceding* Seminar Nasional, Hasil Penelitian Sosial Ekonomi Pertanian "Revitalisasi Kelembagaan Pertanian Untuk Mendukung Swasembada Pangan" (Yogyakarta: Universitas Gadjah Mada) p 154

IOP Conf. Series: Journal of Physics: Conf. Series **1022** (2018) 012041 doi:10.1088/1742-6596/1022/1/012041

Study of waste management towards sustainable green campus in Universitas Gadjah Mada

Mega Setyowati[1]*, Arif Kusumawanto[2] and Agus Prasetya[3]

[1]Master Program of Systems Engineering, Faculty of Engineering, Universitas Gadjah Mada
[2]Architecture and Planning Engineering Department, Faculty of Engineering, Universitas Gadjah Mada
[3]Department of Chemical Engineering, Faculty of Engineering, Universitas Gadjah Mada

*Corresponding email: mega.setyowati@mail.ugm.ac.id

Abstract. Waste management is a part of the green campus achievement program. Universitas Gadjah Mada has a Standard Operating Procedure for managing produced waste. Waste produced by each building or work unit is temporarily accommodated in the waste depot before dumped into the landfill. This research aims to study the waste management system in UGM, in accordance with the concept of a green campus. The concept of green campus to improve the efficiency of waste management needs to be supported by various parties. The success of the green campus program relies on an integrated approach, a sustainable implementation that involves stakeholders of the university. In actualizing the concept of a green campus, the university has its own waste processing system. The organic produced waste is processed into compost, while plastic waste is converted into alternative fuel. Overall, the waste management system that UGM owns is ineffective and inefficient, it was proved by the fact that there is still much waste dumped into the landfill. UGM provides a laboratory that is specialized to process waste that is produced by UGM. It is planned to be able to reduce the amount of waste that is dumped into the landfill. According to the results, vermicomposting technology, the manufacture of liquid fertilizer from leachate, and the manufacture of the composite from a mixture of leaves and paper were offered as solutions.

1. Introduction

The Green campus is the higher education community to increase energy efficiency, conserve resources and enhance the quality of the environment by educating the sustainability and creating a healthier life and learning environment [1]. The green campus continues to be built was the concept of areas of education that pays attention to environmental issues. Not surprising that the university is currently implementing a number of policies, directed to support the effort to become an environmentally friendly campus [2]. The activities of the University staff and students on campus will generate a lot of waste, therefore some of the programs and the handling of waste should be one of the University's attention.

The university already has Standard Operating Procedures for managing the waste generated in the area of the campus. Waste from sources from the Faculty/Unit of work put into Garbage Containers/Container inside Waste which is not far from the area of the Cluster. The staff of the

IOP Conf. Series: Journal of Physics: Conf. Series **1022** (2018) 012041 doi:10.1088/1742-6596/1022/1/012041

Directorate of assets using the truck will take the waste from the appropriate Cluster type for Depo is sent to the laboratory waste recycling (LDUS) located in the center of the Innovation of Agro technology (PIAT) is a locality of Universitas Gadjah Mada in Kalitirto, Berbah, Sleman. The waste has been sent to the LDUS back will be issued in accordance with the type. Organic waste will be processed into Compost so that it would be more beneficial to society and the environment, where the waste residue that cannot be processed will be sent to Landfills in Piyungan [3].

Green campus assessment refers to assessment criteria and standards set by the UI GreenMetric World University Ranking as its base. The criteria used are the recycling programs, toxic waste recycling, organic waste processing, inorganic waste treatment, waste disposal, a policy to reduce the use of paper and plastic on the campus of [4]. GreenMetric a UI can be a framework and guidelines for building Green University standards, and therefore help the universities to Greenwash the whole of their activity. Reflect the results of the ranking, each participating university can test their strengths and weaknesses in promoting green and sustainable University [5].

This research aims to study the system of waste management at the University of Gadjah Mada, in accordance with the concept of a green campus. Waste management is one of the parameters to realize the concept of a green campus. The activity of the processing and recycling of waste is a major factor in creating a sustainable environment.

2. Methods

Research is done in the area of Gadjah Mada University and laboratory waste recycling (LDUS) in Kalitirto, Berbah, Sleman. Research using qualitative and quantitative approaches, where the early stages of the research process in the form of observations or identification of problems surrounding the waste management system in the University. Data collected through interviews with the stakeholders and a review of the literature on waste management and processing systems. The interview was conducted to staff the trash, landscape, and the funeral of the Directorate of assets and staff of the LDUS.

2.1. GreenMetric

The results of the interviews are then given a valuation based on the scoring system issued by the rank GreenMetric a UI. The results of the assessment can be used as a reference in order to realize the concept of a green campus in the waste sector. The activity of the processing and recycling of waste is a major factor in creating a sustainable environment. The activities of the University staff and students on campus will generate a lot of waste, therefore some of the programs and the handling of waste should be one of the University's attention, namely recycling programs, recycling toxic wastes, sewage treatment organic, inorganic, sewage treatment, waste disposal policy to reduce the use of paper and plastic on the campus of [4]. In this study, some of the programs and the handling of waste should be a major concern in the program of recycling, organic waste processing, and inorganic waste processing. Specific indicators and their wastes are given points shown in table 1 [4].

Tabel 1. Specific Indicators and Their Points Awarded of waste

No.	Criteria and indicators	Points	Score
	Waste (WS)		
WS2	Recycling program for university waste	300	
	None		0
	Partial (less than 25% of waste)		0.33x300
	Partial (25%-50% of waste)		0.66x300
	Extensive (more than 50% of waste)		300
WS4	Organic waste treatment	300	
	Open dumping		0
	Partly composted and compost dumped		0.25x300
	Partly composted and compost used		0.5x300

	Fully composted, compost used		0.75x300
	Fully composted, compost used internally and externally		300
WS5	Inorganic waste treatment	300	
	Burned in open area		0
	Taken off campus to a dump site		0.33x300
	Partially recycled (less than 50%)		0.66x300
	Fully recycled (more than 50%)		300
	Total	900	

3. Result and Discussion

Data used in the study is the "Laporan Survey Sampah Kampus" (UGM) in 2016 and Focus Group Discussion "Menuju UGM Sebagai Pelopor Kampus Swakelola Sampah " PIAT-UGM 2016. Data from the 2016 Garbage Survey Report (Gadjah Mada University) provides information on the generation and composition of garbage in Gadjah Mada University and campus waste management in actual conditions. While the data coming from the Focus Group Discussion "Towards UGM as a pioneer of waste self-management campus" PIAT-UGM 2016 provides information on waste processing that has been implemented in LDUS.

Based on the SOP issued by the Asset Directorate, the waste is sorted by type or compost. Waste discharged into the depots are grouped into four types namely; Organic waste consisting of food scraps and trash from landscaping or tree felling, plastic waste comprising bottles and beverages and plastic wrapping, paper waste consisting of paper and used cardboard boxes, as well as other waste which usually consists of canned drinks, Glass bottles, and so forth. Sorting of waste originating from the source as well as garbage collection at one point or depot would be easier for officers to carry out waste transport. The percentage composition of waste generation in table 2 [6]

Table 2. Percentage Composition of Waste Generation

Name of depot	Percentage of Waste Generation (% mass average)			
	Organic	Plastic	Paper	Other
Depo Gelanggang	66.39	15.69	4.73	13.18
Depo Kantor pusat	68.46	11.06	9.68	10.79
Depo Vokasi	61.71	18.00	14.01	6.28
Depo Belimbingsari	88.87	0.88	2.39	7.86
Depo Filsafat	74.86	17.14	3.98	4.01
Depo Pertanian	61.23	8.69	25.10	4.99
Depo Agro	52.07	7.02	21.23	19.68
Depo Purnabudaya	79.03	1.17	15.15	4.65
Depo Teknik	41.53	16.23	4.96	37.27
Depo Biologi	70.74	3.67	1.96	23.63
Average	66.49	9.96	10.32	13.23

Based on table 2, organic waste is the most generated waste in UGM with an average percentage of organic waste disposed at each depot is 66.49%. The organic waste collected in the depot consists of garden waste or felling, and food waste from canteens or food courts and buildings. Waste from Depo Biologi and Depo Teknik is waste sent to LDUS to be processed because the waste from both depots is considered to fulfill the criteria of LDUS incoming waste which is 80% of organic waste and 20% of waste is included. In addition to being located in an overgrown area of trees, biological clusters have two depots where one depot operates filled with street sweeps or tree felling, while the technical depot is a depot where the depot is constructed separately between organic waste, Plastics, paper waste, and other garbage. Other garbage consisting of glass waste, iron, cans, and others have the second largest percentage of 13.23%. While paper waste consisting of used paper lecture and cardboard food wrapper occupy the third position in the percentage that is equal to 10.32%. Plastic waste consisting of the former food wrappers and bottles of drinks is the type of waste with the average percentage of the smallest of 9.96%. Trash that has been sorted and collected at the depot and then transported by truck to LDUS in the PIAT, Berbah Sleman, and landfills in Piyungan. LDUS a waste treatment facility campus with a total area of 2 ha and a capacity of 10 m3. LDUS development goal is to process organic waste from UGM campus and Berbah district community. Table 3 describes the waste management and technology facilities operating in LDUS under existing conditions in managing campus waste in order to achieve the concept of green campus [7].

Table 3. Waste Management Facilities Operating in LDUS under Existing Conditions

Type of waste	Type of Technology	Capacity	Product	Process Time
Organic	Indore heap	32 tub x 6 m^3	Compost	60 days
	Windrow		Compost	45 days
Plastic	Pyrolysis	30 kg/hour	Fuel oil	
Other	Incinerator	20 kg/hour		

The waste produced in UGM is partly sent to UGM waste processing laboratory known as LDUS and located in PIAT, Kalitirto Berbah Sleman. The technology applied in LDUS as seen in table 3 operates more on organic waste processing technology in accordance with the applicable SOP. Organic waste derived from UGM area is processed into compost using 2 kinds of technology ie compost or indore heap and windrow composting kompos using indore heap method takes 60 days process time, while windrow composting takes 45 days process. Indore heap composting method, the pile of garbage is on a box-shaped cement media with aeration through pipes contained in the waste pile. Reversal with this method is done every two weeks to once a month by workers. Bioactivator used is leachate water from compost while composting with windrow method, compost pile is above bamboo building in the form of equilateral triangle. Giving aeration naturally through the hole contained in the windrow tunnel flow through the density of buildings on each side of the windrow. Reversal during the study was conducted twice for one month. The bioactivator used is propucin. Although the windrow method takes a shorter processing time, but this method has its drawbacks. Lack of windrow method compared with indore heap method is the absence of leachate water treatment out during the composting process takes place. Leachate should be returned to the heap with leachate flow into the container so that the nutrients contained in the leachate water is not wasted [8]. Plastic waste can be recycled into fuel oil using pyrolysis technology. Pyrolysis equipment owned by LDUS has a capacity of 30 kg / hour each time the process. On the pyrolysis process, can produce alternative fuel in the form of oil of 1-2 liters. Recyclable waste is then burned by incineration process with a tool capacity of 20 kg / h at a temperature of 1000 C. Unrecycleable waste combustion processes are preferred for waste to be sent to the Piyungan final disposal site can be reduced [7]. In table 4 presented the results of the assessment of campus waste management system on the existing condition.

Table 4. Results of The Assessment of Campus Waste Management System on The Existing Condition

No.	Criteria and indicators	Score GreenMetric	UGM Points
	Waste (WS)		
Recycling program for university waste			
WS2	None	0	**99**
	Partial (less than 25% of waste)	0.33x300	
	Partial (25%-50% of waste)	0.66x300	
	Extensive (more than 50% of waste)	300	
Organic waste treatment			
WS4	Open dumping	0	**150**
	Partly composted and compost dumped	0.25x300	
	Partly composted and compost used	0.5x300	
	Fully composted, compost used	0.75x300	
	Fully composted, compost used internally and externally	300	
Inorganic waste treatment			
WS5	Burned in open area	0	**198**
	Taken off campus to a dump site	0.33x300	
	Partially recycled (less than 50%)	0.66x300	
	Fully recycled (more than 50%)	300	
	Total	1200	**447**

In Table 4, the total value of the three specific indicators of waste UI GreenMetric ranking was 447 out of a maximum score of 900 [4]. Universitas Gadjah Mad has implemented an integrated waste management system, where waste generated has had a treatment site.

Waste generation produced in UGM is not all can be processed in LDUS. Waste sent to LDUS is sewage coming from the depot biology and engineering depot. Based on the waste generation data in table 3, the average amount of waste generation that can be processed in LDUS is about 16% of the total waste. This is the reason for the assessment of indicators Recycling program for university waste has score 99. The waste that can be processed in LDUS is mostly organic waste that is recycled into compost in accordance with the SOP of waste management [3]. Organic waste will be recycled into compost process. The process of recycling organic waste into compost in LDUS using two methods namely Indore heap method and windrow method. Indicator Organic waste treatment on scoring system UI GreenMetric ranking has 150 points with the reason that organic waste produced in UGM area especially leaf waste has been processed into compost in LDUS and the compost product produced is used as fertilizer for plants in UGM and PIAT areas. The inorganic waste produced in UGM area can be through pyrolysis and incineration process to reduce waste disposal to Piyungan disposal site. UI GreenMetric rating indicator Inorganic waste treatment rate has a score of 198 because the inorganic waste generated has been recycled in LDUS even though only a small part. Plastic waste can be recycled into fuel oil through pyrolysis process. Meanwhile, other inorganic waste (especially waste included from biological depots and engineering depots) can be through incineration process. The fuel oil produced in the pyrolysis process can be used as fuel in the incineration process.

Waste processing technology at UGM needs to be improved in order to meet the indicators issued by UI GreenMetric ranking. To meet the criteria and indicators Recycling program for university waste to be effective and efficient, waste produced at least 50% processed. Universities can enact regulations on waste recycling programs starting from the source. In addition to promoting waste recycling programs, universities may encourage academicians to reduce the use of potentially wasteful waste materials such as reducing paper use in lectures, bringing drinking or reusable feeding, and so on. The organic waste generated especially the garden garbage and road sweep can be processed into compost in LDUS. Composting technology that has been run on LDUS can be added using vermicomposting technology that requires faster processing time than indore heap and windrow method that is 30 days [9]. The process of recycling organic waste in composting with indore heap and windrow method using leaves as raw materials. While the vermicomposting method can use the waste of food that comes from canteen, building, and foodcourt. The vermicomposting process has faster compost harvesting time than the other two methods of 30 days. The utilization of food waste can reduce the amount of waste disposal to Piyungan's final dump. And the compost produced from the vermicomposting method can be used for plant fertilizer around the canteen or leach foodcourt resulting from the composting process can be used as liquid fertilizer. With the method of composting with worms or vermicomposting where the process time is faster, it is expected that all organic waste can be processed so that the criteria in organic waste treatment can be maximized (Fully composted, compost used internally and externally) and the compost can be used in the region. The resulting leaf litter can be mixed with paper waste to become composite. Paper waste can be used as raw material for recycled paper or mixed as a composite material together with the leaf litter with a paper waste composition ratio of 60% by weight and leaves 40% [10]. Paper waste in the form of cardboard also can be processed into recycled paper. Other waste that cannot be processed can be burned in the incinerator to reduce waste disposal to the final disposal of piyungan. Incinerator installations require relatively little land when compared to landfills and composting and can operate continuously. The process of recycling inorganic waste such as waste plastic, waste paper, and other debris can increase the score in the ranking GreenMetric UI indicator becomes maximum scores and also can reduce the amount of waste in landfills Piyungan. Additional methods of waste processing can improve score indicator on UI GreenMetric ranking because 50% of waste generated in UGM area can be recycled. In addition, waste generated will increase in trash in the area of the campus there is an activity involving many people. to reduce the rate of waste dumped into the depo, University can provide mandatory regulations are adhered to by the event organizers. the given rule can be done before the event takes place by way of doing the mapping timbulan rubbish, at the time the event took place with, and while the event was taking place.

4. Conclusion s

The waste management system at UGM is an integrated waste management system. Academic community as a source of waste sorting waste according to the type of organic waste, plastic waste, paper waste, and other garbage. In order to meet the parameters of green campus concept that is the effectiveness of waste management, UGM has processed waste generation. Waste processing is done in LDUS PIAT UGM Kalitirto Area Berbah Sleman District. Assessment of the effectiveness of the application of green campus concept using UI GreenMetric ranking system. The indicator used in this research is Recycling program for the waste university, Organic waste treatment, and Inorganic waste treatment. In LDUS, organic waste is recycled into compost, plastic waste is recycled into alternative fuel oil through pyrolysis processes, and another untreated waste is incinerated. In the existing condition, the total score earned on the three indicators used amounted to 447 of the total maximum value of 900. so that the score on every indicator is maximal, additional technology is needed to recycle the waste generated. Organic waste that dominates the type of waste at UGM can apply the vermicomposting method in composting. The composting method with Indore heap and windrow uses leaf litter as raw material for composting, while in the vermicomposting method it can use the waste of food from canteens, buildings, and food court. Plastic waste generated in UGM area other than recycled into fuel oil can be used as handicraft materials by establishing cooperation with the parties component. Other waste that can not be recycled can be burned through the incineration process so as to reduce the amount of waste disposal to the Piyungan final disposal site. With additional waste treatment alternatives, the GreenMetric UI indicator score can be scaled up to the maximum score.

5. References

[1] Humblet E M, Owens R and Roy L P 2010 *Roadmap to a green campus* (washington, D.C.: U.S. Green Building Council 2101 L Street NW Suite 500 Washington, D.C. 20037)
[2] Marwati 2011 The Blue Campus Turns Green. Retrieved from website Universitas Gadjah Mada *ugm.ac.id*
[3] Mada U G 2014 *Bidang Sub Bidang Nomor SOP Nama SOP : Pertamanan Persampahan dan Pemakaman : Pertamanan : Ijin Pemasangan Baliho : Mulai Tahun 2014 : Direktur Aset RUANG LINGKUP Pelayanan pemasangan media informasi luar ruang kepada sivitas akademika UGM . Mewujudkan da* (Indonesia)
[4] Team U G R 2016 Guideline of UI GreenMetric World University Ranking 2016 *http://greenmetric.ui.ac.id/*
[5] Suwartha N and Sari R F 2013 Evaluating UI GreenMetric as a tool to support green universities development: Assessment of the year 2011 ranking *J. Clean. Prod.* **61** 46–53
[6] Asset D and UGM P 2016 *Laporan Survei Sampah UGM* (Yogyakarta)
[7] UGM P 2016 Menuju UGM Sebagai Pelopor Kampus Swakelola Sampah 1–16
[8] Marsiciwati N, Ma'as A and Asmara A A 2016 *Pengaruh Kerapatan Bangunan Windrow dan Perlakuan Awal Bahan Baku Kompos terhadap Kecepatan Pengomposan Sampah Kota* (Universitas Gadjah Mada)
[9] Sucipto C D 2012 *Teknologi Pengolahan Daur Ulang Sampah* (Yogyakarta: Gosyen Publishing)
[10] Masturi 2010 Pemanfaatan kuarsa sebagai penguat pada komposit sampah daun-kertas *J. Pendidik. Fis. Indones.* **6** 138–41

IOP Conf. Series: Journal of Physics: Conf. Series **1022** (2018) 012042 doi:10.1088/1742-6596/1022/1/012042

Simulation and prediction the impact of climate change into water resources in Bengawan Solo watershed based on CCAM (Conformal Cubic Atmospheric Model) data

Sinta B Sipayung[1*], Amalia Nurlatifah[2] and Bambang Siswanto[3]

[123]Pusat Sains dan Teknologi Atmosfer – LAPAN

[*]Corresponding email: s_berlianasipayung@yahoo.com

Abstract. Bengawan Solo Watershed is one of the largest watersheds in Indonesia. This watershed flows in many areas both in Central Java and East Java. Therefore, the water resources condition greatly affects many people. This research will be conducted on prediction of climate change effect on water resources condition in terms of rainfall conditions in Bengawan Solo River Basin. The goal of this research is to know and predict the climate change impact on water resources based on CCAM (Conformal Cubic Atmosphere Model) with downscaling baseline (historical) model data from 1949 to 2005 and RCP 4.5 from 2006 to 2069. The modeling data was validated with in-situ data (measurement data). To analyse the water availability condition in Bengawan Solo Watershed, the simulation of river flow and water balance condition were done in Bengawan Solo River. Simulation of river flow and water balance conditions were done with ArcSWAT model using climate data from CCAM, DEM SRTM 90 meter, soil type, and land use data. The results of this simulation indicate there is (i) The CCAM data itself after validation has a pretty good result when compared to the insitu data. Based on CCAM simulation results, it is predicted that in 2040-2069 rainfall in Bengawan Solo River Basin will decrease, to a maximum of only about 1 mm when compared to 1971-2000. (ii) The CCAM rainfall prediction itself shows that rainfall in Bengawan Solo River basin will decline until 2069 although the decline itself is not significant and tends to be negligible (rainfall is considered unchanged) (iii) Both in the DJF and JJA seasons, precipitation is predicted to decline as well despite the significant decline. (iv) The river flow simulation show that the water resources in Bengawan Solo River did not change significantly. This event occurred because the rainfall also did not change greatly and close to 0 mm/month.

1. Introduction

The water resources is one of the important issues that can affect many things. For example, in America, water resources of the Midwestern United States have a significant collective impact on multiple economic sectors in the US, North America, and the world [1]. Climate change can impact water resources through its impact into quantity, variability, timing, form, and intensity of precipitation [2].

Impacts of climate change are already evident in Indonesia and will likely worsen due to further human-induced climate change. Rising concentrations of GHG (greenhouse gas) will continue to raise the surface and ocean temperatures, change precipitation patterns, increase sea levels, and cause various other impacts from more frequent forest fires to increased health risks. Climate change will also continue to affect 'natural' climate variability, such as El Niño, and may lead to more frequent

IOP Conf. Series: Journal of Physics: Conf. Series **1022** (2018) 012042 doi:10.1088/1742-6596/1022/1/012042

and more intense weather events [3] and global sea-level rise is currently increasing at about 2mm per year and is projected to accelerate to a rate of about 5mm per year over the next century a change of this magnitude will undoubtedly result in significant losses of Indonesia's 80,000km of coastline, affecting thousands of islands and associated marine resources [4].

The Bengawan Solo River is the largest river on the island of Java [5], so spatial and temporal climate variability are key characteristics in hydrologic modelling distributed hydrological models like SWAT, it is designed to predict the impact of management on water, sediment and agricultural chemical yields in ungauged catchments [6]. This research aim to predict and simulate the impact of climate change into water resources in Bengawan Solo River Basin. Most areas in Bojonegoro are located in the Solo River watershed, which is a flood-prone area. Flooding is mostly caused by Bengawan Solo River basin overflow: the higher the intensity of the overflow, the more flooding there is downstream. The districts with the highest potential for flooding are Padangan, Kalitidu, Malo, Trucuk, Bojonegoro, Sumberejo, Kanor and Baureno.

2. Data and Methods

2.1. Data

Research data used in this research include two things:

1) Insitu Data/ Station Measurement
   In situ data is measurement data that obtained from PUSAIR. The data that used is data from 2000-2013.
2) CCAM Modeling Data
   CCAM modeling data is obtained from CCAM model simulation results and have monthly temporal resolution.It is climate data. The data that used is historical data from 1971-2005 and RCP data for 2006-2069.

2.2. Methods: Modelling with CCAM (Conformal Cubic Atmospheric Model)

The Conformal Cubic Atmospheric Model is a global atmospheric model principally developed for atmospheric climate modelling and regional climate downscaling [7][8]. CCAM employs a conformal cubic grid to model the atmosphere without requiring lateral boundaries [9]. CCAM is a hydrostatic atmospheric model with two-time-level semi-implicit time differencing. It uses semi-Lagrangian advection associated with bicubic horizontal interpolation and total-variation-diminishing vertical advection. In a typical case, CCAM runs with 18 vertical levels and employs a C48 grid on which there are 48×48 grid points for each of the six cubic panels (i.e., $N2 = 6 \times 48 \times 48$ grid points in the horizontal) [7].

In this case, RCP 4.5 data is used as the basis for the modelling and prediction of climatic conditions until 2069 by CCAM. Representative Concentration Pathway (RCP) 4.5 is a scenario that stabilizes radiative forcing at 4.5 Watts per meter squared in 2100 [10].

3. Results and Discussion

Bengawan Solo River basin is the largest river basin in Java and have geographical position in 110^0 18' and 112^0 45' E and 6^0 49' and 8^0 08' S. The basin belongs administratively to two provinces, Central Java and East Java. The majority of the Upper Solo River basin belongs to Central Java Province while the majority of the Lower Solo River basins belongs to East Java Province. The basin covers 17 regencies: Boyolali, Klaten, Sukoharjo, Wonogiri, Karanganyar, Sragen, Blora, Rembang, Ponorogo, Madiun, Magetan, Ngawi, Bojonegoro, Tuban. Lamongan, Gresik, Pacitan; and 2 municipalities: Surakarta and Madiun [5]. Figure 1 shows the map of Bengawan Solo River Basin.

IOP Conf. Series: Journal of Physics: Conf. Series **1022** (2018) 012042 doi:10.1088/1742-6596/1022/1/012042

Figure 1. Map of Bengawan Solo Watershed [11]

The majority of riparian areas in the Solo River District in Bojonegoro prone to flooding are flooded 5–6 times annually, covering a total area of 33.59 ha. The areas most susceptible to flooding cover 25.28ha, with an extensive flooding frequency of more than 7 times/year. Meandering rivers cause differences in flooding frequency in each sub-district. Residential areas in the district of Bojonegoro flood 5–7 times each year and 47 per cent of the district is affected. Based on data from the Bojonegoro Regional Disaster Management Agency, there were as many as 16 floods in 2010 [12].

The rainfall pattern in the Java Island area, including Bengawan Solo, generally has a monsunal type [13]. Type monsoonal type means maximum rainfall occurs in December-January-February (DJF) and minimum rainfall occurs in June-July-August (JJA). Based on the analysis of station measurement data (insitu data) it can be seen that in Bengawan Solo Basin area of Ngawi, Bojonegoro, and Ngrambe, the maximum rainfall is around December-January-February. While the minimum rainfall occurs around June-July August. This indicates that rainfall in Bengawan Solo Basin has monsoonal pattern. Rainfall in Bengawan Solo basin area in 2000-2013 has maximum value around 400 mm / month in wet season (December-January-February) and around 50 mm / month in dry season (June-July-August) (Figure 2).

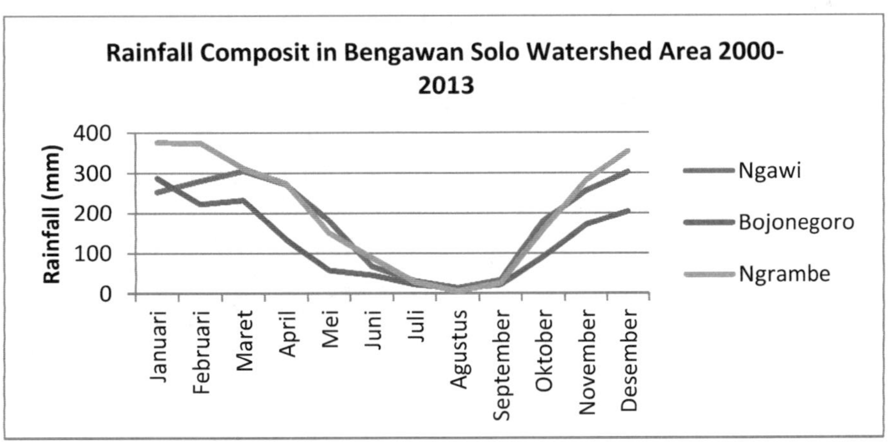

Figure 2. Rainfall Composit in Bengawan Solo Watershed (2000-2013) Based on Insitu Data

Meanwhile, according to CCAM data simulation, rainfall composite in Bengawan Solo Basin in 2006-2013 indicates that maximum rainfall also happened in December-January-February and minimum rainfall happened in June-July-August. However, there is considerable bias between the

IOP Conf. Series: Journal of Physics: Conf. Series **1022** (2018) 012042 doi:10.1088/1742-6596/1022/1/012042

maximum precipitation quantities simulated by CCAM when compared to observational data (in situ measurements / stations). The CCAM data itself shows that the maximum rainfall in Ngawi and Bojonegoro is about 300 mm/month (Figure 3).

CCAM data after being validated by observation data /station data shows good performance in simulating monthly rainfall conditions in Bengawan Solo River Basin. The correlation of CCAM rainfall data and station data in Ngawi reaches 0.82 while in Bojonegoro the value is 0.91. This indicates that CCAM is sufficiently able to simulate rainfall conditions in Bengawan Solo River Basin and its simulation results can be used in the analysis of water resources conditions in Bengawan Solo River basin.

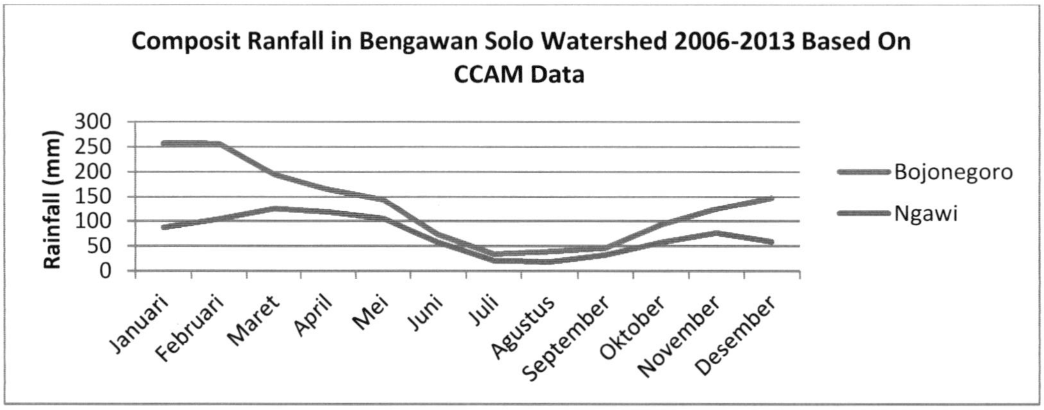

Figure 3. Composit Rainfall in Bengawan Solo Watershed (2000-2013) Based On CCAM Data

The condition of water resources can be influenced by many things. In the present study the condition of water resources to be discussed is viewed from the rainfall condition. This is because rainfall is one of the main factors that can determine the condition of water resources in a region. The rainfall itself is part of weather parameters that can change and be influenced by climate change phenomena.

To determine climate change occurrence in Bengawan Solo River Basin, the temperature simulation plot was conducted in 1971-2000 and CCAM also predict temperature conditions in 2040-2069. From the the that temperature plot, it was observed that in 1971-2000 the temperature in the Bengawan Solo basin is relatively small around 300-302 K or 27-29 °C (Figure 4a). While in 2040-2069, the temperature is predicted to increase reach about 302-304 K or 29-31 °C (Figure 4b). Anomaly temperature plot was also conducted to determine temperature change during the period 2040-2069 when compared with the period 1971-2000. It is seen that the changes occurring during the history (1971-2000) as compared to RCP (2040-2069) are about 1.7-2.7 K (Figure 4c).

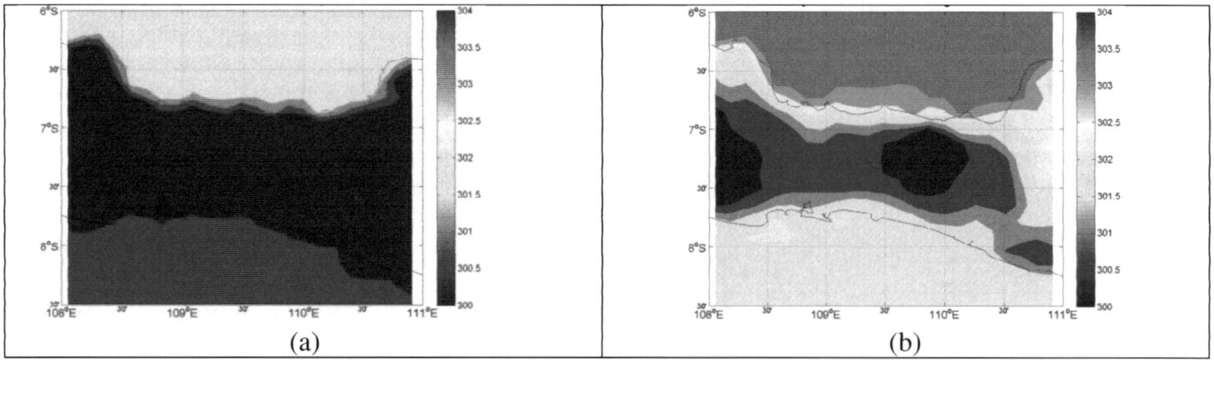

(a) (b)

IOP Conf. Series: Journal of Physics: Conf. Series **1022** (2018) 012042 doi:10.1088/1742-6596/1022/1/012042

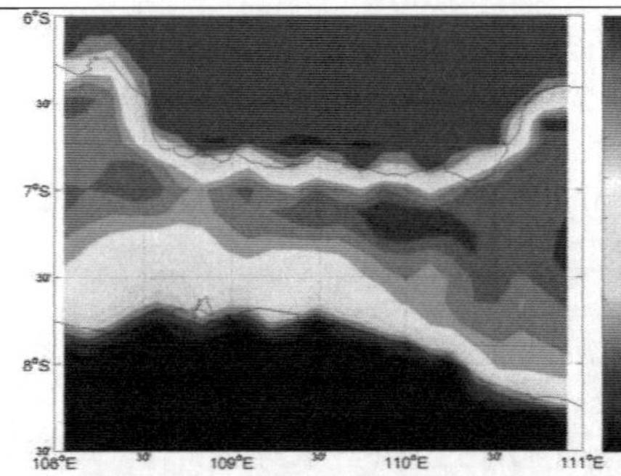

(c)

Figure 4. (a) Monthly Temperature in Bengawan Solo Based on CCAM data in 1971-2000 (history)(in Kelvin), (b) Monthly Temperature in Bengawan Solo Based on CCAM data in2040-2069 (RCP)(in Kelvin), and (c) Anomaly Temperature (RCP-history) (in Kelvin)

Climate change is believed to affect the rainfall conditions in Bengawan Solo River Basin. In the period 1971-2000, rainfall in Bengawan Solo was in the range of 50-250 mm with maximum conditions in the southwestern part of Central Java reaching 400 mm (Figure 5a). Conditions are not much different is predicted to occur in 2040-2069 (Figure 5b). Therefore, the CCAM model shows a decrease in rainfall in 2040-2069 when compared to 1971-2000 only about 0-1 mm.

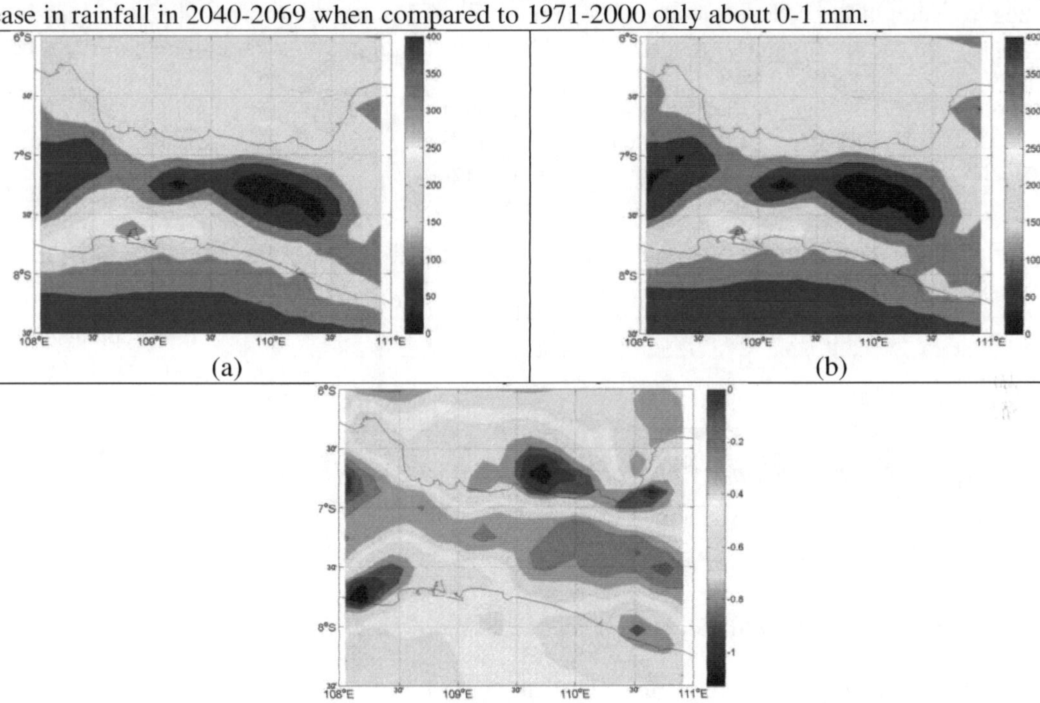

Figure 5. (a) Monthly Rainfall in Bengawan Solo Based on CCAM data in 1971-2000 (history) (in mm), (b) Monthly Rainfall in Bengawan Solo Based on CCAM data in 2040-2069 (RCP) (in mm), and (c) Anomaly Rainfall (RCP-history) (in mm)

Rainfall simulation in Bengawan Solo, especially in Ngawi and Bojonegoro until 2069 are predicted to decrease. It is shown in figure 6 that the linear regression shows the negative value on slope (x coefficient). Nevertheless, the coefficient value of x in linear regression process is very small

IOP Conf. Series: Journal of Physics: Conf. Series **1022** (2018) 012042 doi:10.1088/1742-6596/1022/1/012042

even close to 0. This means that until 2069, there will be a decrease of rainfall in Bengawan Solo basin but the decrease itself will not be too significant.

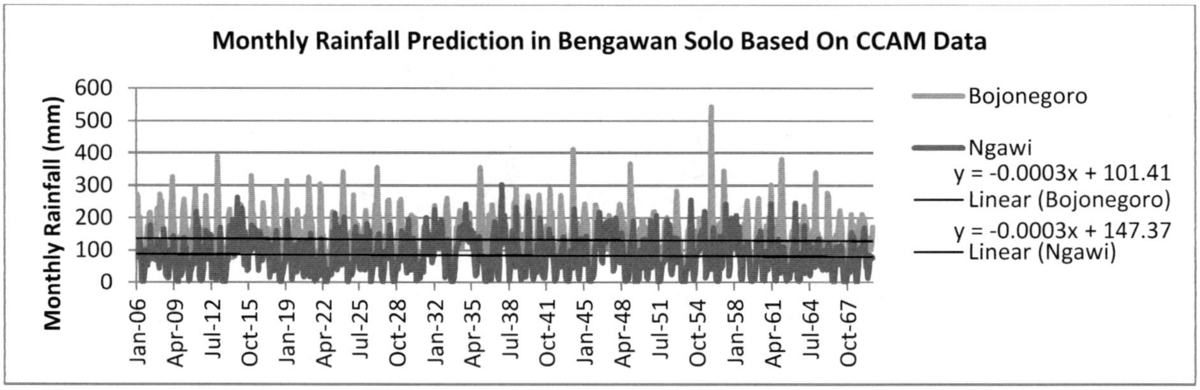

Figure 6. Monthly Rainfall Prediction (2006-2067) in Bengawan Solo Based on CCAM Data

To know how the effect of climate change on rainfall condition in wet season in Bengawan Solo River Basin, there is plot of rainfall simulation until 2069 is for DJF season. It is seen that in Bojonegoro, rainfall is predicted to decrease as evidenced by the slope of the linear regression line decreasing with the coefficient of x is negative. However, the decrease will not be significant because the slope value is near 0 (Figure 7).

In contrast to Bojonegoro, the increase in precipitation is predicted to occur in Ngawi until 2069. This is due to the positive slope of linear regression line. However, the increase will not be too significant because the value of its slope is close to 0 (Figure 7).

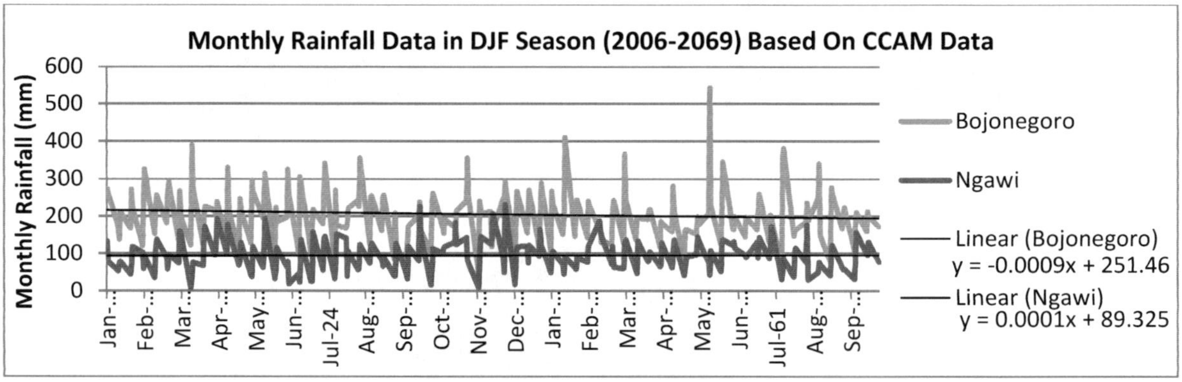

Figure 7. Monthly Rainfall Data in DJF Season (2006-2069) Based on CCAM Data

To find out how the effect of climate change on rainfall conditions in the dry season in Bengawan Solo River Basin, there is plot of rainfall prediction up to 2069 specifically on JJA season. It can be seen that both in Bojonegoro and Ngawi, rainfall is predicted to decrease as evidenced by the slope of the linear regression line that decreases and the coefficient of x is negative. However, the decrease will not be significant because the slope value is close to 0 (Figure 7). The decreases of rainfall in Bojonegoro region itself is predicted to be higher when compared with Ngawi in the JJA season because the coefficient of x in Bojonegoro is much more negative when compared with Ngawi.

If water resources are observed from rainfall conditions, it can be predicted that future water resources conditions in Bengawan Solo River Basin until 2069 will not change much. It will be a

IOP Conf. Series: Journal of Physics: Conf. Series **1022** (2018) 012042 doi:10.1088/1742-6596/1022/1/012042

decline but the value is not significant enough. This is due to by linear regression analysis, the slope of the regression line is close to 0 mm/month which means the decrease is close to 0.

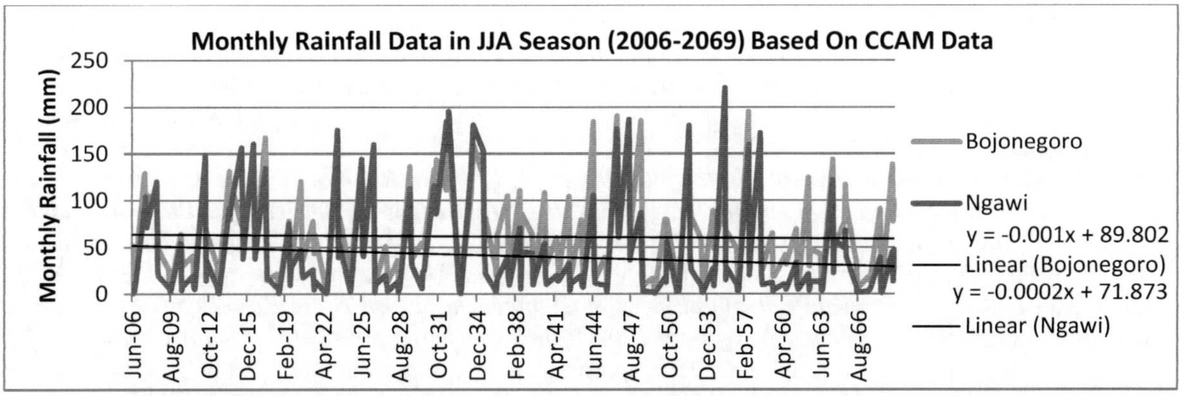

Figure 8. Monthly Rainfall Data in JJA Season (2006-2069) Based on CCAM Data

To better representation about the condition of water availability in Bengawan Solo River, calculation and simulation of the Bengawan Solo River discharge condition was done using ArcSWAT software. The results of this simulation found that the future of Bengawan Solo river discharge will not change much. According to the results of linear regression, it was found that in the Ngawi area, the Bengawan Solo River discharge will decrease by about $3x10^{-5}$ m^3/sec per month and the flow of Bengawan Solo river in Bojonegoro will decrease about $6x10^{-6}$ m^3/sec per month (Figure 9). The magnitude is very small and close to 0 so it can be concluded that the Bengawan Solo River discharge in the Ngawi and Bojonegoro areas is fixed. But different with [14], he has explored different approaches to precipitation input in an alpine area of large topographic relief and concluded that radar, combined with rain gauge data, performed better than gauge-only data. Consequently, these studies have highlighted the need for additional research to improve precipitation input to SWAT simulations.

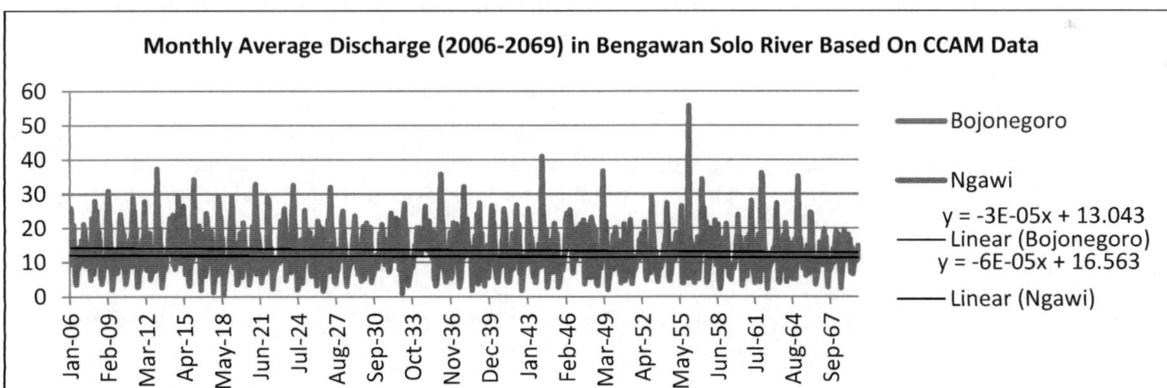

Figure 9. Monthly Average Discharge (2006-2069) in Bengawan Solo River Based on CCAM Data (in m^3/second)

4. Conclusions

CCAM data has good validation results when compared to in situ data. This means that CCAM data can be used for analysis of rainfall conditions in Bengawan Solo basin. Climate change is predicted to occur in Bengawan Solo basin as evidenced by the predicted rise in temperature in Bengawan Solo

River Basin in 2040-2069 compared to temperatures in 1971-2000 based on CCAM data. This resulted in a decrease in rainfall in the period 2040-2069 even though the maximum decrease is only 1 mm when compared with the year 1971-2000.CCAM predicts there will be a decrease in rainfall up to 2069 in Bengawan Solo (Ngawi and Brojonegoro) watersheds although the decrease is not significant and close to 0. This means that the rainfall conditions in Bengawan Solo basin will not change much until 2069. This means the water resources in Bengawan Solo River Basin until 2069 is fixed.

5. References

[1] Lofgren B and A Gronewold 2012 *Water Resources In: U.S. National Climate Assessment Midwest Technical Input Report. J Winkler, J Andresen, J. Hatfield, D. Bidwell, and D. Brown, coordinators.* Available from the Great Lakes Integrated Sciences and Assessments (GLISA) Center, http://glisa.msu.edu/docs/NCA/MTIT_WaterResources.pdf.

[2] Adams R M and Dannele E P 2008 *Effects of Climate Change on Water Resources*

[3] Diposaptono S, Budiman and Agung F 2009 *Menyiasati perubahan iklim di wilayah pesisir dan pulau-pulau kecil* (Bogor: Penerbit Sarana Komunikasi Utama)

[4] Cruz R V, Harasawa H, Lal M, Wu S, Anokhin Y, Punsalmaa, B, Honda Y, Jafari M, Li C and Huu Ninh, N 2007 *Asia climate change 2007: impacts, adaptation and vulnerability. contribution of working group ii to the fourth assessment report of the intergovernmental panel on climate change* M. L. Parry, O. F. Canziani, J. P. Palutikof, P. J. van der Linden and C. E. Hanson (Eds.) Cambridge University Press: UK 469–506 www.ipcc.ch/pdf/assessment-report/ar4/ wg2/ar4-wg2-chapter10.pdf

[5] Hidayat F, Harry Sungguh and Harianto 2008 *Impact of climate change on floods in Bengawan Solo and Brantas River basins, Indonesia* Proceedings of the 11th International River Symposium, Brisbane, Australia

[6] SWAT *Soil and water assessment tool, swat model software. U. S. Department of Agriculture-Agricultural Research Service Grassland* Soil & Water Research Laboratory (Temple, Texas) Available online: http://swatmodel.tamu.edu/software/swat-model/ (accessed on 22 June 2011).

[7] McGregor J L 2005 CCAM: Geometric aspects and dynamical formulation *CSIRO Marine and Atm. Res. Tech. Paper* **70**(43)

[8] McGregor J L and M R Dix 2008 A*n updated description of the conformal cubic atmospheric model high resolution simulation of the atmosphere and ocean* K. Hamilton and W. Ohfuchi (Eds.) Springer 51–76

[9] Thatcher M, J L McGregor 2009 Using a scale-selective filter for dynamical downscaling with the conformal cubic atmospheric model *American Meteorology Society*

[10] Thomson A M, Katherine V C, Steven J S, G Page K, April V, P Patel, S Delgado-Arias, B Bond-Lamberty, M A Wise, L. E Clarke and J A Edmonds 2011 RCP4 5: A *pathway for stabilization of radiative forcing by 2100*

[11] https://id.wikipedia.org/w/index.php?title=Berkas:Lokasi_WS_Bengawan_Solo.jpg&filetimestamp=20160817115943&

[12] Bengawan Solo Water Resources Management Boards 2013 *Bengawan Solo Watershed Tecnical Operation Unit Report*

[13] Aldrian E and Susanto R 2003 Identification of three dominant rainfall region within Indonesia and their relationship to sea surface temperature *Intl. J. of Clim.* (Wiley InterScience)

[14] Yu M, Chen X, Li L, Bao A and de la Paix MJ 2011 Streamflow simulation by SWAT using different precipitation sources in large arid basins with scarce raingauges. *Water Resour. Manag* **25** 2669–81.

Acknowledgments
The authors would like to thank to PSTA-LAPAN (Centre of Atmospheric Science and Technology-National Institute Of Aeronautics and Space) for their support both morally and materially so that the writer can finish this research.

IOP Conf. Series: Journal of Physics: Conf. Series **1022** (2018) 012043 doi:10.1088/1742-6596/1022/1/012043

Behavioral intention and its relationship with gender: a study of green school students in Surakarta, Indonesia

I A Fakhrudin[1]*, P Karyanto[1] and M Ramli[1]

[1] Department of Biology Education, Universitas Sebelas Maret, 36 A Ir. Sutami Street, Surakarta, Central Java 57126, Indonesia

*Corresponding email: azizfakhrudin18@gmail.com,

Abstract. The environmental problems faced today have had a great impact on human beings. The root causes of all environmental problems are related to injudicious human behavior. Since human behavior is determined by Behavioral Intention (BI), it is crucial to examine the role of BI with regard to the study of environmentally friendly behavior. As such, in this research, BI was scrutinized; in particular, this study aimed to identify the factors affecting the particular degree of BI from the perspective of gender. This study was conducted with 334 high school student participants in the district of Surakarta, Indonesia. According to the New Ecological Paradigm (NEP) result, the BI possessed by high school students does not sufficiently promote environmentally friendly behavior. This research also found that there is no significant link between BI and gender.

Key words: Environmentally Friendly Behavior, Behavioral Intention, Gender

1. Introduction

Environmental problems are an ever-increasing global issue and exert adverse effects on human life. Such problems should be dealt with immediately, with respect to both organic and inorganic elements. One organic component that contributes to environmental problems, and is becoming the focus of attention for those who are attempting to preserve the environment, is human behavior [1][2][3]. Ever-changing human behavior requires the consumption of increasingly large amounts of Earth's natural resources [4]. As such, environment and social experts are focusing their efforts on improving human behavior as quickly as possible, in order to realize sustainable development [5][6].

Environmentally friendly behavior is supported by Behavioral Intention (BI) as the main factor in conditioning the emerging behavior [7][8][9][10]. Every individual's BI is different, due to the effects of certain factors, one of which is gender. In psychological research, gender has been found to affect mindset, attitude, and belief in making decisions, impacting on the readiness to perform a behavior [11][12][13]. Another study mentions that gender does not affect an individual's BI [14]. The discussion relates to the effect of gender on BI; generally, significantly different effects have been found in various studies, and this study will attempt to establish the nature of the relationship between gender and BI in Surakarta district, Indonesia.

1.1. Environmentally Friendly Behavior and Behavioral Intention: New Ecological Paradigm Scale Measurement

Human behavior is affected by BI. Ajzen (1991) [7] notes that BI is the main factor assumed to capture the motivation factors affecting human behavior [15]. An individual realizing that the result of performing a behavior is positive will have a positive attitude, belief and perspective regarding the behavior, and similarly, a behavior with a negative result will engender a negative attitude [10][15].

BI on the environment consists of some factors that are assumed to encourage an individual to behave in an environmentally friendly manner. The factors affecting BI are subjective norms, which control attitudes towards the environment [7,16]. Other aspects affecting an individual's behavior regarding the environment are the cognition an individual has of the environment [17-20], beliefs, and the amount of responsibility they feel for the environment [21]. BI represents an individual's cognitive readiness to behave [22] and affects the behavior to appear [14, 17, 18]. Cultural and social differences in every area affect the cognitive level, thereby impacting on different BI [16, 23].

BI can be measured by using the New Ecological Paradigm [24]. The NEP has been used widely as an instrument to determine the relationship between BI and supporting aspects [14, 25-32]. The NEP instrument is one which is adapted to social life in America [33]. The limitations relate to the content existing in the NEP statement, which should be adjusted according to the area of the object of research.

1.2. The Effect of Gender on Behavioral Intention

Gender plays an important part in affecting an individual's BI. The difference between the mindsets and beliefs of men and women results in the difference in values supporting BI. Such differences have resulted from cultural patterns and biological factors. However, cultural factors have a greater potential to cause differences in factors affecting BI than biological factors [34-37].

The differences between men and women are created by sociocultural norms and expectations, behavior, and attitude [11]. These differences are explained in the research conducted by Park & Slaby (1983) which found that parents tend to give boys more freedom than girls to develop their potential and to prevent girls from taking action [38]. Further differences were studied by [11] Eisler et al. (2003), who explained that men's knowledge levels tend to be higher than those of women, while women's motivation levels are higher than those of men. The effect of gender on behavioral attention potentially has a close relationship with and impact on an individual's BI.

2. Methodology

This research was conducted by using an NEP instrument [33]. This instrument is valid and reliable, and has been applied in various countries, including Niger [14], Malaysia [39], and the United State of America [40], to measure NEP and factors affecting environmentally friendly behavior. The NEP instrument consists of 15 revised statements [41] and gender status. The NEP status is categorized using a scale (1–5): 1 = strongly disagree, 2 = disagree, 3 = not sure, 4 = agree, and 5 = strongly agree.

The sample was taken from the Senior High Schools that have been named as green schools in Surakarta district, Indonesia, using a random sampling technique. The questionnaire was distributed to the students of Senior High Schools because they are adolescents. Adolescent students tend to exhibit manners of thinking that can lead to a wise or emotional decision [42-43]. The sample was conducted in Solo Raya because the regencies in that area have achieved the Adipura award as a form of appreciation their care for the environment. In such an area, it is assumed that awareness and understanding of environmentally friendly behavior is encouraged. The NEP instrument was distributed to 352 students, of which 334 questionnaires (89 males and 245 females) were deemed valid; 18 were not valid because some question items had not been answered. In other words, the response rate was 100%, and 94.88% of questionnaires were complete. Data analysis was carried out quantitatively using SPSS 17 application and Iteman version 3.0 in order to determine the validity, reliability, mean score, and relationship between BI and gender.

3. Results and Discussions

3.1 Results

The result of the NEP score, deriving from the mean score of 334 students, was 3.27 (5-scale). The mean NEP obtained indicates low BI among the students [44]. The reliability score was 0.465, indicating low instrument reliability based on Cronbach's alpha criteria. The validity of the question for items 1–15 obtained, respectively, the item-scale correlation scores of: 0.44, 0.48, 0.49, 0.46, 0.37, 0.01, 0.15, 0.41, 0.04, 0.41, 0.43, 0.46, 0.47, 0.26, and 0.19. The data validity shows that items 5, 6, 7, 8, 14, and 15 obtained item-scale correlation coefficient < 0.3 and are thereby stated as invalid. The details of the mean score and standard deviation of the respondents can be seen in Table 1.

Table 1. NEP Items with Frequency, Mean, and Standard Deviation of Respondents

NEP facets	Statement items	Responses (%)			Mean	SD
		A	D	U		
Limit to growth	The earth's ability of supporting the human population number is limited.	62	18	20	3.60	1.03
	Actually, earth has much natural resource to support human life, but human does not sufficient knowledge to utilize it.[a]	11.6	73.8	14.6	2.14	0.88
	Earth has limited space and resources to support human life.	49.5	24.2	26.3	3.42	1.22
Anti-anthropocentrism	Human beings has right to utilize resource and environment completely in sufficing their life.[a]	15.5	69	15.5	2.21	1.09
	Living organism other than human being has equal right to live and to utilize resource and environment.	92.8	1.5	5.7	4.51	0.48
	Human being can master natural resource and environment completely to be utilized for his interest.[a]	40.8	27.5	31.7	3.19	1.09
Fragility of nature	When human being utilizes natural resource and	58.8	15.5	25.7	3.54	0.91

	environment, natural balance becomes damaged frequently.					
	Natural resource and environment will not be harmed by modernization and technology created by human beings.[a]	57.9	10.5	31.6	3.58	0.84
	Natural balance is damaged easily.	66.3	19.1	14.6	3.62	1.35
Rejection of exemptionalism	Human beings with intellectuality and technology they create can always solve natural resource and environment problems.[a]	12.5	40.6	46.9	2.61	0.74
	Despite intellectuality and technology, human beings keep submitted to natural resource.	60	10.2	29.8	3.70	0.84
	Human beings always learn from nature to be able to control natural resource and their environment for their interest.[a]	6.6	68.7	24.7	2.17	0.71
Possibility of an eco-crisis	In utilizing natural resource, human beings always act without mature consideration.	47.7	25.1	27.2	3.28	1.11
	The current condition of natural resource and environment tend to be worse and harmful to human life sustainability.	82.7	4.5	12.8	4.10	0.68
	The story of environment and natural resource damage is actually exaggerated.[a]	51	17.3	31.7	3.44	1.00
	Mean total NEP score				3.27	0.33

[a] Reverse coded

The result of analysis shows that the mean scores of the facets of NEP were obtained as follows: limit to growth result 3.05 ± 1.21, anti-anthropocentrism 3.30 ± 1.33, fragility of nature balance 3.58 ± 1.01, rejection of exemptionalism 2.82 ± 1.08, and possibility of an eco-crisis 3.61 ± 1.02. The comparison of the mean score for every facet can be seen in Figure 1.

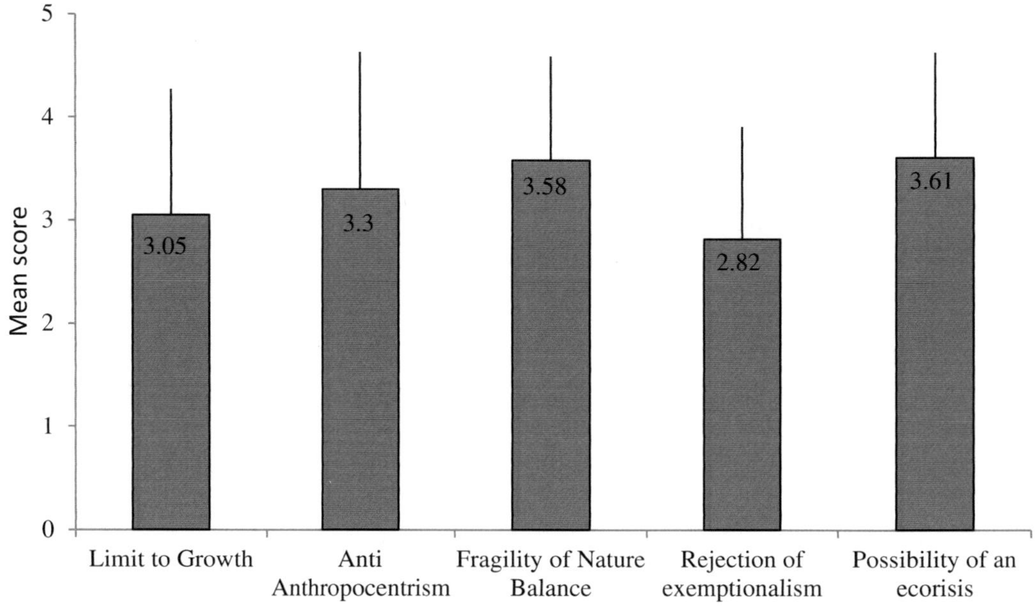

Figure 1. Respondents' Mean Scores and Standard Deviation on NEP Facets

The result of the analysis shows the mean scores of 3.33 and 3.25 for men and women, respectively. The significance level of 0.058 shows that there is no significant difference between women and men. The comparison of NEP mean scores between men and women can be seen in Table 2.

Table 2. Comparison of NEP Mean Scores of Males And Females

		Mean score	SD
Gender	*Male*	3.33	0.33
	Female	3.25	0.33

3.2 Discussion

This research aimed to discover the NEP value of Senior High School students from schools with a green status and to determine the relationship between gender and NEP value as the indicator of BI assessment. The result of the analysis shows that the NEP value of students is not sufficient to support the appearance of environmentally friendly behavior. The score obtained (3.27) is similar to those of a number of samples from areas such as Africa [14] (2.95; Ogunbode, 2013), Australia [45] (3.96; Blaikie, 1992), the United States [46] (3.57; Kortenkamp, & Moore, 2006), Brazil [47] (3.55; Schultz et al., 2000), Turkey [48] (3.50; Erdogan, 2009). The difference in the NEP mean score of women and men is not significant (0.058); this finding is similar to the result of the research conducted by Ogunbode (2013) [14]. The analysis of learning materials utilized in schools shows that, in fact, some of the opinions expressed by students regarding the rejection of exemptionalism items are inconsistent with the content taught.

Environmental education, as an attempt to create sustainable development, should provide output that encourages the establishment of sustainable development, either individually or collectively [49]. The low NEP value was obtained because the individuals have not yet understood the content of the statements proposed [14], and the learning process does not yet support the intention to behave in an environmentally friendly manner [50-51]. Different cultures and social conditions in every area also affect the materials taught. An individual's perspective determines whether or not something is good, and this is highly influenced by culture and social life. For example, in Indonesia's south beach, there is a tradition to dispose of a buffalo's head on the beach as an expression of gratitude to God. This traditional perspective can be viewed differently when related to various disciplines. Culture and social life are inseparable from an individual's perspective in determining behavior and becoming the factors which affect different BI [23, 29, 30, 32, 52].

The different effects of sex on BI play a significant role. Eisler et al. (2003) explain that differences in sex affect mindset, perspective, and behavior, due to the process of establishing maturity by environment and culture [11]. Psychological and biological factors of men and women can also impact behavioral differences [37]. Men tend to be more intelligent, in theory, but women have a greater understanding of feelings in making decisions. The absence of significant differences in this research is due to some factors, such as: age [12], moral development or maturity [53], and education [54]. Further research related to the factors accounting for BI differences should be conducted in the future.

4. Conclusions

The results of the research show that the NEP mean score of Senior High School students based in a green school is not adequate to support the appearance of BI. There is no significant difference between the BI of boys and girls. Varying NEP values in every aspect show that the students have different perspectives on BI, which are affected by cultural, social, education, psychological, and biological factors. More research should be conducted to determine which factors affect the low BI, such that it will be possible to encourage environmentally friendly behavior among students, thereby improving the social and environmental conditions in which Indonesian people live.

5. References

[1] Skinner B F 1965 *Science and human behavior* (New York: The Free Press, a division of Simon & Schuster Inc.)
[2] Schultz P W and Zelezny L C 1999 Values as predictors of environmental attitudes: Evidence for consistency across cultures *J. of Env. Psych.* **19** 255–65
[3] European Environment Agency 1999 *Annual report* Copenhagen: Kongens Nytorv
[4] European Commission 2013 *A healthy and sustainable environment for future generations.* Luxembourg: European Union
[5] Kurz, T 2002 The psychology of environmentally sustainable behavior: Fitting together pieces of the puzzle *Analyses of Social Issues and Policy* **2** 257–78
[6] Mascia B M, Brosius J P, Dobson T A, Forbes B C, Horowitz L, McKean M A, et al 2003 Conservation and the social sciences *Conservation Biology* **17** 649-50
[7] Ajzen I 1991 The theory of planned behavior *Organizational Behavior and Human Decision Processes* **50** 179–211
[8] Christensen R and Knezek G 2015 The climate change attitude survey: Measuring middle school student beliefs and intentions to enact positive environmental change. *International J. of Env. and. Sci.Edu.* **10**(5) 773–88
[9] McGuire N M 2015 Environmental education and behavioral change: An identity-based environmental education model *Intl. J. of. Env. and. Sci. Edu.* **10**(5) 695–715
[10] Yu T Y, Yu T K. and Chao CM 2017 Understanding Taiwanese undergraduate students' pro-environmental behavioral intention towards green products in the fight against climate

change *J. of Cl. Prod.* **161** 390–402
[11] Eisler A D Eisler H and Yoshida M 2003 Perception of human ecology: Cross-cultural and gender comparisons *J. of Env. Psyc.* **23**(1) 89–101
[12] Bart W M, Hokanson B, Sahin I and Abdelsamea M A 2015 An investigation of the gender differences in creative thinking abilities among 8th and 11th grade students *Thinking Skills and Creativity* **17** 17–24
[13] Binngießer J and Randler C 2015 Association of the environmental attitudes "Preservation" and "Utilization" with pro-animal attitudes *Intl. J. of Env. and Sci. Edu.* **10**(3) 477–92
[14] Ogunbode C A 2013 The NEP scale: Measuring ecological attitudes/worldviews in an African context *Environment, Development and Sustainability* **15**(6) 1477–94
[15] Cristea M and Gheorghiu A 2016 Attitude, perceived behavioral control, and intention to adopt risky behaviors *Transportation Research Part F: Psychology and Behaviour* **43** 157–165
[16] Riepe C, Fujitani M, Cucherousset J, Pagel T, Buoro M, Santoul F, Lassus R and Arlinghaus R 2017 What determines the behavioral intention of local-level fisheries managers to alter fish stocking practices in freshwater recreational fisheries of two European countries? *Fisheries Research* **194** 173–87
[17] Kaiser F G, Wolfing S, and Fuhrer U 1999 Environmental attitude and ecological behaviour. *J. of. Env. Psyc* **19** 1–19
[18] Al-dajeh H L 2012 Assessing environmental literacy of pre-vocational education teachers in Jordan *College Student Journal* 492–508
[19] McBride B B, Brewer C A, Berkowitz A R, and Borrie W T 2013 Environmental literacy, ecological literacy, ecoliteracy: What do we mean and how did we get here? *Ecosphere* **4**(5) 67
[20] Yen Y, Wang Z, Shi Y, Xu F, Soeung B, Tayyab M, … Abdalla S 2017 The predictors of the behavioral intention to the use of urban green spaces: The perspectives of young residents in Phnom Penh, Cambodia *Habitat International* **64** 98–108
[21] Shamuganathan S and Karpudewan M 2015 Modeling environmental literacy of Malaysian pre-university students *Intl. J. of Env. and Sci. Edu.* **10**(5) 757–71
[22] Ajzen I, Netemeyer R and Ryn M Van 1991 The theory of planned behavior *Organizational Behavior and Human Decision Processes* **50**(2) 179–211
[23] Malhotra K N and McCort D J 2001 A cross-cultural comparison of behavioral intention models - theoretical consideration and an empirical investigation *International Marketing Review* **18**(3) 235–69
[24] Dunlap R E, Van Liere K D, Mertig A G and Jones R E 2000 Measuring endorsement of the New Ecological Paradigm: A revised NEP scale **56**(3) 425–442
[25] Vining J and Ebreo A 1992 Predicting behavior from global and specific environmental attitudes and changes in recycling opportunities *J. of App. Soc. Psyc.* **22** 1580–607
[26] Blake D E, Guppy N and Urmetzer P 1997 Canadian public opinion and environmental action. *Canadian J. of Pol. Sci.* **30** 451–72
[27] Ebreo A, Hershey J and Vining J 1999 Reducing solid waste: Linking recycling to environmentally responsible consumerism. *Environment and Behavior* **31** 107–135
[28] O'Connor RE, Bord R J and Fisher A 1999 Risk perceptions, general environmental beliefs, and willingness to address climate change *Risk Analysis* **19** 461–71
[29] Rideout B E 2014 The liberal arts and environmental awareness: Exploring endorsement of an environmental worldview in college students *Intl. J. of Env. and Sci. Edu.* **9**(1) 59–76
[30] Li W and Lang G 2015 Effects of green school and parents on children's perceptions of human-nature relationships in China. *Child Indicators Research* **8**(3) 587–604
[31] Maltese C F and Zimmerman T D 2015 A garden-based approach to teaching life science produces shifts in students' attitudes toward the environment *Intl. J. of Env. and Sci. Edu.* **10**(1) 51–66
[32] Chua K B, Quoquab F, Mohammad J and Basiruddin R 2016 The mediating role of New

Ecological Paradigm between value orientations and pro-environmental personal norm in the agricultural context *Asia Pacific J. of Mark. and Log.* **28**(2) 323–49

[33] Dunlap R E and York R 2008 The globalization of environmental concern and the limits of postmaterialist values explanation: Evidence from four multinational surveys. *The Sociological Quarterly* **49** 529–563

[34] Bandura A 1986 *Social foundations of thought and action: A special cognitive theory.* Englewood Cliffs NJ: Prentice-Hall

[35] Beall A E and Sternberg R J 1993 *The psychology of gender* New York: Guilford Press

[36] Epstein C F 1997 The multiple realities of sameness and difference: Ideology and practice *J. of Soc. Iss.* **53** 259–78

[37] Bussey K and Bandura A 1999 Social cognitive theory of gender development and differentiation. *Psychological Review* **106** 676–713

[38] Park R D and Slaby R G 1983 The development of aggression. In P. H. Mussen, & E. M. Hetherington (Eds.) *Handbook of child psychology: Socialization, personality, and social development* pp 547–641 New York: John Wiley

[39] Darner R 2014 Influences on students' environmental self determination and implications for science curricula *Intl. J. of Env. and Sci. Edu.* **9**(1) 21–39

[40] Joseph C, Obrin E N, Tamoi J and Madi N 2013 Environmental literacy and attitudes among Malaysian business educators *Intl. J. of Sus. in Hi. Edu.* **14**(2) 196–208

[41] Darner R 2014 Influences on students' environmental self determination and implications for science curricula *Intl. J. of Env. and Sci. Edu.* **9**(1) 21–39

[42] Kopnina H 2011 Applying the New Ecological Paradigm scale in the case of environmental education: Qualitative analysis of the ecological world view of Dutch children *J. of Peace. Edu. and Soc. Just.* **5**(3) 374–88

[43] Papalia D E, Olds S W, and Feldman R D 2009 Human Development: Perkembangan Manusia. Jakarta: Salemba Humanika

[44] Papalia D E and Feldman R D 2014 Experience Human Development: Menyelami Perkembangan Manusia. Jakarta: Salemba Humanika

[45] Johannsdottir L 2009 *Environmental literacy of business students* Reykjavik: University of Iceland

[46] Blaikie N H 1992 The nature and origins of ecological world views: An Australian study. *Social Science Quarterly* **73** 144–65

[47] Kortenkamp K V and Moore C F 2006 Time, uncertainty, and individual differences in decisions to cooperate in resource dilemmas *Personality and Social Psychology Bulletin* **2** 257–78

[48] Schultz P W, Unipan J B, and Gamba R J 2000 Acculturation and ecological worldview among Latino Americans *J. of Env. Edu.* **31** 22–7

[49] Erdogan N 2009 Testing the new ecological paradigm scale: A Turkish case. *African Journal of Agricultural Research* **4** 1023–31

[50] Galang A P 2010 Environmental education for sustainability in higher education institutions in the Philippines *Intl. J. of Sus. in Hi. Edu.* **11**(2) 173–83

[51] Tuncer G, Tekkaya C, Sungur S, Cakiroglu J, Erterpinar H and Kaplowitz M 2009 Assessing pre-service teachers' environmental literacy in Turkey as a mean to develop teacher education programs *Intl. J. of Edu. Dev* **29** 426–36

[52] Teksoz G, Sahin E and Enterpinar H 2010 A new vision for Chemistry education students: Environmental education *Intl J. of Env. and Sci. Ed.* **5**(2) 131–49

[53] Haslberger A 2005 Facets and dimensions of cross-cultural adaptation: Refining the tools. *Personnel Review* **32**(1) 85–109

[54] Samanci N K 2015 A study on the link between moral judgment competences and critical thinking skills *Intl. J. of Env. and Sci. Edu.* **10**(2) 135–43

[55] Veloo A, Perumal S and Vikneswary 2013 Inquiry-based instruction, students' attitudes and

teachers' support towards science achievement in rural primary schools. *Procedia - Social and Behavioral Sciences* **93** 65–9

Acknowledgments
I would like to thank to Puguh Karyanto and Murni Ramli for being the research team and to my colleagues for helping with this research

IOP Conf. Series: Journal of Physics: Conf. Series **1022** (2018) 012044 doi:10.1088/1742-6596/1022/1/012044

Molecular status of Human Immunodeficiency Virus, Hepatitis B virus, and Hepatitis C virus among injecting drug male commercial sex workers in Surakarta, Indonesia

Afiono Agung Prasetyo[1,2,3*], Marwoto[1,2,3], Zainal Arifin Adnan[1,4] and Hartono[1,5]

[1] A-IGIC (A-Infection, Genomic, Immunology,& Cancer) Research Group, Universitas Sebelas Maret, Jl. Ir. Sutami 36A, Surakarta, 57126, Indonesia.
[2]Center of Biotechnology and Biodiversity Research and Development, Universitas Sebelas Maret, Jl. Ir. Sutami 36A, Surakarta, 57126, Indonesia.
[3] Division of Virology, Department of Microbiology, Faculty of Medicine, Universitas Sebelas Maret, Jl. Ir. Sutami 36A, Surakarta, 57126, Indonesia.
[4] Department of Internal Medicine, Faculty of Medicine, Universitas Sebelas Maret, Jl. Ir. Sutami 36A, Surakarta, 57126, Indonesia.
[5] Department of Physiology, Faculty of Medicine, Universitas Sebelas Maret, Jl. Ir. Sutami 36A, Surakarta, 57126, Indonesia.

*Corresponding author: afie.agp.la@gmail.com; afieagp@yahoo.com; afie@staff.uns.ac.id

Abstract. Male commercial sex workers are one of the high-risk community for blood-borne viruses. However, there are no data concerning the molecular status of Human Immunodeficiency Virus (HIV), Hepatitis B Virus (HBV), and Hepatitis C Virus (HCV) circulated among male commercial sex workers with injecting drug habits in Surakarta, Indonesia. Blood samples obtained from injecting drug male commercial sex workers in Surakarta were examined for HIV antibodies, HBsAg, and HCV antibodies, respectively, by immunological assays. Blood samples were also subjected to viral nucleic acid extraction and molecular detection of HIV, HBV, and HCV by nested (RT) PCRs. The PCR products were purified from agarose gels, and the nucleotide sequences were retrieved and molecular analyzed. HIV, HBV, and HCV were detected in 29.4% (10/34), 17.6% (6/34), and 52.9% (18/34), respectively. HIV CRF01_AE and B were found to be circulating in the community. HBV genotype B3 was predominated, followed by C1. HCV genotype 1a was predominated, followed by 1c, 3a, 1b, and 4a. HIV, HBV, and HCV were found circulating in the male commercial sex workers with injecting drug habits in Surakarta, Indonesia.

1. Introduction

Male commercial sex workers still considered not accepted in Indonesian society, therefore become a neglected population in the community. In a fact, male commercial sex workers represent a key population in Human Immunodeficiency Virus (HIV) epidemic [1]. Injecting drug abused history is a risk factor for blood-borne virus infection; therefore, male commercial sex workers with injecting drug abused history will have a higher risk of blood-borne virus infection, including HIV infection, also for Hepatitis B Virus (HBV) and Hepatitis C Virus (HCV) [2].Previously, we reported the presence of

HIV, HBV, and HCV in male commercial sex workers in Surakarta [3-4]. We then continued the study, to evaluate the circulation of HIV, HBV, and HCV in male commercial sex workers with injecting drug abused history in Surakarta, Indonesia.

2. Materials and Methods

2.1. Study Population

Since 2009, our research group (A-IGIC/A-Infection, Genomic, Immunology,& Cancer research group) has been conducting a molecular epidemiology study of human blood-borne viruses, by collecting epidemiological-clinical data and blood samples from the high-risk communities in Central Java, Indonesia, including that of the male commercial sex workers in Surakarta [3-4]. A 5-ml blood sample was collected from willing participants and then fractionated, aliquoted, and stocked in -80 °C until further study. Written informed consent was obtained from all the individuals who participated in the study. Approval was obtained from the institutional ethics committee review boards of the Faculty of Medicine of Universitas Sebelas Maret and the Dr. Moewardi General Hospital, Surakarta, Indonesia. All blood samples collected from male commercial sex workers with injecting drug abused history in Surakarta in 2010-2014 were used for the present study (n= 34).

2.2. Immunological Assays

Blood samples aliquot obtained were subjected to the immunological assays as described previously [3-6]. Briefly, HIV antibodies were detected using a Determine HIV-1/2 Kit (Abbott Diagnostics Japan, Tokyo, Japan) and positive results were confirmed using Vironostika HIV Uniform II Antigen Ag/Ab (BioM´erieux, Marcy l'´Etoile, France). A SERATEC Hepatitis B Quick Test (Gesellschaftfür Biotechnologie GmbH, Göttingen, Germany) and Ortho HCV PA II (Ortho Diagnostics, Tokyo, Japan) were used for the detection of the HBsAg and anti-HCV, respectively. All assays were performed according to manufacturer's instructions in duplicate.

2.3. Nucleic Acid Extraction and Molecular Detection

The viral nucleic acid extraction and molecular detection techniques were performed as described previously [2,5,8]. Briefly, a portion of the HIV *gag* gene encoding the p24 region was amplified by nested RT-PCR. A portion of the HBV *HBsAg* gene was amplified by nested PCR. A portion of the *E1–E2* region of the HCV genome were amplified by nested RT-PCR. Internal amplification controls were included to exclude any false-negative results. Corresponding positive controls and one negative control (sterile water) were included in each group. The specificity was confirmed by sequencing the amplicons. All samples were tested at least in duplicate.

2.4. Sequencing and Phylogenetic Analysis

Determination of nucleotide sequences and phylogenetic analysis were performed as described previously [2,8]. Briefly, the PCR products were purified from agarose gels, and the nucleotide sequences were determined for the HIV *gag* region(position 1577-2039 on HIV-1 reference HXB2), HBV *HBsAg* region(corresponding to 255-671 nt in HBV ADW GenBank Accession Number V00866), and HCV *E1-E2* region (position 1322-1859 in H77 strain, GenBank Accession Number NC_004102), respectively. Initial genotyping was conducted using the NCBI genotyping tool. The sequences were then submitted to the Blast program to check their similarity to related strains deposited in Genbank/EMBL/DDBJ. The reference strains with the highest homology score to each analyzed strain were retrieved from the GenBank/EMBL/DDJB databases and aligned with the tested sequences. All the HIV, HBV, and HCV sequences isolated in Indonesia deposited in GenBank were also included in the alignment analysis for each tested sequence. The frequency of nucleotide substitution at each base was estimated by the Kimura two-parameter method. A phylogenetic tree was constructed by the neighbor-joining method, and its reliability was estimated by 1000 bootstrap replications. The phylogenetic tree was constructed using the CLC Main Workbench 7.9.1 software package (CLC Bio).

3. Results and Discussion

This is the first molecular epidemiology study of HIV, HBV, and HCV in male commercial sex workers with injecting drug abused history in Surakarta, Indonesia. For the best of our knowledge, prior to this study, there were only limited data about molecular epidemiology profiles of the respected viruses from the community in Indonesia, especially from Surakarta.

All blood samples obtained from injecting drug male commercial sex workers in Surakarta subjected for the HIV, HBV, and HCV immunological assays. Consistent with previous reports [2,3,8], the anti-HIV-1/2, HBsAg, and anti-HCV were found in 29.4% (10/34), 17.6% (6/34), and 52.9% (18/34) samples, respectively. All blood samples were also subjected to molecular assays. HIV, HBV, and HCV were detected in 29.4% (10/34), 17.6% (6/34), and 52.9% (18/34) samples, respectively. None of the respected genome viruses were detected from the negative immunological assays plasma samples.

A portion of the HIV *gag* gene could be amplified by nested RT-PCR in ten anti-HIV samples. The amplified portions of the HIV-1 *gag* region of ten HIV isolates (HIDSKA-31, HIDSKA-32, HIDSKA-33, HIDSKA-34, HIDSKA-35, HIDSKA-36, HIDSKA-37, HIDSKA-38, HIDSKA-39, andHIDSKA-40) were genotyped. The six HIV isolates were clustered together in the phylogenetic tree and shared 96-98% nucleotide homology with HIV-1 CRF01_AE isolated in drug abused inmates in Central Java (09IDSKA-18). The four HIV isolates were genotyped as B, shared 98% homology with HIV-1 B isolate HIVHXB2CG [9] (figure 1). The results obtained were consistent with HIV circulated in drug abused inmates in Central Java's correctional facilities [2] and transgender commercial sex workers in Surakarta [8].

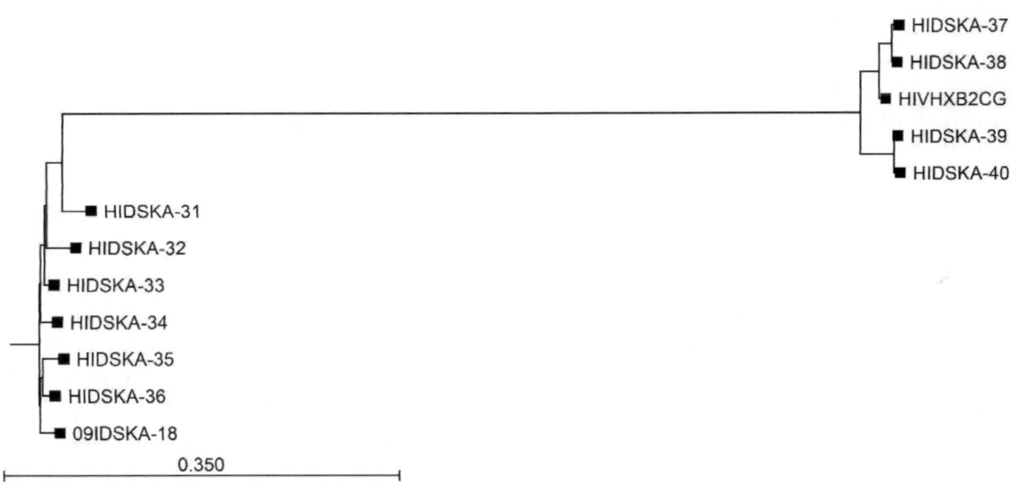

Figure 1. Phylogenetic Analysis of HIV Isolates Obtained from The Male Commercial Sex Workers with Injecting Drug Abused History in Surakarta Based on HIV-*Gag* Nucleotide Sequences

A portion of the HBV preS2/S gene was successfully amplified in six HBsAg positive samples. The HBIDSKA-21, HBIDSKA-22, HBIDSKA-23, and HBIDSKA-24 were clustered together with HBV genotype B3 isolated in Central Java (09IDSKAB-5). The HBIDSKA-25 and HBIDSKA-26 were closely related to HBV genotype C1 isolated in Central Java (09IDSKAB-14) [2] (figure 2). The results obtained were consistent with that of in transgender commercial sex workers in Surakarta, as reported previously [8]. Moreover, the HBV genotypes in the present report were already reported found circulated in drug abused inmates in Central Java's correctional facilities [2].

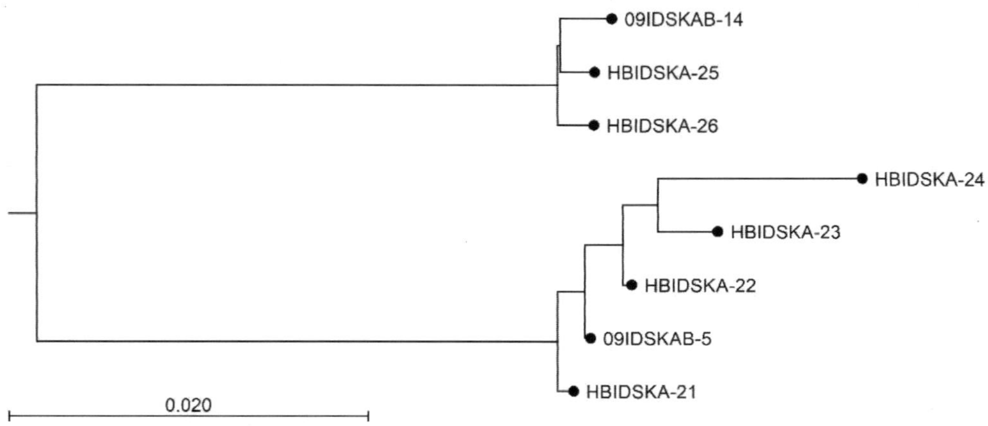

Figure 2. Phylogenetic Analysis of HBV Isolates Obtained from The Male Commercial Sex Workers with Injecting Drug Abused History in Surakarta Based on HBV *HbsAg* Nucleotide Sequences

All blood samples positive for anti-HCV were found positive for HCV RNA by nested RT-PCR (n= 18). The PCR products were then subjected to the determination of the *E1-E2* sequence. HCV genotypes 1a was found in 8 subjects (HCIDSKA-45, HCIDSKA-46, HCIDSKA-47, HCIDSKA-48, HCIDSKA-49, HCIDSKA-50, HCIDSKA-51, and HCIDSKA-52) and were clustered together with HCV 1a isolated in Central Java (09IDSKAC-18) [2]. HCV genotypes 1c was found in four subjects (HCIDSKA-53, HCIDSKA-54, HCIDSKA-55, and HCIDSKAC-56) and were clustered together with HCV 1c isolated in Central Java (09IDSKAC-7) [2]. HCV genotypes 3a was found in three subjects (HCIDSKA-57, HCIDSKA-58, and HCIDSKA-59) and were clustered together with HCV 3a isolated in Central Java (09IDSKAC-2) [2]. HCV genotype 1b was found in two subjects (HCIDSKA-60 and HCIDSKA-61) were clustered together with HCV 1b isolated in Central Java [2]. HCV genotype 4a was found in 1 subject (HCIDSKA-62) and was clustered together with HCV 4a isolated in Central Java (09IDSKAC-13) [2]. The present data indicated all HCV circulated in male commercial sex workers with injecting drug abused history in Surakarta were consistent with HCV circulated in drug abused inmates in Central Java's correctional facilities.

Previously, we proposed that drug abuse prisoner is useful as a surrogate sentinel to gain molecular epidemiology data on human blood-borne viruses for a country such as Indonesia in which it is difficult to perform active surveillance among the general population [2]. In the present study, the molecular data of HIV, HBV, and HCV found in the male commercial sex workers with injecting drug abused history in Surakarta were consistent with the respected viruses in drug abuse prisoner imprisoned in Central Java's correctional facilities. In general, the male commercial sex workers with injecting drug abused history in Surakarta were found at high risk for blood-borne viruses infection, consistent with previous reports [2-8,10-12].People engaging in transactional sex are considered a key population for HIV prevention; therefore, health interventions are needed to improve knowledge, risk perception, and health behaviors in the key population [13]. Moreover, co-infection and polymorphisms status study should be performed for the community, and the pathogen found should be molecularly characterized as previously [14-22].

4. Conclusions

HIV, HBV, and HCV were found circulating in the male commercial sex workers with injecting drug abused history in Surakarta, Indonesia.

5. References

[1] Jin H, Friedman M R, Lim S H, Guadamuz T E and Wei C 2016 *LGBT Health* **3** 465-71
[2] Prasetyo A A, Dirgahayu P, Sari Y, Hudiyono H and Kageyama S 2013 *J. Infect. Dev. Ctries.* **7**

453-67
[3] Prasetyo A A, Ariapramuda E R, Kindi E A, Dirgahayu P, Sari Y, Dharmawan R, Hudiyono, Hartono and Kageyama S 2014 *Southeast Asian J. Trop. Med. Public Health* **45** 1032-47
[4] Sari Y, Haryati S, Raharjo I and Prasetyo A A 2015 *Southeast Asian J. Trop. Med. Public Health* **46** 977-85
[5] Prasetyo A A, Sariyatun R, Reviono, Sari Y, Hudiyono, Haryati S, Adnan Z A, Hartono and Kageyama S 2015 *J. Clin. Virol.* **70** 67–71
[6] Prasetyo A A and Zaini K U N 2015 *Southeast Asian J. Trop. Med. Public Health* **4** 662-8
[7] Prasetyo A A, Nurfitria F B, Adnan Z A, Redhono D, Sari Y and Hudiyono 2015 *Int. J. Interferon Cytokine Mediat. Res.***7** 1-7
[8] Prasetyo A A, Sari Y, Dharmawan R and Marwoto 2017 *AIP Conference Proceedings* **1813** 020011
[9] Berg M G, Yamaguchi J, Alessandri-Gradt E, Tell R W, Plantier J C and Brennan C A 2016 *J. Clin. Microbiol.* **54** 868-82.
[10] Prasetyo A A and Sari Y 2017 *AIP Conference Proceedings* **1813** 020001
[11] Prasetyo A A, Luvi S D, Hartono and Sari Y 2017 *AIP Conference Proceeding* **1813** 020004
[12] Prasetyo A A, Dharmawan R, Raharjo I and Hudiyono 2016 *J. Global Infect. Dis.* **8** 75-81
[13] Bowring A L, Pasomsouk N, Hughes C, van Gemert C, Higgs P, Sychareun V, Hellard M and Power R 2017 *Arch. Sex Behav.* **46** 1047-59
[14] Marwoto, Lismana U H V L, Prasetyo A A, Suradi, Reviono and Harsini 2015 *J. Clin. Diagn. Res.* **9** 10-3
[15] Marwoto, Prasetyo A A, Sariyatun R, Suradi, Reviono and Harsini 2015 *Asian Biomed.* **9** 473-9
[16] Haryati S, Prasetyo A A, Sariyatun R, Sari Y and Murkati 2015 *Asian Pac. J. Trop. Dis.* **5** 798-803
[17] Prasetyo A A and Sariyatun R 2015 *Asian J. Microbiol. Biotechnol. Environ. Sci.***17** 349-55
[18] Prasetyo A A and Sariyatun R 2015 *J. Teknol.***77** 67-70
[19] Prasetyo A A, Desyardi M N, Tanamas J, Suradi, Reviono, Harsini, Kageyama S, Chikumi H and Shimizu E 2015 *Intervirology* **58** 57-68
[20] Prasetyo A A, Dharmawan R, Sari Y and Sariyatun R 2017 *AIP Conference Proceedings* **1813** 020020
[21] Sari Y, Haryati S, Prasetyo A A, Hartono and Adnan Z A 2017 *AIP Conference Proceedings* **1813** 020005
[22] Prasetyo A A 2017 *AIP Conference Proceedings* **1788** 030100

Acknowledgments
This work was supported in part by grants from the Indonesian Directorate of Higher Education-APBN/DIPA UNS (No. 501/UN27.11/PN/2014 and No. 339/UN27.11/PL/2015), and from the Indonesian Ministry of Research, Technology and Higher Education (No 353/ UN 27.21/ PN/ 2016and No 873/ UN27.21/ PP/ 2017).

IOP Conf. Series: Journal of Physics: Conf. Series **1022** (2018) 012045 doi:10.1088/1742-6596/1022/1/012045

Human T-lymphotropic Virus-1/2 detected in drug abused men who have sex with men infected with HIV in Surakarta, Indonesia

Afiono Agung Prasetyo[1,2,3*] **and Yulia Sari**[1,2,4,5]

[1] A-IGIC (A-Infection, Genomics, Immunology, & Cancer) Research Group, Universitas Sebelas Maret, Jl. Ir. Sutami 36A, Surakarta, 57126, Indonesia.
[2]Center of Biotechnology and Biodiversity Research and Development, Universitas Sebelas Maret, Jl. Ir. Sutami 36A, Surakarta, 57126, Indonesia.
[3] Division of Virology, Department of Microbiology, Faculty of Medicine, Universitas Sebelas Maret, Jl. Ir. Sutami 36A, Surakarta, 57126, Indonesia.
[4] Department of Parasitology, Faculty of Medicine, Universitas Sebelas Maret, Jl. Ir. Sutami 36A, Surakarta, 57126, Indonesia.
[5] Doctoral Program of Medical Sciences Faculty of Medicine, Universitas Sebelas Maret, Jl. Ir. Sutami 36A, Surakarta, Indonesia.

*Corresponding author: afie.agp.la@gmail.com; afieagp@yahoo.com; afie@staff.uns.ac.id

Abstract. Human T-lymphotropic virus types 1 and 2 (HTLV-1/2) share similar routes of transmission with human immunodeficiency virus (HIV), and the HTLV-1/2 co-infection may affect the clinical course of HIV infection. The HIV/HTLV-1/2 co-infection risk higher if the patient performing the high-risk activities. This study evaluated the presentation of HTLV-1 and 2 in HIV-infected men who have sex with men with drug abused history in Surakarta Indonesia. Blood samples collected from HIV-infected men who have sex with men with drug abused history in Surakarta were tested using HTLV-1/2 enzyme-linked immunosorbent assays and confirmed by RT-PCR nested addressed the part of HTLV-1 *LTR* and HTLV-2 *LTR* region, respectively. The specificity of the molecular assays was confirmed by sequencing the amplicons. The anti HTLV-1/2 positive rate was 17.4% (8/46). All positive serological samples were confirmed by nested RT-PCR. Of these, three was HTLV-1 positive and five was HTLV-2 positive. Molecular analysis of positive PCR products revealed that all HTLV-1 isolates had a close relationship with HTLV-1 isolated in Japan while all HTLV-2 isolates with that of isolated in the USA. HTLV-1 and HTLV-2 were detected in drug abused men who have sex with men infected with HIV in Surakarta.

1. Introduction

Human T-cell lymphotropic virus type 1 (HTLV-1) was the first human retrovirus to be reported [1]. The retrovirus is endemic to certain regions of the world and infects around 10-20 million people [2]. The virus is associated with neoplastic, neurological, autoimmune, and infectious complications [1-3]. HTLV-2 is associated with a small increased risk of bacterial infections of the chest and bladder, increased cancer risk, and of HTLV- associated myelopathy/tropical spastic paraparesis [4,5].

HTLV-1/2 may affect the clinical course of human immunodeficiency virus-1 (HIV-1) and can lead to increased morbidity [6,7]. Moreover, HTLV-1/2 and HIV-1 mainly infecting CD4(+) T lymphocytes and also antigen-presenting cells such as dendritic cells (DCs) in a lesser extent [8]. Because HTLV-1/2 shares a similar transmission route with HIV, HTLV-1/2 infection may be more prevalent in HIV-infected individuals [6-7].

Drug abused history and high-risk sexual activities have been associated with HIV-1/HTLV-1 and HIV-1/HTLV-2 co-infections [1,2,7,9,10]. Previously, we reported the HTLV-1 and HTLV-2 were detected in the men who have sex with men with drug abused history in Surakarta, Indonesia [11].Since the rates of HTLV-1/2 co-infection among drug abused men who have sex with men HIV-infected individuals have not been studied in Surakarta, we continued the study.

2. Materials and Methods

Since 2009, our research group (A-IGIC/A-Infection, Genomics, Immunology,& Cancer research group) has been conducting a molecular epidemiology study of human blood-borne viruses, by collecting epidemiological-clinical data and blood samples from the high-risk communities in Central Java, Indonesia, including that of the men who have sex with men with drug abused history in Surakarta [10-21]. Written informed consent was obtained from all the individuals who participated in the study. Approval was obtained from the institutional ethics committee review boards of the Faculty of Medicine of Universitas Sebelas Maret and the Dr. Moewardi General Hospital, Surakarta, Indonesia. Blood samples collected from HIV-infected men who have sex with men with drug abused history in Surakarta in 2010-2014 were used for the present study (n=46).

2.1. Immunological Assay

Blood samples aliquot obtained were subjected to the immunological assays as for the presence of specific antibodies to HTLV-1/2 using an ELISA kit that employs HTLV-1/2 viral lysates plus recombinant proteins from HTLV-1/2 as antigens (HTLV-I/II ELISA 4.0, MP Biomedicals, Santa Ana, California). All assays were performed according to manufacturer's instructions in duplicate [10-13].

2.2. Nucleic Acid Extraction and HTLV-1/2 Molecular Detection

All blood samples were confirmed by a molecular assay as previously described [10-13]. Viral nucleic acid was extracted from plasma by using High Pure Viral Nucleic Acid Kit (Roche Life Science, Manheim, Germany) according to the manufacturer's instructions. DNA was extracted from whole blood by using High Pure PCR Template Preparation Kit (Roche Life Science). The nucleic acids were then aliquoted, and one aliquot was reverse-transcribed according to the Transcriptor High Fidelity cDNA Synthesis Kit protocol using random hexamers (Roche Life Science). Molecular detection was performed by nested PCR using the FastStart High Fidelity PCR System (Roche Life Science). *LTR* segments of HTLV-1 and HTLV-2 were amplified to confirm HTLV-1 and HTLV-2 infection. Briefly, the segment of *LTR* regions of the HTLV-1 genome was amplified using primers LTR1/LTR3 in the first round and LTR1/LTR2 in the second round while the segment of *LTR* regions of the HTLV-2 genome was amplified using primers VS1/VS2 in the first round and VS3/VS4 in the second round. Internal amplification controls were included to exclude any false-negative results. Corresponding positive controls and one negative control (sterile water) were included in each group. The specificity was confirmed by sequencing the amplicons. All samples were tested at least in duplicate.

2.3. Sequencing and Phylogenetic Analysis

Determination of nucleotide sequences and phylogenetic analysis were performed as described previously [10-13]. Briefly, the PCR products were purified from agarose gells and then subjected to the determination of nucleotide sequences directly with the primers of LTR1/LTR2 for HTLV-1 *LTR* region and VS3/VS4 for HTLV-2 *LTR* region, respectively. Initial genotyping was conducted using the NCBI genotyping tool. Sequences were then submitted to the BLAST program in order to check their similarity to related strains deposited in Genbank/EMBL/DDBJ. The reference strain with the highest homology score to each analyzed strain was retrieved from the GenBank/EMBL/DDJB database and then was aligned with the tested sequences. Kimura two-parameter method was used to estimate the frequency of nucleotide substitution at each base. A phylogenetic tree was constructed by the neighbor-joining method with 1000 bootstrap replications to estimate its reliability. The phylogenetic tree was constructed using the CLC Main Workbench 7.9.1 software package (CLC Bio).

3. Results and Discussion

Out of 46 HIV-positive blood samples, eight (17.4%) were positive foranti-HTLV-1/2 antibodies. Nested PCR was possible to be performed in all positive samples, confirmed the presence of DNA/HTLV in the plasma and whole blood for the LTR regions of the genome. Of these, three was HTLV-1 positive and five was HTLV-2 positive. Based on 317 nucleotides part of HTLV-1 LTR region, the all HTLV-1 isolates was shared 100 % homology with B1033-2009 (HTLV-1 isolated in Japan; GenBank Accession Number AB513134) and 09IDSKAH-1-3 (HTLV-1 isolated in Central Java, Indonesia, GenBank Accession Number JN247457), consistent with previous results [10-11].Based on 666 nucleotides part of HTLV-2 LTR region, all HTLV-2 isolates had shared 99 % homology with HTLV-2 isolated in USA (GenBank Accession Number AF412314) and HTLV-2 isolated from Central Java, Indonesia (09IDSKAH-2-1, GenBank Accession Number JN247458), consistent with previous results [10-11].

HTLV-1/HIV co-infection is known to elevate the CD4+ T-cell counts of treatment-naïve persons. Also, HTLV-1/HIV co-infected participants continued to have elevated CD4+ T-cell counts after developing virologic failure on ART, despite no difference in their HIV viral load levels when compared with HIV mono-infected participants; therefore, the CD4+ T-cell count testing maybe not be a useful strategy to monitor ART response in the presence of HTLV-1 infection [22]. The HTLV/HIV co-infection was detected in 17.4% of drug abused men who have sex with men infected with HIV in Surakarta, indicated the necessity of HTLV detection in HIV patients with high-risk histories of respected co-infection. Moreover, co-infection with HTLV-1 has been associated with faster progression to AIDS [23].HTLV-1 infection can modify the expression of main functional transcription factors, FOXP3 and GITR, which may lead to immune response deterioration of Tregs [24].

HTLV viral transmission requires cell-to-cell contacts, while cell-free virions are poorly infectious and almost absent from body fluids; therefore, horizontal transmission between adults is likely the main route of HTLV infection in the general population and that this is likely to occur through sexual contact and or sharing blood contain injecting needles as performed by the injecting drug abused users [25,26]. Health intervention like increasing availability of free condoms methadone therapy provided by the government may decline the HTLV seroprevalence [27]. Moreover, other co-infection status study should be performed in the community, as previously [28-33].

4. Conclusions

The HTLV-1 and HTLV-2 were detected in the HIV-infectedmen who have sex with men with drug abused history in Surakarta, Indonesia.

5. References

[1] Stewart J, Heitzinger K, Pollett S, Calderón M, Alarcón J, Ton TG and Zunt J R 2017 *Am. J. Trop. Med. Hyg.* **96** 373-9

[2] Khan M Y, Khan I N, Farman M, Al Karim S, Qadri I, Kamal M A, Al Ghamdi K and Harakeh S 2017 *Curr. Top. Med. Chem.* **17** 1320-30

[3] Ghezeldasht S A, Sadeghian H, Azarpazhooh MR, Shamsian S A, Rafatpanah H, Mahmoodi M and Rezaee S A 2017 *Appl. Biochem. Biotechnol.* doi: 10.1007/s12010-017-2406-7 [Epub ahead of print]

[4] Fox J M, Mutalima N, Molyneux E, Carpenter L M, Taylor G P, Bland M, Newton R and Martin F 2016 *Trop. Med. Int. Health.* **21** 312-24.
[5] Marano G, Vaglio S, Pupella S, Facco G, Catalano L, Piccinini V, Liumbruno G M and Grazzini G 2016 Transfusion **56** 249-60
[6] Kozlowski A G, Matos M A, Carneiro M A, Lopes CL, Teles S A, Vicente C P and Martins R M 2016 *Rev. Inst. Med. Trop. Sao Paulo* **58** 80
[7] Rossheim A E, Cunningham T D and Troy S B 2016 *Am. J. Med. Sci.* **352** 258-60
[8] Dutartre H, Clavière M, Journo C and Mahieux R 2016 *J. Virol.* **90** 7607-17
[9] Campos K R, Gonçalves M G and Caterino-de-Araujo A 2017AIDS Res. Hum. Retroviruses **33** 382-5
[10] Prasetyo A A, Dirgahayu P, Sari Y, Hudiyono H and Kageyama S 2013 *J. Infect. Dev. Ctries.* **7** 453-67
[11] Prasetyo A A and Sari Y 2017 *AIP Conference Proceedings* **1813** 020001
[12] Prasetyo A A, Ariapramuda E R, Kindi E A, Dirgahayu P, Sari Y, Dharmawan R, Hudiyono, Hartono and Kageyama S 2014 *Southeast Asian J. Trop. Med. Public Health* **45** 1032-47
[13] Prasetyo A A and Zaini K U N 2015 *Southeast Asian J. Trop. Med. Public Health* **4** 662-8
[14] Sari Y, Haryati S, Raharjo I and Prasetyo A A 2015 *Southeast Asian J. Trop. Med. Public Health* **46** 977-85
[15] Prasetyo A A, Dharmawan R, Raharjo I and Hudiyono 2016 *J. Global Infect. Dis.* **8** 75-81
[16] Marwoto, Lismana U H V L, Prasetyo A A, Suradi, Reviono and Harsini 2015 *J. Clin. Diagn. Res.* **9** 10-3
[17] Marwoto, Prasetyo A A, Sariyatun R, Suradi, Reviono and Harsini 2015 *Asian Biomed.* **9** 473-9
[18] Haryati S, Prasetyo A A, Sariyatun R, Sari Y and Murkati 2015 *Asian Pac. J. Trop. Dis.* **5** 798-803
[19] Prasetyo A A and Sariyatun R 2015 *Asian J. Microbiol. Biotechnol. Environ. Sci.***17** 349-55
[20] Prasetyo A A and Sariyatun R 2015 *J. Teknol.* **77** 67-70
[21] Prasetyo A A, Nurfitria F B, Adnan Z A, Redhono D, Sari Y and Hudiyono 2015 *Int. J. Interferon Cytokine Mediat. Res.***7** 1-7
[22] Vandormael A, Rego F, Danaviah S, Alcantara L, Boulware D and de Oliveira T 2017 *Curr HIV Res.* doi: 10.2174/1570162X15666170216114917. [Epub ahead of print]
[23] Geddes V E, José D P, Leal F E, Nixon D F, Tanuri A and Aguiar R S 2017 *Virology* **504** 45-51
[24] Ghezeldasht S A, Sadeghian H, Azarpazhooh M R, Shamsian S A, Rafatpanah H, Mahmoodi M and Rezaee S A 2017*Appl. Biochem. Biotechnol.* doi: 10.1007/s12010-017-2406-7. [Epub ahead of print]
[25] Alais S, Dutartre H and Mahieux R2017 *Methods Mol. Biol.* **1582** 47-55
[26] Nunes D, Boa-Sorte N, Grassi M F, Taylor G P, Teixeira M G, Barreto M L, Dourado I and Galvão-Castro B 2017*PLoS One* **12** e0171303
[27] Stewart J, Heitzinger K, Pollett S, Calderón M, Alarcón J, Ton T G and Zunt J R 2017 *Am. J. Trop. Med. Hyg.* **96** 373-9
[28] Prasetyo A A, Desyardi M N, Tanamas J, Suradi, Reviono, Harsini, Kageyama S, Chikumi H and Shimizu E 2015 *Intervirology* **58** 57-68
[29] Prasetyo A A, Sariyatun R, Reviono, Sari Y, Hudiyono, Haryati S, Adnan Z A, Hartono and Kageyama S 2015 *J. Clin. Virol.* **70** 67–71
[30] Prasetyo A A, Dharmawan R, Sari Y and Sariyatun R 2017 *AIP Conference Proceedings* **1813** 020020
[31] Prasetyo A A, Luvi S D, Hartono and Sari Y 2017 *AIP Conference Proceedings* **1813** 020004
[32] Prasetyo A A, Sari Y, Dharmawan R and Marwoto 2017 *AIP Conference Proceedings* **1813** 020011
[33] Prasetyo A A 2017 *AIP Conference Proceedings* **1788** 030100

IOP Conf. Series: Journal of Physics: Conf. Series **1022** (2018) 012045 doi:10.1088/1742-6596/1022/1/012045

Acknowledgments
This work was supported in part by grants from PNBP UNS (No 632/ UN 27.21/ LT/ 2016 and No 623/UN27.21/PP/2017).

Torque Teno Virus in HIV-infected transgender in Surakarta, Indonesia

Hartono[1,2], Afiono Agung Prasetyo[1,3,4*] and Mohammad Fanani[5]

[1]A-IGIC (A-Infection, Genomics, Immunology, & Cancer) Research Group, Universitas Sebelas Maret, Jl. Ir. Sutami 36A, Surakarta, 57126, Indonesia.
[2] Department of Physiology, Faculty of Medicine, Universitas Sebelas Maret, Jl. Ir. Sutami 36A, Surakarta, 57126, Indonesia.
[3] Center of Biotechnology and Biodiversity Research and Development, Universitas Sebelas Maret, Jl. Ir. Sutami 36A, Surakarta, 57126, Indonesia.
[4] Division of Virology, Department of Microbiology, Faculty of Medicine, Universitas Sebelas Maret, Jl. Ir. Sutami 36A, Surakarta, 57126, Indonesia.
[5] Department of Psychiatry, Faculty of Medicine, Universitas Sebelas Maret, Jl. Ir. Sutami 36A, Surakarta, 57126, Indonesia.

*Corresponding author: afie.agp.la@gmail.com; afieagp@yahoo.com; afie@staff.uns.ac.id

Abstract. Torque Teno Virus (TTV) is a circular single-stranded DNA virus that may co-infected with human immunodeficiency virus (HIV), especially in the high-risk community e.g. the transgender performing high-riskbehavior. TTV shows an increased viremia in HIV patients and maybe influence the HIV clinical progression. Blood samples collected from transgender performing high-riskbehavior in Surakarta were tested by serological and molecular assays to detect the presence of HIV infection. The blood samples with HIV positive status were then tested by a nested polymerase chain reaction (PCR) to detect the presentation of TTV DNA. The amplified PCR products were molecularly cloned and subjected to sequence analysis. TTV DNA was detected in 40.0% HIV-positive samples. The molecular characterization revealed that the most prevalent was genogroup 3, followed by genogroup 2 and 1, respectively. TTV was detected in HIV-infected transgender performing high-riskbehavior in Surakarta with high infection rate.

1. Introduction

Torque Teno Virus (TTV), is a circular single-stranded DNA virus member of Anelloviridae family considered a worldwide distributed emergent virus. The pathogenicity of TTV remains unclear, however, it has been constantly associated with concurrent infections [1-4]. TTV has been detected in almost every human tissue type or body fluid reaching near 100% prevalence [5] and frequently found co-infected with human immunodeficiency virus (HIV) [5-13]. TTV showing an increased viremia in HIV patients and TTV load may be used as asurrogate marker of immune function related to B cells exhaustion and impaired viral immune surveillance in HIV patients. TTV might play a role in the development of liver disease in HIV patients, such as the patients coinfected with HIV and HCV, and maybe influence the HIV clinical progression [10-11].

The transgender community is at a higher risk for substance use, sexually transmitted diseases (STDs), bullying, isolation, rejection, anxiety, and depression as compared to the general population. Transgender usually receive poor quality of care due to stigma, lack of healthcare providers' awareness, and insensitivity to the unique needs of this community [12]. Previously, our research group (A-IGIC/A-Infection, Genomics, Immunology, & Cancer research group) reported the circulation of TTV in Indonesia [4,6-8,14]. The circulation of HIV in transgender living in Surakarta also already reported by our research group [6-8,15-18]. The present study describes the circulation of TTV in HIV-infected transgender performing high-risk behavior in Surakarta.

2. Materials and Methods

Since 2009, our group has been conducting a molecular epidemiology study of human blood-borne pathogens in Central Java, Indonesia [4, 6-8,14-24,26-31]. Epidemiological-clinical data and blood samples were collected from the high-risk communities, including that of the transgender performing high-risk behavior in Surakarta. Written informed consent was obtained from all the individuals who participated in the study. Approval was obtained from the institutional ethics committee review boards of the Faculty of Medicine of Universitas Sebelas Maret and the Dr. Moewardi General Hospital, Surakarta, Indonesia. All procedures were conducted according to the principles of the Declaration of Helsinki.

The study participants completed an interviewer-administered questionnaire by a specially trained interviewer that assessed general characteristics associated with blood-borne virus infections, as reported previously [6,14,17,24]. All blood samples were subjected to human blood-bornepathogens detection by immunological and molecular assays, as reported previously [4,6-8,14-31]. All blood samples were screened by immunological and molecular assays for HIV-infection detection.HIV antibodies were detected using a Determine HIV-1/2 Kit (Abbott Diagnostics Japan, Tokyo, Japan) and positive results were confirmed using Vironostika HIV Uniform II Antigen Ag/Ab (Bio M´erieux, Marcy l'´Etoile, France).Viral nucleic acid was extracted from plasma by using High Pure Viral Nucleic Acid Kit (Roche Life Science, Manheim, Germany) according to the manufacturer's instructions. The nucleic acids were then aliquoted, and one aliquot was reverse-transcribed according to the Transcriptor High Fidelity cDNA Synthesis Kit protocol using random hexamers (Roche Life Science). To detect the presence of HIV RNA, a portion of the HIV *gag* gene encoding the p24 region was amplified by nested PCR using the cDNA as template, by the Fast Start High Fidelity PCR System (Roche Life Science). From transgender performing high-risk behavior in Surakarta in 2009-2011, a total of 40blood samples were found positive for HIV and then used for the present study.

The molecular detection techniques for TTV were performed as described previously [14]. TTV DNA was extracted from whole blood by using High Pure PCR Template Preparation Kit (Roche Life Science). TTV DNA was amplified by a nested PCR amplification using the Fast Start High Fidelity PCR System (Roche Life Science). Internal amplification controls were included to exclude any false-negative results. Corresponding positive controls and one negative control (sterile water) were included in each group. To prevent PCR contamination, the reagent preparation, sample processing, and nested PCR assays were performed in rooms separate from where the amplified products were analyzed. Aerosol-resistant pipette tips were used throughout the assays. The PCR products were subjected to electrophoresis in 2% agarose gels, stained with ethidium bromide, and visualized under ultraviolet illumination. The amplicons were purified, subcloned, and sequenced. All samples were tested at least twice.

3. Results and Discussion

During 2009-2011,40transgender performing high-risk behavior in Surakarta were found positive for HIV infection. All participants identified themselves as not married. The majority of the participants (33/40, 82.5%) were informal employees, and 7 participants were work as commercial sex workers. All participants lived in Surakarta and nearby cities (Table 1). None of the participants knew anything about TTV but all participants knew about the prevention and transmission routes of HIV.

Approximately 30.0% (12/40) of participants had had an HIV test, and 25.0% (10/40) had undergone STI screening; however, none of them had been screened for liver function and hepatitis. None of the participants had ever had intercourse with a person whom they had known to have a history of infection with HIV or hepatitis. Unfortunately, all participants never asked about their sex partner's HIV or STI or hepatitis infection status. Regarding other risk behaviors, 10.0% (4/40) of the participants had been an inmate, 70.0% (28/40) had abused non-injected drugs, and 25.0% (10/40) had abused injected drugs. Overall, 15 participants (37.5%) had a tattoo, 40 (100.0%) had been pierced, and 10 (25.0%) had both a tattoo and a piercing. All participants had shared needle history for tattoo and or piercing. All participants had a history of sexual activities with saliva transfer. All participants also had sexual activities (oral and anal) without condom history. There were no reports about contact with blood, being pricked with a needle contaminated with blood, blood transfusions, dialysis, history of traumatic injury, or transplantation. Almost all participants had been a circumcised (35/40, 87.5%) and had a smoking history (33/40, 82.5%). Alcohol intoxication was self-reported by 25 participants (62.5%). The health status was good for all of the participants in the time of data collection.

Table 1. Socio Demographic Data of HIV-Infected Transgender Performing High-Risk Behavior in Surakarta, Indonesia

Factors	Total	
	N (40)	% (100)
Age (years)		
19-29	24	60.0
30-39	16	40.0
40-50	0	0.00
Marital status		
Married	0	0.0
Ever married	0	0.0
Not married	40	100.0
Education		
Undergraduate	0	0.0
Senior high school	15	37.5
Junior high school	15	37.5
Elementary school	10	25.0
Occupation as sex worker	7	17.5
Occupation as non-sex worker		
Unemployed	20	50.0
Employee	13	32.5
Monthly Income		
<100 USD	30	75.0
>100 USD	10	25.0
Ethnicity		
Javanese	32	80.0
Non-Javanese (Ethnic minority)	8	20.0
Residence		
Surakarta	26	65.0
Nearby cities	14	35.0

Note: N, the number of participants; USD, US Dollar.

TTV DNA was detected in 40.0% (16/40) samples. The molecular characterization revealed that the most prevalent was genogroup 3 (50.0%, 20/40), followed by genogroup 2 (30.0%, 12/40), and 1 (20.0%, 8/40), respectively, in line with a previous report [14]. Clinical progression of HIV patients leads increase TTV viral load [32], and the TTV viral load itself may associate with patient's clinical output [10-11]. The susceptibility of TTV co-infection and its pathogenesis role in Indonesian HIV patients may associate with patient's polymorphisms status and other co-infection status or the molecular status of the pathogen itself [4,6-8,14-24,26-31].

HIV disproportionately impacts populations that have traditionally suffered from health disparities as in transgender community[33]. Transgender individuals also often experience significant health disparities [34]. In the present study, all transgender infected with HIV were found had high-risk activities (sharing needle for tattoo and or piercing, saliva transfer, and sexual activities without a condom). Co-infection of TTV was also found, indicated the high-risk activities would increase the risk of other pathogens co-infection, therefore, may influence the patient's clinical output.

4. Conclusions

TTV co-infection was found with high infection rate in HIV-infected transgender performing high-riskbehavior in Surakarta.HIV patient with high-risk activities history should be screened for other blood-borne pathogens.

5. References

[1] Prasetyo A A, Kamahora T, Kuroishi A, Murakami K and Hino S 2009 *Virology* **385** 85-92
[2] Hino S and Prasetyo A A 2009 *Curr. Top. Microbiol. Immunol.* **331** 117-30
[3] Tyschik E A, Shcherbakova S M, Ibragimov R R and Rebrikov D V 2017 *Virol. J.* **14** 92
[4] Prasetyo A A, Desyardi M N, Tanamas J, Suradi, Reviono, Harsini, Kageyama S, Chikumi H and Shimizu E 2015 *Intervirology* **58** 57-68
[5] Cancela F, Ramos N, Mirazo S, Mainardi V, Gerona S and Arbiza J 2016 *Infect. Genet. Evol.* **44** 501-6
[6] Prasetyo A A, Ariapramuda E R, Kindi E A, Dirgahayu P, Sari Y, Dharmawan R, Hudiyono, Hartono and Kageyama S 2014 *Southeast Asian J. Trop. Med. Public Health* **45** 1032-47
[7] Prasetyo A A, Sariyatun R, Reviono, Sari Y, Hudiyono, Haryati S, Adnan Z A, Hartono and Kageyama S 2015 *J. Clin. Virol.***70** 67–71
[8] Prasetyo A A, Dharmawan R, Raharjo I and Hudiyono 2016 *J. Global Infect. Dis.* **8** 75-81
[9] Pirouzi A, Bahmani M, Feizabadi M M and Afkari R 2014 *Rev. Soc. Bras. Med. Trop.* **47** 275-9
[10] García-Álvarez M, Berenguer J, Alvarez E, Guzmán-Fulgencio M, Cosín J, Miralles P, Catalán P, López JC, Rodríguez J M, Micheloud D, Muñoz-Fernández M A and Resino S 2013 *Eur. J. Clin. Microbiol. Infect. Dis.* **32** 289-97
[11] Fogli M, Torti C, Malacarne F, Fiorentini S, Albani M, Izzo I, Giagulli C, Maggi F, Carosi G and Caruso A 2012 *Clin. Dev. Immunol.* **2012** 829584
[12] Devalle S, Rua F, Morgado M G and Niel C 2009 *Arch Virol.* **154** 1285-91
[13] Nasser T F, Brajão de Oliveira K, Reiche E M, Amarante M K, Pelegrinelli Fungaro M H and Watanabe M A 2009 *Microb. Pathog.* **47** 33-7
[14] Prasetyo A A, Luvi S D, Hartono and Sari Y 2017 *AIP Conference Proceedings* **1813** 020004
[15] Sari Y, Haryati S, Raharjo I and Prasetyo A A 2015 *Southeast Asian J. Trop. Med. Public Health* **46** 977-85
[16] Prasetyo A A, Nurfitria FB, Adnan ZA, Redhono D, Sari Y, and Hudiyono 2015 *Int. J. Interferon Cytokine Mediat. Res.***7** 1-7
[17] Prasetyo A A, Dirgahayu P, Sari Y, Hudiyono H, and Kageyama S 2013 *J. Infect. Dev. Ctries.* **7** 453-67
[18] Prasetyo A A, Sari Y, Dharmawan R, and Marwoto 2017 *AIP Conference Proceedings***1813** 020011
[19] Marwoto, Lismana U H V L, Prasetyo A A, Suradi, Reviono, and Harsini 2015 *J. Clin. Diagn. Res.***9** 10-3
[20] Marwoto, Prasetyo A A, Sariyatun R, Suradi, Reviono, and Harsini 2015 *Asian Biomed.* **9** 473-9
[21] Haryati S, Prasetyo A A, Sariyatun R, Sari Y, and Murkati 2015 *Asian Pac. J. Trop. Dis.* **5** 798-803

[22] Prasetyo A A and Sariyatun R 2015 *Asian J. Microbiol. Biotechnol. Environ. Sci.* **17** 349-55
[23] Prasetyo A A and Sariyatun R 2015 *J. Teknol.* **77** 67-70
[24] Prasetyo A A and Zaini K U N 2015 *Southeast Asian J. Trop. Med. Public Health* **4** 662-8
[25] Kageyama S, Agdamag D M D, Alesna E T, Abellanosa-Tac-An I P, Corpuz A C, Telan E F O, Que ER, Leaño P S A, Jereza L D, Emphasis Y E E, Prasetyo A A, Tanimoto T and Ichimura H. 2009 *J. Med. Virol.* **81** 1157-62
[26] Prasetyo A A, Dharmawan R, Sari Y and Sariyatun R 2017 *AIP Conference Proceedings* **1813** 020020
[27] Prasetyo A A and Sari Y 2017 *AIP Conference Proceedings* **1813** 020001
[28] Haryati S, Sari Y, Prasetyo A A and Sariyatun R 2016 *IOP Conf. Series* **107** 012052
[29] Haryati S, Sari Y, Prasetyo A A and Sariyatun R 2016 *IOP Conf. Series* **107** 012055
[30] Sari Y, Haryati S, Prasetyo A A, Hartono and Adnan Z A 2017 *AIP Conference Proceedings* **1813** 020005
[31] Prasetyo A A 2017 *AIP Conference Proceedings* **1788** 030100
[32] Thom K and Petrik J 2007 *J. Med. Virol.* **79** 1-7
[33] Del Rio C 2016 *Top. Antivir. Med.* **24** 86-9
[34] Hayon R 2016 *FP Essent.* **449** 27-36

Acknowledgments
This work was supported in part by grants from the Indonesian Directorate of Higher Education-APBN/DIPA UNS (No. 501/UN27.11/PN/2014 and No. 339/UN27.11/PL/2015), and from the Indonesian Ministry of Research, Technology and Higher Education (No 353/ UN 27.21/ PN/ 2016 and No 873/ UN27.21/ PP/ 2017).

IOP Conf. Series: Journal of Physics: Conf. Series **1022** (2018) 012047 doi:10.1088/1742-6596/1022/1/012047

CCR2 V64I polymorphism in rifampicin resistant tuberculosis patients in Moewardi General Hospital Surakarta, Indonesia

Marwoto[1,2], Afiono Agung Prasetyo[1,2,3*], Reviono[4] and Suradi[4]

[1] A-IGIC (A-Infection, Genomics, Immunology, & Cancer) Research Group, Universitas Sebelas Maret, Jl. Ir. Sutami 36A, Surakarta, 57126, Indonesia.
[2] Department of Microbiology, Faculty of Medicine, Universitas Sebelas Maret, Jl. Ir. Sutami 36A, Surakarta, 57126, Indonesia.
[3]Center of Biotechnology and Biodiversity Research and Development, Universitas Sebelas Maret, Jl. Ir. Sutami 36A, Surakarta, 57126, Indonesia.
[4] Department of Pulmonology, Faculty of Medicine, Universitas Sebelas Maret, Jl. Ir. Sutami 36A, Surakarta, Indonesia.

*Corresponding author: afie.agp.la@gmail.com; afieagp@yahoo.com; afie@staff.uns.ac.id

Abstract. CC chemokine receptor-2 (CCR2) play important roles in inflammation. The *CCR2* V64I polymorphism already reported associated with many diseases; however, the association of *CCR2* V64I polymorphism with tuberculosis is still unknown. Also, there is no report about the presentation of *CCR2* V64I polymorphisms in Indonesian tuberculosis patients with rifampicin-mono resistant status has ever been published, to the best of our knowledge. This study evaluated the presence of *CCR2* V64I polymorphisms in Javanese rifampicin-mono resistant tuberculosis patients. In an ongoing molecular epidemiology study of human genomic polymorphisms and infection, 51 Javanese rifampicin-mono resistant tuberculosis patients in Dr. Moewardi General Hospital in Surakarta were enrolled in the study. The blood samples were aliquoted and fractionated. The nucleic acids were extracted from all blood samples and subjected to the *CCR2* V64I polymorphisms detection by a polymerase chain reaction-sequence-specific primer (PCR-SSP) technique. PCR products were analyzed in 3% agarose. *CCR2* 64V and 64I homozygote were found in 23.5% (12/51) and 23.5% (12/51) blood samples, respectively. The *CCR2* VI genotype was found in 52.9% (27/51) blood samples. The *CCR2* VI genotype was found predominant in Javanese rifampicin-mono resistant tuberculosis patients and may have anassociation with the clinical progression.

1. Introduction

Tuberculosis is a major threat to human health [1]. End tuberculosis strategy aims at ending the tuberculosis epidemic by 2030 [2]. However, treatment failure and resistance amplification are common among patients with rifampicin-resistant tuberculosis [3]. The emergence of drug-resistant tuberculosis, including that of the rifampicin-resistant, compromises global tuberculosis control [4].

Chemokines are low-molecular-weight proteins that play a key role in inflammatory processes [5]. With its receptors, both are crucial for an extensive immune response against infectious diseases such as tuberculosis [6]. Chemokines and its receptors are critical for initiating and coordinating the organized and sequential recruitment and activation of immune cells into Mycobacterium tuberculosis-

IOP Conf. Series: Journal of Physics: Conf. Series **1022** (2018) 012047 doi:10.1088/1742-6596/1022/1/012047

infected lungs [7]. CC chemokine receptor 2 (CCR2), a monocyte chemo attractant protein-1 (MCP-1) receptor, is important in the recruitment of immune cells as well as non-immune cells under pathological condition [8,9]. The genetic polymorphisms of CCR2 V64I may influence the susceptibility of some disease in Asian countries, however, there are no data on the CCR2 genotypes inrifampicin-mono resistant tuberculosis, for the best of our knowledge. We, therefore, aim to investigate the association of CCR2 V64I genetic polymorphism with tuberculosis, focusing on Javanese rifampicin-mono resistant tuberculosis patients in Dr. Moewardi General Hospital in Surakarta.

2. Materials and Methods

In this ongoing study 51 Indonesian Javanese tuberculosis patients with rifampicin-mono resistant status detected by the Gene Xpert system and confirmed by Drug Susceptibility Test at Dr. Moewardi General Hospital, Central Java province, Indonesia were enrolled. The clinical characteristics of these patients were assessed. Peripheral blood samples were obtained from all patients to examine their hematologic profiles. Since the study was a part of our research group (A-IGIC/A-Infection, Genomics, Immunology, & Cancer research group) molecular epidemiology study for blood-borne pathogens, all blood samples collected from the patients also subjected for immunological and molecular assays to find out the HIV, HBV, HCV, HDV, TTV, GBV-C, HTLV-1/2, and *Toxoplasma gondii* infection status as described previously [10-29]. Approval was obtained from the institutional ethical committee review boards of the Faculty of Medicine of Universitas Sebelas Maret and Dr. Moewardi General Hospital, Central Java province, Indonesia. Written informed consent was obtained from all study participants. All procedures were conducted according to the principles of the Declaration of Helsinki.

Genomic DNA was isolated from whole blood using the High Pure PCR Template Preparation Kit (Roche Applied Science, Mannheim, Germany). Then, the genomic DNA was used as a template for amplifications that were conducted using a common primer (5'-TGGAAAATAAGGGCCACAGAC-3') and allele-specific reverse primers (5'-GGGCAACATGCTGGTCG-3' or 5'-TGGGCAACATGCTGGTCA-3') [30]. Fast Start HiFi PCR System dNTPack (Roche Applied Science) polymerase was used in the polymerase chain reaction-sequence-specific primer (PCR-SSP)reaction. For data interpretation, we conducted electrophoresis using a 2% agarose gel (100 V for 30 minutes), which was visualized using a UV trans-illuminator. All samples were tested at least twice.

3. Results and Discussion

A total of 51 Javanese tuberculosis patients with rifampicin-mono resistant status were subjected to genotypic analysis of CCR2 V64I polymorphisms status. The samples were derived from 25 (49.0%) men and 26 (51.0%) women aged 18–65 years old (mean: 41.6 ± 11.6 years). All patient bacterial cultures indicated the presence of *Mycobacterium tuberculosis*. Data for the patient body mass index (BMI) were obtained from 47 patients, with a mean BMI of 16.5 ± 3.4 kg/m^2. Most patients (34/47, 72.3%) were found underweight (BMI <18.5 kg/m^2), whereas 27.7% (13/47) patients were normal weight (BMI = 18.5–24.9 kg/m^2). Smoking history was found in 25.5% (13/51) patients. None of the patients had extrapulmonary tuberculosis. Relapse history was found in 21.6% (11/51) patients. Having a history of treatment failure of I and II category were found in 21.6% (11/51) and 47.1% (24/51) patients, respectively. Most patients had normal CD4+ T cell counts (>500 cells/µL, 36/51, 70.6%) and percentages (≥29%, 37/51, 72.5%).CD4+ T cell counts of 200–500 cells/µL was observed in 29.4% (15/51), and none had CD4+ T cell counts<200 cells/µL. CD4+ T cell percentages of 14%–28% was found in 27.5% (14/51) patients, and none had <14%,

Data for the thorax radiographical examination were obtained for 90.2% (46/51) of rifampicin-mono resistant tuberculosis patients. Lung cavity, infiltration, fibrosis, fibrothorax, pleural effusion, atelectasis, and consolidation were observed in 39.1% (18/46), 89.1% (41/46), 52.2% (24/46), 2.2%

(1/46), 6.5% (3/46), 2.2% (1/46), and 2.2% (1/46) patients, respectively.None of the patients suffered nodule, miliary TB, bulla, pneumothorax, or mass appearance.

During follow-up of MDR-TB patients, sputum conversion was observed mostly after two months of treatment (19/51, 37.3%).Other patients showed sputum conversion after one month (12/51, 23.5%), three months (6/51, 11.8%), and four months (2/51, 3.9%) of treatment. Four (7.8%) patients died before sputum conversion.CCR2 64V and 64I homozygote were found in 23.5% (12/51) and 23.5% (12/51) patients, respectively. The CCR2 VI genotype was found in 52.9% (27/51) patients (Table 1).

Table 1. CCR2 V64I Genetic Polymorphism in Javanese Rifampicin-Mono Resistant Tuberculosis Patients

	CCR2 V64I genotype (%)		
	V/V (n= 12)	V/I (n= 27)	I/I (n= 12)
BMI*			
Underweight (< 18.5 kg/m^2) (n= 34)	58.3 (7)	70.4 (19)	66.7 (8)
Normal weight (18.5–24.9 kg/m^2) (n= 13)	33.3 (4)	22.2 (6)	25.0 (3)
CD4+ T cell counts			
>500 cells/μL (n= 36)	75.0 (9)	70.4 (19)	66.7 (8)
200–500cells/μL (n= 15)	25.0 (3)	29.6 (8)	33.3 (4)
CD4+ T cell percentages			
≥29% (n= 37)	75.0 (9)	74.1 (20)	66.7 (8)
14%–28% (n= 14)	25.0 (3)	25.9 (7)	33.3 (4)
Relapse (n= 11)	0.0 (0)	33.3 (9)	16.7 (2)
Treatment failure of I category (n= 11)	33.3 (4)	14.8 (4)	25.0 (3)
Treatment failure of II category (n= 24)	58.3 (7)	40.7 (11)	50.0 (6)
Thorax radiographically imaging			
Cavity (n= 18)	25.0 (3)	37.0 (10)	41.7 (5)
Infiltration (n= 41)	91.7 (11)	81.5 (22)	66.7 (8)
Fibrosis (n= 24)	58.3(7)	44.4 (12)	41.7 (5)
Fibrothorax (n= 1)	0.0 (0)	3.7 (1)	0.0 (0)
Pleural effusion (n= 3)	0.0 (0)	7.4 (2)	8.3 (1)
Atelectasis (n=1)	0.0 (0)	0.0 (0)	8.3 (1)
Consolidation (n= 1)	0.0 (0)	3.7 (1)	0.0 (0)
Sputum conversion (month)*			
1 (n= 12)	25.0 (3)	7.4 (2)	58.3 (7)
2 (n= 19)	50.0 (6)	44.4 (12)	8.3 (1)
3 (n= 6)	8.3 (1)	18.5 (5)	0.0 (0)
4 (n= 2)	8.3(1)	0.0 (0)	8.3 (1)
Died before sputum conversion (n= 4)	0.0 (0)	14.8 (4)	0.0 (0)

* missing data

Albeit the number patients involved in the present study was inadequate to perform statistical analysis (due to the difficult nature of the study condition), relapse was profound in patients with CCR2 I allele (Odds ratio [OR]=1.702, 95% CI:0.623-4.655, p= 0.299). Sputum conversion more than one month was also more frequent in patients with V allele (OR=1.684, 95% CI:0.685-4.144, p= 0.256). Taking all data together, CCR2 V64I genetic polymorphism may influence the clinical status of Javanese rifampicin-mono resistant tuberculosis patients; need further study to confirm the present finding.

The outcome of the disease is determined by acomplex interaction among pathogen, host genetic variability, and surrounding milieu. Variation in expression or function of chemokines caused by genetic polymorphisms could be associated with attenuated immune responses. Exploration of chemokine genetic polymorphisms in therapeutic response, gene regulation, and disease outcome is important [6]. Correct mononuclear cellular recruitment and localization are essential to ensure control of Mycobacterium tuberculosis growth without the development of diffuse and damaging granulocytic inflammation. Determination of how the key chemokines/cytokines and their receptors are balanced and how the loss of that balance can promote disease is vital to understanding tuberculosis pathogenesis and to identify novel therapies for effective eradication of this disease [7].

4. Conclusions

CCR2 V64I genetic polymorphism may influence the clinical status of Javanese rifampicin-mono resistant tuberculosis patients.

5. References

[1] Cantini F, Nannini C, Niccoli L, Petrone L, Ippolito G and Goletti D 2017 *Mediators Inflamm.* **2017** 8909834
[2] El-Samadony H, Althani A, Tageldin M A and Azzazy H M E 2017 *Expert Rev. Mol. Diagn.* **17** 427-43
[3] Kendall E A, Cohen T, Mitnick C D and Dowdy D W 4 2017 *Int. J. Infect. Dis.* **56** 185-9
[4] Maitre T, Aubry A, Jarlier V, Robert J, Veziris N, and CNR-My RMA 3 2017 *Med. Mal. Infect.* **47** 3-10
[5] Ferreira-Fernandes H, Santos A C, Motta F J, Canalle R, Yoshioka F K, Burbano R R, Rey J A, da Silva B B and Pinto G R 2015 *Genet. Mol. Res.* **14** 11710-8
[6] Qidwai T and Khan MY 2016 *Hum. Immunol.* **77** 961-71
[7] Domingo-Gonzalez R, Prince O, Cooper A and Khader S A 2016 *Microbiol. Spectr.* **4** doi: 10.1128/microbiolspec.TBTB2-0018-2016.
[8] Qidwai T 2016 *Immunol. Lett.* **176** 128-38
[9] Narter F, Bireller E S, Engin C, Catmakas T, Narter F, Ergen A and Cakmakoglu B2. 2015 *Int. J. Clin. Exp. Med.* **8** 14075-9
[10] Prasetyo A A, Sariyatun R, Reviono, Sari Y, Hudiyono, Haryati S, Adnan Z A, Hartono and Kageyama S 2015 *J. Clin. Virol.* **70** 67–71
[11] Prasetyo A A, Dirgahayu P, Sari Y, Hudiyono H and Kageyama S 2013 *J. Infect. Dev. Ctries.* **7** 453-67
[12] Prasetyo A A, Ariapramuda E R, Kindi E A, Dirgahayu P, Sari Y, Dharmawan R, Hudiyono, Hartono and Kageyama S 2014 *Southeast Asian J. Trop. Med. Public Health* **45** 1032-47
[13] Prasetyo A A, Desyardi M N, Tanamas J, Suradi, Reviono, Harsini, Kageyama S, Chikumi H and Shimizu E 2015 *Intervirology* **58** 57-68
[14] Sari Y, Haryati S, Raharjo I and Prasetyo A A 2015 *Southeast Asian J. Trop. Med. Public Health* **46** 977-85
[15] Prasetyo A A, Nurfitria FB, Adnan ZA, Redhono D, Sari Y and Hudiyono 2015 *Int. J. Interferon Cytokine Mediat. Res.* **7** 1-7
[16] Marwoto, Lismana U H V L, Prasetyo A A, Suradi, Reviono and Harsini 2015 *J. Clin. Diagn. Res.* **9** 10-3
[17] Marwoto, Prasetyo A A, Sariyatun R, Suradi, Reviono and Harsini 2015 *Asian Biomed.* **9** 473-9
[18] Haryati S, Prasetyo A A, Sariyatun R, Sari Y and Murkati 2015 *Asian Pac. J. Trop. Dis.* **5** 798-803
[19] Prasetyo A A and Sariyatun R 2015 *Asian J. Microbiol. Biotechnol. Environ. Sci.* **17** 349-55
[20] Prasetyo A A and Sariyatun R 2015 *J. Teknol.* 77 67-70
[21] Prasetyo A A and Zaini KUN 2015 *Southeast Asian J. Trop. Med. Public Health* **4** 662-8
[22] Prasetyo A A, Dharmawan R, Raharjo I and Hudiyono 2016 *J. Global Infect. Dis.* **8** 75-81
[23] Haryati S, Sari Y, Prasetyo A A and Sariyatun R 2016 *IOP Conf. Series* **107** 012052
[24] Haryati S, Sari Y, Prasetyo A A and Sariyatun R 2016 *IOP Conf. Series* **107** 012055

[25] Prasetyo A A 2017 *AIP Conference Proceedings* **1788** 030100
[26] Prasetyo A A, Sari Y, Dharmawan R and Marwoto 2017 *AIP Conference Proceedings* **1813** 020011
[27] Sari Y, Haryati S, Prasetyo A A, Hartono and Adnan Z A 2017 *AIP Conference Proceedings* **1813** 020005
[28] Prasetyo A A, Dharmawan R, Sari Y and Sariyatun R 2017 *AIP Conference Proceedings* **1813** 020020
[29] Prasetyo A A and Sari Y 2017 *AIP Conference Proceedings* **1813** 020001
[30] Chatterjee K, Dandara C, Hoffman M and Williamson A L 2010 *BMC Cancer* **10** 278

Acknowledgments

This work was supported in part by grants from the Indonesian Directorate of Higher Education-APBN/DIPA UNS (No. 501/UN27.11/PN/2014 and No. 339/UN27.11/PL/2015), Indonesian Ministry of Research, Technology and Higher Education (No 353/ UN 27.21/ PN/ 2016 and No 873/ UN27.21/ PP/ 2017), and PNBP UNS (No 632/ UN 27.21/ LT/ 2016 and No 623/UN27.21/PP/2017).

Molecular analysis of *Toxoplasma gondii Surface Antigen 1* (*SAG1*) gene cloned from *Toxoplasma gondii* DNA isolated from Javanese acute toxoplasmosis

Sri Haryati[1,2], Afiono Agung Prasetyo[1,3,4*], Yulia Sari[1,2,3,5] and Ruben Dharmawan[1,2]

[1] A-IGIC (A-Infection, Genomics, Immunology, & Cancer) Research Group, Universitas Sebelas Maret, Jl. Ir. Sutami 36A, Surakarta, 57126, Indonesia.
[2] Department of Parasitology, Faculty of Medicine, Universitas Sebelas Maret, Jl. Ir. Sutami 36A, Surakarta, 57126, Indonesia.
[3]Center of Biotechnology and Biodiversity Research and Development, Universitas Sebelas Maret, Jl. Ir. Sutami 36A, Surakarta, 57126, Indonesia.
[4] Division of Virology, Department of Microbiology, Faculty of Medicine, Universitas Sebelas Maret, Jl. Ir. Sutami 36A, Surakarta, 57126, Indonesia.
[5] Doctoral Program of Medical Sciences Universitas Sebelas Maret, Jl. Ir. Sutami 36A, Surakarta, 57126, Indonesia.

*Corresponding author: afie.agp.la@gmail.com; afieagp@yahoo.com; afie@staff.uns.ac.id

Abstract. *Toxoplasma gondii* Surface Antigen 1 (SAG1) is often used as a diagnostic tool due to its immunodominant-specific as antigen. However, data of the *Toxoplasma gondii* SAG1 protein from Indonesian isolate is limited. To study the protein, genomic DNA was isolated from a Javanese acute toxoplasmosis blood samples patient. A complete coding sequence of *Toxoplasma gondii SAG1* was cloned and inserted into an *Escherichia coli* expression plasmid and sequenced. The sequencing results were subjected to bioinformatics analysis. The *Toxoplasma gondii SAG1* complete coding sequences were successfully cloned. Physicochemical analysis revealed the 336 aa of SAG1 had 34.7 kDa of weight. The isoelectric point and aliphatic index were 8.4 and 78.4, respectively. The N-terminal methionine half-life in *Escherichia coli* was more than 10 hours. The antigenicity, secondary structure, and identification of the HLA binding motifs also had been discussed. The results of this study would contribute information about *Toxoplasma gondii* SAG1 and benefits for further works willing to develop diagnostic and therapeutic strategies against the parasite.

1. Introduction

Toxoplasma gondii is an obligate intracellular protozoan parasite with a broad host range including humans [1]. The parasite is of great importance to public health, highlighted by the high diversity of hosts [2]. The parasite could cause toxoplasmosis, a disease that can create serious health problems in humans [3]. Not only humans which are immune compromised, the immune competent humans are also susceptible to toxoplasmosis [4]. Because of the clinical importance of toxoplasmosis, it's timely and accurate diagnosis has a major impact on disease-fighting strategies [3].

Toxoplasma gondii surface antigen 1 (SAG1) is a well-characterized *Toxoplasma gondii* antigen [1]. SAG1 is an immunodominant-specific antigen, therefore is often used as a diagnostic tool [3]. However, data of the SAG1 protein from Indonesian isolate is limited.

2. Materials and Methods

2.1. *Toxoplasma gondii Sample Selection*

Since 2009, our research group (A-IGIC/A-Infection, Genomics, Immunology,& Cancer Research Group) has been conducting a molecular epidemiology study of human blood-borne pathogens in Central Java, Indonesia [5-24]. To select the *Toxoplasma gondii* sample for cloning, all blood samples were subjected for immunological and molecular assays to find out the HIV (Human Immunodeficiency Virus), HBV (Hepatitis B Virus), HCV (Hepatitis C Virus), HDV (Hepatitis D Virus), TTV (Torque Teno Virus), GBV-C (GB Virus C), HTLV-1/2 (Human T-cell Lymphotropic Virus-1/2), and *Toxoplasma gondii* infection status as described previously [5-24]. All samples were tested at least in duplicate.

2.2. *Cloning and Sequencing*

A complete coding sequence of *Toxoplasma gondii SAG1* was cloned from a Javanese acute toxoplasmosis blood samples by nested PCR. The PCR product was inserted into Vari Flex N-terminal SBP-SET Vector (an *Escherichia coli* expression plasmid). For cloning, each primer was attached with a start or stop codon sequences. The recombinant plasmid was then transformed into competent cells, propagated, purified, and sequenced. The sequencing step was performed in both orientations for confirmation, three times for each orientation.

2.3. *BLAST Analysis*

A BLAST analysis was performed to confirm the sequencing results. The sequences were submitted to the BLAST program to check their similarity to related strains deposited in GenBank/EMBL/DDBJ. Multiple alignmentsof reference sequences were reconstructed using ClustalW as implemented in CLC Main Workbench 7.9.1 software. The frequency of nucleotide substitution at each base was estimated using the Kimura 2-parameter method. Synonymous and non-synonymous amino acid substitution events were computed using the Nei-Gojobori method with the Jukes-Cantor correction implemented in CLC Main Workbench 7.9.1 software.

2.4. *Predicted Protein Analysis*

Protein analysis was performed using CLC Main Workbench 7.9.1 software. The parameters for hydrophobicity plot was set by Kyte-Doolittle, Eisenberg, Engelman, Hopp-Woods, Janin, Rose, and Cornette hydrophobicity scale, respectively. To create the antigenicity plot, both Welling and Kolaskar-Tongaonkar antigenic scale was used. The Immune Epitope Database (IEDB) (http://tools.immuneepitope.org/mhcii/) was used to predict the peptide binding of SAG1 sequences to HLA molecules.

3. Results and Discussion

Blood samples collected by our research group were always screened for HIV, HBV, HCV, HDV, TTV, GBV-C, HTLV-1/2, and *Toxoplasma gondii* infection status by immunological and molecular assays. Blood sample positive for IgM anti-*Toxoplasma gondii* and *Toxoplasma gondii* DNA (acute toxoplasmosis), but negative for anti-HIV, HIV RNA, HBsAg, HBV DNA, anti-HDV, HDV RNA, TTV DNA, GBV-C RNA, anti-HTLV-1/2, HTLV-1 RNA, HTLV-2 RNA, and IgG anti-*Toxoplasma gondii* was then used for cloning the *Toxoplasma gondii* interest gene, including the *Toxoplasma gondii SAG1*. A complete coding sequence of *Toxoplasma gondii SAG1* was successfully cloned, and the BLAST and nucleotide analysis were performed to confirm the sequences. Based on the analysis of 1011 nucleotides bases, the cloned was confirmed as *Toxoplasma gondii SAG1*.

We then perform the predicted protein analysis. The predicted protein had336 amino acid sequence length with 34.731kDa in weight. The isoelectric point was 8.4 and the aliphatic index was 78.423. The predicted protein had methionine N-terminal half-life in *Escherichia coli* more than 10 hours, however, shorter than in mammals (30 hours) and in yeast (>20 hours), respectively. The extinction coefficient at 280 nm for non-reduced cysteines was 21,750 (absorption at 280nm 0.1%= 0.626) while for reduced cysteines was 20,910 (absorption at 280nm 0.1%= 0.602).The count of positively and negatively charged residues was 0.074 and 0.068, respectively, in frequency.

Toxoplasma gondii SAG1 is an appropriate antigen with high specificity and sensitivity for the detection of *Toxoplasma gondii* [25]. The predicted SAG1 protein had antigenic positions (figure 1).Beta alpha helix secondary structures were frequent in our predicted SAG1. Only three alpha helix secondary structures were found, at aa (amino acids) 21-26, 37-39, and 243-249, respectively. The induction of CD8+ T cell responses is important in the formulation of successful immune reaction against *Toxoplasma gondii* [26]. The CD8+ T cell response is initiated by the MHC-I presentation resulting in the death of infected cells [27]. Further analysis of the possible antigenic regions presented in figure 1, nine epitopes were found had high probability of eliciting an immune response (score >0.1), i.e. IPEAEDSWWT (0.3516), LTPTENHFTL (0.27859), KAVRRAVTAG (0.24846), SDNGATLTIN (0.20798), FAMVTGLIGS (0.17702), IKLTVPIEKF (0.14864), PVTTQTFVVG (0.14582), SSVVNNVARC (0.14446), and SENPWQGNAS (0.13894), respectively.

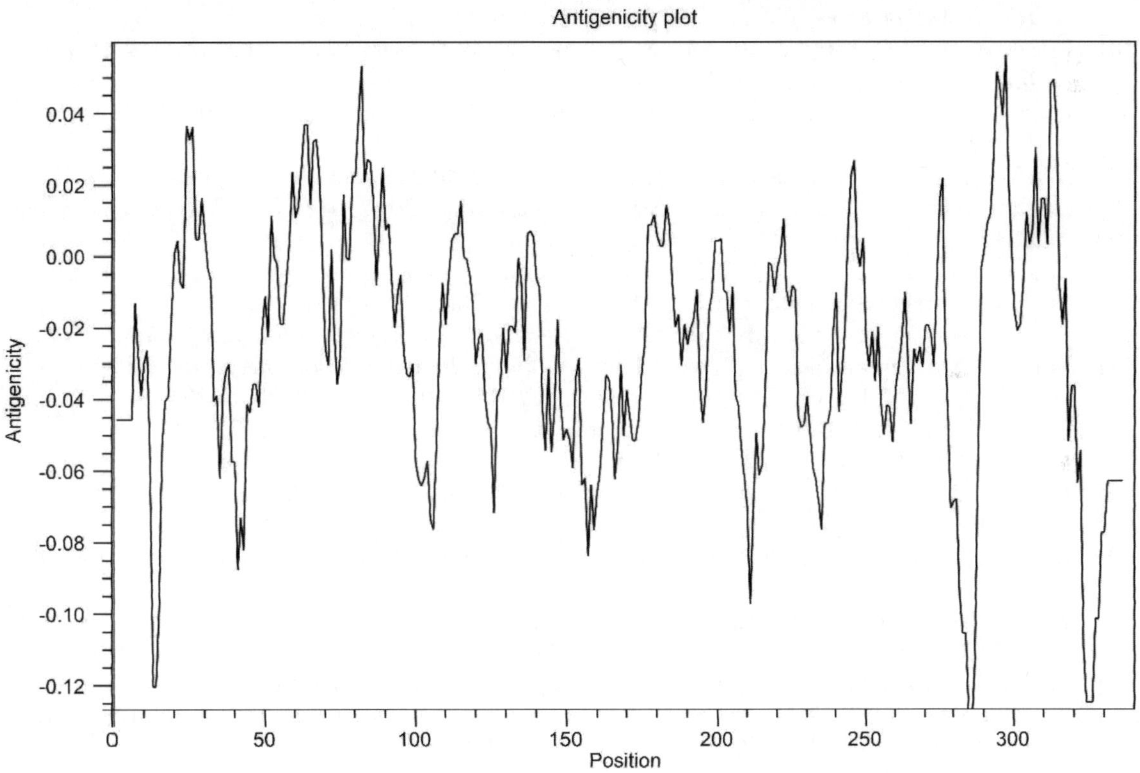

Figure 1. Antigenicity Plot of *Toxoplasma Gondii* SAG1 Predicted Protein

4. Conclusions

Finally, we concluded that our clone was *Toxoplasma gondii SAG1*. Our predictive protein should be a mature protein, tended to be positively charged, had both α-helix and β-sheet secondary structures, and 9 possible antigenic regions high probability of eliciting an immune response. Taken all data together, it will increase our understanding of our *Toxoplasma gondii* SAG1 and benefits for further works willing to develop diagnostic and therapeutic strategies against the parasite.

5. References

[1] Han Y, Zhou A, Lu G, Zhao G, Wang L, Guo J, Song P, Zhou J, Zhou H, Cong H and He S2017 *BMC Infect. Dis.* **17** 59
[2] da Silva R C and Langoni H 2016 *Vet. Parasitol.* **225** 73-80
[3] Sert M, Cakir Koc R and Budama Kilinc Y 2017 *Sci. Rep.* **7** 852
[4] Rather H A, Din M M, Sheikh A A, Tewari A K and Maharana B R 2017 *J. Parasit. Dis.* **41** 551-6
[5] Prasetyo AA, Dirgahayu P, Sari Y, Hudiyono H and Kageyama S 2013 *J. Infect. Dev. Ctries.* **7** 453-67
[6] Prasetyo A A, Ariapramuda E R, Kindi E A, Dirgahayu P, Sari Y, Dharmawan R, Hudiyono, Hartono and Kageyama S 2014 *Southeast Asian J. Trop. Med. Public Health* **45** 1032-47
[7] Prasetyo A A, Sariyatun R, Reviono, Sari Y, Hudiyono, Haryati S, Adnan Z A, Hartono and Kageyama S 2015 *J. Clin. Virol.* **70** 67–71
[8] Prasetyo A A, Desyardi M N, Tanamas J, Suradi, Reviono, Harsini, Kageyama S, Chikumi H and Shimizu E 2015 *Intervirology* **58** 57-68
[9] Sari Y, Haryati S, Raharjo I and Prasetyo A A 2015 *Southeast Asian J. Trop. Med. Public Health* **46** 977-85
[10] Prasetyo A A, Nurfitria F B, Adnan Z A, Redhono D, Sari Y and Hudiyono 2015 *Int. J. Interferon Cytokine Mediat. Res.* **7** 1-7
[11] Marwoto, Lismana U H V L, Prasetyo A A, Suradi, Reviono and Harsini 2015 *J. Clin. Diagn. Res.* **9** 10-3
[12] Marwoto, Prasetyo A A, Sariyatun R, Suradi, Reviono and Harsini 2015 *Asian Biomed.* **9** 473-9
[13] Haryati S, Prasetyo A A, Sariyatun R, Sari Y and Murkati 2015 *Asian Pac. J. Trop. Dis.* **5** 798-803
[14] Prasetyo A A and Sariyatun R 2015 *Asian J. Microbiol. Biotechnol. Environ. Sci.* **17** 349-55
[15] Prasetyo A A and Sariyatun R 2015 *J. Teknol.* 77 67-70
[16] Prasetyo A A and Zaini K U N 2015 *Southeast Asian J. Trop. Med. Public Health* **4** 662-8
[17] Prasetyo A A, Dharmawan R, Raharjo I and Hudiyono 2016 *J. Global Infect. Dis.* **8** 75-81
[18] Haryati S, Sari Y, Prasetyo A A and Sariyatun R 2016 *IOP Conf. Series* **107** 012052
[19] Haryati S, Sari Y, Prasetyo A A and Sariyatun R 2016 *IOP Conf. Series* **107** 012055
[20] Prasetyo A A 2017 *AIP Conference Proceedings* 1788 030100
[21] Prasetyo A A, Sari Y, Dharmawan R and Marwoto 2017 *AIP Conference Proceedings* **1813** 020011
[22] Sari Y, Haryati S, Prasetyo A A, Hartono and Adnan Z A 2017 *AIP Conference Proceedings* **1813** 020005
[23] Prasetyo A A, Dharmawan R, Sari Y and Sariyatun R 2017 *AIP Conference Proceedings* **1813** 020020
[24] Prasetyo A A and Sari Y 2017 *AIP Conference Proceedings* **1813** 020001
[25] Sharifdini M, Mohebali M, Keshavarz H, Hosseininejad M, Hajjaran H, Akhoundi B, Rahimi Foroushani A and Zarei Z 2016 *J. Parasit Dis.* **40** 1278-82
[26] Alexander J, Oseroff C, Dahlberg C, Qin M, Ishioka G, Beebe M, Fikes J, Newman M, Chesnut R W, Morton PA, Fok K, Appella E and Sette A 2002 *J. Immunol.* **168** 6189–98
[27] Da Silva S A, Mazini P S, Reis P G, Sell A M, Tsuneto L T, Peixoto P R and Visentainer J E 2009 *BMC Infect. Dis.* **9** 134

IOP Conf. Series: Journal of Physics: Conf. Series **1022** (2018) 012048 doi:10.1088/1742-6596/1022/1/012048

Acknowledgments
This work was supported in part by grants from the Indonesian Directorate of Higher Education-APBN/DIPA UNS (No. 339/UN27.11/PL/2015), and from the Indonesian Ministry of Research, Technology and Higher Education (No 353/ UN 27.21/ PN/ 2016 and No 873/ UN27.21/ PP/ 2017).

IOP Conf. Series: Journal of Physics: Conf. Series **1022** (2018) 012049 doi:10.1088/1742-6596/1022/1/012049

Buccal swab, a minimally invasive method for the screening of oral cancer in active smokers

Suyatmi[1,2*], P Subiyantoro[3] and S Indrakila[4]

[1]Doctoral Program of Medical Science Universitas Sebelas Maret, Indonesia
[2]Dept. of Histology, Faculty of Medicine, Universitas Sebelas Maret, Indonesia
[3]Dept.of Dentistry, Faculty of Medicine, Universitas Sebelas Maret, Indonesia
[4]Dept. of Ophtalmology, Faculty of Medicine Universitas Sebelas Maret, Indonesia

*Corresponding email: suyatmi72@staff.uns.ac.id

Abstract. Smoking is the main risk factor for developing oral cancer. The previous study showed that there was a strong correlation between the length of smoking with the risk to develop oral cancer. Early detection of epithelial changes of oral mucosa will be a good prevention of the incidence of oral cancer among active smokers. This study evaluated the potential use of buccal swab for the screening of early signs of malignancy in active smokers. This study involved 80 participants including those who were smokers and non smokers. The buccal swab was conducted using sterile cytobrush. An epithelial smear was made from the buccal swab and stained with Papanicolaou's technique. An cytomorphometric analysis was conducted by comparing the ratio of nuclear cell to cytoplasmic diameter (ND/CD) between the two groups. The mean of ND observed in this study were 8.963μ for active smokers and 7.991μ for non smokers groups. While the mean of CD were 58.249μ and 63.473μ for active smoker and non-smoker respectively. The mean of ND/CD ratio were 0.156 for active smokers and 0.129 for non smokers groups. This study detected a significant difference on the ND/CD ratio among active smokers vs non smokers ($p<0.0001$; 95% CI = -0.040 – -0.014). In conclusion buccal swab could be a routine procedure to obtain sample for identification of changes in cells morphology to screen an early development of oral cancer.

1. Introduction

Oral cancer is cancer which occurs on mouth, tongue, and oropharynx. The incidence of oral cancer is quite high with 300,286 cases worldwide were reported in 2012. Compared to other regions, the incidence and mortality rate of oral cancers in South East Asia Region is the highest. In fact, 34.46% (103,464/300,286) of oral cancer cases worldwide are in South East Asia Region [1]. Among the South East Asia countries, Indonesia reported the highest incidence of oral cancer. A number of 5329 cases were reported in Indonesia in 2012 and this incidence is predicted to increase year by year [2]. Although these data showed a serious problem, there is a lack of awareness of health society toward oral cancer. A study in Dharmais National Cancer Hospital of Indonesia showed that most of the patients with oral cancer seek medical help in severe conditions, with the majority of patients were first diagnosed in stage 4; this condition explains the high mortality rate of oral cancer patients in Indonesia [3].

The efforts to prevent oral cancer incidence should include programs to educate the society about the risk factors of oral cancer and to detect early cancer development. There are several risk factors for oral cancer development including, smoking, alcohol consumption, and HPV infection [1,4]. It has been shown that tobacco smoking is strongly correlated to oral cancer incidence . In India, 74% of oral cancer patients had history of long term tobacco used [4]. As part of detection oral cancer, biopsy followed by histopathology examination of a clinically pathologic lesion is recommended as an early detection of oral cancer [5]. However, this effort is quite late, considering that the process of cancer development has been started at a long time before the clinical lesion can be observed.This study aims to assess the potential of buccal swab as a minimally invasive method to collect a sample source for detection of early oral cancer development in active smokers.

2. Material And Methods

This study involved 80 male participants; 40 active smokers and 40 nonsmokers. The inclusive criteria for an active smoker are smoking over 10 years and consuming at least 10 cigarettes per day. Participants with diabetes mellitus, a history of radiotherapy and or chemotherapy were excluded from this study. The nonsmokers participants were those who were not smoking nor being passive smokers, not diabetic patient and not having a history of radiotherapy or chemotherapy. Research were conducted in District Surakarta, Central Java, Indonesia. Participants were selected by convenient sampling method; a person who meets the subject criteria are asked to be research participants. All participants were given written informed consent and the research protocol has been approved by Ethics Committee of Muwardi Hospital and Faculty of Medicine Universitas Sebelas Maret, Indonesia.

2.1. Buccal Swab

Buccal swab for a collection of oral epithelial samples was conducted using a sterile cytobrush. The participants were instructed to brush the teeth and do mouth washing thoroughly with water prior to epithelial swab. The epithelial swab was taken from buccal and tongue areas by stirring the cytobrush for 5 times on the areas. An epithelial smear was made from the buccal swab and then the preparation was stained with Papanicolaou's technique for cytomorphometric measurement.

2.2. Cytomorphometry

The cytomorphometry was conducted by measuring the ratio of nuclear cell diameter (ND) vs cytoplasmic cell diameter (CD) on 40 cells per individual Papanicolaou's smear. Measurement of ND and CD were made using Optilab Image Raster software® under 1000x microscopic magnification, as can be shown in figure 1. The difference in ratio of ND to CD was analyzed statistically using Independent t-test.

3. Result

Characteristics of participants are distributed according to age, length of smoking, amount of cigarette smoked, daily habit of oral hygiene, and alcohol consumption. These characteristics are shown in Table 1. The age of participant involved in this study range from 21 to 73 years. The average of age among active smokers and nonsmokers groups (43.05 vs 43.95 years) is not significantly different ($p = 0.743$). The data on oral hygiene habit which is categorized based on the frequency of tooth brushing also showed that there were no differences in oral hygiene habit between active smokers and nonsmokers groups ($p = 0.817$). However, there is a different number of participant consuming alcohol between the two groups. The number of participant who consumes alcohol is higher in active smokers group compared to nonsmokers group (14/40 vs 2/40).

3.1. Cytomorphometry

The mean of CD and ND of epithelial cells collected from active smokers are 58.249μ and 8.963μ respectively. On the other hand, the mean of CD and ND are 63.473μ and 7.991μ for epithelial cell collected from nonsmokers. On average epithelial cells collected from active smokers have smaller cells size with bigger nuclear size compared to those collected from nonsmokers. The means of ND/CD ratio are 0.156 and 0.129 for active smokers and nonsmokers respectively. The ratio is higher

IOP Conf. Series: Journal of Physics: Conf. Series **1022** (2018) 012049 doi:10.1088/1742-6596/1022/1/012049

for active smokers compared to nonsmokers. A significant difference of the ND/CD ratio was observed between the active smoker and non-smoker p<0.0001 (95% CI = -0.040 to -0.014) (Table.1).

Table 1. The Characteristic of Participants. Participants are Distributed According to Age, Frequency of Toothbrushing, Alcohol Consumption, Length of Smoking and The Number of Cigarette Consumption Perday. The Cytomorphometry Result is Also Shown

	Participant	
	active smoker	Non-smoker
Age (p=0.743)		
Min	25	21
Max	73	66
Mean	43.050	43.950
STDev	10.561	13.679
Toothbrush frequency (p = 0.817)		
1x	7	6
2x	28	31
3x and more	5	3
Alcohol use		
Frequently	8	0
Rarely	6	2
Never	26	38
Smoking years		
Min	10	
Max	54	
Mean	25.875	
Cigarette amount		
Min	12	
Max	40	
Mean	18	
Cytomorphometry (p < 0.001)		
Mean CD	58.249μ	63.473μ
Mean ND	8.963μ	7.991μ
Mean ND/CD	0.156	0.129

4. Discussion

Smoking has been widely considered as a risk factor for cancer [6]. The effects of smoking on oral epithelial mucous include the increase of micronuclei, changes in keratin pattern and dysplasia. Examination of epithelial morphology for investigating the effect of smoking on buccal epithelial cells showed the increase in the formation of DNA adduct in heavy smokers [7]. Another parameter to measure the changes in epithelial cells lining the oral mucosa is by cytomorphometric analysis. This measurement gives information on the changes in average cells size from buccal epithelial smear. A cytomorphometric study revealed that the decrease in cytoplasmic area and the increase in the nuclear

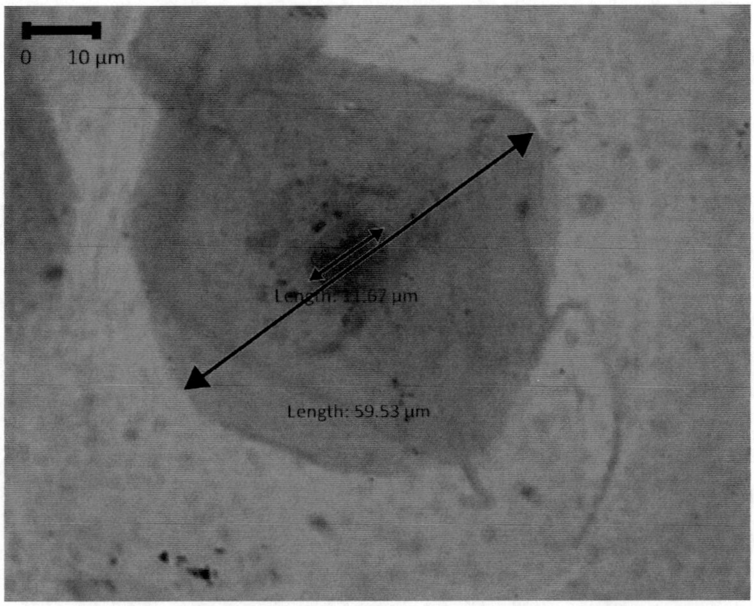

Figure 1. Measurement of Nuclear Diameter and Cytoplasmic Diameter on Epithelial Smear Stained with Papanicolaou's Technique Under 1000x Microscopic Magnification

Area of epithelial cells are associated with the development of malignancy [8]. The increase of nuclear size is one of the histological signs of epithelial dysplasia. Cellular changes including the increase ratio of nuclear to cytoplasmic size can be an important finding in the detection of epithelial dysplasia prior to the clinical manifestation of early cancer lesion. Other cellular appearances of epithelial dysplasia are hyperchromatism and prominent nucleoli [1,8].

Using the sample collected from the buccal swab, this study shows a valuable information on the morphological finding of oral epithelial cells exposed to cigarette smoking. The cytomorphometric measurement in this study shows that the mean of ND of epithelial cells is higher in active smokers compared to the mean in nonsmokers. On the other hand, the mean of CD is lower in active smokers compared to the mean of CD in nonsmokers. These data indicate that the average cell size of epithelial cells collected from active smokers is smaller than the average size of cells collected from nonsmokers. The bigger nuclear cells size in active smokers than the size in nonsmokers predicted the effect of smoking on epithelial cells morphology. The significant difference of ND/CD ratio between active smokers and nonsmokers in this study gives a stronger suggestion on the contribution of smoking to epithelial dysplasia. In conclusion, the result of this study shows that a simple procedure like buccal swab can be an easy way to screen early morphological changes on oral epithelial mucosa to prevent the development of oral cancer. Further study is required to establish this method for oral cancer screening.

5. References

[1] César R 2015 Essentials of oral cancer *Int. J. Clin. Exp. Pathol.* **8**(9) 11884-11894

[2] Cheong S C, Vatanasapt P, Yi-Hsin Y, Zain R B, Kerr A R and Johnson N W 2017 Oral cancer in South East Asia. *Translational Research Oral Oncology*. **2**(1) 2057178X1770292

[3] Sutandyo N, Ramli R, Sari L and Soeis D S 2014 Profile and survival of tongue cancer patients in "Dharmais" cancer hospital, Jakarta *Asian Pac. J. Cancer. Prev.* **15**(5) 1971–5

[4] Goyal S, Tiwari V K, Nair K S and Raj S 2014 Risk factors and costs of oral cancer in a

Tertiary Care Hospital in Delhi *Asian Pac. J. Cancer. Prev.* **15**(4) 1659–65
[5] Rashid A and Warnakulasuriya S 2015 The use of light-based (optical) detection systems as adjuncts in the detection of oral cancer and oral potentially malignant disorders: a systematic review *J. Oral Pathol. Med.* **44**(5) 307–28
[6] Gandini S, Botteri E, Iodice S, Boniol M, Lowenfels A B, Maisonneuve P and Boyle P 2008 Tobacco smoking and cancer : A meta-analysis *Int. J. Cancer* **122** 155–64
[7] Verma R, Singh A, Badni M, Chandra A and Gupta S 2015 Evaluation of exfoliative cytology in the diagnosis of oral premalignant and malignant lesions: A cytomorphometric analysis *Dent. Res. J.* **12** 83–8
[8] Shirani S, Kargahi N, Razavi S M and Homayoni S 2014 Epithelial dysplasia in oral cavity *IJMS* **39**(5) 406–417

Acknowledgement
The authors would like to thank the Ministry of Research and Higher Education of Republic Indonesia for research funding.

IOP Conf. Series: Journal of Physics: Conf. Series **1022** (2018) 012050 doi:10.1088/1742-6596/1022/1/012050

Combating Hepatitis B and C through immunological approach

Tri Nugraha Susilawati[1,5,7*], **Sigit Setyawan**[2,5], **T Y Pramana**[3], **Ambar Mudigdo**[4] **and Afiono Agung Prasetyo**[1,5,6]

[1] Department of Microbiology, Faculty of Medicine, Universitas Sebelas Maret, Surakarta, Indonesia
[2] Department of Parasitology, Faculty of Medicine, Universitas Sebelas Maret, Surakarta, Indonesia
[3] Department of Internal Medicine, Dr. Moewardi Hospital/ Faculty of Medicine, Universitas Sebelas Maret, Surakarta, Indonesia
[4] Department of Anatomic Pathology, Dr. Moewardi Hospital/ Faculty of Medicine, Universitas Sebelas Maret, Surakarta, Indonesia
[5]A-IGIC (A-Infection, Genomic, Immunology, & Cancer) Research Group, Universitas Sebelas Maret, Surakarta, Indonesia
[6] Center of Biotechnology and Biodiversity Research and Development, Universitas Sebelas Maret, Surakarta, Indonesia
[7] Corresponding author

*Corresponding email: tri.susilawati@staff.uns.ac.id

Abstract. Infections with hepatitis B and C viruses are the main factors contributing to the development of chronic liver disease and have been known as the major global health problems. This paper examines evidence that demonstrates the involvement of host immune responses in hepatitis B and C, particularly in the protection against immune-mediated liver injury. The proposed mechanisms of protection range from T cell responses that facilitate spontaneous resolution during acute infection and prevent persistent infection to immunoregulatory cytokines that inhibit destructive immune responses. Regulatory T cells (Tregs), TGF-β1, IL-4, and IL-10 are the main components of the immune system that play an important role in the protection mechanisms against the detrimental effects of hepatitis B and C viruses in liver tissues. Thus, factors contributing to increased Tregs activity and immunoregulatory cytokines should be elaborated. Recent studies reported factors that facilitate the development of Tregs during hepatitis C viral infection include HCV epitope, expression of miR 146a in monocytes and the Tim-3/Gal-9 pathway. On the other hand, the generation of Tregs is inhibited by IL-6 produced during inflammation. These findings suggest that immunomodulation strategy should be further developed and applied in the management of hepatitis B and C.

1. Introduction

Hepatitis B virus (HBV) and hepatitis C virus (HCV) are the blood-borne viruses with the potential to cause chronic infection and severe liver damage, including cirrhosis and cancer, which lead to liver failure. Currently, there are 257 million and 71 million of people living with chronic liver disease

IOP Conf. Series: Journal of Physics: Conf. Series **1022** (2018) 012050 doi:10.1088/1742-6596/1022/1/012050

caused by HBV and HCV, respectively, leading to more than a million of fatality per year due to liver failure [1, 2]. Until now, hepatitis B and C cannot be cured as the antiviral therapy can only slow down the disease progression and the vaccine is available only for hepatitis B. At the end stage of liver failure, transplantation is the only treatment for life-saving and this is practically difficult even though in developed countries. The nature of persistent infection, a few available treatment, and poor treatment responses complicate the management of hepatitis B and C. Hence, a new strategy for combating hepatitis B and C is urgently needed.

The host defense responses to HBV and HCV viruses mainly involve subpopulation of white blood cells named T lymphocytes (T cells). T cells responses are critical for spontaneous resolution of HBV and HCV acute infection as well as preventing persistent infection [3]. Unfortunately, T cells also contribute to the immune-mediated liver injury through unknown mechanisms [4]. It is likely that liver damage in patients with chronic hepatitis is caused by pro-inflammatory cytokines produced by non-specific T cells during the viral clearance in liver tissue [5] so that regulatory/inhibitory cytokines are needed to prevent liver destruction. Therefore, from an immunological perspective, restoring the effective functions of T cells and controlling cytokines levels may be the key to preventing liver damage caused by HBV and HCV.

Regulatory T cells (Tregs) are a group of T cells that suppress the activity, proliferation, differentiation, and effector function of immune cells, including T and B-lymphocytes, natural killer (NK) cells and dendritic cells (DCs) [6, 7, 8, 9]. In patients with HBV and HCV infections, Tregs protect liver tissues from the damage by suppressing the immune responses against viral replication [10]. Although there is an increased number of studies focusing on Tregs amongst hepatitis B and C patients within the last decade, the exact mechanisms of protection and the best strategy to modify Tregs' roles in the course of hepatitis B and C infections remain unclear. This review aims to deepen our understanding on the roles of Tregs and various cytokines involved in HBV and HCV infections so that the best strategy for modulating immune system for the management of hepatitis B and C can be developed.

2. Regulatory T Cells and Immunity

Immune system develops mechanisms for protecting the host from pathogen invasion. Unfortunately, such mechanisms may cause excessive immune responses or target self-antigens and cause tissue damage and death. Tregs have an important role in regulating immune responses to specific self and non-self antigens, specifically for maintaining tolerance. This role is conducted by limiting the undesired immune responses that could be harmful to the host; for instance: severe inflammation, autoimmunity, and allergy [11]. Thus, the presence of Tregs is important to provide appropriate immune responses with sufficient reactivity against pathogen and assure self-tolerance.

2.1. Development of Tregs

Regulatory T cells are produced by the normal immune system and specialized for immune suppression. These cells are being part of $CD4^+$ T cells population and express high levels of IL-2 receptor α chain (CD25), cytotoxic T-lymphocyte antigen-4 (CTLA-4), and the transcription factor Foxp3 [12]. Foxp3 is the main regulator of Treg development and function. In human, Foxp3 is highly expressed in Tregs but up-regulated transiently and in lower quantities in activated conventional T cells. Foxp3 deficiency or mutation can cause autoimmune disease in various tissues and the manifestations include massive lymphoproliferation, diabetes, exfoliative dermatitis, thyroiditis and enteropathy [13].

Activated nonregulatory T cells produce IL-2 that is critical for maintaining immune homeostasis through negative feedback mechanism of immune responses. This substance is important for the development of natural Tregs and maintaining their survival and function whilst limit the development of nonregulatory T cells. This negative feedback control is essential for Tregs suppressive activity as Tregs do not produce IL-2 themselves and rely on the use of IL-2 secreted by nonregulatory T cells.

IOP Conf. Series: Journal of Physics: Conf. Series **1022** (2018) 012050 doi:10.1088/1742-6596/1022/1/012050

Reduced IL-2 production interferes Tregs' development and function of which consequently may precipitate autoimmunity and promote the development of inflammatory diseases [14, 15, 16].

Tregs are divided into two major subgroups: naturally occurring Treg (nTreg) cells and peripherally induced Tregs.

2.1.1. Naturally Occurring Treg Cells or Thymus-Derived Tregs

This type of Tregs is developed during maturation process of T cells in thymus. Central T cell tolerance in thymus involves the process of "negative selection" during which self-reactive T cells are destroyed (Figure 1). However, some developing $CD4^+$ T cells that recognize self-peptides survive and these are naturally occurring regulatory T cells (nTregs). These cells account for 5–10% of peripheral T cells and are present in healthy individuals from birth [11].

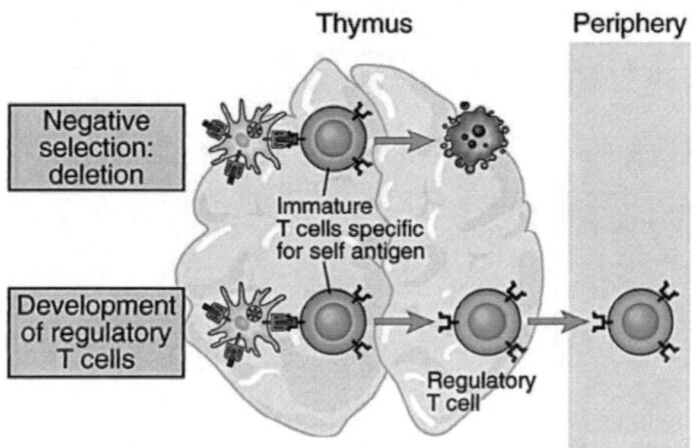

Figure 1. The Development of Regulatory T Cells in Thymus
Abbas & Lichtman: Basic Immunology, 3rd Edition. Copyright © 2008 by Saunders, an imprint of Elsevier, Inc. All rights reserved.

2.1.2. Peripherally Induced Treg Cells

These cells are generated from naive T cells under the stimulation of appropriate antigens and cytokines either naturally (e.g., during inflammatory processes) or experimentally generated. These cells are essential in peripheral self-tolerance when there is a lack of self antigen-specific nTreg. A major substance that can induce the transformation of $CD4^+CD25^-$ T cells into Tregs is transforming growth factor-β (TGF-β) either alone or in conjunction with retinoic acid, or 2,3,7,8-tetrachlorodibenzo-p-dioxin, a ligand for the aryl hydrocarbon [11, 14]. Retinoic acid plays an important role in TGF-β–dependent immune responses by inhibiting IL-6. Interleukin-6 poses detrimental effects to the host as it can induce the differentiation of proinflammatory Th17 cells and inhibit the differentiation of anti-inflammatory Tregs during chronic inflammation [17, 18, 19].

2.2. Mechanisms of Tregs-Mediated Suppression

Regulatory T cells migrate to and become activated in inflamed tissues, infectious sites, tumours, and in regional lymph nodes where the specific antigens are presented. Following their activation and proliferation, antigen-activated Tregs contact DCs then modulate DC function to prevent the activation of other T cells (Figure 2). After this, Tregs may then secrete granzyme/perforin or immunosuppressive cytokines [15]. As the results, Tregs maintain peripheral tolerance by suppressing the expansion and function of conventional T cells. Tregs can also suppress antigen presenting cells (APCs), macrophages, DCs, NK, and B cells [14]. From a functional perspective, it has been suggested that Tregs' suppression activities use four mechanisms: (a) Suppression by inhibitory cytokines (i.e., IL-10, IL-35 and TGF-β); (b) Suppression by cytolysis or killing mechanisms that depend on granzyme A, granzyme B and perforin; (c) Suppression by metabolic disruption mediated by CD25, cyclic AMP (cAMP), CD39 and/or CD73 and adenosine receptor 2A; and (d) Suppression by modulation of DC maturation or function which involves lymphocyte-activation gene 3 (LAG3 or CD223), major histocompatibility complex (MHC) class I, CTLA4, CD80/CD86 and indoleamine 2,3-dioxygenase (IDO) [20].

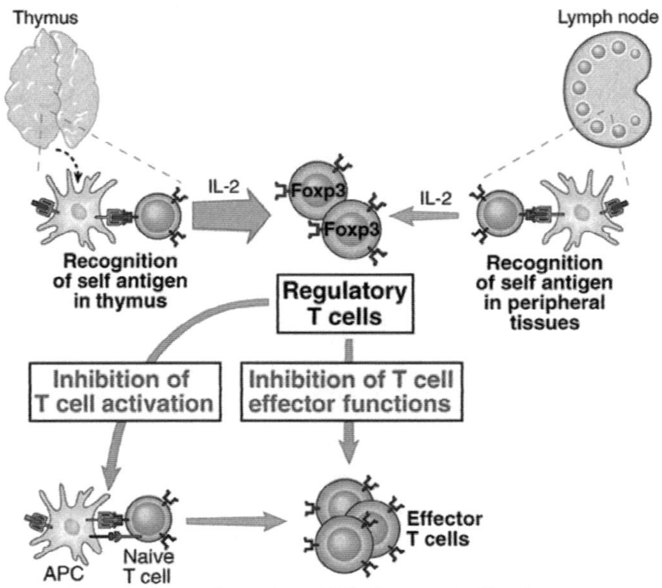

Figure 2. Tregs-Mediated Suppression
Abbas & Lichtman: Basic Immunology, 3rd Edition. Copyright © 2008 by Saunders, an imprint of Elsevier, Inc. All rights reserved.

3. The Roles of Tregs in Responses to HBV and HCV Infections

When the microbes entering host, cells of immune system are recruited to the site of infection and is often resulting various degree of inflammation. Regulation of immune response is required to contain or eliminate the pathogen as well as develop tolerance to non-pathogenic agents. A dysregulation of Tregs-mediated suppression mechanisms is harmful because it can lead to excessive inflammation causing severe tissue damage.

It has been reported that during HCV infection, several factors contribute to the development of Tregs, including HCV epitope [21], expression of miR 146a in monocytes [22], the Tim-3/Gal-9 pathway [23, 24] and decline of miR-124 in myeloid cells [25]. Tregs can suppress hepatitis antiviral responses [26] so it can be hypothesized that the decrease of Tregs numbers may increase host's capacity in viral clearance, which is important during the acute stage of hepatitis. In contrast, increasing the numbers of Tregs in chronic hepatitis may be advantageous to hamper immune-mediated liver damage [27].

4. The Roles of Cytokines in Hepatitis B and C

Cytokines are important mediators of immune responses. They are secreted by immune cells and give feedback effects to immune cells. For example, T cells switch to Tregs in peripheral tissues as a result of exposure to TGF-β, and then the resulting Tregs produce TGF-β and Foxp3. There are immunoregulatory/inhibitory cytokines that regulate/inhibit immune responses (e.g., TGF-β1, IL-4 and IL-10) and inflammatory or effector cell-associated cytokines (e.g., IFN-γ, IL-17 and TNF-α) [28]. In the course of illness, however, the functions of cytokines are often ambiguous and need to be further investigated.

Hepatitis B and C viruses are both non-cytopathic, meaning that the viruses themselves cannot cause liver injury. Instead, the liver injury during infections is the result of specific immune responses

IOP Conf. Series: Journal of Physics: Conf. Series **1022** (2018) 012050 doi:10.1088/1742-6596/1022/1/012050

controlled by cytokines. A previous study reported that there was a strong correlation between viral load and IL-4 or IL-6 in hepatitis B and IL-6 or IL-10 in hepatitis C, indicating the association of these cytokines with persistent infection [29].

Some cytokines are useful for combating HBV and HCV infections. Interleukin-4 has been shown to have direct antiviral effect on HBV [30] while IFN-α, IL-1β and IL-6 play an important role in inhibiting HBV entry [31, 32, 33]. Other studies demonstrated that IL-6, IL-1β and TGF-β repress expression of essential HBV transcription factors [34, 35, 36, 37]. Interferon-α disrupts transcription process through epigenetic control of HBV DNA minichromosomes [38, 39, 40] thus inhibit HBV replication as well as inhibit HCV protein translation [41]. Furthermore, IFN-γ, TNF-α and TGF-β could induce HBV DNA degradation [42, 43]. At post-transcriptional level, HBV replication could be inhibited by IFN-γ, IFN-β and TNF-α [44, 45, 46]. Finally, IFN-α has been shown to block virion secretion from an inducible HBV replicating cell line [47].

Infection with HCV activates both innate and adaptive immune responses [48]. Innate immune response is initiated by the synthesis of type I interferon (IFN-α/β) which induces NK and NK T cells (NKT) to release perforin, granzyme β, IFN-γ and TNF-α. This is followed by the expression of Fas ligand (FasL or CD95L) and causes the death of infected hepatocytes. The adaptive immune response is mediated by interactions between DC and $CD4^+$ T cells, resulting in the release of various cytokines (i.e., IFN-γ, TGF-β, TNF-α, IL-2, IL-4, IL-10 and IL-13). Currently, IFN-α is the only cytokine used for the treatment of hepatitis C.

5. Conclusions

Management of hepatitis B and C is challenging since currently available treatments can only control HBV and HCV replication but cannot eliminate the infection, leading to chronic inflammation and liver failure. Tregs' are essential to maintaining peripheral tolerance and regulating effector cells in the immune system so that immune-mediated liver injury can be prevented. Various cytokines have also been shown to control viral replication and contribute to the cure of hepatitis B and C. Therefore, in addition to modulating Tregs, the roles of cytokines associated with Tregs should be considered in order to provide adequate treatment to the patients. In the future, modulation of Tregs and cytokines should be further developed for management of hepatitis B and C in conjunction with antiviral therapy and vaccination.

6. References

[1] World Health Organization. Accessed: 10 July 2017. URL: http://www.who.int/mediacentre/factsheets/fs204/en/

[2] World Health Organization. Accessed: 10 July 2017. URL: http://www.who.int/mediacentre/factsheets/fs164/en/

[3] Schmidt J, Blum H E and Thimme R 2013 *Emerg. Microbe.s Infect.* **2** e15

[4] Park S and Rehermann B 2014 *Immunity* **40** 13–24

[5] Rehermann B 2013 *Nat. Med.* **19** 859–68

[6] Shevach E, DiPaolo R, Andersson J, Zhao D, Stephens G and Thornton A 2006 *Immunol. Rev.* **212** 60–73

[7] Zhao D, Thornton A, DiPaolo R and Shevach E 2006 *Blood* **107** 3925–32

[8] Ghiringhelli F, Menard C, Martin F and Zitvogel L 2006 *Immunol. Rev.* **214** 229–38

[9] Mahnke K and Enk A 2005 *Curr. Top. Microbiol. Immunol.* **293** 133–50

[10] Jung M K and Shin E 2016 *Immune Netw* **16** 330–36

[11] Roncarolo M G and Battaglia M 2007 *Nat. Rev. Immunol.* **7** 585–98

[12] Fransisco L M, Sage P T and Sharpe A H 2010 *Immunol. Rev.* **236** 219–42

[13] Zheng Y and Rudensky A Y 2007 *Nat. Immunol.* **8** 457–62

[14] Piccirillo C A, D'Hennezel E, Sgouroudis E and Yurchenko E 2008 *Curr. Opin. Immunol.* **20** 655–62

[15] Sakaguchi S, Yamaguchi T, Nomura T and Ono M 2008 *Cell* **133** 755–87

[16] Yamanouchi J *et al* 2007 *Nat. Genet.* **39** 329–37
[17] Mucida D, Park Y, Kim G, Turovskaya O, Scott I, Kronenberg M and Cheroutre H 2007 *Science* **317** 256–60
[18] Kimura A and Kishimoto T 2010 *Eur. J. Immunol.* **40** 1830–35
[19] Scheller J, Chalaris A, Schmidt-Arras D and Rose-John S 2011 *Biochim. Biophys. Acta* **1813** 878–88
[20] Vignali D A A, Collison L W and Workman C J 2008 *Nat. Rev. Immunol.* **8** 523–32
[21] Losikoff P T *et al* 2015 *J. Hepatol.* **62** 48–55
[22] Ren J P, Ying R S, Cheng Y Q, Wang L, El Gazzar M, Li G Y, Ning S B, Moorman J P and Yao Z Q 2016 *J. Viral Hepat.* **23** 755–66
[23] Ji X J, Ma C J, Wang J M, Wu X Y, Niki T, Hirashima M, Moorman J P and Yao Z Q 2012 *Eur. J. Immunol.* **43** 458–67
[24] Moorman J P, Wang J M, Zhang Y, Ji X J, Ma C J, Wu X Y, Jia Z S, Wang K S and Yao Z Q 2012 *J. Immunol.* **189** 755–66
[25] Ren J P, Wang L, Zhao J, Wang L, Ning S B, El Gazzar M, Moorman J P and Yao Z Q 2016 *Immunology* **150** 213–20
[26] Xu D *et al* 2006 *J. Immunol.* **177** 739–47
[27] Claassen M A, de Knegt R J, Tilanus H W, Janssen H L and Boonstra A 2010 *J. Hepatol.* **52** 315–21
[28] Imai A, Sugita S, Kawazoe Y, Horie S, Yamada Y, Keino H, Maruyama K and Mochizuki M 2012 *Invest. Ophthalmol. Vis. Sci.* **53** 7299–309
[29] Khan S, Bhargava A, Pathak N, Maudar K K, Varshney S and Mishra P K 2011 *Indian J Clin Biochem* 161–68
[30] Lin S J, Shu P Y, Chang C, Ng A K and Hu C P 2003 *J. Immunol.* **171** 4708–16
[31] Xia Y, Blossey C, Esser K, Lucifora J and Protzer U 2014 *J. Hepatol.* **60** S122-123
[32] Lee Vee M, Gripon P, Stieger B and Fardel O 2008 *Drug Metab. Dispos.* **36** 17–22
[33] Le Vee M, Lecureur V, Stieger B and Fardel O 2009 *Drug Metab. Dispos.* **37** 685–93
[34] Hosel M *et al* 2009 *Hepatology* **50** 1773–82
[35] Isorce N, Testoni B, Locatelli M, Fresquet J, Rivoire M, Luangsay S, Zoulim F and Durantel D 2016 *Antiviral Res.* **130** 36–45
[36] Isorce N, Lucifora J, Zoulim F and Durantel D 2015 *Antiviral Res.* **122** 69–81
[37] Hong M H, Chou Y C, Wu Y C, Tsai K N, Hu C P, Jeng K S, Chen M L and Chang C 2012 *PLoS ONE* **7** e30360
[38] Belloni L, Allweiss L, Guerrieri F, Pediconi N, Volz T, Pollicino T, Petersen J, Raimondo G, Dandri M and Levrero M 2012 *J. Clin. Invest.* **122** 529–37
[39] Liu F *et al* 2013 *PLoS Pathog.* **9** e1003613
[40] Tropberger P, Mercier A, Robinson M, Zhong W, Ganem D E and Holdorf M 2015 *Proc. Natl. Acad. Sci. U.S.A* **112** E5715–24
[41] Gupta R, Kim S and Taylor M W 2012 *J Transl Med.* **10** 54
[42] Xia Y *et al* 2016 *Gastroenterology* **150** 194–205
[43] Qiao Y, Han X, Guan G, Wu N, Sun J, Pak V and Liang G 2016 *FEBS Lett.* **590** 419–27
[44] Guidotti L G, Ishikawa T, Hobbs M V, Matzke B, Schreiber R and Chisari F V 1996 *Immunity* **4** 25–36
[45] Pasquetto V, Wieland S F, Uprichard S L, Tripodi M and Chisari F V 2002 *J. Virol.* **76** 5646–53
[46] Biermer M, Puro R and Schneider R J 2003 *J. Virol.* **77** 4033–42
[47] Yan R, Zhao X, Cai D, Liu Y, Block T M, Guo J T and Guo H 2015 *J. Virol.* **89** 9200–12
[48] Fierro NA, Castro-Garcia F P and Panduro A 2013 *Adv Biosci Biotechnol* **4** 13–18

IOP Conf. Series: Journal of Physics: Conf. Series **1022** (2018) 012051 doi:10.1088/1742-6596/1022/1/012051

Study of cultivation and growth rate kinetic for mixed cultures of local microalgae as third generation (G-3) bioethanol feedstock in thin layer photobioreactor

Wahyu Prihastuti Yuarrina[1], Yano Surya Pradana[2], Arief Budiman[1,2,5], Akmal Irfan Majid[4,5], Indarto[4,5] and Eko Agus Suyono[3,5*]

[1] Master Program of Systems Engineering, Universitas Gadjah Mada, Jalan Teknika Utara 3, Yogyakarta, Indonesia
[2] Chemical Engineering Department, Universitas Gadjah Mada, Jalan Grafika No. 2, Yogyakarta, Indonesia
[3] Faculty of Biology, Universitas Gadjah Mada, Jalan Teknika Selatan, Yogyakarta, Indonesia
[4] Mechanical and Industrial Engineering Department, Universitas Gadjah Mada, Jalan Grafika No. 2, Yogyakarta, Indonesia
[5] Center for Energy Studies, Universitas Gadjah Mada, Sekip K1a, Yogyakarta, Indonesia

* Corresponding email: eko_suyono@ugm.ac.id

Abstract. The increasing use of fossil fuels causes the depletion in supply and contributes to climate change by GHG emissions into the atmosphere. Microalgae indicate as renewable and sustainable energy sources as they have a high potential for producing large amounts of biomass for third-generation biofuels (bioethanol and biodiesel) feedstock. However, there are several parameters which should be considered for microalgae cultivation, such as environmental conditions, medium composition and microalgae species. The aim of this research was to study cultivation of mixed microalgae cultures (Glagah consortium and *Arthrospira maxima*) in a thin layer photobioreactor. Farmpion medium, Bold's Basal Medium (BBM) and Thoriq Eko Arief (TEA) medium were investigated as cultivation medium for bioethanol feedstock for 7 days. The results showed that the highest dry weight concentration of microalgae was in Farmpion medium (0.35 mg/ml) and the highest carbohydrate concentration of microalgae was in BBM (0.14 mg/ml). Thus, the optimum medium of microalgae cultivation for bioethanol feedstock was BBM because of the highest carbohydrate-dry weight ratio (0.88). In addition, mathematical approach by using Contois model was used to find out the growth rate of microalgae cultivation in each medium.

1. Introduction

Nowadays, renewable energy (RE) is one of hot issues in the world, which related research on this topic is always attractive to be conducted every year [1]. It is caused by the high demands of fossil-based energy which is not followed by the discovery of new reserves in similar capacity [2]. In addition, environmental issues related to accumulation of carbon emission in the atmosphere is also powering on development of RE [3]. In this case, Indonesia, the tropical and top-biodiversity country, has a lot of raw materials that can be utilize in developing RE. One of prominent feedstocks in developing RE is microalgae, which is potentially converted into new alternative energy [4].

Microalgae are single cell organisms contain cellulose, lipid, protein and other components which is consuming CO_2 for photosynthesis [5]. Based on these components, microalgae can be cultivated in order to produce biofuels, such as bioethanol, biodiesel, bio crude oil, called third generation (G-3) biofuels [6]. Microalgae have been proposed as G-3 biofuel feedstock because they have high biomass production per unit of light, can cultivate at any condition of land and water, and have no restriction with other sectors, such as food and pharmaceutical [7]. Moreover, microalgae can be cultivated in single culture, such as *Arthrospira sp.* and *Botryococcus sp.*, or mixed cultures, such as local strain, to reach optimum production.

Generally, microalgae can be cultivated in two methods, i.e. closed and open. In closed system, microalgae are grown in closed pond by bubbling it with air or CO_2. It will produce dry microalgae with high specification because of less contaminant during cultivation. Meanwhile, the open system will cultivate microalgae in pond which is freely contacting with air. The open system is easier to operate compared to the closed ones. In the other hand, dry microalgae product is less in quality and need to be treated for resulting in food grade or higher.

The aim of this research was to study the effect of culture medium on carbohydrate and dry weight concentration of mixed cultures microalgae which contain *Arthrospira maxima* and Glagah Consortium in thin layer photobioreactor (TL-PBR) with an open system cultivation for bioethanol feedstocks. Furthermore, the kinetics of microalgae growth rate in all medium were also investigated.

2. Material and Methods

2.1. Microalgae

This study used a mixed culture microalgae containing *Arthrospira maxima* and Glagah Consortium. *Arthrospira maxima* was one of the common microalgae species for mass production, which has high growth rate and biomass productivity by converting light energy into biomass efficiently. Otherwise, this species can survive both in fresh and salt water. Glagah Consortium was a natural mixed culture taken from Glagah Beach, Yogyakarta, Indonesia. This culture was then isolated in the Laboratory of Biotechnology, Faculty of Biology, Universitas Gadjah Mada. The species composing Glagah Consortium were *Cyclotella polymorpha, Cylindrospermopsis raciborskii, Corethron criophilum, Golenkinia radiata, Chlamydomonas sp.* and *Syracosphaera pirus* [8]. These cultures grew on the wide range of salinity, dry weight concentration achieving 3.42 g/L and lipid content reaching 13.58 % in cultivation medium using salt water. The aim of using mixed culture of *Arthrospira maxima* and Glagah Consortium was for mutual interaction beyond organism to complete the activity metabolism of the mixed cultures. All microorganisms in the cultures could defend as well as mutual support for growth [8].

2.2. Culture medium

Microalgae need culture medium, such as carbon source, vitamins, salts and nutrients, for growing [9]. Optimization of medium composition is necessary to enhance the productivity of microalgae. The main nutrient in the culture medium is nitrogen (N), which is available in inorganic matter, such as nitrite (NO_2) and nitrate (NO_3), or in organic matter, such as ammonium (NH_4^-). Phosphorus (P) is also the primary nutrient for culturing microalgae. This matter is found in inorganic phosphates, such as potassium phosphate monobasic (KH_2PO_4) and potassium phosphate dibasic (K_2HPO_4), but its use cannot be assimilated with organic phosphate simultaneously [10,11]. Besides nitrogen and phosphorus, potassium is

IOP Conf. Series: Journal of Physics: Conf. Series **1022** (2018) 012051 doi:10.1088/1742-6596/1022/1/012051

the other primary nutrient supplied from potassium oxide (K_2O), KH_2PO_4 and K_2HPO_4. This study used three algal cultivation medium, i.e. Farmpion Green 63 Medium, Bold's Basal Medium (BBM) and Thoriq Eko Arief (TEA) Medium, for comparing their performance.

2.2.1. Farmpion Green 63 Medium
Farmpion Green 63 Medium is a agricultural fertilizer which can be used as a medium culture microalgae. It is containing macronutrients, micronutrients and trace metals which were required for microalgae cultivation. The elements in this medium are nitrogen (N) in the form of nitrate (NO_3), phosphate (P) in the form of P_2O_5, potassium (K) in the form of K_2O, calcium (Ca), magnesium (Mg), sulfur (S) and microelements (B, Mn, Fe, Zn, Mo and Cu) [12]. This medium is affordable and easy to obtain in the fertilizer market.

2.2.2. Bold's Basal Medium (BBM)
Bold's Basal Medium (BBM) is the most medium used for cultivating green algae division [8], with fresh water. It uses two sources of nitrogen, i.e. sodium nitrate and ammonium molybdate. Phosphorus and potassium source is supplied from potassium phosphate monobasic (KH_2PO_4) and potassium phosphate dibasic (K_2HPO_4) [11].

2.2.3. Thoriq Eko Arief (TEA) Medium
TEA Medium is a self-generated algal culture medium by modifying of Walne's medium and BBM as common algal culture medium. This modification was expected to increase the growth rate of microalgae and and biomass productivity [13]. As comparison, the summary of N, P, and K concentration at several medium is shown in Table 1.

Table 1. Concentration (g/L) of Nitrogen (N), Phosporus (P) and Potassium (K) at Several Medium

Medium	N	P	K	N/P ratio
Farmpion Green 63 Medium	21	21	21	1.00
Bold's Basal Medium (BBM)	75	25	31	3.00
Thoriq Eko Arief (TEA) Medium	250	175	-	1.43

2.3. Photobioreactor
Various cultivation methods of microalgae have been developed, one of them is photobioreactor. This study used open photobioreactor by flowing the medium forming thin layer, called thin layer photobioreactor (TL-PBR). TL-PBR was designed by referring to Doucha [14] with modification, as shown in Figure 1. The goals of thin layer photobioreactor (TL-PBR) were to avoid sedimentation of microalgae when cultivated and to increase the productivity of microalgae by optimizing light and CO_2 penetration into the medium. The installed tank capacity at this bioreactor was 600 L. To distribute the microalgae suspension properly, TL-PBR was also equipped by a pump (capacity of 75 L / min).

IOP Conf. Series: Journal of Physics: Conf. Series **1022** (2018) 012051 doi:10.1088/1742-6596/1022/1/012051

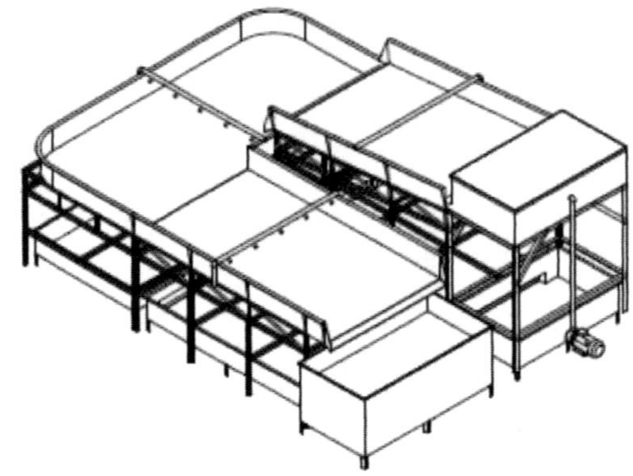

Figure 1. Scheme of TL-PBR

2.4. *Analysis of Carbohydrate*
Carbohydrate content was analyzed by using phenol-sulfuric acid method and spectrophotometry at wavelength of 490 nm [15].

2.5. *Kinetics of Growth Rate*
The relationship between substrate concentration (S) in mg/ml and growth rate of microalgae could be illustrated using Contois Equation, as shown in Equation 1. This kinetics led in facilitating future mass scale development.

$$\mu = \frac{\mu_{max}\, S}{(K_s\, X + S)} - k_d X \qquad (1)$$

To determine the kinetics of growth rate in Equation (1), we needed the differential equation of substrate concentration with time (t) in Equations (2) and carbohydrate concentration with time (t) in Equation (3) [6].

$$\frac{dS}{dt} = -\frac{1}{Y_{X/S}}\frac{dX}{dt} \qquad (2)$$

$$\frac{dC_{carbo}}{dt} = Y_{carbo/x}\frac{dX}{dt} \qquad (3)$$

where, μ was specific growth rate biomass (day^{-1}); μ_{max} was maximum specific growth rate biomass (day^{-1}); K_s was half-saturation constant; k_d was death-rate constant (day^{-1}); X was dry weight concentration (mg/ml); $Y_{x/s}$ was yield of biomass per mg substrate reduction; C_{carbo} was carbohydrate concentration (mg/ml); $Y_{carbo/x}$ was yield of carbohydrate per mg biomass. Committee

3. Results and Discussions
This research was conducted at Biotechnology Laboratory, Faculty of Biology, Universitas Gadjah Mada, Yogyakarta, Indonesia. The cultivation process of microalgae was investigated for 7 days. Figure 2 shows the effect of culture medium on carbohydrate concentration, dry weight concentration and carbohydrate-dry weight ratio of microalgae. Figure 2(a) shows that the highest dry weight concentration was microalgae cultivated in Farmpion, followed by BBM and TEA Medium. The dry weight concentration of microalgae in Farmpion was significantly higher than BBM and TEA Medium. The composition of main components (nitrogen, phosphorus and potassium) in balanced condition, such as in Farmpion medium, would be driven dry weight concentration of microalgae faster [16].

IOP Conf. Series: Journal of Physics: Conf. Series **1022** (2018) 012051 doi:10.1088/1742-6596/1022/1/012051

Meanwhile, Figure 2(b) shows the continuous increasing of carbohydrate concentration in microalgae for 7 days. The results indicated that BBM was the best medium for synthesizing carbohydrate in microalgae, but it was slightly difference with Farmpion. The highest carbohydrate concentration of microalgae cultivated in MBM was 0.1353 mg/ml, followed by Farmpion (0.1290 mg/ml) and TEA Medium (0.760 mg/ml). Carbohydrate, as the product of photosynthesis, required nutrients in the form of nitrogen (N), phosphorus (P) and potassium (K) [17,18]. Based on these results, the high nitrogen-phosphorus ratio (N/P ratio) in culture medium led to produce more carbohydrate and less lipid in microalgae. Nitrogen was one of important elements in driving chlorophyll formation in microalgae [12]. The formation of chlorophyll would uptake more carbon dioxide (CO_2) for synthesizing carbohydrate [19]. Figure 2(c) shows the effect of culture medium on carbohydrate-dry weight ratio of microalgae. This figure indicated that BBM was the best medium of microalgae cultivation for bioethanol feedstock because of the highest carbohydrate-dry weight ratio compared to other medium.

(a)

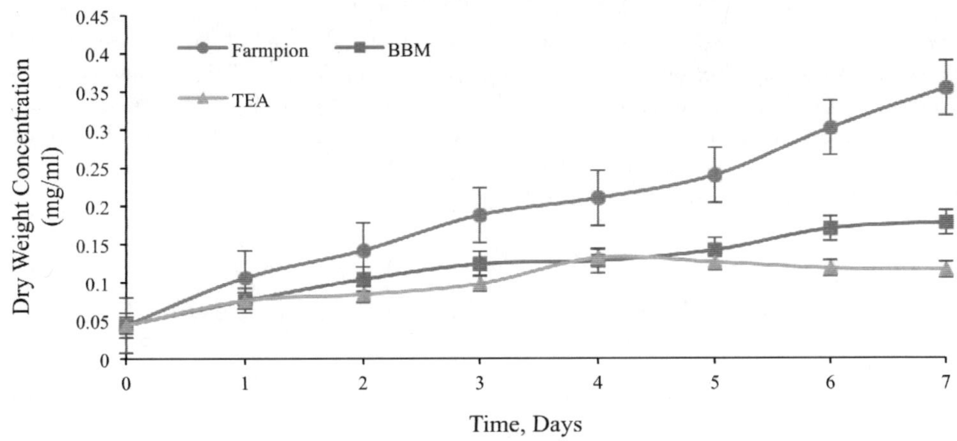

(b)

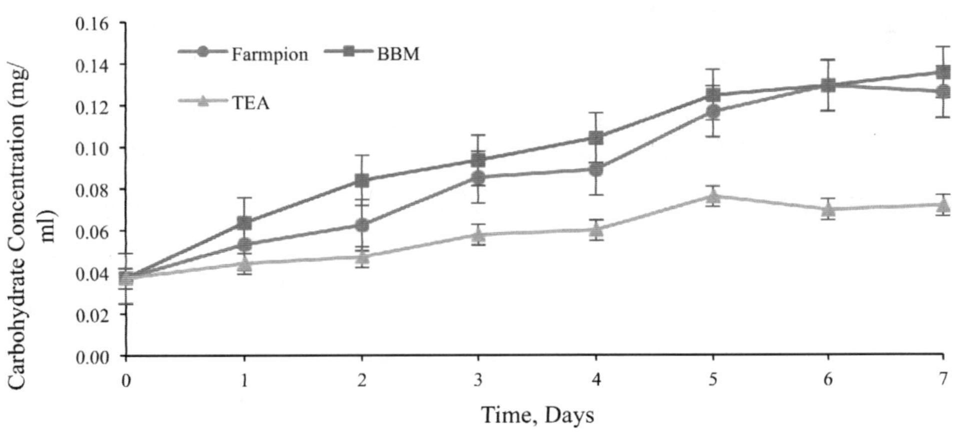

(c)

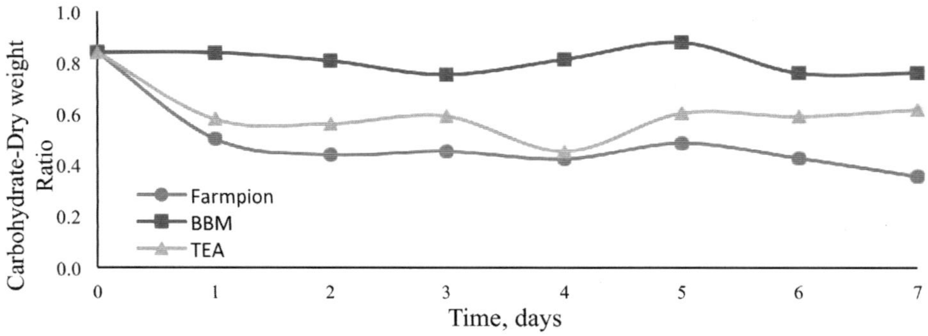

Figure 2. Effect of Culture Medium on: (a) Dry Weight Concentration; (b) Carbohydrate Concentration; (c) Carbohydrate-Dry Weight Ratio

Furthermore, kinetic study was conducted to identify growth rate of mixed cultures microalge of Glagah Consortium and *A. maxima* on each medium using Contois Model. This mathematical approach was used to predict growth rate of microalgae and design mass cultivation of microalgae. The relation between substrate and dry weigh concentration was expressed as in Equation 2. Meanwhile, the relation between carbohydrate and dry weigh concentration was declared as in Equation 3. The results of numerical calculation using MATLAB are shown in Figure 3, 4 and 5.

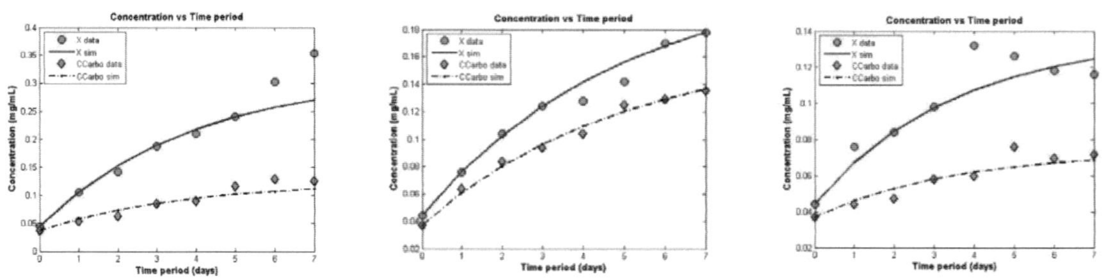

(a) (b) (c)

Figure 3. Relation Between Dry Weight and Carbohydrate Concentration with Time Period Resulted from Kinetic Simulation in: (a) Farmpion Medium; (b) BBM; (c) TEA Medium

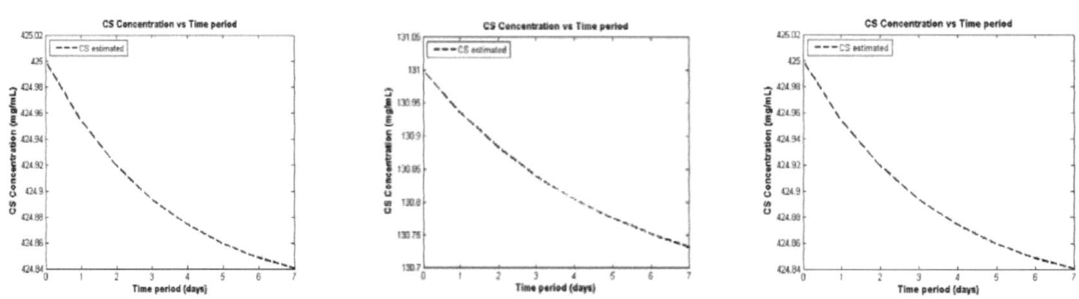

(a) (b) (c)

Figure 4. Relation between Substrate Concentration and Time Period Resulted Ffom Kinetic Simulation in: (a) Farmpion Medium; (b) BBM; (c) TEA Medium

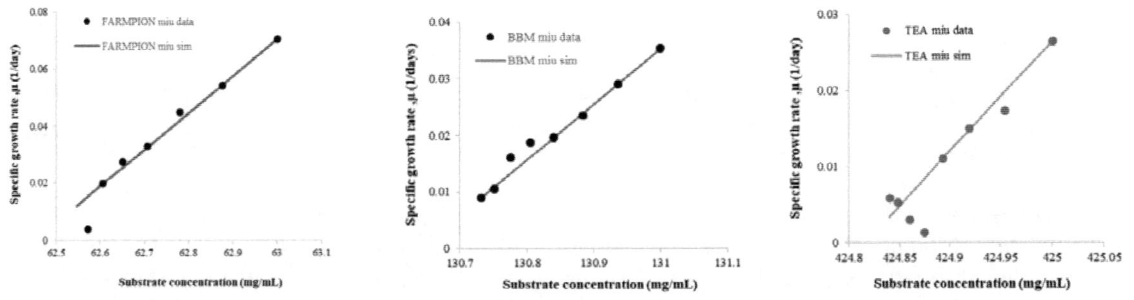

(a) (b) (c)

Figure 5. Relation between Substrate Concentration and Specific Growth Rate Resulted from Kinetic Simulation in: (a) Farmpion Medium; (b) BBM; (c) TEA Medium

Figure 3 shows the relation between dry weight and carbohydrate concentration with time period resulted from kinetic simulation. As comparison of observed data and calculated data, there were deviation in all medium. In Farmpion medium, the deviation of dry weight and carbohydrate concentration was occurred in time period of 6 and 7 days. Meanwhile, the deviation of dry weight and carbohydrate concentration appeared in time period of 4 and 5 days for BBM and TEA medium. Figure 4 shows relation between substrate concentration and time period resulted from kinetic simulation. The substrate concentration was decrease everyday because it was consumpted by microalgae for reproduction and/or biomass, carbohydrate, protein and other components production. However, the different composition of component in the substrate would give different composition of component in microalgae. Furthermore, Figure 5 shows relation between substrate concentration and specific growth rate resulted from kinetic simulation. From this figure, the slope of spesific growth rate (dry weight concentration) was slighly differences in all medium. However, the value of spesific growth rate was totally different. Spesific growth rate of microalgae in Farmpion medium was the highest, followed by BBM and TEA medium respectively.

Parameters calculated from mathematical modeling and simulation of the kinetic of microalgae growth rate are shown in Table 2. From this table, Farmpion was the highest maximum specific growth rate biomass (0.082 day^{-1}). However, the yield of carbohydrate production per biomass production of microalgae in Farmpion medium was the lowest. Meanwhile, BBM had the highest yield of carbohydrate production per biomass production of microalgae because of the high N/P ratio.

Table 2. Kinetics constants of mix culture microalgae on several medium by Contois Model generated from MATLAB simulation

Parameters	Farmpion	BBM	TEA
μ_{max}	0.082	0.044	0.039
S	63	131	425
K_s	31.732	30.13	15.495
$Y_{x/s}$	0.498	0.5	0.504
$Y_{carbo/x}$	0.3310	0.743	0.394
k_d	0.223	0.187	0.284
SSE	30.7151	22.4482	16.9514

4. Conclusions

The results showed that the highest dry weight concentration of microalgae was in Farmpion medium (0.35 mg/ml) and the highest carbohydrate concentration of microalgae was in BBM (0.14 mg/ml). Thus, the optimum medium of microalgae cultivation for bioethanol feedstock was BBM because of the highest carbohydrate-dry weight ratio (0.88). In addition, mathematical approach by using Contois model was used to find out the growth rate of microalgae cultivation in each medium.

5. References

[1] Nafis G A, Mumpuni P Y, Indarto and Budiman A 2015 Combination pulsed electric field with ethanol solvent for Nannochloropsis sp. extraction *AIP Conference Proceedings* **1699** 030021

[2] Daniyanto, Sutijan, Deendarlianto and Budiman A 2016 Reaction kinetic of pyrolysis in mechanism of pyrolysis-gasification process of dry torrified-sugarcane bagasse *ARPN J. Eng. Appl. Sci.* **11**(16) 9974-9980

[3] Chen C Y, Zhao X Q, Yen H W, Ho S H, Cheng C L, Lee D J, Bai F W and Chang J S 2013 Microalgae-based carbohydrates for biofuel production *Biochem. Eng. J.* **78** 1-10

[4] Kim G, Mutjaba G and Lee K 2016 Effects of nitrogen sources on cell growth and biochemical composition of marine chlorophyte Tetraselmis sp. for lipid production *Algae* **31**(3) 257-266

[5] Nigam S, Rai M P and Sharma R 2011 Effect of Nitrogen on Growth and Lipid Content of Chlorella pyrenoidosa *Am. J. Biochem. & Biotech.* **7**(3) 124-129

[6] Sudibyo H, Pradana Y S, Samudra T T, Budiman A, Indarto and Suyono E A 2017 Study of Cultivation under Different Colors of Light and Growth Kinetic Study of Chlorella zofingiensis Dönz for Biofuel Production *Energy Procedia* **105** 270–276

[7] Pradana Y S, Sudibyo H, Suyono E A, Indarto and Budiman A 2017 Oil Algae Extraction of Selected Microalgae Species Grown in Monoculture and Mixed Cultures for Biodiesel Production *Energy Procedia* **105** 277–282

[8] Suyono E A, Fahrunnida, Nopitasari S and Utama I V 2016 Identification of microalgae species and lipid profiling of Glagah consortium for biodiesel development from local marine resource *ARPN J. Eng. Appl. Sci.* **11**(16) 9970–9973

[9] Pradana Y S, Kusumastuti Y, Rahma F N and Effendy N 2017 Chitosan Flocculation-sedimentation for Harvesting Selected Microalgae Species Grown in Monoculture and Mixed Cultures *Chem Eng. Trans.* **56** 1549-1554

[10] Barsanti L and Gualtieri P 2014 *Algae Anatomy, Biochemistry, and Biotechnology* (Pisa, Italy: CRC Press)

[11] Zheng C 2015 *The Influence of Media Ingredients and Ph on the Growth of Chloromonas rosae var. psychrophila* (University of Kentucky)

[12] Suyono E A, Muavatun U, Husna F, Khotimah H, Pratiwi I, Husna R, Cahyani F, Purwanti Y and Samudra T T 2016 The effect of nitrogen stress in mediumfor increasing carbohydrate as a bioethanol source and carotenoid as an antioxidant from Chlorella zofingiensis culture *ARPN J. Eng. Appl. Sci.* **11**(40) 2698-2701

[13] Wahyuni N E 2016 *Produktivitas Biomassa Dan Lipid Mikroalga Kultur Campuran Konsorsium Glagah Dan Arthrospira maxima Pada Thin-Layer Photobioreactor* (Universitas Gadjah Mada)

[14] Doucha J and Lívanský K 2009 Outdoor open thin-layer microalgal photobioreactor: Potential productivity *J. Appl. Phycol.* **21** 111–117

[15] Suyono E A, Nopitasari S, Zusron M, Khoirunnisa P, Islami D A and Prabeswara C B 2016 Effect of silica on carbohydrate content of mixed culture Phaeodactylum sp. and Chlorella sp *Biosci. Biotechnol. Res. Asia* **13** 109–14

[16] Nirmala N and Dawn SS 2016 Environmental Stress Control to enhance lipid content in Oleaginous Microalgae for Biodiesel production *Int. J. ChemTech. Res.* **9**(2) 237-241

[17] Barsanti L and Gualtieri P 2014 *Algae Anatomy, Biochemistry, and Biotechnology* (Pisa, Italy: CRC Press)
[18] Isnansetyo A and Kurniastuty 1995 Teknik Kultur Phytoplankton Zooplankton: Pakan Alam untuk Pembenihan Organisme Laut. (Kanisius: Yogyakarta)
[19] Richardson B, Orcutt D M, Schwertner H A, Martinez C L and Wickline H E 1969 Effects of Nitrogen Limitation on the Growth and Composition of Unicellular Algae in Continuous Culture *Appl. Microbiol.* **18**(2) 245-250

Acknowledgement
The authors would like to thanks to Laboratory of Biotechnology (Faculty of Biology), Process System Engineering research group (PSErg) and Algae Biorefinery Research Group, Universitas Gadjah Mada for financial and technical support in this work.

IOP Conf. Series: Journal of Physics: Conf. Series **1022** (2018) 012052 doi:10.1088/1742-6596/1022/1/012052

Nutrient uptake from liquid digestate using ornamental aquatic macrophytes (*Canna indica*, *Iris pseudacorus*, *Typha latifolia*) in a constructed wetland system

W Ediviani[1], C R Priadi[2] and S S Moersidik[3]

[123]Environmental Engineering Program, Faculty of Engineering, University of Indonesia, Depok, Indonesia

Abstract. Indonesia has implemented energy recovery from organic (food) waste by anaerobic digestion method, but the digestate was commonly treated only by composting, and still as a separated treatment (not integrated into a resource recovery system). Whilst not getting any pretreatment, the digestate was disposed to the environment and then act as a pollutant. Yet it contains nutrients which could be recovered as a nutrient source for plants. The study was about how ornamental aquatic macrophytes could uptake nitrogen from liquid digestate in a constructed wetland method. *Canna indica*, *Iris pseudacorus*, and *Typha latifolia* were the experimented ornamental aquatic macrophytes used to uptake the nutrient (nitrogen—N) from liquid digestate. The study showed that the highest N uptake was done by *C. indica* (25.1%) which has the highest biomass increment as well (80.5%). Effluent quality improvement also shown by N removal by *C. indica* (68.5—76.4% TN), *I. pseudacorus* (61.8—71.3% TN), and *T. latifolia* (61.6—74.5%). This research proved that *C. indica* has the performance for the N uptake, best N removal efficiency, with a great growth rate as well. This system using *C. indica* could also improve the water quality of the effluent and add the aesthetic of environment.

1. Introduction

Organic (food) waste has been treated for recovering energy using the technology of *anaerobic digestion* (AD). This technology is quite well-known in developed countries such as Indonesia, which is also one of the problem solvers of food waste generation. Despite the role of treating waste and recovering energy, AD even puts another problem on the environment by producing end-of-waste, known as digestate [1]. Digestate in both forms, solid (fibre) and liquid (liquor), still contains high nutrient content which makes it as a pollutant if it's not treated before disposal. Liquid fraction/ liquor has higher nutrient content than solid fraction/ fiber; Total Nitrogen (TN) of solid digestate range from 8.8—13.0 kg/t dry solids (DS), while the liquid fraction contains 4,000—6,000 mg/l [2]. Due to the high nutrient content of digestate, it must be treated before disposing to prevent unwanted matter on the environment.

The characteristic of digestate made it as a valuable nutrient source, yet not many technologies are friendly enough to the user. One way to treat digestate is by using a nutrient recovery method, which not only could recover nutrients from digestate by plant uptake, but also create a cleaner environment by making a friendly post-treatment product (low nutrient, BOD, and COD content). Nutrient recovery of digestate that has been done was only up to composting (biofertilizer), mostly on solid form, but it was not good enough to compete with synthetic fertilizers [3] unless it is integrated with an agriculture industry which uses AD technology as well [2].

Another method of nutrient recovery is by using plants, particularly aquatic macrophytes. Aquatic macrophytes are the type of plants which has a macro (can be seen; big) form, and lives in a wet ecosystem [4]. It is known that this vegetation has a reputation in nutrient removal system, such as a constructed wetland. The vegetations that are quite popular in here is *Canna indica*, *Iris pseudacorus*, and *Typha latifolia*. Each of these ornamental aquatic macrophytes recovered the nutrients from liquor state as nutrient uptake which will be the source of their development.

C. indica is one of the most popular ornamental aquatic macrophyte used in the wastewater treatment in a wetland system. As for treating domestic wastewater, *C. indica* could be used as an efficient nutrient removal up to 84% for nitrogen removal and 92% for phosphorus removal, which could create a healthier environment at the end of the treatment [5]. With those capabilities, *C. indica* could be more likely used as a nutrient recovery agent for substances such as liquid products of organic waste treatment which has similarities with domestic wastewater; high nutrient content. In this study, *C. indica* was compared to two other ornamental aquatic macrophytes, *I. pseudacorus* and *T. latifolia*, to see if how much does the ornamental aquatic macrophytes could make use liquid digestate as a nutrient source, especially nitrogen

2. Methods

2.1 Operation

The operation flow started from the AD reactor to the influent storage, and the influent later will be distributed to three beds (Figure 3). The operational time was done in 70 days; acclimatization (A) and feeding (F) stage. Both stages (A and F) started in the morning (10 AM); flowing the influent inside the bed. Sampling was done three times a week (every two days) after the acclimatization stage. As for laboratory test needed 14 days for liquid samples, and 20 days for plants and sediments (count on work days).

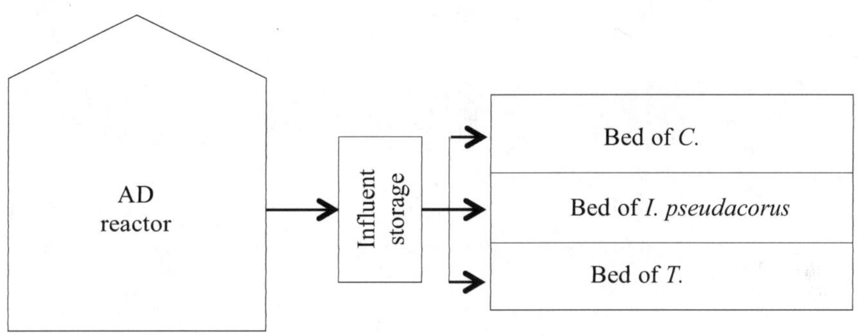

Figure 1. Operation Flow

2.2 Influent Characteristic

The constructed wetland influent was from an anaerobic digester which treats food waste [6, 7]. Effluent from the digester was pretreated by dilution to provide an influent with a BOD_5 characteristics $\pm$ 400 mg/L (**Table 2**). This characteristic must be reached for the survival of the macrophyte itself [8].

2.3 CW Design

The constructed wetland for the research was designed in a horizontal subsurface flow [9]. This system was chosen since the constructing area are located near the campus so it could prevent the smell from the influent (flows in the subsurface), near from water and energy source, and fairly manageable. The site was also for adding the aesthetic view. As for the land used for each bed is 170 × 80 × 70 cm (length : width : depth), with planting area 120 × 60 cm (length : width). *C, indica*, *I. pseudacorus* and *T. latifolia* were the experimented object in the study. The selection of the vegetation was based on literature (the most effective vegetation in nutrient removal), availability in the country, price and aesthetic [10]. Sediments were put in layers with clay as the top layer (20 cm), sandy loam as the middle layer (15 cm), and coarse sand at the bottom (15 cm). Each inlet and outlet were given gravel (15 cm).

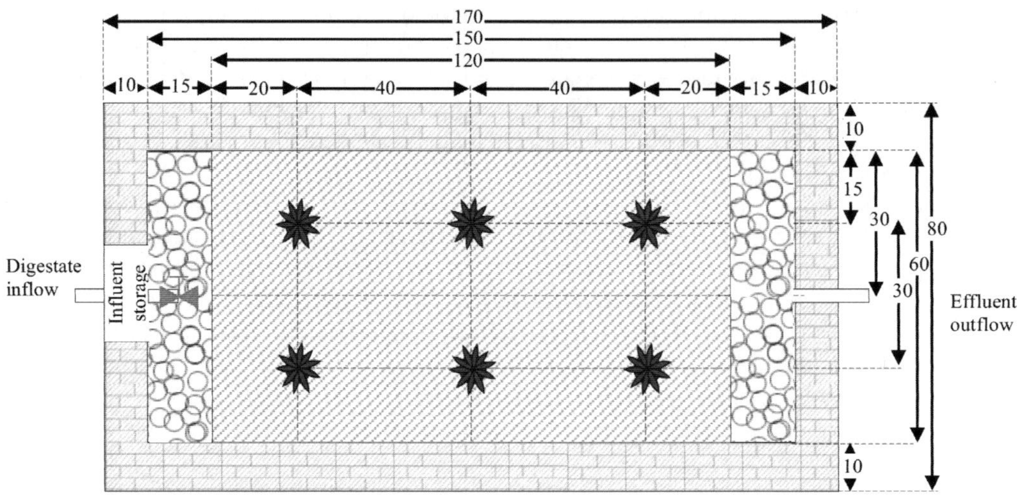

Figure 2. Top View of The HSSF CW Design (scale in centimeter)

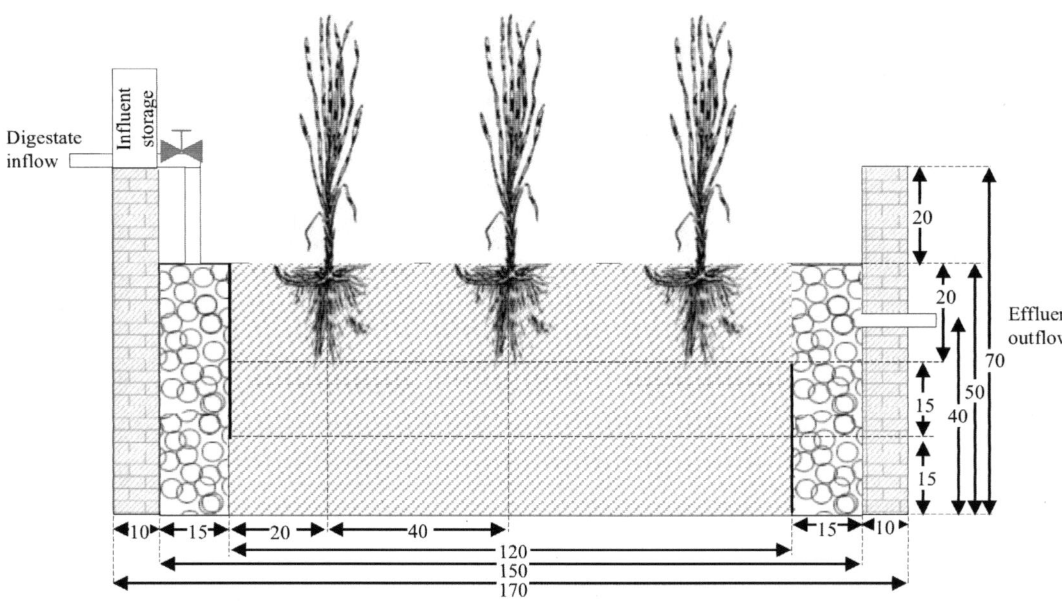

Figure 3. Side Layer of The HSSF CW Design (scale in centimeter)

2.4 Data Analysis

The data which were gained from the study will be analyzed later on by statistical analysis using ANOVA and *Duncan Multiple Range Test* (DMRT) to compare the effect of digestate on plant growth (height, shoots, and leaves). Calculating the nutrient removal by each plant was using an efficiency equation:

$$\%\text{Efficiency}=\left(\frac{\text{input-output}}{\text{input}}\right)\text{mg/L x }100\% \quad (1)$$

TN content by each plant used the equation as followed:

$$\%\text{N uptake}=\left(\frac{\text{total nutrient content}_F - \text{total nutrient content}_A}{\text{total nutrient content}_F}\right)\text{g/kg x }100\% \quad (2)$$

Calculating total biomass increment using equation is:

$$\text{Total biomass (g/m}^2)=\frac{\text{biomass (F-A) (g)}}{\text{planting area (m}^2)} \quad (3)$$

and measuring the growth rate of each plants using kinetics from the order of 0 to 2.

3. Result and Discussion

The statistic shows that the type of the macrophyte which used digestate as the nutrient source has a significant effect for each variable (**Table 1**). *C. indica* was the most improved macrophyte in plant height, and the least improvement of it was shown by *T. latifolia*. *T. latifolia* had a stagnant height situation on the feeding stage; barely add a milimeter of height, while *C. latifolia* could rise 1—4 cm height per week, and *I. pseudacorus* could add 0.5—2 cm per week.

Table 1. Summary of Statistical (DMRT) Analysis

Variable	p<0.005		
	C. indica	*I. pseudacorus*	*T. latifolia*
Plant height	131.500[a]	59.972[b]	14.199[c]
Number of leaves	9.222[a]	4.278[b]	4.833[b]
Number of shoots	1.1111[a]	-	0.5556[b]

Letters (a-c) are indicate homogeneous subsets ($p < 0.05$)

Number of leaves also improved the best on *C. indica* which has lush leaves, while the other two had the same growth performance being in the same group that explain how small the growth of leaves of each macrophytes. *I. pseudacorus* didn't have shoots for growth, so the variable was compared between *C. indica* and *T. latifolia*. *C. indica* leads the performance by having lots of shoots growth every weak since acclimatization until feeding ended. The parent plant of *T. latifolia* withered several days after acclimatization started, but showing shoots number increment two weeks before the study ended.

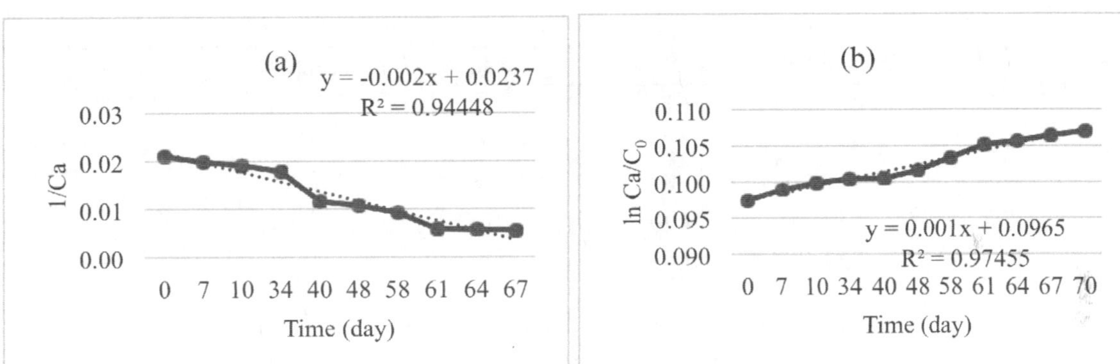

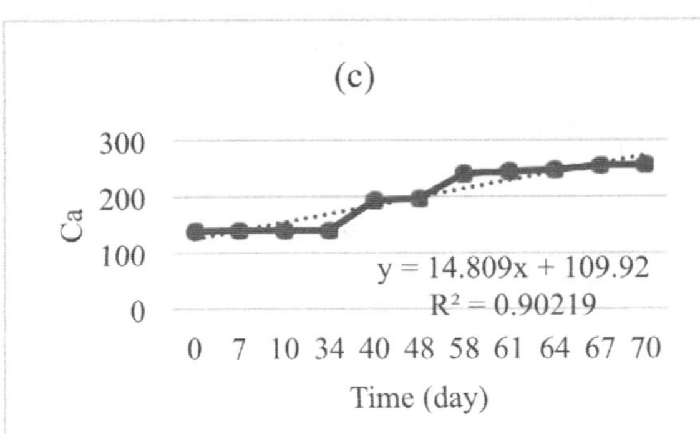

Figure 4. Growth Rate Kinetics of *C. indica* using order 2 (a), *I. pseudacorus* using order 1, and *T. latifolia* (c) using the order 0 which the R^2's are the closest to 1

As for the growth rate, each macrophytes growth rate has shown the result as followed.

C. indica 2nd Order

$$\frac{1}{Ca} = -0.002\ x + 0.0237 \tag{4}$$

I. pseudacorus 1st Order

$$\frac{\ln Ca}{C_0} = 0.001\ x + 0.0965 \tag{5}$$

T. latifolia 0th Order

$$Ca = 14.809\ x + 109.92 \tag{6}$$

Ca is the height of the plant and x is number of the day.

N removal by each macrophyte on the digestate was proved by the TN content on the effluent from every bed are lower than the influent (**Table 2)**. However, there was an anomaly result on period A3; TN content of *C. indica*'s bed effluent was too high compared to other samples. The efficiency calculation shows that *C. indica* could remove nitrogen up to 72.19%, the highest N removal efficiency among the other macrophytes in the study. Even though the influent's nitrogen content wasn't as high as the literature mentioned, but *C. indica* has the performance to improve the quality of the effluent. *I. pseudacorus* and *T. latifolia* show similar efficiency performance, lower than *C. indica*. However, it still shows improvement in the effluent quality.

Table 2. BOD and N Concentration in Influent and Each Effluent (mg/L)

Periode	BOD_5 of the Influent	TN content of the Influent	TN content of the Effluent		
			C. indica	*I. pseudacorus*	*T. latifolia*
A1	420	617	195	224	236
A2	275	659	177	223	242
A3	255	581	1205	222	188
F1	530	670	158	192	257
F2	520	649	189	238	165

Table 3. N Removal Efficiency by Each Macrophyte (%)

Periode	Bed		
	C. indica	*I. pseudacorus*	*T. latifolia*
A1	68.5	63.8	61.8
A2	73.1	66.1	63.3
A3	-	61.8	67.7
F1	76.4	71.3	61.6
F2	70.8	63.3	74.5
Min-Max	68.5—76.4	61.8—71.3	61.6—74.5
Average	72.2	65.3	65.8

C. indica shows the best result in N uptake and biomass increase. *C. indica* gains 25.1% TN during the study, and increase 80.5% of the plant biomass. Meanwhile, *I. pseudacorus* lost 9.7% of TN content and *T. latifolia* lost 62% of TN content, while both still having an increment in their biomass (gain 68.8% and 72.4% biomass, respectively). This result matches the situation presented in **Table 1**, when *C. indica* has the fastest height growth and the most number of leaves.

Table 4. N Uptake and Biomass Increase by Each Macrophyte

Periode	TN content (g/kg)			Biomass (mg)		
	C. indica	*I. pseudacorus*	*T. latifolia*	*C. indica*	*I. pseudacorus*	*T. latifolia*
A	1,652	1,319	1,319	12,430	11,110	12,100
F	2,207	1,203	920	63,590	35,630	43,860
%	+25.1%	-9.7%	-62%	80.5%	68.8%	72.4%

4. Conclusions

Ornamental aquatic macrophyte in a constructed wetland system could be a solution for treating liquid digestate. Using a horizontal subsurface flow constructed wetland, the highest nutrient uptake performance from liquid digestate was done by *C. indica* which has 25.1% N uptake to the plant during the study, and has the highest biomass increase for 80.5%, which was shown by the growth performance as well to be significantly high than the two other experimental macrophytes. *I. pseudacorus* and *T. latifolia* had a lower number of TN content in the plant (loss 9.7% and 62%, respectively), but the biomass shows increment number up to 68.8% for *I. pseudacorus* and 72.4% for *T. latifolia*. Effluent quality improvement also shown by N removal by *C. indica* (68.5—76.4% TN), *I. pseudacorus* (61.8—71.3% TN), and *T. latifolia* (61.6—74.5%). With this result, all of the experimented plants could use liquid digestate as a nutrient source, but the best performance goes to *C. indica*. Besides of the nitrogen uptake value, it could also improve the quality of the effluent before disposal. Nonetheless, the ornamental aquatic macrophyte also has an aesthetic value to the environment.

5. References

[1] Saveyn H 2012 EC *'End-of-Waste' Criteria for Biodegradable Waste Subject to Biological Treatment (Compost and Digestate)* (Rennes: Joint Research Centre, Institute for Prospective Technological Studies) p 39

[2] Anaerobic Digestion & Bioresources Association 2013 *Digestate* **3** chapter 6 p 81–92

[3] Frischmann P 2012 *Enhancement and Treatment of Digestates from Anaerobic Digestion* (United Kingdom: Pell Frischmann Consultants Ltd) p 122

[4] Dhir B 2013 *Phytoremediation: Role of Aquatic Plants in Environmental Clean-up* (India: Springer) p 111

[5] Calheiros C S C, Bessa V S et al 2015 Constructed wetland with a polyculture of ornamental plants for wastewater treatment at a rural tourism facility *Ecological Engineering* **79** pp 1–7

[6] Kurnianingsih S Priadi CR 2017 Pilot scale anaerobic digestion of food waste process performance under various mixing intensities *The 1st International Conference on Science, Mathematics, Environment and Education 2017 Solo* presented on 16—17 September 2017 (ID: 3108) p 10

[7] Wijayanti D F, Suwartha N and Priadi C R 2017 Effect of addition of fat oil and grease (fog) towards performance of dry anaerobic digestion food waste reactor *The 2nd International Tropical Renewable Energy Conference 2017 Bali* presented on 3—4 Oktober 2017 (ID: 12732) p 9

[8] Tousignant E 1999 *Guidance Manual for The Design, Construction and Operations of Constructed Wetlands for Rural Applications in Ontario* (Ontario: Stantec Consulting Ltd) p 61

[9] Environment Protection Agency 1993 *Subsurface flow constructed wetlands for wastewater treatment: a technology assessment* (USA: EPA) p 87

[10] Hasan M R and Chakrabarti R 2009 Use of algae and aquatic macrophytes as feed in small-scale aquaculture: a review (Rome: *FAO Fisheries and Aquaculture Technical Paper*) No. 531 p 123

Acknowledgment

This study is funded by the grant of Green FT Universitas Indonesia and *Publikasi Internasional Terindeks Untuk Tugas Akhir Mahasiswa* (PITTA) 2017 with a number of grant contract 756/UN2.R3.1/HKP.05.00/2017. The influent material was supplied from Septiana Kurnianingsih and Dwita Fitriani Wijayanti research, influent preparation was helped by Mislan, and location of the CW was provided by Faculty of Engineering, University of Indonesia.

IOP Conf. Series: Journal of Physics: Conf. Series **1022** (2018) 012053 doi:10.1088/1742-6596/1022/1/012053

Nutrients content and quality of liquid fertilizer made from goat manure

Yacobus Sunaryo[1, 2], Djoko Purnomo[1], Maria Theresia Darini[2] and Vita Ratri Cahyani[1*]

[1] Graduate School, Sebelas Maret University, Indonesia.
[2] Faculty of Agriculture, University of Sarjanawiyata Tamansiswa, Indonesia.

[*]Corresponding author: vitaratri@staff.uns.ac.id

Abstract. Quality of liquid fertilizer is determined by the content of nutrients and other chemical factors such as pH and EC. This research aimed to examine nutrient contents and dynamic of pH and EC of liquid fertilizer made from goat manure in combination with sugar and ammonium sulfate (ZA) and using Effective Microorganisms (EM) as the decomposer. This research was conducted by employing 3 x 3 factorial experiment with three replications. Each treatment combination was applied in 20 L of water. The first factor was the quantity of sugar which consisted of 3 levels: 12.5, 25, and 50 g L^{-1} of water. The second factor was the quantity of ZA which consisted of 3 levels: 25, 37.5, and 50 g L^{-1} of water. All combinations were added by 100 g of air dried goat manure L^{-1} of water and EM solution 1 ml L^{-1} of water, and incubated for five months. Results of the experiment indicated that the increasing concentration of ZA resulted in the significantly increase of N total and S total. Increasing concentration of sugar resulted in decreasing pH and increasing lactic acid; whereas, increasing concentration of ZA followed by increasing Electrical Conductivity (EC). There was no significantly change of pH and EC of the liquid fertilizer during five months incubation.

Key words: goat manure, liquid fertilizer, nutrients content, quality

1. Introduction

The availability of high quality fertilizers made of materials which are cheaper and easier found by farmers is very important. The quality of the liquid fertilizer is determined by the value of pH and EC, as well as the nutrient elements content of the fertilizer. Fertilizers as source of nutrients are very important for plant growth.

In geoponic cultivation, fertilizers are generally given to the soil as inorganic fertilizers so they can be absorbed quickly by plants. Nevertheless, the application of inorganic fertilizers can reduce the soil productivity. The soil productivity is determined by soil condition, previously fertilizing, organic matters addition, and kinds of plants cultivated [1]

Recently, many kinds of high quality of solid organic fertilizers are offered to the farmers, i.e. vermicomposting and fine compost, can result in higher yield of plants [2, 3]. Furthermore, now a quantity of liquid organic fertilizers is made by either farmers or factories [4, 5, 6]. Organic matters used as raw materials to make liquid organic fertilizer are seaweeds [7, 8, 4, 5], *Eichhornia crassipes, Symphytum officinale* and *Amaranthus retroflexus* [9], and sludge [10]. Moreover, vermicomposting can also result in high quality of liquid organic fertilizer [6].

Organic farming, especially the use of organic fertilizers, in one hand it can result in lower quantity, but in another hand it can result in higher quality of plant yield. The lower yield quantity is due to the lower of macro nutrients content of the organic fertilizers [2, 3], so to get higher quantity of plant yield is necessary to fertilize the plant either by the high quality organic fertilizer or by higher dose organic fertilizer application.

To get the high quality organic fertilizer the use of effectiveness microorganisms (EM) during fermentation of raw materials is recommended. In order to give nutrients and energy sources for EM during fermentation, it can be added urea, NPK, as well as molasses in making liquid organic fertilizer [11]. EM is inoculant microorganisms used to ferment organic materials applied to increase the soil fertility, plant growth, and plant yield [12]. The quality of liquid fertilizer is determined by not only the pH but also nutrient elements content of the fertilizer. The quality of organic liquid fertilizer is also determined by content of another chemical substances such as growth regulator and organic acid [13, 14]. The production of high quality organic liquid fertilizer is very important to support the development of hydroponic cultivation in several countries. The liquid organic fertilizer can substitute inorganic chemical substances that more expensive and more difficult to be found in the market. This research was objected to examine nutrient contents (total of N, K_2O, S, Fe, and Mn) and dynamic of pH and EC of liquid fertilizer made from goat manure in combination with sugar and ammonium sulfate (ZA) and using Effective Microorganisms (EM) as the decomposer.

2. Materials and Methods

2.1. Date and Place of The Experiment

The fermentation and incubation of the liquid fertilizers was conducted in the laboratory of University of Sarjanawiyata Tamansiswa in condition of air room temperature. The liquid fertilizer fermentation was carried out from the first week until the fourth week of February 2016. Whereas, the incubation of the liquid fertilizers was conducted from March until July 2016.

2.2. Preparation of The Materials

Goat manure used in this experiment was the feces of grass fed goat. The goat manure was in air dried condition, and it was grinded to become fine granules. Samples of the fine granules of goat manure were analyzed to get the data of pH and nutrient elements content based on Govere *et al.* [9].

2.3. Formulation of The Materials

Fertilizers were formulated by 3 x 3 factorial experiment with three replications. Each treatment combination was applied in 20 L of water. The first factor was the quantity amount of sugar consisted of 3 levels: 12.5, 25, and 50 g L^{-1} of water noted as G1, G2, and G3 respectively. The second factor was the quantity of ZA which consisted of 3 levels: 25, 37.5, and 50 g L^{-1} of water noted as Z1, Z2, and Z3 respectively. Nine treatment combinations were gained: G1Z1, G2Z1, G3Z1, G1Z2, G2Z2, G3Z2, G1Z3, G2Z3, and G3Z3. All combinations were added by 100 g air dried goat manure L^{-1} of water and EM solution 1 ml L^{-1} of water.

2.4. Fermentation of The Materials

Fermentation of the materials consisted of several steps. The steps were as follows: (1) goat manure, sugar, and ZA were mixed in ratio to 20 L of water. Sugar + ZA were diluted by 20 L of water in the plastic bucket, then it was added by 20 ml EM and 2 kg goat manure. (2) The mixed materials were stirred manually around 5 minutes using wood stick. (3) All of mixed combination materials were fermented in the plastic bucket of 25 liter volume. All of the fermented plastic buckets were capped properly and they were put inside the room in condition of air room temperature. (4) Every day, in the morning and afternoon, all of the mixed fermented materials were stirred manually around 2 minutes using wood stick. (5) During 3 weeks while fermentation, the pH and EC of the solution were observed. The pH and EC were observed using portable pH/EC/TDS/ Temperature meter Hanna Instruments H19811-5. In order to get the real value of EC and TDS, and due to the fertilizer solution which was very high concentrated, it was diluted by adding sterile water in ratio of 1:20 (v/v). (6) In the fourth week of fermentation, both macro and micro elements, and organic acids content of the fertilizer solution were observed [9, 15, 16]. The acid organics observed were citric, acetic, and lactic acid on formulation of G1Z1, G3Z1, G1Z3, and G3Z3 using HPLC method [15, 16].

Each formula of treatment combinations of the liquid fertilizers was taken 5 liters to put inside plastic containers which were capped properly. Those plastic containers were incubated in the room for five months. Every month the pH and EC of the solution were observed.

3. Results and Discussions

Results of the analysis listed in the Table 1 indicated that the nutrient content of the goat manure was in the range of standard as recommended by the Regulation Minister of Agriculture Republic Indonesia [17]. The Regulation states that the solid organic fertilizer must have the content of macro element ($N + K_2O + P_2O_5$) ≥ 4 %, Fe total ≤ 9000 ppm, and Mn total ≤ 5000 ppm, and Zn total ≤ 5000 ppm. The results were also appropriate with the research by Karim *et al.* [18] stating that N content of goat manure is 1.02 %. The nutrient element content of animal husbandry manures is depended on their food and animal species [19]. The results of macro and micro elements analysis of the goat manure are listed in Table 1.

Results of the experiment (Figure 1) indicated that there was no interaction between the quantity of sugar (G) and ZA (Z) on the pH value of liquid fertilizers. The highest quantity of sugar (G3) could result in the lowest pH (the three lowest lines) and the lowest quantity of sugar (G1) could result in the highest pH (three upper lines). There was no effect of ZA on pH of the liquid fertilizers.

Figure 2 indicates that there is no interaction between the quantity of sugar (G) and ZA (Z) on the EC value of liquid fertilizers. The highest quantity amount of ZA (Z3) could result in the highest EC (three upper lines) and the lowest quantity of ZA (Z1) could result the lowest EC (the three lowest lines). There was no effect of sugar on EC of the liquid fertilizers. The neutral pH and the highest EC could be resulted in the formulation of G1Z3.

Table 1. The Average Content of Macro and Micro Elements of The Goat Manure

No.	Nutrient elements	Unit	Content
1	N total	%	1.15 ± 0.11
2	P_2O_5 total	%	1.10 ± 0.14
3	K_2O total	%	2.79 ± 0.16
4	Ca total	%	1.56 ± 0.09
5	Mg total	%	0.42 ± 0.06
6	S total	Ppm	2050 ± 16.09
7	Fe total	Ppm	7214 ± 29.46
8	Mn total	Ppm	512 ± 13.45
9	Zn total	Ppm	924 ± 15.87

$\pm$ indicates the value of standard deviation of mean (n = 3)

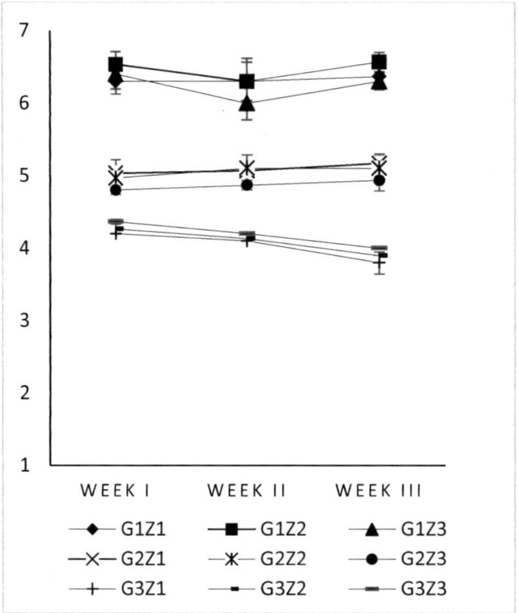

Figure 1. Effect of Sugar and ZA Levels on pH of Liquid Fertilizer during Fermentation. Error Bars Represent The Standard Deviations

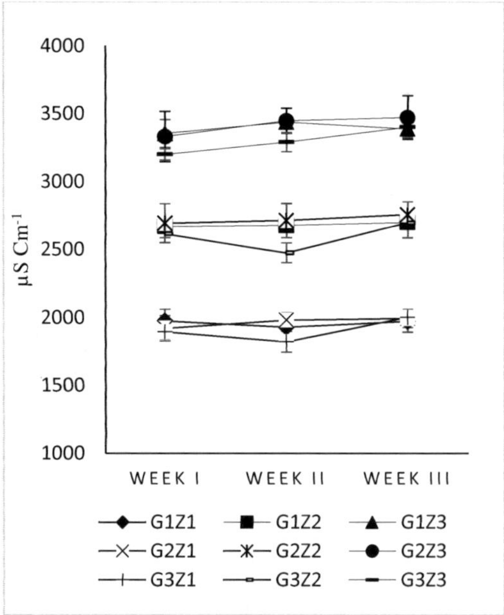

Figure 2. Effect of Sugar and ZA Levels on EC of Liquid Fertilizer during Fermentation. Error Bars Represent Standard Deviations

IOP Conf. Series: Journal of Physics: Conf. Series **1022** (2018) 012053 doi:10.1088/1742-6596/1022/1/012053

Results indicated that there was interaction between sugar (G) and ZA (Z) on the content of N, K, S, Fe, and Mn of the liquid fertilizers. The increasing of ZA could result the increasing of N total content as well as S total content of the liquid fertilizers (Table 2). Both macro and micro element content of the liquid fertilizers is lower than the macro and micro element content of the liquid organic fertilizer as stated in the Regulation Minister of Agriculture Republic Indonesia [17].

Table 2. The Average Content of N, K, S, Fe, and Mn of The Liquid Fertilizers

Formula	N total (%)	K2O total (%)	S total (%)	Fe total (ppm)	Mn total (ppm)
G1Z1	0.44 d	0.17 b	0.06 e	26.71 g	21.34 e
G1Z2	0.66 c	0.13 cd	0.19 b	31.80 f	24.50 d
G1Z3	0.85 a	0.11 d	0.26 a	30.28 f	23.71 d
G2Z1	0.39 d	0.14 c	0.09 d	47.94 c	30.49 b
G2Z2	0.79 b	0.12 cd	0.24 a	45.58 d	37.54 a
G2Z3	0.81 ab	0.12 cd	0.20 b	39.31 e	27.95 c
G3Z1	0.44 d	0.19 a	0.06 e	82.47 a	29.55 cb
G3Z2	0.66 c	0.19 ab	0.15 c	82.33 a	38.03 a
G3Z3	0.85 a	0.21 a	0.20 b	76.68 b	29.44 cb
G * Z (ρ > F)	<.0001	0.0103	<.0001	<.0001	<.0001

Numbers in columns followed by same letter are not significantly different at the 5% level of significance by DMRT

Table 3. The Average Content of Organic Acid of Liquid Fertilizers

No.	Treatment combination	Organic acid content (mg L^{-1})		
		Citric acid	Lactic acid	Acetic acid
1	G1Z1	41.09 a	49.37 d	2,512.17 a
2	G3Z1	35.80 b	7,582.52 a	373.89 d
3	G1Z3	29.39 c	17.2 d	915.38 b
4	G3Z3	30.02 c	7,270.21 b	439.55 c
G * Z (ρ > F)		0.0012	< 0.0001	< 0.0001

Numbers in columns followed by same letter are not significantly different at the 5% level of significance by DMRT

Results of the experiment indicated that there was interaction between sugar (G) and ZA (Z) on the content of organic acid of liquid fertilizers. Results indicated that the highest of quantity amount of sugar (G3) could result the highest of lactic acid content in combination with both the lowest ZA (Z1) and the highest ZA (Z3): 7,582mg L^{-1} and 7,270mg L^{-1} respectively. The highest acetic acid content could be resulted from the combination of G1 and Z1: 2,512 mg L^{-1} (Table 3).

The higher content of lactic acid due to the higher amount of sugar could result in the lower pH of the liquid fertilizer while reducing the quality of liquid fertilizer. Nevertheless, several studies show that lactic acid can accelerate the plant growth. L-lactic acid can stimulate the growth of duckweed plant (*Lemna minor* L.) and corn (*Zea mays* L.) as it was indicated at increasing plant biomass [20]. L-(d)-lactic acid can exhibit typical growth promotor activity at very low concentrations and dosage application [21].

Acetic acid as well lactic acid can make plants healthty. These organic acids can decrease the number of seeds infected by fungi, however at the highest concentration it negatively affected germination parameters and declined the seed vigor of Zinnia seeds [22]. Compared to the content of lactic acid as well acetic acid, the content of citric acid was very low. Citric acid can accelerate the plant growth. Talebi *et al.* [23] report that citric acid environmentally sounds chemical, in concentration 300 mg L^{-1}, is effective on the various aspects of growth and development of Gazania plants.

Results of the experiment indicated that the combination of 100 g goat manure + 12.5 g sugar + 50 g ZA had the normal pH (around 6.0 - 6.5), the higher EC (around 3200 - 3400 μS cm^{-1}, after diluted by the water in ratio 1:20), and the higher content of N total and S total. Otherwise, it had the lowest lactic acid content.

Results of the experiments indicated that there was no significantly change of pH (Figure 3) and EC (Figure 4) of the liquid fertilizers during five months incubation. These results indicated that the storage during five months did not change the quality of the liquid fertilizers.

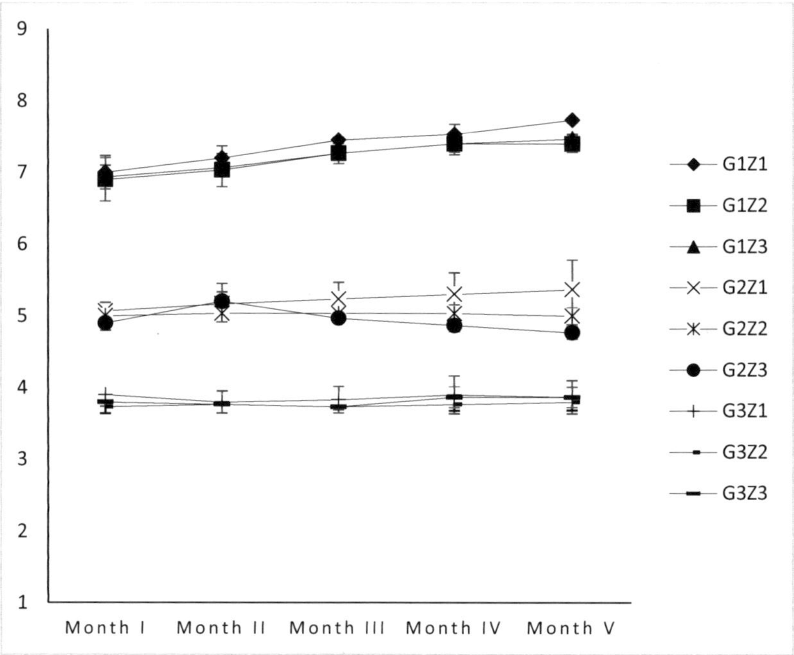

Figure 3. The pH Dynamic of Liquid Fertilizers During Storage. Error Bars Represent Standard Deviations

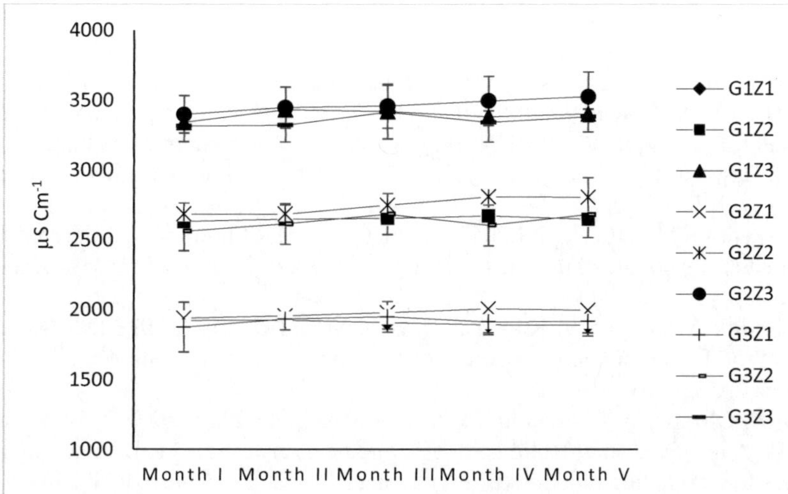

Figure 4. The EC Dynamic of Liquid Fertilizers During Storage. Error Bars Represent Standard Deviations

4. References

[1] Koenig R and Johnson M 2011 *Selecting and using organic fertilizers* Utah State University. Cooperative Extension, extention.usu.edu pp 6

[2] Chanda G K, Bhunia G and Chakraborty SK 2011 The effect of vermicompost and other fertilizers on cultivation of tomato plants *J. of Hort. and Fore.* **3**(2) 42-45

[3] Joshi R, Vig A P and Singh J 2013 Vermicompost as soil supplement to enhance growth, yield and quality of Triticum aestivum L.: a field study *Intl. J. of Recyc. of Org. Wast. in Agri.* **2**(16) 1-7

[4] Sridhar S, and Rengasamy R 2010 *Studies on the effect of seaweed liquid fertilizer on the flowering plant tagetes erecta in field trial* Reseacch Paper, Advances Bioresearch **1**(2) 29-34

[5] Sasikumar K, Govindan T and Anuradha C 2011 effect of seaweed liquid fertilizer of dictyota dischotoma on growth and yield of Abelmoschus esculantus L. *Euro. J. of Exp. Biol.* **1**(3) 223-27

[6] Quaik S and Ibrahim M H 2013 A review on potential of vermicomposting derived liquids in agricultural use *Intl. J. of Scie. and Res. Publication* **3**(3) 1-6

[7] Zodape S T, Kawarkhe V J, Patolia J S and Warade A D 2008 Effect of liquid seaweed fertilizer on yield and quality of okra (Abelmoschus esculentus L.). *J. of Sci. & Indus. Res.* **67** 1115 – 17

[8] Thirumaran G, Arumugam M, Arumugam R and Anantharaman P 2009 Effect of Seaweed Liquid fertilizer on Growth and Pigment Concentration of Cyamopsis tetrogonolaba (L) Taub. *Amer.-Eura. J. of Agro.* **2**(2) 50-56

[9] Govere S, Madziwa B and Mahlatini P 2011 The nutrient content of organic liquid fertilizers in Zimbabwe. *Intl. J. of Mod. Eng. Res.* **1**(1) 196-202

[10] Jambaldorj G, Takahashi M and Yoshikawa K 2007 *Liquid fertilizer production from sewage sludge by hydrothermal treatment* Proceedings of International Symposium on Eco Topic

Science 605-608.
[11] Desyane H K and Wiyana A F 2012 Proposed quality improvement of liquid fertilizers "Herbafarm" to meet national standards in Indonesia *The Ind. J. of Bus. Adm.* **1**(6) 343-352.
[12] Higa T and Parr J F 1994 *beneficial and effective microorganism for a sustainable agriculture and environment*. International Nature Farming Research Center Atami, Japan pp 16.
[13] Arancon N Q, Edwards C A, Lee S and Byrne R 2006 Effects of humic acids from vermicomposts on plant growth *Eur. J. of Soil Biol.* **42** S65-S69.
[14] Campitelli P, Velasco M and Ceppi S 2012 Characterization of humic acids derived from rabbit manure treated by composting-vermicomposting process *J. of Soil Scie. and Plan. Nutr.* **12**(4) 875-91
[15] Kishore G, Karthik A, Gopal S V, Kumar A R, Bhat M and Udupa N 2013 Development of RP-HPLC method for simultaneous estimation of lactic acid and glycolic acid *Der Pharma Chemica* **5**(4) 335-340
[16] Zamanova M K, Glotova V N, Izhenbina T N, Krutas D S and Novikov V T 2014 P Simultaneous HPLC-UV determination of lactic acid, glycolic acid, glycolide, lactide and ethyl acetate in monomers for producing biodegrsdsble polymers *Procedia Chemistry* **10** 244-251.
[17] Peraturan Menteri Pertanian Nomor 70/**Permentan**/SR.140/10/**2011** *Tentang. **Pupuk Organik**, Pupuk Hayati dan Pembenah Tanah* pp 88. *The Regulation Minister of Agriculture Republic Indonesia No.70/Permentan/SR.140/10/2011 About, Organic Fertilizer, Biofertilizer and Soil Ameliorant pp 88*. http://perundangan.pertanian.go.id/admin/file/Permentan-70-11.pdf
[18] Karim T H, Qadir K S and Salahaddin A A 2013 Rate of soil aggregates disintegration as affected by type and rate of manure application and water quality *Intl. J. of Plan. Anim. and Envi. Scie.* **3**(2) 94-102
[19] Schoenian S 2012 *Nutrient management on sheep farm* http://www.sheep101.info/201/nutrientmgt.html
[20] Kinnersley A M, Taylor C S, Yopp J H, and Whitten G H 1990 Promotion of plant growth by polymers of lactic acid *Plant Growth Regulation* **9**(2) 137-49
[21] Young D C 1989 *Methods for regulating the growth of plants and growth regulant compositions*. Patents: US4863506 A
[22] Szopińska D 2013 The effects of organic acids treatment on germination, vigour and health of Zinnia (Zinnia elegans Jacq.) seeds *Acta Sci. Pol. Hortorum Cultus* **12**(5) 17-29.
[23] Talebi M, Hadavi E and Jaafari N 2014 *foliar sprays of citric acid and malic acid modify growth, flowering, and root to shoot ratio of gazania (Gazania rigens L.): A comparative analysis by ANOVA and structural equations modeling*. Advances in Agriculture. Article ID 147278 pp 6 http://dx.doi.org/10.1155/2014/147278

Acknowledgement

The present study is supported with the fund from the Ministry of Research Technology and Higher Education of Indonesia

IOP Conf. Series: Journal of Physics: Conf. Series **1022** (2018) 012054 doi:10.1088/1742-6596/1022/1/012054

A new candidate of calcium channel blocker in silico from *Tectona grandis* for treatment of gestational hypertension

A Azizah[1], Y H Suselo[2*], M Muthmainah[3] and D Indarto[2,4]

[1]) Medical Student, Faculty of Medicine, Universitas Sebelas Maret
[2]) Department of Physiology, Faculty of Medicine, Universitas Sebelas Maret
[3]) Department of Anatomy, Faculty of Medicine, Universitas Sebelas Maret
[4]) Biomedical Laboratory, Faculty of Medicine, Universitas Sebelas Maret

Corresponding email : yulianaheri@staff.uns.ac.id

Abstract. Gestational Hypertension is one of the three main causes of maternal mortality in Indonesia. Nifedipine which blockes the Cav1.2 calcium channel has frequently been used to treat gestational hypertension. However the efficacy of nifedipine has not been established yet and the prevalence of gestational hypertension is still high (27.1 %). Indonesian herbal plants have potential to be developed as natural drugs. Molecular docking, a computational method, is very often used to depict interaction between molecules and target receptor This study was therefore to identify Indonesian herbal plants that could inhibit the calcium channel *in silico*. This was a bioinformatics study with molecular docking approach. Three-dimensional structure of human calcium channel Cav1.2 was determined by modelling with rabbit calcium channel (ID:5GJW) as template and using the SWISS MODEL software. Nifedipine was used as a standard ligand and obtained from ZINC database with the access code ZINC19594578. Active compounds of Indonesian herbal plants were registered in HerbalDB database and their molecular structure was obtained from PubChem. Binding affinity of human Cav1.2 model-ligand complexes were assesed using AutoDock Vina 1.1.2 software and visualization of molecular conformation used Chimera 1.10 and PyMol 1.3 softwares. The Lipinsky's rules of five were used to determine active compounds which fullfilled drug criteria. The human Cav1-2 model had 72.35% sequence identity with rabbit Cav1.1. Nifedipine bound to the human Cav1.2 model with -2.1 kcal/mol binding affinity and had binding sites at Gln^{1060}, Phe^{1129}, Ser^{1132}, and Ile^{1173} residues. A lower binding affinity was observed in 8 phytochemicals but only obtusifolin 2-glucoside (-2.2 kcal/mol) had similar binding sites as nifedipin did. In addition, obtusifolin 2-glucoside met the Lipinsky criteria and the molecule conformation was similar with nifedipine. From the HerbalDB database, obtusifolin 2-glucoside is found in Tectona grandis. Obtusifolin 2-glucoside computationally becomes a potensial candidate of calcium channel blocker. In vitro assays should be performed to evaluate the antagonist effect of obtusifolin 2-glucoside on calcium channel Cav1.2.

1. Introduction

Gestational hypertension is one of the three main causes of maternal death in Indonesia. Based on Indonesian Health Profile (2014), the prevalence of gestational hypertension is 21.5% in 2010 and increases by 5.6 % in 2013 [1]. In short time, untreated gestational hypertension can cause pre-eclampsia, placental abruption, and cerebrovascular complications like intravascular coagulation [2-3].

Subsequently, premature birth, low birth weight, intrauterine fetal death and mental retardation are the most common complication that is found in babies who are delivered by pregnant women with gestational hypertension [3].

Pharmacological regiments which are currently available for treatment of gestational hypertension are methyldopa and nifedipine. Methyldopa is the first line drug for gestational hypertension but this drug has many side effects [5]. Although nifedipine is the second choice for gestational hypertension, it is commonly used in Indonesia. Nifedipine blocks calcium (Cav1.1 and Cav1.2) channels which results in systemic arterial dilation, increases oxygen transport to myocardial tissue and reduces total peripheral resistance, systemic blood pressure and after load [6, 7]. Senadheera and co-workers (2013) have reported that the expression of Cav1.2 increases in uterine artery during pregnancy [8]. So, inhibition of Cav1.2 expression by nifedipine may have beneficial effects to treat gestational hypertension due to its tocolytic property and placental perfusion [2].

In some cases, extensive use of nifedipine, however, does not reduces the incidence of gestational hypertension, which leads to high morbidity and mortality rates. In addition, several side effects of nifedipine treatment has been reported including gingival hyperplasia, peripheral edema, vomitus, fatigue, constipation, bradycardia, atrioventricular block, and heart failure [9]. Therefore, it is required for development of new drugs as an alternative treatment for gestational hypertension.

Herbal plants have been used for human remedy either their natural products or derivatives for thousand years ago in the world including Indonesia [13]. Indonesia has more than 30.000 plants and around 9.600 of them are known to have pharmacological properties [14]. *Digitalis purpurea* and *Allium sativum* are identified to have anti-hypertension properties [15] but these plants have not been developed yet as an anti-hypertension drug.

Virtual screening is one of the common method for drug development which is started by searching small molecules [10]. Molecular docking is a computational method that is widely used to describe interaction between molecules and target receptor [11]. This method can certainly be used to investigate phytochemicals which are able to interact with target receptors [12]. Thus, the aim of this study was to explore Indonesian phytochemicals which can act as calcium channel blocker for development of gestational hypertension drug.

2. Methods

Three dimensional (3D) structure of Cav1.2 calcium channel, nifedipine, and Indonesian phytochemicals were used in this bioinformatics study with a molecular docking approach. Sequence target of human Cav1.2 protein was downloaded from Uniprot with ID: Q13936. Nifedipine was a standard ligand which was obtained from ZINC database with the access code ZINC19594578. Phytochemicals derived from Indonesian herbal plants should be registered in HerbalDB database and have 3D structures which were obtained from PubChem NCBI. Selected phytochemicals had to meet the *Lipinski's rule of five*. Because 3D structure of human (h) Cav1.2 protein has been not established yet, this protein structure was made by modeling with the rabbit (r) Cav1.1 channel (ID: 5GJW) using SWISS-MODEL (http://swissmodel.expasy.org/). PyRx 0.8 and AutoDock Tools 1.5.6 sofwares were used to molecularly dock phytochemicals with the hCav1.2 model while Chimera 1.10 and PyMol 1.3.softwares were used to visualize phytochemical- Cav1.2 binding complexes.

To optimize interaction between hCav1.2 model and nifedipine as the standard ligand, water molecules were removed from the hCav1.2 model and it was then added with hydrogen molecules. Molecular docking was run in the grid box position at X center = ±148.378, Y center = ±184.750 and Z center = ±174.310 to restrict area interaction between hCav1.2 model and nifedipine. Nifedipine was molecularly docked with the hCav1.2 model at least three times to get the root mean square deviation (RMSD) <2 Å as a valid conformation. Nifedipine- hCav1.2 binding complexes were then visualized using Chimera 1.10 and PyMol 1.3 and their binding sites were compared with the existing binding sites of rCav1.1 (Gln^{949}, Phe^{1008}, Ser^{1011}, and Ile^{1052}). Phytochemicals were finally docked with the hCav1.2 model and considered as calcium channel blocker when the docking score was lower or equal to the docking score of nifedipine and they similarly bound to Cav1.2 at four residues after visualization with Chimera and PyMol softwares [16].

3. Results and Discussion

3.1 Human Cav1.2 Model

In this study, we have firstly demonstrated 3D structure of hCav1.2 which was generated from rCav1.1 as calcium channel template. The superposition showed that the hCav1.2 model had 72.35% similarity with rCav1.1 (Figure 1).

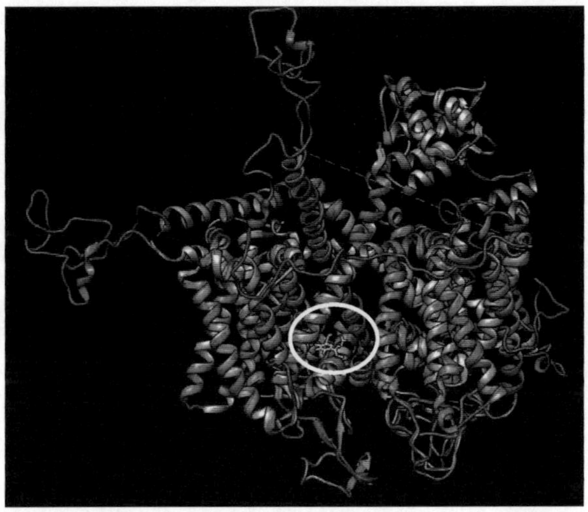

Figure 1. Superposition of 3D structure between rCav1.1 and h Cav1.2. Homology modeling was performed using SWISS-MODEL and superposition was visualized using Chimera 1.10 software. Blue ribbon indicated 3D rCav1.1 structure. Red ribbon was hCav1.2 model, yellow sticks were nifedipine structure and yellow circle was calcium channel binding site.

Figure 2 indicated sequence alignment between hCav1.2 and rCav1.1 which were comparable from residu 74 to 2.221. The binding sites of hCav1.2 were as same as binding sites of rCav1.1 and located at residues Gln[1060], Phe[1129], Ser[1132], dan Ile[1173]. The hCav1.2 model was then analyzed using homology modeling curve [17]. Our hCav1.2 model contained 2.221 amino acids with 72.35% sequence identity so that the created hCav1.2 model was located in a safe and represented the 3D hCav1.2 (Figure 3).

```
Q13936     940 FTTIFTIEIALKILGNADYVFTSIFTLEIILKMTAYGAFLHKGSFCRNYF   989
                                  .|||:||:||:||||.|||||||||||||||
5GJW       838 -------------------AFTSVFTVEIVLKMTTYGAFLHKGSFCRNYF   868

Q13936     990 NILDLLVVSVSLISFGIQSSAINVVKILRVLRVLRPLRAINRAKGLKHVV  1039
               |||||||:|||||.|::||.|:||||||||||||||||||||||||||||
5GJW       869 NILDLLVVAVSLISMGLESSTISVVKILRVLRVLRPLRAINRAKGLKHVV   918

Q13936    1040 QCVFVAIRTIGNIVIVTTLLQFMFACIGVQLFKGKLYTCSDSSKQTEAEC  1089
               ||||||||||||||:||||||||||||||||||||.::|:|.||.||.||
5GJW       919 QCVFVAIRTIGNIVLVTTLLQFMFACIGVQLFKGKFFSCNDLSKMTEEEC   968

Q13936    1090 KGNYITYKDGEVDHPIIQPRSWENSKFDFDNVLAAMMALFTVSTFEGWPE  1139
               :|.|..||||:.....::||.|.::.|.|||||:|||:|||||||||||:
5GJW       969 RGYYYVYKDGDPTQMELRPRQWIHNDFHFDNVLSAMMSLFTVSTFEGWPQ  1018

Q13936    1140 LLYRSIDSHTEDKGPIYNYRVEISIFFIIYIIIIAFFMMNIFVGFVIVTF  1189
               ||||:|||:.||.||:||.|||::||||||||:|||||||||||||||||
5GJW      1019 LLYRAIDSNEEDMGPVYNNRVEMAIFFIIYIILIAFFMMNIFVGFVIVTF  1068
```

Figure 2. Sequence alignment between hCav1.2 (Q13936) and rCav1.1 (5GJW) primary structures. The binding sites of hCav1.2 were localized at Gln^{1060}, Phe^{1129}, Ser^{1132}, dan Ile^{1173} (black boxes) and red boxes designated binding sites of rCav1.1 at Gln^{939}, Phe^{1008}, Ser^{1011}, and Ile^{1052}. Residue deletion or insertion were indicated by dash. Vertical bar designated identical amino acids and colon was amino acid with similar property whilst dot was different amino acids.

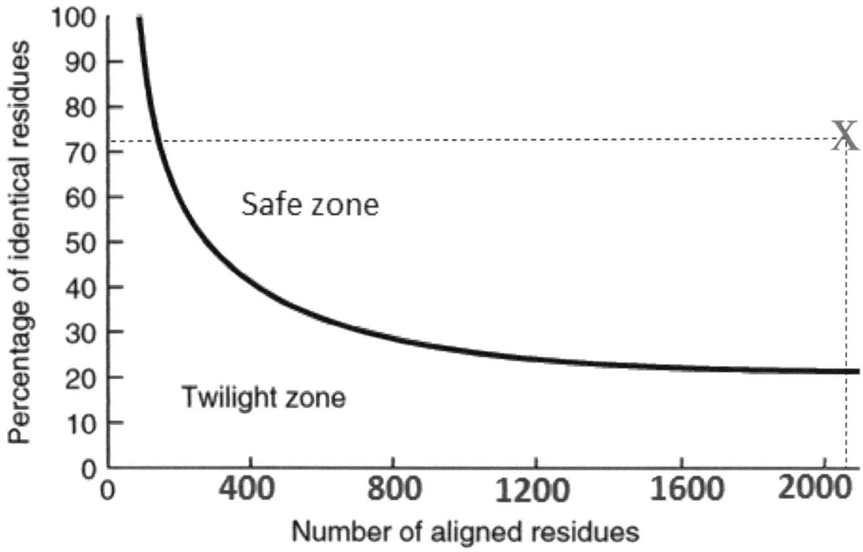

Figure 3. The homology modeling threshold (curved line) of h Cav1.2 model. The red-cross mark showed the plot of residue number of hCav1.2 model and its sequence identity.

3.2 Validation of Nifedipine-hCav1.2 Binding Complexes

After five times docking, nifedipine bound to hCav1.2 model with -2.1 kcal/mol binding energy and RMSD value = 0 (Figure 4). The localization of binding sites was the same as the binding sites in the sequence alignment. Hydrogen bond was observed at residue Gln^{1060} and Van der Waals bond was at Phe^{1129}, Ser^{1132}, and Ile^{1173}.

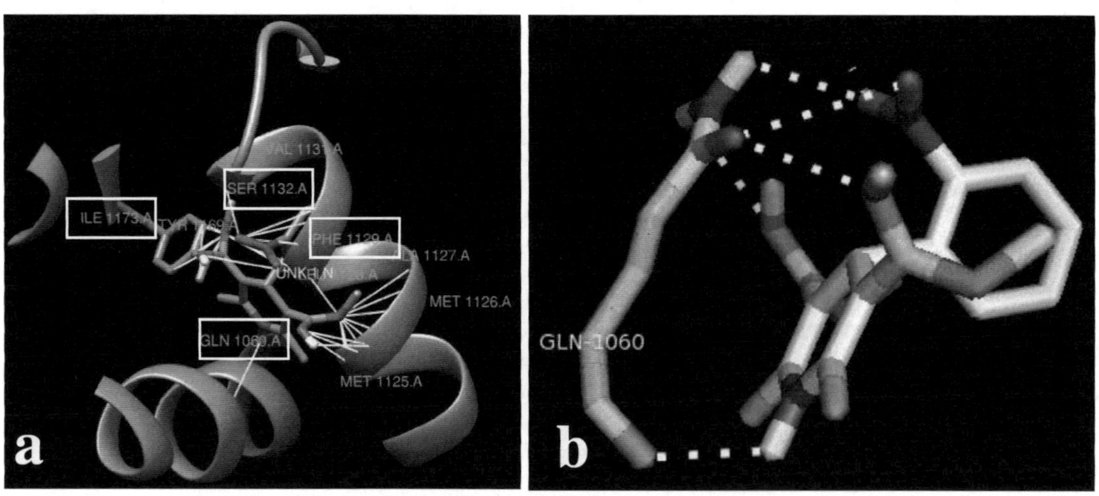

Figure 4. a) Visualization of nifedipine-hCav1.2 binding complexes with Chimera 1.10. White boxes showed interaction between nifedipine (grey bar) and hCav1.2 model (blue ribbon) at Gln^{1060}, Phe^{1129}, Ser^{1132}, and Ile^{1173} residues. Yellow lines were atom interaction between nifedipine and hCav1.2 model. b) Nifedipine bound to hCav1.2 model at Gln^{1060} which was visualized using PyMol 1.3. Green: carbon (C), red: oxygen (O), white: hydrogen, blue: nitrogen (N), yellow dashes: detailed interaction between nifedipine and hCav1.2 model at Gln^{1060} residue.

4. Docking of Indonesian Phytochemicals and hCav 1.2

A total of 6,776 phytochemicals registered in the HerbalDB database, 517 met the Lipinski's rule of five and had their 3D structures. After molecularly docked with hCav1.2 model, it revealed 66 compounds with lower binding energy, compared to nifedipine. There were only 8 phytochemicals that bound to hCav1.2 at Gln^{1060}, Phe^{1129}, Ser^{1132}, Ile^{1173} residues (Table 1). Among these phytochemicals, morindone, obtusifolin 2-glucoside, actinodaphnine, oxonantenine, and gibberellin A44 had optimal lipophilicity. Obtusifolin 2-glucoside had five H-bond donors and ten H-bond acceptors. The conformation of obtusifolin 2-glucoside fited with nifedipine (Figure 5a).This compound had Van der Waals bond at Gln^{1060}, Phe^{1129}, and Tyr^{1169} residues while the hydrogen bonds were at Gln^{1060}, Met^{1126}, Tyr^{1130}, and Ser^{1132} (Figure 5b). Additional binding sites of obtusifolin 2-glucoside to hCav1.2 was observed at Met^{1126}, Thr^{1130}, and Tyr^{1169} residues but the specific function is unknown [18].

In contrast to obtusifolin 2-glucoside, actinodaphnine had two H-bond donors and five H-bond acceptors. It had Van der Waals bond at Gln^{1060}, Phe^{1129}, and Leu^{1128} residues and hydrogen bond at Ser^{1132} residue. Oxonantenine did not have H-bond donor but had six H-bond acceptors. This compound bound to Cav1.2 at Gln^{1060}, Phe^{1129}, and Ser^{1132} residues with hydrogen bond and at Phe^{1129} residue with Van der Waals bond. Morindone had three H-bond donor and five acceptors. It interacted with hCav1.2 at Phe^{1129} residue with Van der Waals bond, Ser^{1132} residue with H-bond and Gln^{1060} with both bond types. Gibberellin A44 had two H-bond donors and five acceptors. It interacted with hCav1.2 at Phe^{1129} residue with Van der Waals bond and at Gln^{1060} and Ser^{1132} residues with hydrogen bond. However, their conformation did not fit with nifedipine.

Obtusifolin-2-glucoside which is a secondary metabolite belongs to glycoside group and is found in *Tectona grandis* (teak). Some studies has showed that this compound has antiviral effect against Chikungunya virus, antiplasmodial property, antibacterial effect, analgesics, and anti-inflamation [19-22].

IOP Conf. Series: Journal of Physics: Conf. Series **1022** (2018) 012054 doi:10.1088/1742-6596/1022/1/012054

Table 1. Binding energy, sites and Lipinski's criteria of phytochemicals and hCav1.2 model

No	Pubchem Code	Compound Name	Average Docking Score	*Binding site* (with Pymol)	Bond Type
1	**4485**	**Nifedipine**	**-2,1 ± 0,000**	**Gln-1060 Phe-1129, Ser-1132, Ile-1173**	**Hydrogen Van der Waals**
2	442756	*Morindone*	-2,2 ± 0,000	**Gln-1060, Ser-1132** Met-1125, **Phe-1129**	Hydrogen Van der Waals
3	442761	*Obtusifolin 2-glucoside*	-2,2 ± 0,000	**Gln-1060**, Met-1126, **Phe-1129, Tyr-1130, Ser-1132,**Tyr-1169	Hydrogen Van der Waals
4	72276	*(-)-Epicatechin*	-2,2 ± 0,173	**Gln-1060 Phe-1129, Ser-1132**	Hydrogen Van der Waals
5	13964005	*BR-xanthone A*	-2,3 ± 0,000	**Gln-1060,** Leu-1128**, Ser-1132** Met-1125, Phe-1129	Hydrogen Van der Waals
6	160502	*Actinodaphnine*	-2,3 ± 0,000	**Ser-1132 Gln-1060, Phe-1129**	Hydrogen Van der Waals
7	3084224	*Oxonantenine*	-2,3 ± 0,000	**Gln-1060, Ser-1132 Phe-1129**	Hydrogen Van der Waals
8	392169	*Thwaitesixanthone*	-2,3 ± 0,000	**Ile-1173 Gln-1060,** Ala-1124, Met-1126, Leu-1128	Hydrogen Van der Waals
9	46173798	*Gibberellin A44*	-2,3 ± 0,000	**Gln-1060, Ser-1132 Phe-1129**	Hydrogen Van der Waals

No	Pubchem Code	Compound Name	Molecule Formula	*Lipinski's Rule of Five*			
				Molecular weight (<500g/mol)	*H-Bond Donor* (<5)	*H-Bond Acceptor* (<10)	*Compound lipophilicity* (Log P<5)
1	**4485**	**Nifedipine**	**C17H18N2O6**	**346.339**	**1**	**7**	**2.2**
2	442756	*Morindone*	C15H10O5	270.24	3	5	3.3
3	442761	*Obtusifolin 2-glucoside*	C22H22O10	446.408	5	10	1.2
4	72276	*(-)-Epicatechin*	C15H14O6	290.271	5	6	0.4
5	13964005	*BR-xanthone A*	C23H24O6	396.439	2	6	4.8
6	160502	*Actinodaphnine*	C18H17NO4	311.337	2	5	2.4
7	3084224	*Oxonantenine*	C19H13NO5	335.315	0	6	3.3
8	392169	*Thwaitesixanthone*	C23H20O5	376.408	1	5	5
9	46173798	*Gibberellin A44*	C20H26O5	346.423	2	5	1.6

IOP Conf. Series: Journal of Physics: Conf. Series **1022** (2018) 012054 doi:10.1088/1742-6596/1022/1/012054

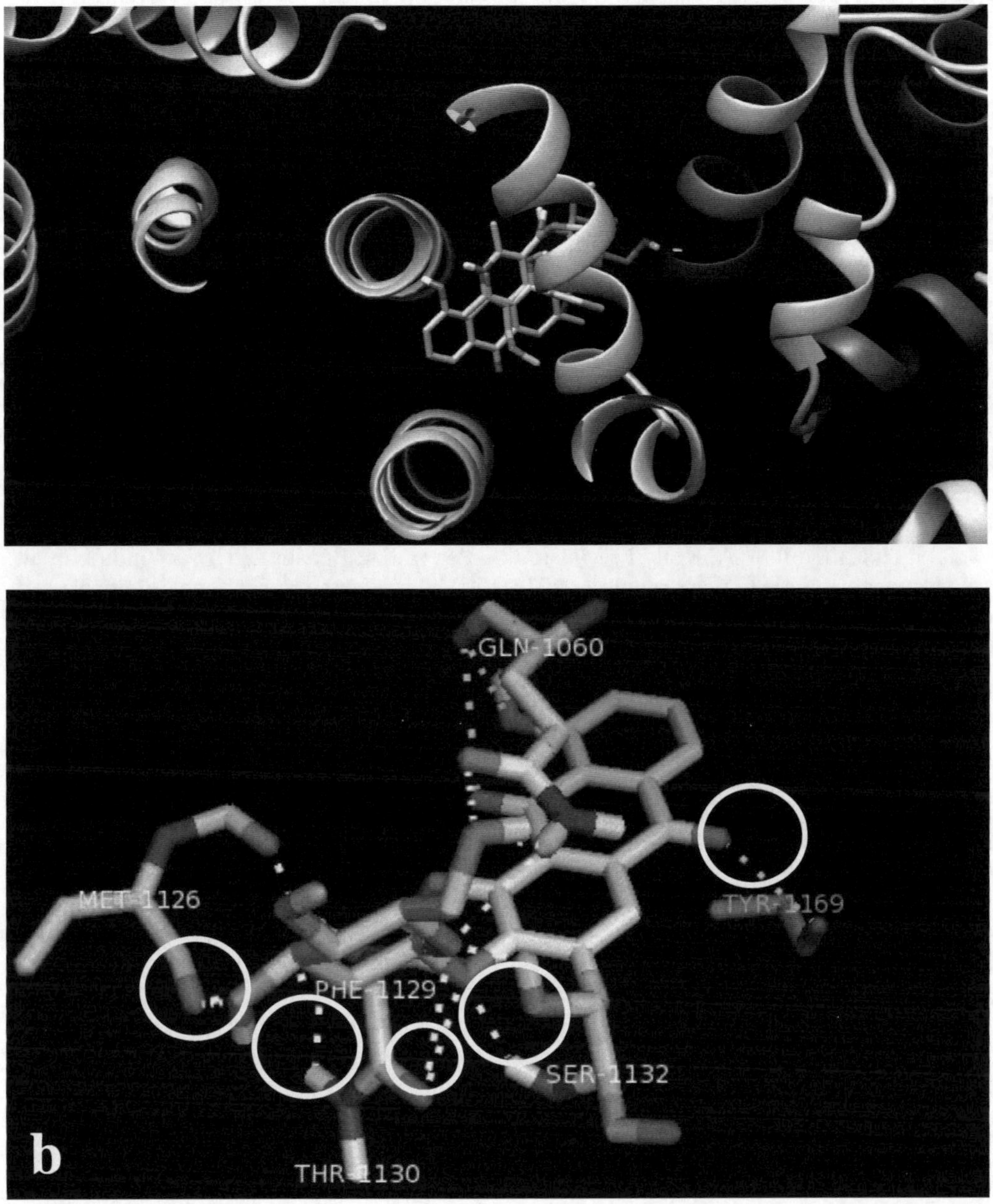

Figure 5. Visualisation of obtusifolin 2-glucoside- hCav1.2 binding complexes. (a) Superposition of obtusifolin 2-glucoside (cream-coloured) and nifedipine (purple) in hCav1.2 model with Chimera *1.10*; (b) Binding sites at Gln^{1060}, Met^{1126}, Phe^{1129}, Thr^{1130}, Ser^{1132}, Tyr^{1169} residues (yellow dashes). Green: carbon (C), red: Oxygen (O), white: Hydrogen (H), blue: Nitrogen (N), yellow dashes: interaction between atom

5. Limitation of The Study

Because biochemical properties of human Cav1.2 channel are limited, it is likely that the endogenous structure and function of calcium channels in both species differ. Secondly, AutoDockVina software is a rigid receptor-flexible ligand program which is unable to describe receptor-macromolecule interaction inside the human body since endogenous receptors have flexibility to interact with macromolecules.

6. Conclusion

Obtusifolin 2-glucoside becomes a potensial candidate of calcium channel blocker in silico. In vitro assays should be performed to evaluate the antagonist effect of obtusifolin 2-glucoside on hCav1.2 channel.

7. References

[1] Ministry of Health Republic of Indonesia 2015 Indonesian Health Profile 2014 Jakarta
[2] Moussa H N, Arian S E and Sibai B M 2015 Management of hypertensive disorders in pregnancy *Women's Health* **10**(4) 385-404
[3] Kintiraki E, Papakatsika S, Kotronis G, Goulis D G and Kotsis V 2015 Pregnancy-Induced hypertension *Hormones* **14**(2) 211-23
[4] Mudjari N S and Samsu N 2015 Management of Hypertension in Pregnancy *Acta Med. Indo. – The Indo. J. of Inte. Med.* **47**(1) 78-86
[5] Royal College of Obstetricians and Gynecologists (RCOG) 2011 Hypertension in pregnancy: the management of hypertensive disorders during pregnancy. London: Royal College of Obstetricians and Gynaecologists 111–47
[6] Sojitra R, Dungrani M, Virani P, Raj H and V S R 2015 A review on Nifedipine co-administered with Metoprolol succinate for the treatment of hypertension *Intl. J. of Adv. in Scie. Res.* **1**(3) 129-31
[7] Alexander S P H, Catterall W A, Kelly E, Marrion N, Peters J A, Benson H E, Faccenda E et al. 2015 The concise guide to pharmacology 2015/16: Voltage-gated ion channels *British Journal of Pharmacology* **172** 5904–41
[8] Senadheera S, Bertrand P P, Grayson T H, Leader L, Tare M, Murphy T V and Sandow S L 2013 Enhanced contractility in pregnancy is associated with augmented TRPC3, L-type, and T-type voltage-dependent calcium channel function in rat uterine radial artery *Amer. J. of Phys. Reg. Intg. and Comp. Phys.* **305** 917-26
[9] Vekaria A, Sheth T, Shah S and Shah M 2015 Nifedipine-induced gingival enlargement - a case report *J. of Advd. Oral Res.* **6**(3) 49-52
[10] Atanasov A G, Waltenberger B, Pferschy-Wenzig E M, Linderd T, Wawroscha C, Uhrine P, Temm V et al. 2015 Discovery and resupply of pharmacologically active plantderived natural products: A review *Biotechnol Advances* **33**(8) 1582–614
[11] Yanuar A, Mun'im A, Lagho A B A, Syahdi R R, Rahmat M and Suhartanto H 2011 Medicinal plants database and three dimensional structure of the chemical compounds from medicinal plants in Indonesia *Intl. J. Comp. Scie.* **8**(5) 180–83
[12] Shouk R, Abdou A, Shetty K, Sarkar D and Eid AH 2013 Mechanisms underlying the antihypertensive effect of garlic bioactives *Nutrition Research* **34** 106-15
[13] Hughes J P, Rees S, Kalindjian S B and Philpott K L 2011 Principles of early drug *Brit. J. of Phar.* **162** 1239-49
[14] Yanuar A 2012 Penambatan Molekular: Praktek dan Aplikasi pada Virtual Screening. Fakultas Farmasi Universitas Indonesia Jakarta.
[15] Meng XY, Zhang HX, Mezei M and Cui M 2011 Molecular docking: a powerful approach for structure-based drug discovery *Current Computer-Aided Drug Design* **7**(2) 146-57
[16] Xu L, Li D, Tao L, Yang Y, Li Y and Houa T 2016 Binding mechanisms of 1,4-dihydropyridine derivatives to L-type calcium channel Cav1.2: Molecular modeling study. *Molecular Biology System* **2**(2) 379-90
[17] Krieger E, Nabuurs S B and Vriend G 2003 Homology modeling *Stuctural Bioinformatics* **44**(25) 507-21
[18] Wu J, Yan Z, Li Z, Qian X, Lu S, Dong M, Zhou Q et al. 2016 Structure of the voltage-gated calcium channel Cav1.1 at 3.6 Å resolution *Nature* **537** 191-96
[19] Sangeetha K, Purusthothaman I, and Rajarajan S 2016 Spectral characterization, antiviral activities, in silico ADMET and molecular docking of compounds isolated from Tectona grandis to chikungunya virus *Biomedicine & Pharmacotherapy* **87** 302-10
[20] Komlaga G, Cojean S, Dickson R A, Beniddir M A, Suyyagh-Albouz S, Mensah MLK, Agyare C, Champy P and Loiseau PM 2016 Antiplasmodial activity of selected medical plants used

to treat malaria in Ghana *Parasitology Research* **115**(8) 3185–95

[21] Bitchagno G T M, Fonkeng L S, Kopa T K, Tala M F, Wabo H K, Turne C B, Tane P et al. 2015
Antibacterial activity of ethanolic extract and compounds from fruits of Tectona grandis (Verbenaceae) *BioMed Central Complementary and Alternative Medicine* **15** 265

[22] Giri S P and Varma SB 2015 Analgesic and anti-inflammatory activity of Tectona grandis Linn stem extract *J. of Bas. and Clin. Phys. and Phar.* **26**(5) 479-8

IOP Conf. Series: Journal of Physics: Conf. Series **1022** (2018) 012055 doi:10.1088/1742-6596/1022/1/012055

The influence of discovery learning model application to the higher order thinking skills student of Srijaya Negara Senior High School Palembang on the animal kingdom subject matter

F Riandari[1,a], R Susanti[2,b] and Suratmi[3,c]

[1,2,3]Department of Biology Education, Faculty of Teacher Training and Education,Universitas Sriwijaya, Jl. Raya Palembang-Prabumulih Indralaya Ogan Ilir 30662, Palembang, Indonesia

Corresponding email: [a)]feni.riandari@yahoo.com, [b)]mamahabnur@yahoo.co.id, [c)]ami_suratmi@ymail.com

Abstract. This study aimed to find out the information in concerning the influence of discovery learning model application to the higher order thinking skills at the tenth grade students of Srijaya Negara senior high school Palembang on the animal kingdom subject matter. The research method used was pre-experimental with one-group pretest-posttest design. The researchconducted at Srijaya Negara senior high school Palembang academic year 2016/2017. The population sample of this research was tenth grade students of natural science 2. Purposive sampling techniquewas applied in this research. Data was collected by(1) the written test,consist of pretest to determine the initial ability and posttest to determine higher order thinking skills of students after learning by using discovery learning models. (2) Questionnaire sheet, aimed to investigate the response of the students during the learning process by using discovery learning models. The t-test result indicated there was significant increasement of higher order thinking skills students. Thus, it can be concluded that the application of discovery learning modelhad a significant effect and increased to higher order thinking skills students of Srijaya Negara senior high school Palembang on the animal kingdom subject matter.

1. Introduction

The 2013 Curriculum is applied in developing Indonesian education to improve the quality of students' potential to face the challenges of the future [1]. This expectation can be realized by changing the mindset of students through student centered learning and scientific approach to build students' creativity and critical thinking. Students are required to be able to think critically and creatively in solving a problem to develop the capacity of thinking. Critical and creative thinking processes of students included in the higher order thinking skills [2].

The ability to think critically is higher thinking processes. Ability to think is not just memorizing the facts, but it takes an understanding and mastering of students to connect knowledge and experience in solving the problem [3-5]. Higher order thinking skills of students can be seen from cognitive test result. Related with Graduates Competency Standards of the 2013curriculum, the knowledge is gained through the remembering, understanding, applying, analysing, evaluating, and creating activities.

IOP Conf. Series: Journal of Physics: Conf. Series **1022** (2018) 012055 doi:10.1088/1742-6596/1022/1/012055

These knowledge activities refer to Bloom's Taxonomy revision by Anderson & Krathwohl (2001) [6]. Higher order thinking skillsis needed for students in preparation to solve the daily real life problems. Higher order thinking skills consist of the three top cognitive processes: analysing, evaluating, and creating [7-8].

Based on the observation and communication with biology teacher is known that higher order thinking skills of students are not optimally trained. The students are not accustomed to analyze issues so they think low level. This is indicated by the results of the pretest that 70% of students have a low level critical thinking category. Low level thinking skills include three cognitive processes: remembering, understanding, and applying. As a result, students can not apply their knowledge in the real because higher order thinking skills are not trained [5]. This is due to the implementation of learning model have not facilitatedstudents to think higher level. Regarding to that problem, learning model required that students' facilitates to improve higher order thinking skills. In relation to this problem in the Curriculum of 2013 the Ministry of Education and Culture suggests some learning model that are Project Based Learning, Problem Based Learning and Discovery Learning.

The discovery learning model is a learning which students by actively engage finding their own concepts and principles analysing process to solve their problems, so they have learnt meaningfully [9-11]. This model has six stages, such as stimulation, problem statement, the data collection, the data processing, verification and generalization [12]. At the problem statement stage the students are emphasizedto identify problems to this the students have to analyze problems solving. Furthermore, at the data collection stage and data processing, students work together in groups to gather information and process it to make hypotheses. Afterward, at the verification stage, the students prove the truth of their hypothesis. Then at the generalization stage, the students make conclusion. Thus with the model students are trained to identify problem by analysing, hypothesis making, hypothesis proving, and conclusion making. These activities are higher order thinking of cognitive process [13]. Related to this, discovery learning model has advantages in helping students to increase their cognitive processes skills, and problem solving ability [10]. After participating in the learning process by this model, the students are expected able to solve their daily real life problems with higher order thinking skills. Real life problems are the main study in biology. One of the studies in Biology is the animal kingdom. The real life problems associated with the animal kingdom such as elephantiasis, worm infections in cattle, tapeworms and hookworms as a parasite.

Based on the explanation above, this study tries to present how the effect of applying the discovery learning model to higher order thinking skills of the tenth grade Srijaya Negara senior high school Palembang students on the animal subject matter.

2. Methodology

Pre-experimental research was applied because this design is not yet a pain staking experimentation and persistence of external variables that take effect on the dependent variable [14]. The research design was one-group pretest-posttest design. This study used the measurement of initial tests before being treated and the measurement of the final test after being treated. The treatment in question of the study was the application of discovery learning model.

The study population was tenth grade students of natural science at Srijaya Negara senior high school Palembang academic year 2016/2017.The sample of this study was selected by using technique purposive sampling namely sampling technique with a certain consideration [14], in order to obtain tenth grade students of natural science 2 which includes 33 students as an experimental class.

The judgments made by researchers using the average value of the highest class as to help researchers determine the improvement in higher order thinking skills significantly by using discovery learning model.

Data collection techniques are written test form multiple-choice test with five possible answers are 20 questions that pertain to Bloom's taxonomy revision of cognitive levels to analyze, to evaluate, and to create. Higher order thinking skills of students was seen from the cognitive test result to use Bloom's Taxonomy revisions by Anderson & Krathwohl (2001). In addition, the researchers used a questionnaire sheet directly response and closed by using a Linkert scale. Questionnaires were administered in this study is the student questionnaire responses in the form of statements to evaluate the response after being given the student learning by applying the discovery learning model. The student questionnaire responses contain of 19 statements with 15 positive and 4 negative statements.

The data analysis answer scores of initial tests and test scores end of each student each added together to determine the value of the initial test and final test the students with the following formula [15] :

$$\text{Value} = \frac{score\ obtained}{score\ maximum} \text{x } 100$$

Furthermore, the test scores were grouped by ranges of values used with 80-99,9 criteria (excellent), 60-79,9 (good), 40-59,9 (enough), 20-39,9 (poor), and < 19,9 (very poor). The increase in higher order thinking skills of students after learning gain is calculated using the normalized values (n-gain). For the calculation of the normalized gain and level categories used formula of Hake [16], as follows:

gain = value end-value tests initial tests

$$\text{n-gain} = \frac{posttest\ -\ pretest}{score\ maximum\ -\ pre}$$

Increased higher order thinking skills of students categorized gain index value which is divided into three categories of Hake [16], among others high (g> 0.7), moderate (0.3 <g <0.7), and low (g <0.3).

Then to calculate or analyze an average of data questionnaire has been responded by the students is done by changing each item in terms of percent later described. Test for normality using the assistance program SPSS 22 is by test Shapiro Wilk if the data are normally distributed then using statistical parametric and vice versa, if the data were not normally distributed then using statistical parametric [14]. After that, proceed with the hypothesis test is performed using sample t test to determine the significance of the increase in higher order thinking skills by discovery learning model.

3. Results and Discussion

Students are provided the pretest before the beginning of learning process at a meeting to determine the readiness and initial ability of students, then students are given posttest after the application of the learning process with the treatment discovery learning model is completed at the last meeting. Students have two types of test value, pretest value and posttest value. The percentage increase of higher order thinking skill on pretest and posttest are presented in Figure 1.

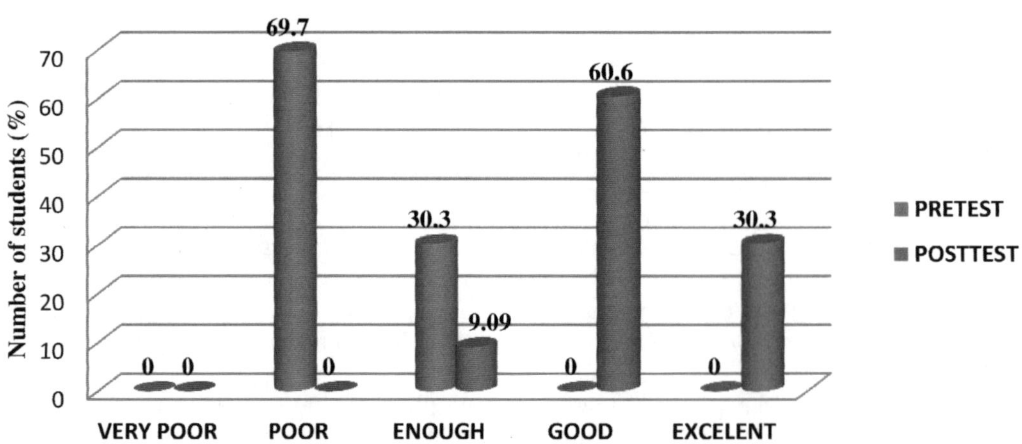

Figure 1. Number of Students (%) Based on The Category of Higher Order Thinking Skills

Based on the results on Figure 1, show that the percentage of pretest valueinhigher order thinking skills of student for " poor" and "enough"criteria as much 69.7% and 30.3%, while for "good" and "excelent" criteria is none. Then the percentage of posttest valuein higher order thinking for"enough, very good, and excelent" criteria as much 9.09%, 60.6%, and 30.3%, while for "poor" is none. So, after discovery learning model was applied in learning, higher order thinking skills of student increased for "good" criteria from 0% to 60.6% and "excellent" criteria from 0% to 30.3%. Whereas, " poor"and "enough" criteria decreased from 69.7% to 0% and from 30.3% to 9.09%.

Higher order thinking skills of students can be seen through a different test mean value of pretest and posttest that should be normally distributed. However, due to the results of the initial test and final test is not normal, then the statistical analysis used is nonparametric statistics. Different test mean value of pretest and posttest using one samplet test with SPSS22**.** The number of students, the mean value of the pretest and posttest value, distribution, and significance are shown in Table 1.

Table 1. Normality Test and Mean Value Difference Between Pretest And Posttest

N	Pretest		Posttest		*P*(sig.)
	The mean value	Distribution	The mean value	Distribution	
33	34.39	Normal	67.73	normal	0.000 (significant)

Based on the data in Table 1 show that the value of the pretest and posttest were normally distributed after the results of testing for normality using non parametric statistics. The results analysis of sample t test derive significant value of 0.000. The significant value less than the value of α (0.05). It showed that the mean value of pretest and posttest of the higher order thinking skills students have significant differences. Thus, the higher order thinking skills students have increased on the animal kingdom subject matter. The magnitude of the increase in higher order thinking skills of students can be seen by calculating the normalized gain. The scores of pretest and posttest, mean pretest and posttest value, the gain and the n-gain in Table 2 below.

Table 2. Pretest, Posttest, Gain, and N-Gain Data

Pretest		**Posttest**		**Gain**	**N-gain**
Scores	**Mean**	**Scores**	**Mean**		
6.88	34.39	13.55	67.73	33.33	0.5

The results in Table 2, explain that the mean value of pretest (34.39), posttest (67.73)and was normalized (n-gain) of0.5.It show that the increase in higher order thinking skills of students on the animal kingdom subject matter included the medium category.

In addition, the students were given a response questionnaire at the end of the lesson to determine the students' response while the learning process. The students' responses were obtained through the direct and closed response questionnaire with Linkert scale. The questionnaire consisted of 19 statements with 15 positive statements and 4 negative statements. The results of students' responses learning process are presented in Table 3 below.

IOP Conf. Series: Journal of Physics: Conf. Series **1022** (2018) 012055 doi:10.1088/1742-6596/1022/1/012055

Table 3. Response Students of Discovery Learning Model

No.	Statement	The nature of State	Response of Student SD[a]	D[b]	A[c]	SA[d]
1.	Learning by using a model of discovery learning improve my learning spirit.	Positive	2	18	11	2
2.	Learning to use the model of discovery learning makes me lazy to learn.	Negative	3	20	7	3
3.	Learning by using a model of discovery learning improve my thinking skills.	Positive		7	23	3
4.	Learning to use the model of discovery learning improve my ability to solve problems.	Positive		8	20	5
5.	Learning through discovery learning models increased my understanding.	Positive	1	14	15	3
6.	Learning to use the model of discovery learning makes my learning process becomes more meaningful.	Positive	3	17	10	3
7.	The Animal subject matter very tedious to learn.	Negative	8	21	3	1
8.	Learning by using a model of discovery learning improve my analysis.	Positive	3	7	21	2
9.	Learning to use the model of discovery learning improve my creativity.	Positive	1	10	21	2
10.	Learning to use the model of discovery learning improve my work.	Positive		13	17	3
11.	Learning to use the model of discovery learning boost my confidence.	Positive	2	11	17	3
12.	Learning to use the model of discovery learning increases my curiosity.	Positive		9	21	3
13.	Learning through discovery learning models makes me difficult to understand the subject matter.	Negative	3	17	7	6
14.	How to learn to use the model of discovery learning makes me more active in learning.	Positive	7	11	13	2
15.	Learning to use the model discovery learning increase the sense of my responsibility.	Positive	3	14	12	4
16.	Learning to use the model of discovery learning makes the subject matter easier to understand.	Positive	8	15	9	1
17.	Learning by using a model of discovery learning more fun.	Positive	6	11	14	2
18.	The Animal subject matter very interesting to learn.	Positive	2	6	17	8
19.	Learning by using a model of discovery learning makes me bored.	Negative	4	17	10	2

[a.] SD = Strongly Disagree.
[b.] D = Disagree.
[c.] A = Agree.
[d.] SA = Strongly Agree.

The results of data response questionnaire in Table 3 above then changed in percentage form, the percentage is presented in Table 4.

Table 4. The Percentage of Amount Student in The Response Questionnaire

No.	Statements	Strongly Disagree		Disagree		Agree		Strongly Agree	
		Amount students	%	Amount students	%	Amount students	%	Amount students	%
1.	Positive	38	7.68	171	34.55	241	48.69	46	9.29
2.	Negative	18	13.6	75	56.82	27	20.45	12	9.09

Based on a student responses questionnaire in Table 4 above for a positive statement obtained as much as 9.29% the percentage who responded strongly agreed (SA), 48.69% responded agree (A), while the response who disagree (D) as much as 34.55%, and the response strongly disagree (SD) as much as 7.68%. Data questionnaire responses were obtained for the total percentage of negative statements as much as 13.6% and 56.82% student response stated strongly disagree (SD) and disagree (D), while students who responded agree (A) of 20.45 %, as well as responding strongly agreed (SA) as much as 9.09%.

Based on the explanation above, the higher order thinking skills students have increased the applicationof discovery learning model on the animal kingdom subject matter. This is indicated by the differences in mean values of pretest and posttest by using discovery learning model. Increased higher order thinking skills can be seen in Table 2 shows that the mean value of posttest is higher than the mean value of pretest. During the learning process, students are given a worksheet that presents real problems in life. The problems presented are emphasize studentsable to classify animals into phyla based on body shape, the body symmetry, the body cavity, and reproduction. It requires high level thinking skills to analyze problems in resolving the issue. Student worksheets were suitable with the discovery learning syntax model.

Discovery learning model provides the chance to engage actively acquire unknown knowledge to find their own concepts and principles through scientific activities. Teachers design the students activities to using mental process activities such as observing, classifying, measuring, suspect, analysing, and conclusion making to find concepts or principles [17]. Through this model, students find something they learned themselves [18]. This model emphasizes the students do various activities such as observing, asking questions, gathering information from various sources, conduct experiments, analysing and interpret data [19]. These activities are higher order thinking of cognitive process [7]. These activities train the learners to develop thinking skills to determine the problem, evaluate the problem, make a hypothesis, test hypothesis, making inferences, and making a solution in resolving the problem. This activity is to train students to solve problems. Problem solving is done through various activities to determine the problems, evaluate them, make a hypothesis, test hypothesis, making conclusion, and creating solutions. This is supported by the research results Chinedu & Kamin [13] and the results of research Yuliani & Saragih [20] which states that the discovery learning model are trained students to solve the problem by giving problems through analysing the information to determine the problem, evaluate the problem and create a solution. Troubleshooting is also building knowledge and skills of students from their own sources of knowledge gained [21]. This model has six stages, among others stimulation, problem statement, the data collection, the data processing, verification and generalization [12].

Stimulation stage, teachers provide stimulus by reading, questions, pictures, and videos of the issues presented. These stimulus desire to bring learners to find out the problems, so the students get the learning experience in observing the conceptual knowledge by reading and watching a situation or viewing pictures. Furthermore, at the identification problems stage students are required to identify problem by analysing and evaluating problems. In this activity, students are given the experience to ask questions, seek information and formulate the problem. This stage train students to resolve the problem by finding the concepts of the lessons. Then in the data collection stage students are given experience to find and collect data / information.It can be used to find the problems solutions. This activity will also train them work together in groups to process information and to solve problems solutions. The data processing activities will train students to try and explore the competence of their conceptual knowledge to apply in the real life. At the verification stage directing students prove the truth data processing byactivities in the preceding stage such as asking questions to classmates, discuss, when seeking information from relevant sources, and to connect itmaking conclusion. The conclusion last stage led students making conclusions from the problems presented. Students can also decide the facts obtained from the hypothesis then making conclusions [22].

In connection with this, the students are able to think critically to find theirown knowledge of the concepts and principles by discovery learning model to solve the problems. That is, students are able higher order thinking to classify animals into phyla based on body shape, the body symmetry, the body cavity, and reproduction by using discovery learning model. Other than that, the students become more active and mutual cooperation among individuals to find knowledge related to the problems. It is also supported by the results data analysis of questionnaire responses showed that the discovery learning model bring up self-confidence, and high curiosity. In addition, students can improve the ability to think critically, analysis, and creativity in solving the problems. This shows that the discovery learning model increase high order thinking skills. In line with the research results Tondang & Sahyar [23] and Yuliani & Saragih [20] shows that the model of discovery learning increase to higher order thinking skills.

4. Conclusions

Based on the research and discussion above, it can be concluded that the discovery learning model increase the high order thinking skills of students on the animal kingdom subject matter. The increase in higher order thinking skills of students on the animalkingdom subject matter included in the medium category.

The model of discovery learning is able to bring up a positive attitude of students as it builds knowledge, self-confidence, cooperation between individuals, and a high curiosity.

5. References

[1] Kemendikbud 2014 *Materi pelatihan guru implementasi kurikulum 2013* (Jakarta: Kementerian Pendidikan dan Kebudayaan)

[2] Johnson E B 2014 *Contextual teaching and learning: menjadikan kegiatan belajar mengajar mengasyikkan dan bermakna* Translated by Ibnu Setiawan (Bandung: Kaifa)

[3] Rofiah E, Nonoh S A and Elvin Y E 2013 Penyusunan instrument tes kemampuan berpikir tingkat tinggi fisika pada siswa SMP *Jurnal Pendidikan Fisika* **1**(2) 17

[4] Istiyono E, Mardapi D and Suparno 2014 Pengembangan tes kemampuan berpikir tingkat tinggi fisika (pysthots) siswa SMA *J. Penel. dan Eval. Pend.* **18**(1) 1-12

[5] Saputra H 2016 *Pengembangan mutu pendidikan menuju era global penguatan mutu pembelajaran dengan penerapan HOTS (High Order Thinking Skills)* (Bandung: SMILE's Publishing)

[6] Kemendikbud 2016 *Permendikbud 22 tahun 2016 tentang standar proses pendidikan dasar dan menengah* (Jakarta: Kementrian Pendidikan dan Kebudayaan)

[7] Anderson L W and Krathwohl D R 2015 *Kerangka landasan untuk pembelajaran, pengajaran, dan asesmen revisi taksonomi pendidikan bloom* Translated by A. Prihantoro (Yogyakarta: Pustaka Pelajar)

[8] Wilson L O 2016 A succinct discussion of the revisions to bloom's classic cognitive taxonomy by lorin anderson and david krathwohl and how to use them effective

[9] Ilahi M T 2012 *Pembelajaran discovery strategy & mental vocational skill* (Yogyakarta: Diva Press)

[10] Hosnan M 2014 *Pendekatan saintifik dan kontekstual dalam pembelajaran abad 21* (Bogor:

Ghalia Indonesia)

[11] Nurdin S and Adriantoni 2016 *Kurikulum dan pembelajaran* (Jakarta: PT Raja Grafindo Persada)

[12] Syah M 2016 *Psikologi pendidikan dengan pendekatan baru* (Bandung: PT Remaja Rosdakarya)

[13] Chinedu C C and Kamin Y 2015 Strategies for improving higher order thinking skills in teaching and learning of design and technology education *J. of Techni. Edu.Train.* **7** 2229-8932 pp 35-43

[14] Sugiyono 2015 *Metode peneltian pendidikan pendekatan kuantitatif, kualitatif, dan r&d.* (Bandung: Alfabeta)

[15] Sudijono A 2013 *Pengantar evaluasi pendidikan* (Jakarta: Raja Grafindo Persada)

[16] Hake R R 1998 Interactive engagement versus traditional methods: A six-thousand-student survey of mechanics test data for introductory physics courses *Americ. J. of Phys.* **66** pp 164-74

[17] Ulumi F D et al 2015 The influence of guided discovery learning model on biology result study at SMA N 2 Sukoharjo academic year 2013/2014 *J Bio Edu* **7** pp 68-79

[18] In'am A and Hajar S 2017 Learning Geometry through Discovery Learning Using a Scientific Approach *Intl. J. of Instr.* **7** 1694-609X pp 55-70

[19] Nuryakin and Riandi 2017 Improving middle school students' critical thinking skills through reading infusion-loaded discovery learning model in the science instruction *J.of Phys.: Conf. Series* **812** 012003

[20] Yuliani K and Saragih S 2015 The development of learning devices based guided discovery model to improve understanding concept and Critical thinking Mathematically ability of students at Islamic Junior High School of Medan *J. of Educ. and Pract.* **6** 2222-288X pp 116-128

[21] Marzoan *et al* 2016 Learning styles, learning strategies and learning outcomes of science in primary school *Academic Research International* **7** 2223-9944 pp 183-196

[22] Haris F et al 2015 The influence of guided discovery learning model toward critical thinking skills of x grade students at SMA Negeri Karangpandan in academic year 2013/2014 *J Bio Edu* **7** pp114-122

[23] Tondang K N and Sahyar 2016 The effect of discovery learning model toward student higher order thinking skills in dynamic electricity subject matter at SMA Raksana Medan Academic year 2014/2015 *J. Inpafi* **4** pp 54-61

Acknowledgement

The authors recognize that many of those involved in helping to resolve this journal. Therefore, on this occasion the authors thank profusely to the supervisor who has guided me in this study. I would also like to thank the principal, a biology teacher, and the tenth grade of natural science 2 Srijaya Negara senior high school Palembang students.

IOP Conf. Series: Journal of Physics: Conf. Series **1022** (2018) 012056 doi:10.1088/1742-6596/1022/1/012056

Misconception of biology education student of teacher training and education of Sriwijaya University to the concept of photosynthesis and respiration

Rahmi Susanti

Department of Biology Education Sriwijaya University,
Jln. Raya Palembang-Prabumulih Indralaya, Km 32 Ogan Ilir – Indonesia

Corresponding email: mamahabnur@yahoo.co.id

Abstract. This study aimed to gain an overview of misconceptions on the concept of photosynthesis and respiration. The study involved 58 students from Biology Education of Sriwijaya University. Collecting data used written test of 16 questions, which are 10 questions of multiple choice and 6 of choice with reason. The results showed that:photosynthesis occurs continuously (37.9%), energy used for photosynthesis are light and heat energy (34.5%), plants take $CO_2$to respiration (47%), plants carry on respiration in the absence of light for photosynthesis (22.4%), respiration in plants occurs only in leaf cells (76.4%),and only animals that take O_2 of photosynthesis to respiration (68.9%). The conclusion: 1) on the concept of photosynthesis is still prevailing misconceptions about the concept of the place and time of the occurrence of photosynthesis in plants, the role of the sun in photosynthesis, energy is required in the form of photosynthesis, and the role of photosynthesis for the plant. 2) on the concept of respiration is still prevailing misconceptions about the place of the respiration in plants, gas necessary for respiration of plants, and the plants perform respiration time, as well as the cycle of CO_2 and O_2 that occurs in nature.

1. Introduction

Study on misconceptions in science has often been studied by many experts. Misconceptions can be interpreted as a mismatch someone description of a concept. This discrepancy refers to the scientific understanding or definition accepted by experts. Some research indicates that there are a lot of misconceptions that happen to students, ranging from primary school level to the college level in the various fields of study. Some research suggests a common misconception among students, among others misconceptions on the concept of photosynthesis, respiration, diffusion, osmosis, ecology, genetics, classification, and human circulation system [1]; [2]; [3]; [4]; [5]; [6];[7]. Most learners experiencing misconceptions on the concept of diffusion and osmosis [8]. Results of a study involving 200 high school students and college students, showed that more than 60% of high school students and college students have misconceptions on the concept of diffusion [9]. In addition to the concept of diffusion and osmosis which are prevalent misconceptions, another concept also undergone many misconceptions is the concept of photosynthesis.

Photosynthesis is a process that is very important in life. Photosynthesis provided food for almost the entire life of the world, either directly or indirectly [10]. Food gained by living things, will be used as a source of energy to do all the activities of life. This energy can be produced by living things by way of oxidizing the food with the use of oxygen through the process of respiration. Photosynthesis not only provide food, but also provide the oxygen for respiration or breathing all living creatures on earth. Thus the process of photosynthesis and respiration are two processes that are essential for life. Once the importance of these two processes for life, so that it can be said that there is no life if there is no process of photosynthesis and respiration [11].

Research on the misconception that specifically examine in depth the concept of photosynthesis and respiration have been done [12]; [13]; [14]; [15]. Things to cause difficulties in understanding the concept of photosynthesis, and occurs misconceptions because this concept is abstract and complex, and involves many chemical reactions [16]; [17]. The results showed that 99% of students do not understand the role of chlorophyll in photosynthesis and 46% of students do not understand the changes of light energy into chemical energy [18].

The process of photosynthesis is closely related to respiration. CO_2gas that results respiration of living organisms (plants and animals) will be used by plants for photosynthesis. O_2gas that results of photosynthesis in plants used by living things (plants and animals) for respiration. Given the importance of the concept of photosynthesis and respiration, as well as common misconceptions on both the concept, then in this paper will be presented the results of a descriptive study of misconceptions on photosynthesis and respiration in biology education student.

2. Research Method

This study is a descriptive study that revealed the understanding of concepts and misconceptions biology student teachers about the basic concept of photosynthesis and respiration. Subjects were first-year students of Biology Education of the Faculty Teacher Training and Education University of Sriwijaya consisting of 58 people. Collecting data using diagnostic tests that refers to Haslam and Treagust [15]. The test consists of 16 questions, which are divided into two groups, 10 questions multiple choice test on photosynthesis, and 6 a matter of choice on the grounds of photosynthesis and respiration. Research data in the form of qualitative and quantitative data were processed and presented descriptively.

3. Results and Discussion

Questions posed to gain knowledge about the concept of photosynthesis, consisting of 10 questions in the form of multiple choice, namely questions related to: 1) the site of photosynthesis, 2) the role of solar energy in photosynthesis, and 3) the compound is necessary and produced of photosynthesis. Two questions of 6 questions with answers reason is related to the concept of respiration which occurs in plants. The results of the analysis of student answers to questions about the site of photosynthesis is presented in Table 1.

Table 1. Percentage of Student Answers to Questions About The Site of Photosynthesis

No	Question	Answer	%
1	In which part of the plant photosynthesis to occur?	Photosynthesis occurs in all green plant parts	67.2
		occurs in leaf photosynthesis	32.8
2	Which parts of plants containing chlorophyll?	Chlorophyll contained in the green part of the plant	68.9
		contained chlorophyll in the leaves	31.1
3	In which part of the plant photosynthesis to occur, and where there is chlorophyll?	Photosynthesis occurs in the green part of the plant and chlorophyll contained in the green section	60.4
		Photosynthesis occurs in the leaves and leaf chlorophyll contained	39.6

IOP Conf. Series: Journal of Physics: Conf. Series **1022** (2018) 012056 doi:10.1088/1742-6596/1022/1/012056

Based on analysis of student answers to questions about the site of photosynthesis and the presence of chlorophyll in plants can be explained that most students (>60%) have understood that photosynthesis occurs in the green part of the plant. Their green color is caused by the presence of chlorophyll. Nevertheless, it is still large enough that 39.6% occurred on student misconceptions, which occurs in leaf photosynthesis and chlorophyll contained in leaves. The concept of photosynthesis occurs in leaves already owned by students from an elementary school level. Results of previous studies show that 69% of elementary school students stated that photosynthesis occurs in the leaves, the rest in stomata, root and leaf stalk. Further results showed there were 22% of primary school students expressed photosynthesis occurs in the spines of cactus plants, because they assume cactus spines are leaves [19]. This occurs because the green leaves caused by the presence of chlorophyll, whereas chlorophyll plays an important role in photosynthesis. Thus the students chose the answer.

The concept of the right of where photosynthesis and the presence of chlorophyll in plants are all green parts of the plant contain chloroplasts in their cells. However, most plant organs contain chloroplasts of leaves. Chlorophyll contained in the chloroplasts of green that plays an important role in the process of photosynthesis. Therefore, it is often stated that a key site of photosynthesis in plants is the leaves [10]. However, in some plants whose leaves are modified, for example, be a thorn (cactus) or the stem is green, the process of photosynthesis does not occur in the leaves, but it occurs in the trunk. The results of the analysis of student answers to questions about the role of solar energy in plants is presented in Table 2.

Table 2. Percentage of Students Answers to Questions About The Role Of Solar Energy in Photosynthesis

No	Question	Answer	%
1	What is the role of the sun in photosynthesis	Sun provides energy for the plant.	100
2	When did this to occur	Photosynthesis occurs at noon.	32.8
		Photosynthesis occurs when there is light.	34.4
		Photosynthesis occurs continuously (day and night).	32.8
3	Time photosynthesis and the role of the sun in photosynthesis	sun provides energy for plants perform photosynthesis and photosynthesis take place during the daytime.	25.9
		The sun provides the energy for photosynthesis and plants perform photosynthesis occurs when there is light.	36.2
		The sun provides energy for plants perform photosynthesis and photosynthesis take place continuously (day and night).	37.9
4	Forms of energy needed for photosynthesis	Energy sunlight	65.5
		Energy sunlight and heat energy	34.5

Based on the analysis presented in Table 2 can be explained that 100% of the students answered questions about the role of the sun is to provide energy for photosynthesis. However, when combined with questions about the timing of photosynthesis, the energy spread evenly student answers to three different answers that photosynthesis occurs during the day (25.9%), photosynthesis occurs when there is no light (36.2%), and photosynthesis occurs continuously constantly day and night (37.9%). The shape of the energy needed for photosynthesis is the solar light energy and heat energy is answered by the students as much as 34.5%. Misconception that happens is photosynthesis occurs continuously day and night. The right thing is photosynthesis occurs only when there are sufficient light. The

IOP Conf. Series: Journal of Physics: Conf. Series **1022** (2018) 012056 doi:10.1088/1742-6596/1022/1/012056

misconception is also the case that plants require heat energy for photosynthesis. The energy used by plants for photosynthesis is light energy instead of heat energy.

The results of the analysis to the question of the necessary compounds produced by photosynthesis and are presented in Table 3.

Table 3. Percentage of Student Answers to The Question of The Necessary Compounds and Generated Photosynthesis

No	Question	Answer	%
1	Pair of compounds which is needed for photosynthesis?	CO_2 and water	92.8
		O_2 and Water	3.2
		CO_2 and oxygen	4.0
2	Pairs where the resulting compound of the photosynthetic	Compounds produced from photosynthesis is glucose and O_2	94.5
		Compound is produced from photosynthesis is glucose and CO_2	2.9
		The compound resulting from photosynthesis are CO_2 and water	2.6
3	Pairs of compounds needed in photosynthesis, and the pair compound resulting fromphotosynthesis	The process of photosynthesis plants require CO_2 and water, and produce glucose and O_2	93.8
		in the process of photosynthesis in plants need glucose and O_2, produces glucose and CO_2	6.2

Based on the analysis presented in Table 3 can be expressed energy 100% of the students already understand the concept of compound needed and the compounds produced in photosynthesis. The process of photosynthesis in plants require gas CO_2 and water, and produce glucose and O_2gas. The results of the analysis of student answers to questions about the role, factors affecting, and the equation of photosynthesis is presented in Table 4.

Table 4. Percentage of Answers to Students About The Role Of Photosynthesis, and The Factors Affecting The Photosynthesis Reaction

No	Question	Answer	%
1	What is the most important role for plants green to do photosynthesis?	Convert light energy into chemical energy, because during photosynthesis solar light energy is converted and stored in a glucose molecule	29.3
		Moving the CO_2 from the air, because the CO_2 taken up by the leaves through the stomata during photosynthesis	25.7
		Generating energy for photosynthesis to provide energy for plant growth	45
2	Factors what is not important in the process of photosynthesis?	Factors that are not important in photosynthesis is the amount of O_2 because O_2 is not required for photosynthesis, O_2 is the result of photosynthesis.	100
3	photosynthesis equation right?	Photosynthesis correct equation is: $CO_2 + H_2O \xrightarrow[\text{Energy of sunlight}]{\text{chlorophyll}} \text{glucose} + O_2$ Reason: The energy of sunlight is used plants containing chlorophyll to combine CO_2 and H_2O to form glucose and O_2	82.7
		$CO_2 + H_2O \xrightarrow[\text{Energy of sunlight}]{\text{chlorophyll}} \text{glucose} + O_2$ Chlorophyll joining the CO_2 in the presence of light and produce glucose and water.	17.3

IOP Conf. Series: Journal of Physics: Conf. Series **1022** (2018) 012056 doi:10.1088/1742-6596/1022/1/012056

Based on the results presented in Table 4 can be explained that as much as 45% of the students still have the misconception that an important role for plant photosynthesis is to produce energy. Another misconception is happening is the absence of light, chlorophyll joined with CO_2 to form glucose and O_2 (17.3%). The process of photosynthesis in plants do not produce energy, but photosynthesis is the process of converting light energy into chemical energy stored in chemical bonds form sugars that are made from carbon dioxide and water [10].Analysis of the results of student answers on respiration in plants and its relation to photosynthesis are presented in Table 5.

Table 5. Percentage of Student Answers on Respiration in Plants and Its Relation to Photosynthesis

No	Question	Answer	%
1	Gas what more taken by greenery when the sun's light energy is less ? (CO_2 or O_2),provide a reason?	CO_2 gas, because gas is used by green plants for photosynthesis happens all the time.	13.8
		Gas CO_2, because this gas is used by green plants for respiration which occurs when light is less	46.6
		Gas O_2, because this gas is used in respiration which only occurs in green plants, when there is no light for photosynthesis	22.4
		Gas O_2,because the gas is used for respiration occurs continuously in green plants.	17.2
2	Where is the place of the respiration in plants? (only in the root cells, every cell of the plant, or only in leaf cells), provide a reason.	Respiration occurs in every cell of the plant, because of all the living cells need energy to live	23.6
		respiration occurs only in cells of the leaf, because only the leaves that have a special pores (stomata) for gas exchange.	76, 4
3	Figure out which is right for the show cycle CO_2 and O_2 in the wild? Give a reason	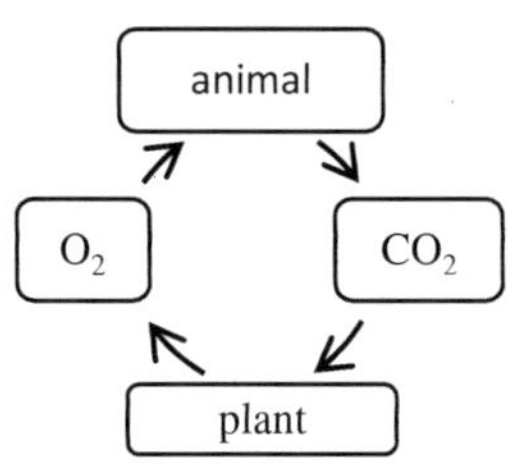 O_2 results from photosynthesis of plants used by animals to breathe, and CO_2 results from animal breathing taken by plants for photosynthesis.	68.9
		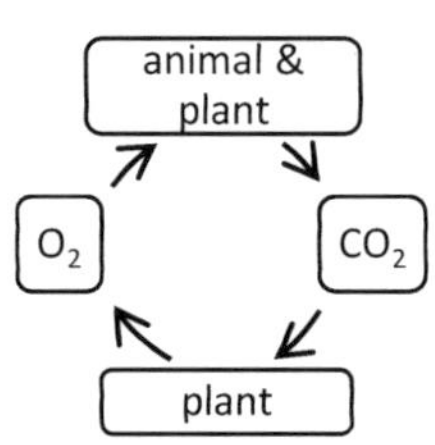 O_2 results from photosynthesis of plants used by animals and plants to breathe, and CO_2 results from respiration of animals and plants taken plants for photosynthesis.	31.1

Based on the analysis presented in Table 5 can be explained that 46.6% of the students stated gas is needed in plant respiration is gas CO_2. Misconception that occurs is respiration in plants requireCO_2.Misconceptions also happens to students that respiration in plants occurs only when there is no light (22.4%). This shows that when there is no light, and plants that are doing photosynthesis does not perform respiration. 76.4% of students answered that respiration only occurs in leaf cells, since only the leaves that have a special pores (stomata) for gas exchange. 68.9% of students answered that O_2 gas resulting photosynthesis of plants used by the animal to breathe, and CO_2 resulting respiration of animals taken by plants for photosynthesis. Misconception that happens is that the only animals that require O_2 gas for respiration.

The concept that is specifically related to the results of the analysis in Table 5 is the O_2 gas produced from the photosynthesis of plants used for respiration not only by animals but also used by plants. Green plants are living things. Therefore, plants must perform respiration to produce energy (in the form of ATP) needed in the metabolic processes that occur in all living cells. Respiration in plants occurs in all living cells, and occurs continuously day and night.

4. Conclusions

Based on our research and discussions described above can be summarized as follows. On the concept of photosynthesis is still prevailing misconceptions about the concept of the place and time of the occurrence of photosynthesis in plants, the role of the sun in photosynthesis, energy is required in the form of photosynthesis, and the role of photosynthesis for the plant. On the concept of respiration is still prevailing misconceptions about the place of the respiration in plants, gas necessary for respiration of plants, and the plants perform respiration time, as well as the cycle of CO_2 and O_2 that occurs in nature.

5. References

[1] Boo H K 2007 Primary science assessment item setter's misconceptions concerning biological science concept *Asia-Pacipic Forum on Learning and Teaching* **8**1

[2] Tekkaya C 2003 Remediating high school students misconceptions concerningdiffusions and osmosis through concept mapping and conceptual change text *Res.in Sci.Tech.Educ* **21**

[3] Sanger M J, Brecheisen D M and Hynek B M 2001 Can computer animations affect college biology students' conceptions about diffusion and osmosis? *The Americ. Biol. Teach.* **63**(2) 104-09

[4] Odom A L and Barrow L H 1999 The development and application of a two-tier diagnostic test measuring college biology students' understanding of diffusion and -osmosis after a course of instruction *J. of Res. in Sci. Teach.* **32**45-61

[5] Odom L 1995 Secondary and college biology students' misconceptions about diffusion and osmosis *The Am. Biol.Teac.* **57**(7) 409-415

[6] Zuckerman T T 1994 "Problem Solvers" conception about osmosis *The Am. Biol. Teac.* **56** 22-25

[7] Friedler Y Amir R and Tamir P 1987 High students' difficulties in understanding osmosis *Int. J.Sci. Edu.* **9** 541-51

[8] Marek E 1986 Understanding and misunderstanding of biology concepts *The Am. Biol .Teac.* **48**(1) 37-40

[9] Westbrook S and Marek E A 1991 A cross-age study of student understanding of the concept of diffusion *J. of Resin Sci. Teac.* **28** 649-60

[10] Campbell N A, Reece J B, Taylor M R, Simon E J and Dickey J 2009 *Biology: Concept and Connections.* Sixth Edition. San Francisco: Pearson Benjamin Cummings

[11] Susanti R 2013 Effect of application of problem based learning in practice to improve photosynthesis and respiration generic capability science. *Papers.* Presented at the Seminar Functional hike to Associate Professor at the Level Faculty- on May 8, 2013

[12] Anderson C W, Sheldon T H and Dubay J 1990 The effect of instruction on college non majors'conception of respiration and photosynthesis *J. of Res.in Sci.*Teac. **27**(8) 761-76

[13] Amir R and Tamir 1994 In depth analysis of misconception as a basis for developing research based remedial instruction: The case of photosynthesis *The Am. Biol .Teac.* **56**(2) 94-100

[14] Eisen Y and Stavy R 1988 Students' understanding of photosynthesis *The Am. Biol. Teac.* **50**(4) 208-12
[15] Haslam F and Treagust D F 1980 Diagnosing secondary students "misconception of photosynthesis and respiration in plants using a two-tier instrument multiple choice *J. Biol. Edu.* **21**(3) 203-10
[16] Barker M and Carr M 1989 Teaching and learning about photosynthesis. part 1: an assessment in terms of student prior knowledge *Int .J. Sci. Edu.* **11** (1) 49-56
[17] Susanti R, Nuryani Y R and Sri R 2010 Profile material difficulty level of plant physiology according to prospective biology teacher *Prooced. of the 4th Inter. Seminar on Sci. Edu.* ISBN: 978-979-99232-3-3, 30 October 2010, the Science Education Program School of Postgraduate Studies, Indonesia University of Education, Bandung, West Java, Indonesia
[18] Simpson M and Arnold B 1982 Availibility of prerequisite concepts for learning biology at the certificate level *J. Bio. Edu.* **16**(1) 65-72
[19] Susanti R 2009 Early knowledge of photosynthesis: Descriptive study on school elementary s students *Proceedings of the National Seminar on Biology: Biology and Biology Education Innovation in Human Resources Development.* ISBN: 987-602-95207-0-5, July 15 to 16, 2009. UPI Bandung, Indonesia

IOP Conf. Series: Journal of Physics: Conf. Series **1022** (2018) 012057 doi:10.1088/1742-6596/1022/1/012057

Analysis of high school students' environmental literacy

R A K Wardani[1,a], **P Karyanto**[2,b] **and M Ramli**[3,c]

[1,2,3]Department of Biology Education, Faculty of Teacher Training and Education Universitas Sebelas Maret, Jl. Ir. Sutami No. 36 A Kentingan, Surakarta, Indonesia

Corresponding email: [a)]rikaastutikw@student.uns.ac.id, [b)]puguhkaryanto@staff.uns.ac.id, [c)]mramlim@staff.uns.ac.id

Abstract. The student's environmental literacy (EL) is a vital component to improve the awareness of student on environmental issues. This research aims to measure and analyse the EL of high school students, and how the topic of environment has been taught in high school. The research was conducted in February to April 2017. The EL was measured on three aspects, i.e. knowledge, attitude and concern. The participants were sixty-five (21 boys, 44 girls) purposively selected from students of grade X, XI and XII of one Senior High School in Karanganyar Regency, Indonesia. The knowledge of students on concepts of environmental issues was tested by fourteen main questions followed by supported questions. The result showed that 80% of students were classified as inadequate category. The attitude of students was measured by New Ecological Paradigm (NEP) consisted of fifteen items, and students' average score was 46.42 (medium). The concern was measured by fifteen statements about environment, and it was ranged from 2.58 to 4.18. EL of students may low due to students' lack understanding of the environment concepts, the limited theories and concepts transferred to students, inappropriate lesson plan to meet the EL components.

1. Introduction

Environmental issues, such as deforestation, forest degradation, water pollution and air pollution are the world-wide problem that causes serious impacts on life. Deforestation has threatened three continents in the world; Africa, America and Asia. Deforestation has led to the loss of biodiversity and soil degradation [1]. Rapid population growth and infrastructures development in the coastal area of Mumbai resulted in water contamination. The water of Mumbai Beach has been contaminated by anthropogenic heavy metals that caused adverse effects on human health, biodiversity, marine ecosystem productivity, and aesthetic [2]. Air pollution has been affected nine out of ten people living in urban areas around the world [3]. Exposure to air pollution in the form of ozone and NO_2 increases risk of death, asthma, lung cancer and respiratory infections.

Humans and the environment are closely linked and influencing one another. Either directly or indirectly, human activity has impact to sustainable living [4]. The declining of environmental quality was influenced by human excessive exploitation of natural resources without regard to the environmental carrying capacity and ecological functions [5]. Management of the natural resources should be done wisely, appropriately and sustained so that the continuity of living beings and preservation of the ecosystems can be maintained [6].

Education has an important role in addressing the complexity of environmental issues [7] [8]. Environmental education is an effort to manage the environment through a formal approach by

IOP Conf. Series: Journal of Physics: Conf. Series **1022** (2018) 012057 doi:10.1088/1742-6596/1022/1/012057

providing insight knowledge to students [9]. Environmental education is a commitment of the international government and community and implemented through the school curriculum.

Based on the results of the 1997 UNESCO's Tbilisi Declaration, environmental education aims to foster learners' environmental awareness and ability to provide solutions to the environmental problems. Environmental education attempts to involve learners in imparting knowledge to change beliefs, attitudes and behaviors towards the environment [10]. One of the ultimate goals of environmental education is to increase students' EL [11]. The implementation of environmental education integrates EL on it, so that the students understand the environmental and natural relationships with human being [12].

EL is the ability to care for the environment and ability to solve environmental problems [13] [14]. EL, affirmed by the Environmental Education and Training Partnership is the ability to overcome the existing environmental problems and avoid the emergence of the new problems. The concept of EL is emphasized on three aspects; nature, environmental issues, and suitable and sustainable solutions to existing problems [15].

EL was measured based on three aspects, i.e. knowledge, attitude and concern [16]. Knowledge is measured using fourteen multiple choice questions. Meanwhile, attitude was measured using a New Ecological Paradigm scale (NEP) consisted by fifteen items of statements with a likert scale. Concern was measured using fifteen statements with a Likert scale.

The questions of knowledge based on indicators of lesson plan, especially in the chapter of environmental changes and waste recycle. The fourteen questions have been tested for validity and reliability.

The EL of attitude was measured using NEP Scale. NEP is new paradigm in ecology, first NEP stands for New Environmental Paradigm. The conversion of New Environmental Paradigm to New Ecological Paradigm as a confirmation of ecological basis for society. The statements of NEP are air pollution, water pollution, loss of aesthetic value and conservation of natural resources. The NEP has been revised, so statements in the next article are environmental issues, such as waste, ozone depletion, deforestation, biodiversity extinction, global climate changes and global warming. The NEP was consisted by five components, each containing three statements, namely limits to growth, balance of nature, anti-anthropocentrism, anti-exemptionalism, and eco crisis [17]. Limits to growth is the belief that the earth's limitation to provide natural resources. Limits to growth was listed number 1, 6 and 11. Balance of nature is the belief that humans activity affects natural resources. Balance of nature was listed number 3, 8 and 13. Anti-anthropocentrism is the belief that humans may could change and control the environment. Anti-anthropocentrism was listed number 2, 7 and 12. Anti-exemptionalism is the belief that humans were responsible to preserve the nature. Anti-exemptionalism was listed number 4, 9 and 14. Eco crisis is the belief that humans cause environmental damage. Eco crisis was listed number 5, 10 and 15. The statements of concern about environment, especially perception and action on environmental issues. The fifteen statements have been tested for validity and reliability.

Students' EL can be seen through a learning process of biology in the classroom, especially in the chapter of environmental changes and waste recycle. Environmental education in classroom should actively involve teachers to build students' knowledge about the environment [12]. The teachers should understand the characteristics of the topic of environment by using contextual learning model, provides relevant-daily life examples to create meaningful learning [18].

The result of students' EL was important issues to be analyzed. This study aims to figure and analyze the students' EL and how the learning process run to nurture the EL of students, especially in the chapter of "Environmental changes and waste recycle".

2. Methods

The research was a quantitative-qualitative research conducted from February to April 2017 in one of the non-*Adiwiyata* (not-yet-green) high school in Karanganyar. The participated school has 18 science classes, and the topic of environment is taught in Grade tenth. The participants of this research were three classes selected purposively from three graders. The first class was Grade tenth, consisted of 17 students, while the second and the third class were grade eleventh and twelfth, each consisted of 24 students. Regarding the gender, there were twenty boys and forty-four girls of total joined as participants.

The data of EL was measured by testing knowledge, attitude and concern. The knowledge of students on concepts of environmental issues was tested by fourteen main questions followed by supported questions. The attitude of students was measured by New Ecological Paradigm (NEP) consisted of fifteen items. The concern was measured by fifteen statements about environment, which was modified from Al-dajeh (2011). Details of the instruments is figured out in Table 1.

Table 1. The Instrument for EL

EL Components	Types of Question	Number of Items	Scoring	Category
Knowledge	Multiple choice	14	Correct answer: 1 Wrong answer: 0	Category based on score of total correct answer Correct answer 10-14: adequate Correct answer 8-1: inadequate
Attitude	Likert scale	15	Odd numbers: SA (5), A (4), U (3), D (2), SD (1) Even numbers: SA (1), A (2), U (3), D (4), SD (5)	Category knowledge based on accumulative score: 16-30: very low 31-45: low 46-60: medium 60-70: high Category of attitude based on score of NEP components 4-6: very low 7-9: low 10-12: medium 13-15: high
Concern	Likert scale	15	Number 3, 8, 15: SA (5), A (4), U (3), D (2), SD (1) Other numbers: SA (1), A (2), U (3), D (4), SD (5)	Uncategorized

Notes: SA (Strongly Agree), A (Agree), U (Uncertain), D (Disagree), SD (Strongly Agree)

The knowledge of students was categorized into two groups. First category is "adequate", covers the score from 10 to 14 (maximum score of correct answer), while the second category is "inadequate", covers the lower range of score. The overall attitude of students was categorized into four groups, i.e. "very low", "low", "medium" and "high". While, the attitude can be categorized based on NEP components, i.e. limits to growth (question number 1, 6, 11); balance of nature (question number 3, 8, 13); anti-anthropocentrism (question number 2, 7, 12); anti-excemptionalism (question number 4, 9, 14); and eco-crisis (question number 5, 10, 15). The NEP components were categorized into four groups, i.e. "very low", "low", "medium" and "high". The concern only to know range mean on answer statements about environment (uncategorized).

The practices of EL in the learning process were observed in one month or 9 meetings in the Grade X, when students learned about environmental changes and waste recycle. Observations were focused on students and teacher behavior, the learning content, and the activities of students during the learning process.

The interviews to grasp the information on what the factors that affecting EL, were conducted on a sample of 15 students which selected purposively based on the distribution groups of students' scores.

IOP Conf. Series: Journal of Physics: Conf. Series **1022** (2018) 012057 doi:10.1088/1742-6596/1022/1/012057

Two biology teachers who taught the classes and one vice principal of the curriculum were also interviewed. The results of the EL interview about the attitude and concern, and answers for knowledge questions were analyzed using some coding to understand the patterns and the number of similar answers [19].

3. Result and Discussion

The results of environmental knowledge measurement presented that the percentages of students' knowledge are in the range 3.08% to 93.85%. The results of environmental knowledge measurement (Figure 1).

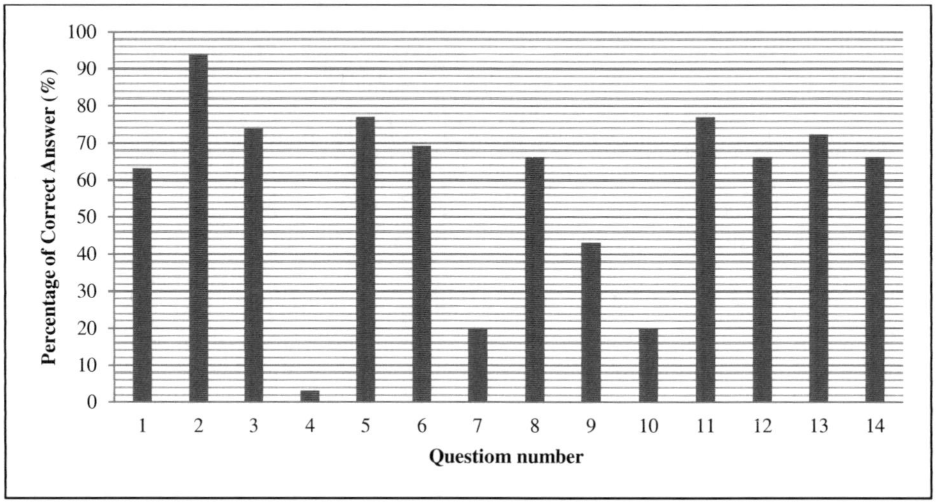

Figure 1. The Results of Environmental Knowledge Measurement

Note: Question number 1 factual issues about environmental problems.
Question number 2 impact of environmental problems.
Question number 3 factors that causing environmental problems.
Question number 4 the threshold values of water pollution.
Question number 5 the water pollution.
Question number 6 how to deal with water pollution.
Question number 7 the factors that causing water pollution.
Question number 8 the fact of environmental change.
Question number 9 recycling.
Question number 10 the factors that cause, impacts, and ways to prevent tornado.
Question number 11 the impact of noise pollution.
Question number 12 and 13 how to handle noise pollution.
Question number 14 the impact of noise pollution.

The results of EL questionnaire as environmental knowledge show that 80% of students were categorized as inadequate and the rest classified as adequate. The questions number 4, 7, 9 and 10 have percentage of correct answer lower than 50%. The lowest percentage of correct answer is question number 4 (3.08%), which asked about threshold values of water pollution.

Question number 4 (Q4) is about the threshold values of water pollution. Environmental knowledge measurement shows that Q_4 has the lowest percentage of correct answer. Most of the students indicated lack of knowledge on threshold values of water pollution, showed by choosing the wrong answer that the students do not understand the concepts of BOD, COD and DO and fabricated inappropriate reasons. Other students partially understood about threshold values of water pollution, showed by choosing the wrong answer, and the lack of proper reasoning.

Question number 7 (Q7) is about the factors that causing water pollution. Some students did not understand the material on eutrophication, some students' even incapable to give reasons why eutrophication occurred. Most of the students partially understood the material, showed by the wrong answer, and give inappropriate reason. There are few students who already understood the material about eutrophication.

Question number 10 (Q10) is about the factors that cause, impacts, and ways to prevent tornado. Many students have not understood the cause, the impacts, and ways to prevent the tornado is showed by choosing the wrong answer and fabricate inappropriate reason. There are more students who have not understood the material than those who understood.

Close to 50% students had already understood the concept of recycling, shown by Question number 9 (Q9), compared with number 4, 7 and 10. Most of the students did not understand the concept of recycling because they could not distinguish between reuse, recycle, reduce, recovery and replace.

The result of the students' attitude measured by the New Ecological Paradigm Scale (NEP) questionnaire shows that overall average score was 46.42. The NEP consisted by five components, three statements each, that form the basis of environmental caring attitudes; limits to growth, balance of nature, anti-anthropocentrism, anti-exemptionalism and eco-crisis. The average students' score for each of NEP Components (Figure 2).

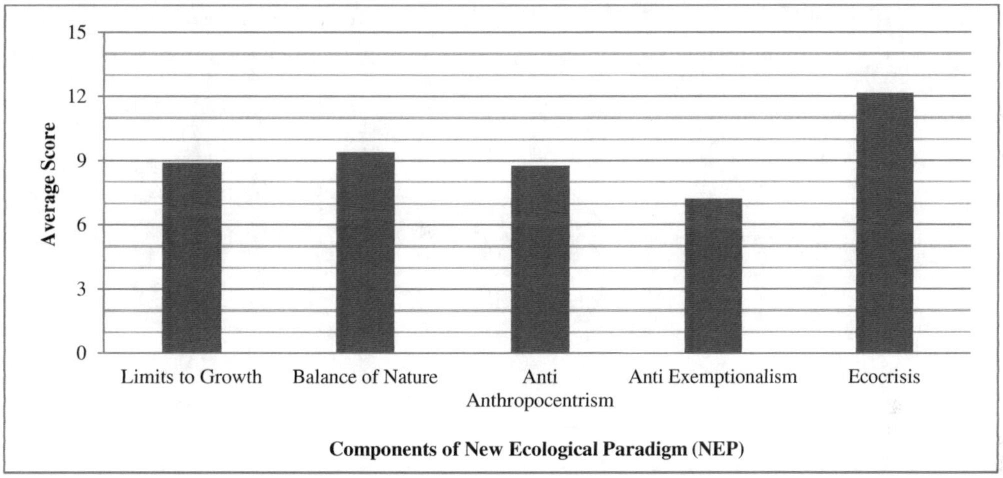

Figure 2. The Average of Students' Score for Each Of NEP Components

Note: Statements about limits to growth were listed on questions number 1, 6, and11
Statements about balance of nature were listed on questions number 3, 8, and 13
Statements about anti-anthropocentrism were listed on questions number 2, 7, and 12
Statements about anti-excemptionalism were listed on questions number 4, 9, and 14
Statements about eco-crisis were listed on questions number. 5, 10, and 15

The average score of some components of New Ecological Paradigm Scale (NEP) were still categorized as low, i.e. limits to growth, anti-anthropocentrism, and anti-exemptionalism. The lowest score was the anti-exemptionalism (which is 7.22). The result of the students' attitude measured by the NEP questionnaire shows that most of students have not understood the concepts about the environment.

Anti-exemptionalism contains beliefs that humans were responsible to preserve the nature. Most students have not grasped the concepts of natural resources use and real-life examples, explanations of natural laws and the ways the nature works. The student still answers incorrectly and gives an inappropriate reason for concepts that belong to anti-exemptionalism.

Anti-anthropocentrism contains beliefs that humans may could change and control the environment. Most students have not yet understood about the management of the natural resources, environmental conservation and environmental ethics. Some students cannot give a reason to the facts about environmental issues and have not yet able to give examples of the application of environment conservation in daily life.

Limits to growth contains beliefs about the earth's limitation to provide natural resources. Most of the students have not understood the concept of natural resources, could not able to connect the concepts about the pyramid of the ecosystem with the Robert Malthus' theory. Most of the students also could not relate the facts of environmental damage to the factors that cause environmental damage and could not give a concrete example of proper use of the natural resources. Some of the students also have not understood that the earth has limitations as a resources provider.

Environmental concern can be seen based on the average score of each question. The result of environmental concern measurements using 15 questions shows that the average score was between 2.58 to 4.18. The result of environmental concern measurement (Figure 3).

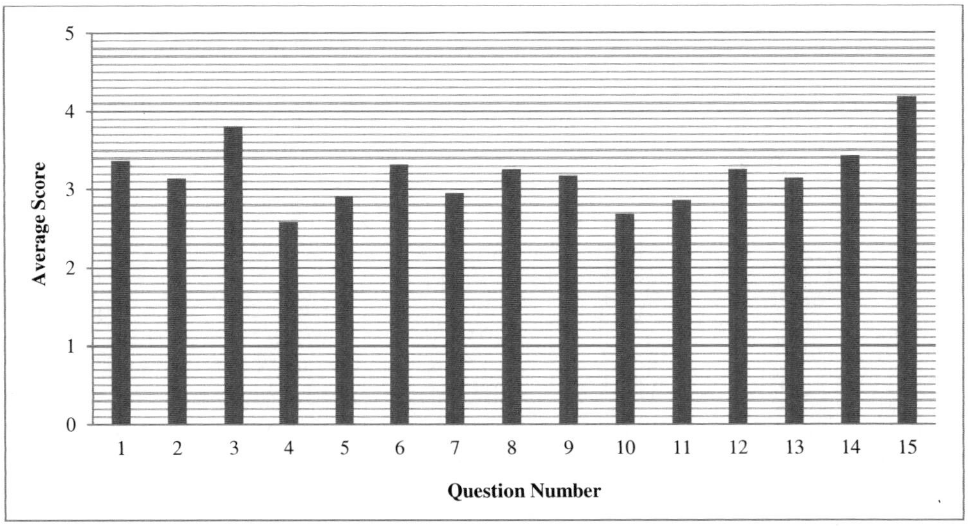

Figure 3. The Average Results of The Environmental Concern Questionnaire

Note: Questions number 1, 2, 4, 6, 9, 10, 11, 12, 13, 14 and 15 were about the perceptions of the environment.
Questions No. 3, 5, 7, 8 were about the action on environmental issues.

The lowest average score for the environmental concern was 2.58 (question number 4). Question number 4 contains the perception about the use of plastic for shopping bags in everyday life. The results of environmental concern questionnaire show that some of the students were not have determined perception about the use of plastic for shopping bags. Some of the students have not fully understood the environmental impact of plastics usage. Students could not make opinion about the fact that the decomposition of plastics that can be sped up by the bacteria. There are more students who have understood about plastic recycling than those who have not understood, so that most of the students can decide their perceptions, but the reasons given were less detailed.

The result of observation of environment education learning in the classroom shows that the learning was dominated by teacher's continuous lecture (direct instruction learning) based on power point slides and the book. From the interviews, the teachers state they were unwilling to make variation for their teaching model. Their lesson plans show that they used discovery and inquiry. The teachers explained that the learning models besides the direct instruction were takes a long time and make hard to assess the students' ability.

The numbers of indicators and learning materials that teachers delivered at the classroom were 16 the students' understanding. The teachers also give no real-world and complex examples of environmental problems through video or pictures, so students only know the theories. Teachers should understand the characteristics of the learning material by using contextual learning model, provides relevant-daily life examples to create meaningful learning [18].

The analysis of learning instrument shows that the lesson plans were unsuitable for the needs of EL. The analysis was conducted based on indicators of each NEP. The indicators in the teachers' lesson plans have included limits to growth, balance of nature, anti-anthropocentrism, and eco-crisis components, but do not contain the anti-exemptionalism component.

Students' activity during environmental education learning activities were: a few students were listening to teacher and making note. Some of students talk to other, daydreaming, yawning, playing alone, using their gadget, reading to brochures, doing other tasks and leaning to friends' shoulder. Most of the students were interested to pay attention to learning when the teacher provides video, pictures and gives a real example of environmental issues. Environmental education was not only dealing with the theories and concepts, but also know the real problems so that the students were involved in solving problems and encourage the students to take action on the environmental issues [20].

Based on the interview with the vice principal of the curriculum, shows that the school has not yet integrated the environmental education into the curriculum, because of some shortcomings and limitations. The students have not able to implement what they learn in school, for example they still did not care about the trash in the classroom and school zone and the picket team shift were not going well. Observations show that the distance between the one trash can with another was 14.4 meters and not all classes have the trash can.

The implementation of environmental education in the classroom was still constrained and has not yet establishing high EL. The weakness of environmental education was the material and the lesson model applied in the school is not adequate and unsuitable so that the learners were not have complete understanding about environmental conservation. The proper action to overcome this weakness was to make learning plan that suitable to the characteristics of the learning materials [20].

4. Conclusions

About 80% of students' EL were categorized as inadequate and the rest were classified as adequate. The students' attitude measured using New Ecological Paradigm scale (NEP) shows the average students' score was 46.42 (medium). Meanwhile, students' concern score was between 2.58 to 4.18. Observation during the learning process showed that the students did not understand well the learning content as the teachers taught only the theories and concepts without demonstration or application of those concepts through a series of experiment or investigation. Furthermore, teachers' lesson plan and the learning process conducted by the teachers have not fulfilled the EL components. The school, also has not yet integrated the environmental education through the curriculum.

5. References

[1] Olagunju and Ebenezer T Impacts of human-induced deforestation, forest degradation and fragmentation on Food Security 2015 *New York Science J.* **8**(1) 4-16

[2] Sawant S N and Bhave P P 2014 Impact of pollution on marine environment-a case study of coastal Mumbai *J. of Advanced Technology in Engineering and Science* **2**(7) 182-88

[3] Ozlem Kar K, Jingjing, Pinkerton and Kent E 2016 Pulmonary health effects of air pollution *J. of Current Opinion* **22**(2) 138-43

[4] Sullivan S 2012 Banking nature? The spectacular financialisation of environmental *A Radical J. of Geo.* **45**(1) 1-20 http://dx.doi.org/10.1111/j.1467-8330.2012.00989.x

[5] Hawcroft L J and Milfont T L 2010 The use (and abuse) of the new environmental paradigm scale over the last 30 years: a meta-analysis *J. of Environ. Pscyc.* **30** 143-58 http://doi.org/10.1016/j.jenvp.2009.10.003

[6] Larson S, Freitas, D M De and Hicks C C 2013 Sense of place as a determinant of people's attitudes towards the environment: implications for natural resources management and planning in the great barrier reef, Australia *J. of Environmental Management* **117** 226-34 http://dx.doi.org/10.1016/j.jenvman.2012.11.035
[7] Orlins S and Guan D 2016 China's toxic informal e-waste recycling: local approaches to a global environmental problem *J. of Cleaner Produc.* **114** 71-80 http://dx.doi.org/10.1016/j.jclepro.2015.05.090
[8] Jickling B and Wals A E J 2008 Globalization and environmental education: looking beyond sustainable development *J. of Curriculum Studies* **40**(1) 1-21 http://dx.doi.org/10.1080/00220270701684667
[9] Carleton-hug A and Hug J W 2010 Challenges and opportunities for evaluating environmental education programs *J. of Evaluation and Program Planning* **33** 159-64 https://doi.org/10.1016/j.evalprogplan.2009.07.005
[10] Frantz C M and Mayer F S 2014 The importance of connection to nature in assessing environmental education programs *J. Studies in Educational Evaluation* **41** 1-5 http://dx.doi.org/10.1016/j.stueduc.2013.10.001
[11] Srbinovski M, Erdogan M and Ismaili M 2010 Environmental literacy in the science education curriculum in Macedonia and Turkey *J. of Procedia Social and Behavioral Sciences* **2** 4528-32 https://doi.org/10.1016/j.sbspro.2010.03.725
[12] Locke S and Russo R O 2013 Environmental education and eco-literacy as tools of education for sustainable development *J. of Sustainability Education* **4**
[13] Pe'er S, Goldman D and Yavetz B 2007 Environmental literacy in teacher training: attitudes, knowledge and environmental behavior of beginning students *J. of Environ. Educ.* **39**(10) 45-59 http://dx.doi.org/10.3200/JOEE.39.1.45-59
[14] Yavetz B, Goldman D and Pe'er S 2009 Environmental literacy of pre-service teachers in Israel: a comparison between students at the onset and end of their studies *J. Environ. Educ. Research* **15**(4) 393-415 http://dx.doi.org/10.1080/13504620902928422
[15] Mcbeth W and Volk T L 2010 The national environmental literacy project: a baseline study of middle grade students in the United States *J. of Environ. Educ.* **41**(1) 55-67 http://dx.doi.org/10.1080/00958960903210031
[16] Al-dajeh H L 2011 Assessing environmental literacy of pre-vocational education teachers in Jordan *College Student J.* **46**(3) 492-507
[17] Dunlap R E, Liere, K D Van, Mertig A G, and Jones R E 2000 Measuring endorsement of the New Ecological Paradigm: a revised NEP scale *J. of Soc. Issues* **56**(3) 425-42 http://dx.doi.org/10.1111/0022-4537.00176
[18] Stuckey M, Heering P, Mamlok-Naaman R, Hofstein A and Eilks I 2015 The philosophical works of ludwik fleck and their potential meaning for teaching and learning science *J. of Science and Educ.* **4**(3) 281-98 https://doi.org/10.1007/s11191-014-9723-9
[19] Miles M and Huberman A 1994 *Qualitative Data Analysis* (California: SAGE Publication)
[20] Stevenson R 2007 Schooling and environmental education: contradictions in purpose and practice *J. of Environ. Educ. Reseacrh* **13**(2) 139-53 http://dx.doi.org/10.1080/13504620701295726

IOP Conf. Series: Journal of Physics: Conf. Series **1022** (2018) 012058 doi:10.1088/1742-6596/1022/1/012058

The development of the learning video for the flipped classroom model on student of open university on human skeletal system and muscles

V S Andrini

STKIP PGRI Nganjuk, Jl.A.R. Saleh 21 Nganjuk 64411, Indonesia

Corresponding email: vera@stkipnganjuk.ac.id

Abstract. The objectives of the research are to develop the learning video for the flipped classroom model for Open University's student and to know the effectiveness of the video. The development of the video used Research and Development ADDIE design (Analyses, Design, Development, Implementation, Evaluation). The sampling used purposive sampling was 28 students in Open University of Nganjuk. The techniques of data collection were the observation data to know the problems of the students, and learning facilities, the test (pre-test and post-test) to know a knowledge aspect, a questionnaire to know advisability of video learning, a structured interview to confirm their answer. The result of the expert of matter and media showed that the average product score was 3.75 of 4 or very good, the small-scale test showed that the average score was 3.60 of 4 and the large-scale test showed that the average score was 3.80 of 4, it had a very good category. The t-test with paired sample test showed that sig. (2-tailed) < 0.05. The N-gain score of pre and post test was 0.55, it had the medium category. It can be concluded that the development of the learning video for flipped classroom was effective to be implemented.

1. Introduction

The new paradigm that appears related to the learning process, it is no longer describes a meeting face to face in the classroom even though the social interaction is maintained. The internet technology makes it easy for people to interact without being bound by space and time. Discussing in the group by email, chatting and, the others internet media is the part of innovation in this post-modern [2].

The learning process in Open University always uses a module to reach out and to communicate with student and the lecturer. The module is designed in order to the student can understand and learn independently. If the student can't understand the matter from the module, the Open University provides tutorial learning by direct and online. The tutorial learning is implemented during 8 times in one semester.

The Open University is dominated by the students have 25-29 years old of Teaching and Learning Faculty who is a teacher. This condition raised limitations and problem of interaction between the student and lecturer. So that, the students don't understand the matter. To solve this problem shall implement the innovative learning concept as Flipped Classroom. According to Johnson [7], the lecturer can give a flipped classroom model minimizing the number of direct learning in the classroom. This model utilizes technology which supports the matter by online for the students. This section gives free up the time for class previously.

The flipped classroom model supports the students to get more matter not only in the classroom but outside the classroom course also. The students can access repeatedly the matter through the internet or video learning which are given by lecturer. Before the flipped classroom learning begins, the students study the topic with video which is created by lecturer and the other. The study is supported by research that has been done by Holzinger [5], the flipped classroom learning begins watching instructional videos that contain material which is accompanied by worksheet. The student must confirm with worksheet that they had been watching the video. Then, learning in the classroom is used with discussion to solve a problem relating the material.

Flipped classroom learning, the students watch the video learning in the home and they find the matter concept. In the classroom, the students had the concept of the matter so they more quick to accept the lesson. Beside it, they can do task, project, and discussion easily in the classroom. In this case, the lecturer has to plan, direct, evaluate the lesson. According to Basal [1], the level of learning is (1) the lecturer plan the learning; (2) to choose the activities which conform with the topic; (3) to integrate the assignment with activities; (4) to manage the lesson and to present all process an organized manner.

2. Research Methods

The research used the Research and Development (R&D) ADDIE Design (Analyses, Design, Development, Implementation and, Evaluation). The matter in the learning process is the human skeletal system and muscles. The research was done at the Open University. The population of the research was the Elementary Education's students in The Open University. The sampling technique was a purposive sampling. They were 28 students who are the experimental group. The techniques of data collection were the observation data to know analyses level that is the problems of the students, lecturer and learning facilities, test (pre-test and post-test) to know a knowledge aspect, a questionnaire with the Likert scale to know advisability of video learning, a structured interview to confirm their answer on the test and a questionnaire.

The validation of video was developed and done through review by expert and trial test to the students. The conversion score of validation by Suyanto[10] is:

Table 1. The Conversion Score of Validation of Video Development

Score	Category
3.26 – 4.00	Very Valid/Very Proper/Very Good
2.51 – 3.25	Valid/Proper/Good
1.76 – 2.50	Less Valid/ Less Proper / No too Good
1.00 – 1.75	No Valid/ No Proper/ Bad

Meanwhile, the effectiveness of learning video was known by the paired t-test with one group pre-test and post-test design. The difference of pre-test and post-test score used N-Gain score. The normalisation of N-Gain could be classification with: (1) if $g \geq 0.7$ so N-gain score on a high category; (2) if $0.7 > g \geq 0.3$ so N-gain score on a medium category and, (3) if $g < 0.3$ so N-gain score on low category [4].

3. The Result And Discussion

3.1. The Result of Analysing Level

At this level, the needs analysis to the students was done through the observation. The analysing was gotten which included all of the problems of the students, lecturer and learning facilities. The observation data showed that the number of the Open University's students was 406.027 students until 2015 years. 291.641 students had working be the teacher in the school, 31.975 students have not worked yet. The students have 25-29 years old as many as 96.561 and the under of 25 years old have

IOP Conf. Series: Journal of Physics: Conf. Series **1022** (2018) 012058 doi:10.1088/1742-6596/1022/1/012058

91.393 people. After that, the researcher was done the interview. It has 75.55% students that find the difficulties to do the assignment and 65.80% students that find the difficulties about the understanding of the matter.The limited in the classroom caused the difficulties of the lesson. Beside it, the students who be the teacher has more the task in the school and their college distance was large from their home, so they didn't understand the matter clearly. The solution of this problem is the researcher make a video. This video consists of the explaining of the lecturer about the human skeletal system and muscles. Rebowo [10] stated that the development of an instructional video to improve learning outcomes. So, the ability of students to listen and analyse is well done, they can observe their problem about the human skeletal system and muscles in daily life and find its solution.

3.2. The Result of Design Level
Design level of learning video was to determine the video format to create a storyboard that was adapted to syllabus and lesson plans. The video was also equipped with the worksheet. The formats of video that was applied in the flipped classroom model were: (1) the title of the matter; (2) making the synopsis; (3) making the storyboard; (4) taking the picture by storyboard; (5) the editing process; (6) the multiplication and dissemination. There were two video titles which were The Human Skeletal System and Muscles. This matter consisted of the parts of the human skeletal and muscles, characteristic, function and the assignment. The making procedures of the video were: (1) to determine the technology was used; (2) to determined how to publish video on the students; (3) to make a video; (4) to create a rule that they really watch the videos.

Implementation of the flipped classroom learning model by Jamaludin et al [6] included three core activities: (1) the students had to watch the video before coming to the classroom; (2) when learning process, the student had to discuss; (3) after that, the students had to do the consolidation and prepare the next matter.

3.3. The Result of Development Level
The development of video has two main matters. They are The Human Skeletal and Muscles. The video of The Human Skeletal contains (1) bone structures; (2) shape and structure of bone; (3) the relationships between bones; (4) the disorders of the joints. The video of The Human Muscles contains (1) type of muscles; (2) type of muscles movement; (3) how the muscles work. The examples of the videos like figure 1 an figure 2.

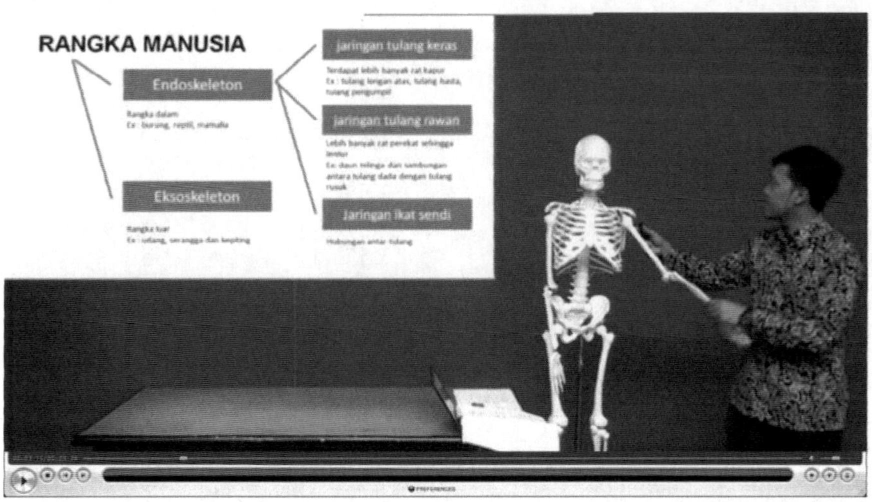

Figure 1. The Print Screen of Content Display on Human Skeletal System

Figure 2. The Print Screen of Content Display on Human Muscles

At this level, the expert judgment, the result of the small-scale test, the result of the large-scale test suggested that the video clarifies the voice of the actors, brighter lighting, zoom in the object of the human skeletal and muscles, the actor use shorter and clearer language, adding the instrumental music.

The making procedure of video used Vegas Program to editing video. The Human Skeletal and Muscles were divided be 2 videos with duration 15-25 minutes. The video published with veraseptiandrini@blogspot.com. The controlling that the student watched the video was the students had to do the assignment in the worksheet so that the student had the responsibility. Sales [9] stated that the flipped classroom model try the students to respond the video and organize the learning process.

Table 2. The Result of Expert Judgment's Validation, The Small-Scale Test and, The Large-Scale Test

Aspect		The Average of Expert's Score	The Average of The Small-Scale Test	The Average of The Large-Scale Test
The Picture				
1	The appearence of the actors.	3,58	3,78	3,72
2	Suitability of shooting location and the matter.	3,77	4,48	3,77
3	Lighting of the video and the picture	3,85	3,50	3,85
The Appeal				
1	Systematic presentation	3,55	3,55	3,66
2	Continuity of the concept	3,80	3,56	3,80
3	Suitability of visualization with the matter presented	3,78	3,58	3,78
4	Clarity the matter	3,70	3,70	3,86
5	Interesting of the content video	3,76	3,76	3,84
The Sound				
1	Clarity of the sound	3,88	3,68	3,88
2	Compatibility of the musical accompaniment	3,82	3,42	3,82
The Average Score		3.75	3.60	3.80
The Category		Very Good	Very Good	Very Good

The validation and advisability of video were done by the expert judgment. They were the expert of matter and media. The small-scale test was given to 6 students and the large-scale test to 28 students. The achievement aspect consisted of the picture clarity, the attraction and, the sound clarity. The average achievement of the expert judgment was 3.75 of 4 (very good category), the result of the small-scale test was 3.60 of 4 (very good category) and the result of the large-scale test was 3.80 of 4 (very good category).

3.4. The Result of Implementation Level

Before the learning process, the students got pre-test in the experiment class. After the learning process, they got post-test. The result like as Table 3.

Table 3. The result of Pre-Test and Post-Test

	Minimum Score	Maximum Score	The Average Score
Pre-test	55,29	84,71	68,46
Post-test	70,29	96,47	85,14

The average of pre-test show that 68.46. It means that the students had low knowledge aspect so it had to be improved. After they accepted the video and also watched it with the flipped classroom model, they had 85.14 of the average of post-test. It means that had the very good category. The effectiveness of the video was determined from the difference of pre-test and post-test like as Figure 3.

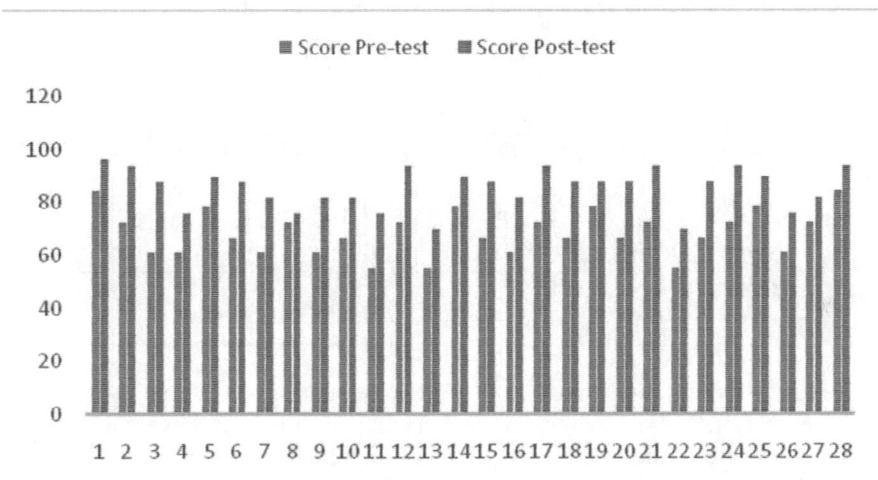

Figure 3. The Difference of Pre-test and Post-test Score.

The prerequisites test with the normality test, its result like as Table 4 and homogeneity test in Table 5.

Table 4. The Normality Test Results

Tests of Normality							
		Kolmogorov-Smirnov[a]			Shapiro-Wilk		
	Test	Statistic	df	Sig.	Statistic	df	Sig.
Result	Pre-test	.139	28	.174	.943	28	.132
	Post-test	.209	28	.200	.917	28	.030
a. Lilliefors Significance Correction							

The normality test used Kolmogorov-Smirnov test that was obtained the significant of pre-test 0.174 > 0.05 and post-test 0.200 > 0.05, so the experimental group was normally distributed.

Table 5. The Homogeneity Test Result

Test of homogeneity of variance					
		Levene Statistic	DF1	DF2	Sig.
Result	Based on Mean	.506	1	54	.480
	Based on Median	.706	1	54	.404
	Based on median and with adjusted df	.706	1	53 994	.404
	Based on the trimmed mean	.560	1	54	.458

The homogeneity test with Kolmogorov-Smirnov test was obtained a significant value of pre-test and post-test for the experimental group of 0.480 which was greater than 0.05. It could be concluded that the homogeneous data. The result of the prerequisites test showed that the normal and homogeneous data so it could be done the next test with paired sample t-test.

Table 6. The Results of The Paired Sample t-test

Paired Samples Test									
		Paired Differences							
					95% Confidence Interval of the Difference				
		Mean	Std. Deviation	Std. Error Mean	Lower	Upper	t	df	Sig. (2-tailed)
Pair 1	Pretest-Postest	4.5589 3E1	29.88449	3.99348	53.59240	37.58617	11.416	55	.000

Based on Table 6, it was found that 0.000 of significant (2-tailed) was lower than 0.05. It was concluded that the learning video in the flipped classroom could improve the learning outcomes. Based on the N-Gain score of the improving pre-test to post-test score showed that 0.55 the medium category. It means that the development of learning video was effectively implemented in the flipped classroom model. Enfield [3] stated that the flipped classroom model was effective to help to understand the matter. it improved the ability of to solve the problem about the matter. The student could interact in the class so the students had independently learned.

3.5. The result of Evaluation Level

The product in the research was learning video about The Human Skeletal and Muscles. This product had to be evaluated that the size of the video was too much. So, the downloading process needed a long time and a good connection. The evaluation of learning process used the knowledge aspect of the pre-test, post-test and presentation. According to Zhou [13] stated that the comprehensive assessment in the flipped classroom was the understanding aspect, implementing aspect, analysing at presentation, test and self- evaluation.

The flipped classroom was supported by video can be applied anytime and anywhere. This model can improve the learning outcomes. The students can repeat the video and lesson again. Love *et al* [8]

IOP Conf. Series: Journal of Physics: Conf. Series **1022** (2018) 012058 doi:10.1088/1742-6596/1022/1/012058

stated that the students can more study than in the classroom. They can discuss with their group and another, present, and solve the assignment.

The result of observation showed that the students have more confidence and do not find the difficulties during has the task and discuss. Velegol *et al* [12] stated that the students have more confidence to learn and accomplish the necessary tasks when able to work with a lecturer or a friend in the class.

4. Conclusion and Suggestion

The development of video for flipped classroom model was valid and advisable to implemented. The evaluation of product is picture clarity, the appeal, the sound clarity. The result of the expert judgment of matter and media showed that the average product score was 3.75 of 4 or very good, the result of the small-scale test showed that the average score was 3.60 of 4 and the result of the large-scale test showed that the average score was 3.80 of 4 it had a very good category. The t-test with paired sample test showed that sig.(2-tailed) of $0.000 < 0.05$. The N-gain score of pre-test and post-test was 0.55, it had the medium category.

Based on learning process for flipped classroom, the researcher recommends the development not only make the video but also integrating the learning device with another learning model which makes a student centre. The publishing of video can do through Youtube, Instagram, and another media social.

5. References

[1] Basal A 2015 The implementation of a Flipped Classroom in Foreign LanguageTeaching *Turkish Online Journal of Distance Education* **16**(4) 28-37Another reference

[2] Cuthell J P 2002 *Virtual learning: The impact of ICT on the way young people work and learn* (Ashgate Pub Limited) p 32

[3] Enfield J 2013 Looking at TheIimpact of The Flipped Classroom Model of Instruction on Undergraduate Multimedia Students at CSUN *TechTrends* *57*(6) 14-27

[4] Hake R R 1999 *Analyzing change/gain scores.* (http://www.physics.indiana.edu/~sdi/AnalyzingChange-Gain.pdf)

[5] Holzinger A C 2016 The Flipped Classroom Model For Teaching Vectors *Thesis* Johannes Kepler University Linz Austria p.6

[6] Jamaludin R, Osman S Z M, Yusoff W M W and Jasni N F A 2016 FLIPPED: A Case Study in Fundamental of Accounting in Malaysian Polytechnic Journal of *Education and e-Learning Research* **3**(1) 23-31

[7] Johnson G B 2013 Student perceptions of the Flipped Classroom *Doctoral Dissertation* University of British Columbia p.20

[8] Love B, Hodge A, Grandgenett N and Swift A W 2014 Student learning and perceptions in a flipped linear algebra course *International Journal of Mathematical Education in Science and Technology* **45**(3) 317-324

[9] Sales N 2013Flipping the classroom: Revolutionising legal research training. *Legal Information Management* **13**(4) 231-235

[10] Rebowo W A 2014 Pengembangan Media Video Pembelajaran Berbasis Masalah Materi Pecahan Pada Siswa Kelas Iv Sekolah Dasar. *Pelangi Pendidikan* **21**(2) 94-106

[11] Suyanto E 2009 Pengembangan Contoh Lembar Kerja Fisika Siswa dengan Latar Penuntasan Bekal Awal Ajar Tugas Studi Pustaka dan Keterampilan Proses Untuk SMA Negeri 3 Bandar Lampung *In Prosiding Seminar Nasional Pendidikan* p. 20

[12] Velegol S B, Zappe S E and Mahoney E M I L Y 2015 The Evolution of a Flipped Classroom: Evidence-Based Recommendations *Advances in Engineering Education* **4**(3) n3

[13] Zhou G Q and Jiang X. F 2014 Theoretical research and instructional design of the flipped classroom *In Applied Mechanics and Materials* (Trans Tech Publications) **543** 4312-4315

IOP Conf. Series: Journal of Physics: Conf. Series **1022** (2018) 012059 doi:10.1088/1742-6596/1022/1/012059

Enhancing the prospective biology teachers' Pedagogical Content Knowledge (PCK) through a peer coaching based model

Yenny Anwar

Biology Department, Sriwijaya University

Corresponding email: yenny_anwar@fkip.unsri.ac.id

Abstract. This paper presents the results of implementation Peer Coaching Based Model that was implemented in development and Packaging Learning Tool program aimed at developing a Pedagogical Content Knowledge prospective teachers' capabilities. Development and Packaging Learning Tool is a training program that applies various knowledge, attitude, and skill of students in order to form professional teacher. A need assessment was conducted to identify prospective teachers' professional needs, especially PCK ability. Tests, questionnaires, interviews, field notes and video recordings were used in this research. The result indicated that the ability of Prospective teachers' PCK has increased. This can be shown from the N-Gain that included in the medium category. This increase shows that there is integration of pedagogy and content; they have used varied strategies and can explain the reasons for its used. This means that the pattern belongs to the lower limit of the growing- PCK category. It is recommended to use peer coaching model during peer teaching.

1. Introduction

Education is increasingly important to the success of both individual and nations. This has an impact on the demands on teachers to be able to keep order and provide useful information to students, enabling a diverse group of students to learn ever more complex material and to develop a wider range of skills are increasing. Teachers' abilities are especially crucial contribution to students learning. Teachers need to know how to take the steps necessary to gather information that will allow them to make more grounded judgement about what is going on and what strategies may be helpful. The teachers should understand and be able to integrate knowledge of the content into lessons on curriculum, teaching, learning, and students. Such knowledge can guide teachers to assemble learning situations based on the needs of individual and group of students. That knowledge known as Pedagogical Content Knowledge.

The person who is called the teacher is the person who has the ability to design a learning program and it is able to organize and manage the class so that learners can learn and ultimately can reach the level of maturity as the ultimate goal of the educational process. The teacher is a profession, which means a position that can require special skills as a teacher and cannot be done by any person who do not come from the field of education. For that preparation of a professional teacher must be prepared from the beginning that is when they are still a prospective teacher. A prospective teacher should be

IOP Conf. Series: Journal of Physics: Conf. Series **1022** (2018) 012059 doi:10.1088/1742-6596/1022/1/012059

able to plan and incorporate appropriate science teaching strategies for learners with diverse backgrounds and learning styles [1;2].

Professional competence of biology teacher in high school is required to master the materials, structure, concepts and mindset of scholarship that support the subjects that he or she is recieving in school[3] Quoting from Lawrence D. Hazkew and Jonathan: Teacher is a professional person who conducts classes ". Meanwhile, according to Jean D. Grambs and C. Morris Mc Claire in the Foundation of Teaching An Introduction to Modern Education," The teacher is someone who has the ability to organize and manage the class". To meet expectations that they face now, Programs that prepare teachers need to consider the demands of school nowadays in concert with the growing knowledge base about learning and teaching if we want to produce the professional teachers.

Prospective teachers as one of the most important elements that will be responsible to teaching biology with good and right, need to be equipped with a mastery of the basic concepts (content) is strong as well as the ability to teach content well and correctly. This means that prospective teachers should have good knowledge and skills about PCK, which hope they will be able to do the learning well too. PCK is a knowledge of experience and expertise gained through classroom experiences. PCK is an important knowledge and must be owned by a teacher[4;5]. PCK is a form of teachers' professional knowledge that is highly topic, person, and situation specific [6;7]. Based on the results of several studies show that PCK is important to introduce to prospective students and hoping that they will be able to develop their PCK. The research about PCK has not been deep. No research has developed a teacher's PCK with an explicit approach, in the sense that it is still limited to the search for what is needed, not to the extent of the implications. Because of the importance of PCK that a teacher must have, PCK needs to be explicitly introduced to prospective students, hoping they can continue to develop their PCK skills after becoming a teacher. In other words, provide briefing about the PCK to prospective teachers before they do the real teaching at the school.

In teacher preparing program at Sriwijaya University in Indonesia, there is the Development and Packaging Learning Tool program, it is a training program that applies various knowledge, attitude, and skill of student in order to forming professional teachers. But that program did not make prospective teaches learn how to work, discuss, reflect on, and evaluate in peer communities to improve their Pedagogical Content Knowledge. Therefore, Peer coaching models is a model that prepare prospective teachers to have a command of critical ideas and skills, equally important, the capacity to reflect on, evaluate and learn from their friend and teachers so that it continually improves, hoping their PCK ability will be increase. Based on the reasons above, this study tried to use peer coaching model to develop the pedagogical content knowledge of prospective biology teacher.

2. Method

This paper based on a two-year research project that consisted of two main step. The first step was identifying teachers' professional knowledge especially about PCK. It was conducted by distributing needs assessment to teachers and prospective teachers. The second step was to develop the peer coaching models and do face to face training sessions

This study used Concurrent Triangulation Design[8]. In this design the researcher collected quantitative and qualitative data concurrently, then compares these two databases to determine whether there is convergence, differences, or some combination. This research was conducted at the Biology Education Program at a State University in South Sumatra, Indonesia. The activity of this study involved 20 students who are joined the teaching training program.

The researcher used tests, questionnaires, field notes, video recordings, and interviews to collect the data. To ensure genuine responses from the prospective teachers, interviews were conducted in small groups, though they were individually assessed and did three times test. All interviews recorded and then transcribed and careful noted while interviewing. In data analysing, the researcher used grounded theory, begun with open coding to identify concepts. Then, give labels to each unit of text from transcripts, field notes, test and interviews, forming the basis for identifying concepts throughout the data set, then compare the categories according to Anwar[9]

3. Results

This research was aimed to describe how the ability of student PCK prospective teachers who have done peer teaching and real teaching through peer coaching. in Previously, the assessment instrument and assessment rubrics were developed. Then, measure the ability of prospective biology teacher PCK before and after using peer coaching model and real teaching.

3.1 Perform peer teaching with peer coaching approach

Peer Coaching is an approach applied when implemented peer teaching. In the first stage, a prospective teacher acts as a teacher while the other teacher and lecturer as an observer. In the second stage, after a prospective student teaches he or she will evaluate himself or do self-reflection. Third stage, prospective teacher who acts as an observer and lecturer will provide insert in the form of suggestions and criticism of prospective teachers who have taught earlier. This approach emphasizes the shared learners, while at the same time they learn and evaluate themselves and their peers. Each student displays her teaching planning, while the other student acts as a student as well as an observer who assesses her students' teaching ability. When they assess the ability to teach their friends, in accordance with the planning that has been made, then they also do self-evaluation, do self-reflection of the ability of his friend. It was then that they indirectly developed their PCK capabilities. As Pollard points out, reflective activity often helps the development and maintenance of professional skills[10].

Figure 1. Peer Coaching Approach

3.2. Prospective teachers' content representation skills

The ability of the content representation can be seen from how prospective students describe material content related to learning in class[11]. How can they be able to determine why students need to learn ideas / concepts? Why is it important for students? What things should be explained and not the time explained? What difficulties are usually faced by students? How students think about the concept? What factor that affect the teaching of the concept? How to teach the procedure? How do students understand or confusion learn the concept? Essential concepts raised by prospective teachers can be seen in the following table:

Table 1. The Important Concepts That Appear Before and After Using Peer Coaching

No.	Concepts	Test		
		I	II	III
1	The history of Virus	√	√	√
2	The Structure of Virus	√	√	√
3	The characteristic of Virus	√	√	√
4	Virus replication	√	√	√
5	The Role of virus	√	√	√
6	Lytic Cycle	√	√	√
7	Lysogenic Cycle	√	√	√

8	The Harmful of Virus	√	-	-
9	The Profitable of Virus	√	-	-

There are differences in the number of concepts raised by prospective teachers who used peer coaching approach (Table 1). After peer teaching fewer essential concepts / ideas are raised by prospective teachers, after peer teaching using peer coaching models students bring more important concepts to be taught. Teachers need flexible understanding of subject matter link to an appreciation for how students learn, to make content accessible to learners. The CoRes of pedagogical content knowledge is knowing how students understand their particular subjects and having a repertoire of strategies to help student engage ideas central to discipline.

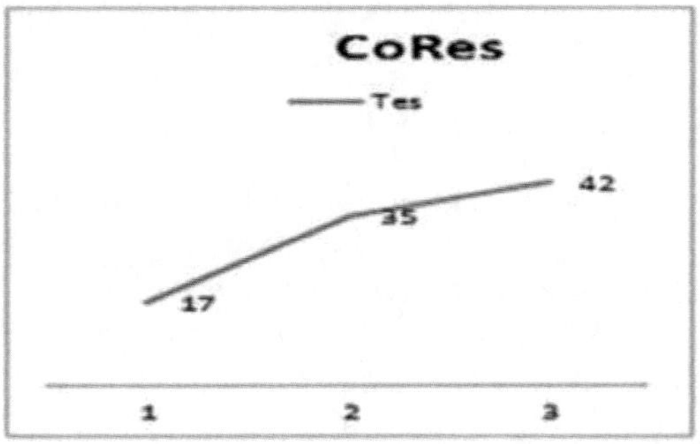

Keterangan :
0 ≥ 50 = less
51 ≥ 70 = enough
80 ≥ 90 = good
100 ≥ 120 = very good

Figure 2. CoRes Value

The value of the representation of the content shows improvement before and after do peer teaching used peer coaching approach and after doing practice teaching in the field. Improvement after peer teaching used coaching approach showed a good improvement. In peer coaching, prospective teachers get feedback on their performance, which shows they have or have not understood what is taught and how to teach it. In addition, they also learn to give feedback to their friends, who in that way also indirectly they self-correct. As expressed by Langs and Evan that giving and receiving feedback will have a profound effect on teacher development in teaching practice[12].

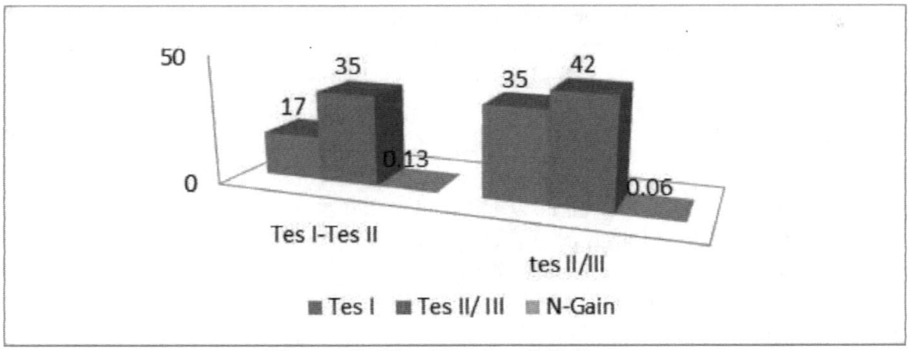

Figure 3. The N-gain of CoRes

From the Scores of N-Gain there is an increase too from the test I to the Test II (after peer coaching) and Tests III after PPL (real teaching). Enhancement after peer coaching was medium category N-Gain: 0,13), while in the third test was classified as low category (N-Gain: 0,06). It shows that peer coaching model is able to improve the ability of prospective teachers PCK.

Table 2. Prospective Teachers' Pedagogical Content Knowledge

No.	Idea/concept	Before peer coaching	After peer coaching
1	Understanding of pedagogy media	There are prospective teachers who have not been able to distinguish between methods and media, such as prospective teachers call the animation as a method.	All prospective teachers are able to distinguish between methods and media
2	Variation of strategy	The methods and media used tend to be less varied. The methods used are more varied	There are use of models and methods not mentioned by prospective teachers
3	Consideration of selection of teaching strategies	Has not considered the efficiency of time and practicality in its use in the classroom.	Less consideration of practicality.
4	The development of mastery of teaching strategies	Initially there are prospective teachers who have not been able to determine a strategy to teach a concept, because knowledge of their pedagogy is still limited.	Have started to have an idea how to teach a concept

The Scores of pedagogy of prospective teachers also increased quite well, which means the ability of teacher pedagogy was getting better. Before peer coaching, the PCK category are still in pre-PCK, after used peer coaching models increase to growing-PCK. This increase has begun shows the integration of pedagogy and content, they have used varied strategies and can explain the reasons for its used. This implies that this pattern belongs to the lower limit of the growing- PCK category [13].

4. Discussion

After peer coaching, the used of variable methods was increasingly and the used of this method has been associated with material characteristics. Each content has certain characteristics and this will be related to how the content should be presented, for example declarative knowledge presented with a proposition, whereas the procedural knowledge by production. Declarative knowledge can be communicated verbally, procedural knowledge is not. The content of science is simple or complex, concrete or abstract and open or closed for general experience, this difference will certainly differ in the way to learn it and teach it[14;15].Peer coaching models prepare prospective teachers need to have a command of critical ideas and skills, equally important, the capacity to reflect on, evaluate and learn from their friend and teachers so that it continually improves, and their PCK ability was increase.

Improvement after peer teaching used coaching approach showed a good improvement. It's reached the upper limit of the pre- PCK category. The term pre PCK describe the early stages of the interaction between pedagogy with content. In *the pre-PCK* prospective teachers have not been able to find a link between knowledge and pedagogical content knowledge so that no slice / integration between the two. At this stage of *Growing PCK,* teacher candidates begin to integrate content and pedagogy that has begun to form a wedge between content and pedagogy. In the maturing stage of PCK, it has been increasingly mature teacher and teachers are able to integrate content and pedagogy in a flexible and rational[13].

In peer coaching, prospective teachers get feedback on their performance. The purpose of providing feedback is to narrow the gap between existing understanding and learning goals. Reflection is fundamental to assessment, decision making, and a deeper understanding of the teaching practice. The act of reflection is primarily concerned with developing insights and discovering solutions to difficulties, or what might be described more correctly as learning opportunities to revise the next topic instructional strategies[16;17]. Feedback can be a reinforcement on students for a successful performance. Improvement also occurs after the student candidates do teaching practice in the field. At the time of the prospective teachers do the real teaching is faced with real state in the classroom, so this experience helps many prospective teachers in developing their abilities. This real teaching helps teachers in improving the planning and learning processes that occur.

The improvement education is depended on teachers PCK, to improve theirs PCK depends on the preparing program. The prospective teachers must work in settings where they can use what they know, work with other friend and teachers to provide a coherent. Hammond and Bransford, similarly reported that the focus is not on the format, length or location of teacher education but on its substance: what prospective teachers need to learn and how they may best be enabled to learn it[18].

5. Conclusions

The ability of prospective teachers' PCK has increased. After peer teaching used peer coaching approach, the improvement of students' PCK ability has improved quite well, including in medium category. In peer coaching prospective teachers get feedback on their performance, which shows they have or have not understood what is taught and how to teach it. In addition, they also learn to give feedback to their friends, who in that way also indirectly they self-correct. As expressed by Langs and Evan that giving and receiving feedback will have a profound effect on teacher development in teaching practice [12]. The purpose of providing feedback is to narrow the gap between existing understanding and learning goals. Feedback can be a reinforcement on students for a successful performance. Peer coaching models is a model to prepare prospective teachers need to have a command of critical ideas and skills, equally important, the capacity to reflect on, evaluate and learn from their friend and teachers so that it continually improves their PCK ability will be increase. Based on it, I argue that the prospective teacher education curriculum should help prospective teacher learn how to work on the improvement of practice as members of such collaborative communities. They can learn how they can engage in partnership in community. From this analysis of the peer coaching approach I recommended the prospective teachers' preparing program better to used peer coaching models to improve the pedagogical content knowledge and skill.

6. References

[1] NSTA 1998 *Standards for science teacher preparation* National Science Teachers Association in collaboration with the association for the Education of Teachers in Science

[2] National Committee of Science Education 1996 *National Science Education Standards* Washington DC. National Academy Press

[3] Depdiknas 2005 Peraturan Pemerintah Republik Indonesia Nomor 19 Tahun 2005 *Standar nasional pendidikan* Jakarta Fokusmedia

[4] Shulman L S 1986 Those who understand knowledge growth in teaching *Educational Researcher* **15**(2) 4-14

[5] Shulman L S 1987 Knowledge and teaching: Foundations ofthe new reform. *Harvard Educational Review* **57** 1—22.

[6] Van Driel J H and Berry A 2010 The teacher education knowledge base: Pedagogical Content Knowledge . In P.L Peterson, E Baker, and B. McGaw (Eds.), Third international encyclopedia of education **9**(7) 656-61 Amsterdam The Netherlands: Elsevier

[7] Loughran J, Mulhall P and Berry A 2008 Exploring Pedagogical Content Knowledge in Science Teacher Education *Int.l J. of Sci. Edu.* **30** 1301-20

[8] Creswell John W, Clark and Vicki L P 2007 *Designing and Conducting Mixed Methods Research* London Sage Publications

[9] Anwar Y 2016 Perkembangan kemampuan pedagogical content knowledge calon guru biologi pada pendekatan konkuren. *Jurnal Cakrawala Pendidikan* **3** 349-56

[10] Pollard A 2005 *reflective teaching* New York: Continuum

[11] Loughran J, Berry A and Mulhall P 2012 *Understanding and developing science teacher's*

pedagogical contentknowledge Rotterdam Sense Publishers

[12] Lang H and Evans D 2006 *Models, strategies, and methods for effective teaching* Boston pearson education, Inc.

[13] Anwar Y 2014 Hypothetical Model to Developing Pedagogical Content Knowledge (PCK) Prospective Biology Teachers in Consecutive Approach. *International Journal of Science and Research* **3** 138-1

[14] Fensham P, Gunstone R, and White R 1994 *The content of science* London The Falmer Press.

[15] Child A McNicholl J 2007 Investigating the relationship between subject content knowledge and pedagogical practice through the analysis of classroom discourse *Intl. J. of Sci. Edu.* **29** 1629-53

[16] Jong S and Chuan S 2009 Develoing in-service science teachers' PCK through a peer coaching-based model. *Education Research* **3** 87-108

[17] Vidmar D J 2006 Reflective peer coaching: Crafting collaborative self-assessment in *teaching. Research Strategies* **20** 135-48

[18] Hammond L D and Bransford J 2005 *Prearing teachers for a changing world* San Francisco Jossey Bass

Institute of Physics
Dirac House, Temple Back
Bristol BS1 6BE UK

ISSN: 1742-6588
ISBN 978-1-5108-6322-4

9 781510 863224

CURRAN ASSOCIATES INC.
proceedings.com